Handbook of Research on Prenatal, Postnatal, and Early Childhood Development

Neriman Aral
Ankara University, Turkey

IGI Global
DISSEMINATOR OF KNOWLEDGE

A volume in the Advances in Medical Diagnosis, Treatment, and Care (AMDTC) Book Series

Published in the United States of America by
IGI Global
Medical Information Science Reference (an imprint of IGI Global)
701 E. Chocolate Avenue
Hershey PA, USA 17033
Tel: 717-533-8845
Fax: 717-533-8661
E-mail: cust@igi-global.com
Web site: http://www.igi-global.com

Library of Congress Cataloging-in-Publication Data

Names: Aral, Neriman, editor.
Title: Handbook of research on prenatal, postnatal, and early childhood development / Neriman Aral, editor.
Description: Hershey, PA : Medical Information Science Reference, [2020] | Includes bibliographical references and index. | Summary: "This book explores children's developmental periods, characteristics, and the principles that affect their development in every stage of life"-- Provided by publisher.
Identifiers: LCCN 2019046394 (print) | LCCN 2019046395 (ebook) | ISBN 9781799829522 (hardcover) | ISBN 9781799829546 (ebook)
Subjects: MESH: Child Development | Prenatal Care | Postnatal Care
Classification: LCC QP84 (print) | LCC QP84 (ebook) | NLM WS 105 | DDC 612.6/5--dc23
LC record available at https://lccn.loc.gov/2019046394
LC ebook record available at https://lccn.loc.gov/2019046395

This book is published in the IGI Global book series Advances in Medical Diagnosis, Treatment, and Care (AMDTC) (ISSN: 2475-6628; eISSN: 2475-6636)

British Cataloguing in Publication Data
A Cataloguing in Publication record for this book is available from the British Library.

For electronic access to this publication, please contact: eresources@igi-global.com.

Advances in Medical Diagnosis, Treatment, and Care (AMDTC) Book Series

ISSN:2475-6628
EISSN:2475-6636

Mission

Advancements in medicine have prolonged the life expectancy of individuals all over the world. Once life-threatening conditions have become significantly easier to treat and even cure in many cases. Continued research in the medical field will further improve the quality of life, longevity, and wellbeing of individuals.

The **Advances in Medical Diagnosis, Treatment, and Care (AMDTC)** book series seeks to highlight publications on innovative treatment methodologies, diagnosis tools and techniques, and best practices for patient care. Comprised of comprehensive resources aimed to assist professionals in the medical field apply the latest innovations in the identification and management of medical conditions as well as patient care and interaction, the books within the AMDTC series are relevant to the research and practical needs of medical practitioners, researchers, students, and hospital administrators.

Coverage

- Patient-Centered Care
- Medical Procedures
- Internal Medicine
- Chronic Conditions
- Alternative Medicine
- Disease Prevention
- Critical Care
- Disease Management
- Cancer Treatment
- Patient Interaction

IGI Global is currently accepting manuscripts for publication within this series. To submit a proposal for a volume in this series, please contact our Acquisition Editors at Acquisitions@igi-global.com or visit: http://www.igi-global.com/publish/.

The Advances in Medical Diagnosis, Treatment, and Care (AMDTC) Book Series (ISSN 2475-6628) is published by IGI Global, 701 E. Chocolate Avenue, Hershey, PA 17033-1240, USA, www.igi-global.com. This series is composed of titles available for purchase individually; each title is edited to be contextually exclusive from any other title within the series. For pricing and ordering information please visit http://www.igi-global.com/book-series/advances-medical-diagnosis-treatment-care/129618. Postmaster: Send all address changes to above address.

Titles in this Series

For a list of additional titles in this series, please visit: http://www.igi-global.com/book-series/advances-medical-diagnosis-treatment-care/129618

Advanced Pharmacological Uses of Medicinal Plants and Natural Products
Ajeet Singh (National Dairy Research Institute, India) Padam Singh (Vaccine and Infectious Disease Research Centre, Translational Health Science and Technology Institute, India) and Navneet Bithel (Gurukul Kangri Universit, India)
Medical Information Science Reference • © 2020 • 540pp • H/C (ISBN: 9781799820949) • US $425.00

Quality Control of Cellular Protein in Neurodegenerative Disorders
Md. Sahab Uddin (Southeast University, Bangladesh) and Ghulam Md. Ashraf (King Abdulaziz University, Saud Arabia)
Medical Information Science Reference • © 2020 • 260pp • H/C (ISBN: 9781799813170) • US $295.00

Multidimensional Perspectives and Global Analysis of Universal Health Coverage
Yeter Demir Uslu (Istanbul Medipol University, Turkey) Hasan Dinçer (Istanbul Medipol University, Turkey) and Serhat Yüksel (İstanbul Medipol University, Turkey)
Medical Information Science Reference • © 2020 • 499pp • H/C (ISBN: 9781799823292) • US $265.00

Examining Biological Foundations of Human Behavior
Barre Vijaya Prasad (Dharwad Institute of Mental Health and Neurosciences, India)
Medical Information Science Reference • © 2020 • 266pp • H/C (ISBN: 9781799828600) • US $285.00

Handbook of Research on Medical Interpreting
Izabel E.T. de V. Souza (Osaka University, Japan) and Effrossyni (Effie) Fragkou (National and Kapodistrian University of Athens, Greece)
Medical Information Science Reference • © 2020 • 511pp • H/C (ISBN: 9781522593089) • US $345.00

Quality Assurance in the Era of Individualized Medicine
Anastasius S. Moumtzoglou (P. & A. Kyriakou Children's Hospital, Greece)
Medical Information Science Reference • © 2020 • 346pp • H/C (ISBN: 9781799823902) • US $285.00

Diagnosing and Managing Hashimoto's Disease Emerging Research and Opportunities
Mahmoud Sakr (Alexandria University, Egypt)
Medical Information Science Reference • © 2020 • 263pp • H/C (ISBN: 9781522596554) • US $245.00

IGI Global
DISSEMINATOR OF KNOWLEDGE

701 East Chocolate Avenue, Hershey, PA 17033, USA
Tel: 717-533-8845 x100 • Fax: 717-533-8661
E-Mail: cust@igi-global.com • www.igi-global.com

List of Contributors

Table of Contents

Detailed Table of Contents

some other periods. Development, which is considered as a whole together with the physical, cognitive, linguistic, social, and emotional growth areas, can be affected by many factors. These factors are expressed as heredity, hormones, and environment. This chapter, which moves from the understanding that a healthy society comprises of healthy individuals and determination of the factors affecting development, is of crucial importance in this regard. It is aimed to address and examine the factors affecting development.

Recognizing this period and the features in this period is regarded very important in terms of evaluating, following, and supporting child development. The concept of a developmental task should be approached in each period in order to understand the development. An individual should be biologically ready to carry out the developmental tasks belonging to the period he is in. However, in addition to biological readiness, the environmental conditions like the education environment should be appropriate in order to carry out the developmental tasks. Especially in the early years, schools have crucial roles in children's fulfillment of the developmental tasks and developing. Development periods and development duties should be known in order to observe and understand the development process better. Based on this view, the development periods and the development duties are to be explained in this section.

Children's exposure to many risk factors such as; need for protection, living on the streets, working, abuse and neglected, pushed into crime, exposed to violence, obliged to immigrate due to war, living under socio-economic disadvantageous conditions, having chronic diseases, being a disabled child and living in divorced families can arise from both their families and from the social structure. Throughout the world, many children live at risk due to various reasons. Irrespective of the reasons, which risk group the children enter and the factors causing this situation should be examined. According to the obtained data, the factors causing to such situations should be determined and necessary precautions should be taken. Thus, the negative conditions, under which the children are, can be improved and the children can be reintegrated into society.

In addition to the children who show typical development, there are also children who show atypical development. Individuals whose development varies compared to their peers and whose differences are identified by experts are called individuals with special needs. Individuals with special needs can be children with intellectual disabilities, sight, hearing impairment, language and speech disorders, physical disability and chronic illness, specific learning difficulties, children with autism spectrum disorder and gifted children. The causes of special needs are generally prenatal causes, which can often be hereditary and can be classified as natal and postnatal. The identification of individuals with special needs from

Living things start life with fertilization. Human offspring develop rapidly in approximately 38-42 weeks. Birth is a transition process where human offspring meets life. After a healthy birth, a new period begins for the baby and the mother. The first years following birth, infancy is a process in which the baby tries to adapt to life and gains basic skills. In this section, childbirth and the birth process are examined. In the postnatal period, it continued with the characteristics of the mother and newborn and the evaluation of the newborn. Development in infancy; physical development, cognitive development, language development and social and emotional development has analyzed.

Early childhood is a period that is associated with the development process of children. It was determined that physical, motor, cognitive, language, emotional and social development, which was created in accordance with the characteristics of the developmental period of children in the early childhood, did not change much direction in later ages, the improvement in the same direction was higher. In the early childhood, children begin to learn the rules of society limited to their own environment. They understand language and show great development. Socialization and gender awareness begin to form. The researches revealed that behaviors and attitudes gained during this period shape the habits, attitudes, beliefs, value judgments and personality traits of the individual when they are adults. In this section, the changes in physical, motor, cognitive, language, emotional-social development of the first childhood are examined.

Child development is segmented into five periods and the bridge between early childhood and adolescence is named as "middle childhood." One of the milestones of this period is schooling. Middle childhood children start learning about the world; their roles, responsibilities and how to participate in this world by communicating with others besides the family members. These abilities are gained mostly at school with peer relations. In this chapter, the developmental areas of middle childhood children will be discussed with examples from recent studies.

Development expresses the process that starts from impregnation and continues till the end of life. Man is a living creature that is born with a number of hereditary features and can improve them. The features of development areas should be known in order to recognize man as a whole. In the classification regarding the development periods of people, the late childhood period is between the ages of eleven to eighteen and it is also expressed as puberty. In this period the process of transition to adulthood is observed.

Section 3
Current Issues in Child Development

Early intervention refers to programs developed for infants and young children with developmental delays or inadequacies or who are at risk group, and their families. The aim of the early intervention is providing services for children with disabilities, minimizing their developmental delay and disability levels and guiding parents for positive development of children. Early intervention includes extremely important studies for children in need of developmental support and their families. The effective early intervention treatment requires a gradual progress, teamwork with a multidisciplinary approach, and family should be considered as the most important element of the treatment and participate in the decision-making at every phase, alongside effective use of the Individualized Family Service Plan and updating the plan when it is necessary. It should be noted that early intervention is an ongoing process. In this process, the family and the child should be considered as a whole and supported in a holistic way.

Family support programs are defined as the programs which are organized to provide parents to take part in all the stages of their child's development and education periods. Through the family support programs, it is aimed for parents to gain their self-confidence about their child's development and education and increase their knowledge and skills they need. Therefore, it is very important to consider social and cultural views of the parents, prioritize the voluntariness and privacy, and establish a clear communication within the frame of respect while organizing programs aimed at this field. Many family support programs are being applied both in our country and abroad.

The natural environment has generally a structure to support all areas of development of children. It can be said that the interior spaces are deprived of naturalness and have less stimulation compared to outdoor spaces. Playgrounds outdoors have forms, textures, heights, and mobility in a natural or artificial landscape have the energy to encourage children to play and move. The first years of the early childhood period are of a vital importance in terms of all areas of development. Pedagogues or parents should provide the activities and environments that will enable the infants and toddlers to get pleasure out of life and that can create a learning skill and give new skills to the children at same time. Therefore, in this study, outdoor and outdoor play activities that will support the development areas of 0-36-month-old children will be discussed.

Preface

Development is a dynamic process starting from prenatal period and lasting for life. Child developments involves cognitive skills, language skills, motor, social and emotional development, skills of communication and self care of the children who are between the ages of 0-18, grow normally, need special education, need protection, work, are refugees, are guilty and are in hospital. It is regarded important to give information about prenatal period, postnatal period and early childhood development to children, parents, educators, society and researchers in order to support all those areas of skills and development. Further, having information about the prenatal period, the postnatal period and the early childhood development is necessary to gain a global perspective, to know children, to create an effective learning environment, to carry out education at the earliest and on the top level and to interfere in the cases when it is required. Knowing the developmental periods of the children, the characteristics of those periods, and principles, which are effective in the development of the children in all areas, supports the heathy and timely development of the children. Accordingly, basic information about prenatal period, postnatal period and early childhood development are to be given in this book and the developmental areas of the children are to be explained in detail and up-to-date researches are to be mentioned. It is hoped that, thanks to this book, the child developers, educators, students and researchers can use their knowledge about the related area by knowing the developmental characteristics of the children. Global Perspectives on Prenatal, Postnatal, and Early Childhood Developmentbook consists of 3 sections and 21 chapters.

In the first section entitled "Introduction to Development," the following subjects are included; foundations of development, factors affecting development, developmental tasks, children at risk, exceptional children, children in hospital and children with adaptations and behavior problems. The chapters within this section are described below.

The name of the first chapter is "Foundations of Development/Introduction to Development." In this chapter it is aimed to give basic information about development in order tocomprehend development, what is more, some basic terms such as development, growth, progress, change, maturation, learning, readiness, period, critical period, time and historical time are discussed. It is required to know the principle of development in order to understand the development. Biological and many other environment factors influence the development. It is acknowledged that development is a product of heredity and environment. Developmental periods which take part in an individual's life and are defined with certain characteristics, are classified as "prenatal, infancy, early childhood, middle childhood, adolescence, adulthood and old age." Development is examined in some branches like "physical, motor, cognitive, language, social and emotional, personality, moral, perceptual development and sexual development". Families and educators play an important role in supporting the development of children. Ethics is a very sensitive issue in child development. Considering this thought, the first chapter entitled "Foundations of Development/Introduction to Development" serves as an introduction.

Second chapter is entitled "Factors Affecting Development: Development," which is regarded as a whole together with the physical, cognitive, linguistic, social and emotional growth areas, can be influenced by many factors. These factors are expressed as heredity, hormones and environment. In this chapter, which starts from the understanding that a healthy society consists of healthy individuals and determination of the factors affecting development is of great importance, it was aimed to address and examine the factors affecting development.

The name of the third chapter is "Developmental Tasks." The concept of developmental tasks should be approached in each period in order to understand the development. It is because of the fact that there are developmental tasks to be fulfilled in each period of development. An individual should be biologically ready to carry out the developmental tasks belonging to the period he is in. However, in addition to biological readiness, the environmental conditions like the education environment should be appropriate in order to carry out the developmental tasks. Especially in the early years, schools have crucial roles in children's fulfillment of the developmental tasks and developing healthily. Development periods and development duties should be known to observe and comprehend the development process better. Based on this view, the developmental periods and the developmental duties were explained in this section.

The fourth chapter is entitled "Children at Risk." Children at risk should be dwelt on in child development because children exposed to many risk factors such as being the children in need of protection, living on the streets, being working children, being abused and neglected, being pushed into crime, being exposed to violence, being obliged to immigrate due to war, living under socio-economic disadvantageous conditions, having chronic diseases, being a disabled child and living in divorced families can arise from both their families and from the social structure. Throughout the world, many children live at risk due to various reasons. Irrespective of the reasons, which risk group the children enter and the factors causing to this situation should be studied. According to the obtained data, the factors causing to such situations should be determined and necessary precautions should be taken. Thus, the negative conditions, under which the children are, can be improved and the children can be reintegrated into society. Based on this view, the children at risk were discussed in this chapter.

The name of the fifth chapter is "Exceptional Children." Besides the children showing typical development, there are also some other children who show atypical development. Therefore, individuals with special needs are one of the important subjects in child development. Individuals whose development varies comparing to their peers and whose differences are identified by experts are called individuals with special needs. The individuals with special needs can be defined as children with intellectual disabilities, sight, hearing impairment, language and speech disorders, physical disability and chronic illness, specific learning difficulties, children with autism spectrum disorder and gifted children. The causes of special needs are generally prenatal causes, which can often be hereditary and can be classified as natal and postnatal. The identification of individuals with special needs from early ages and the use of suitable intervention methods are vital for these children. Children who are happy and are a part of society via early recognition and accurate education methods, are directly related to the quality of the intervention to be performed.

The name of the sixth chapter is "Children in Hospital." Children who are different than healthy children and are in the hosptial due to their diseases, have an important place in child develoment. It is because of the fact that visiting a hospital for hospitalization or treatment is one of the foremost factors triggering stress for children. Under this stress, lie both the restriction of children's freedom and independence due to getting into an unknown environment and the rapidly changing events getting into an unknown around the child. The child's reactions may become even more aggressive, especially if they feel the fear and anxiety experienced by their parents. Based on this importance, this chapter focused on the responses of children to hospital life according to their developmental stages and on what can be done about their adaptation in hospitals.

The seventh chapter is entitled "Children with Adaptation and Behavior Problems." In this chapter, adaptation and behavior problem is explained conceptually, and detailed information is presented about causal factors related to biological, family, and school. Especially in the preschool period, adaptation and behavioral problems such as thumb-sucking, nail-biting, enuresis, encopresis, masturbation, lying, aggression, stubbornness, jealousy, fear, anxiety, stealing, sleep disorders, and eating disorders are explained in this chapter, and precautions to reduce/prevent these problems were mentioned.

In the second section entitled "Periods of Life; child development" is approached starting from the prenatal period, also childbirth and development in infancy, early childhood development, middle childhood development, development in late childhood are explained in detail. The chapters within this section are described below.

The name of the eighth chapter is "Prenatal Development." The prenatal period is an important process for the parents to become conscious about the development of the baby. Particularly in this process, in which mental and physical changes are observed in the mother, both the mother and father should be supported in order for the baby in the mother's womb to have a healthy period and to prevent a risky pregnancy period. In this regard, especially with the impact of recent technological innovations and with the increasing awareness of parents through increasing literacy, it is reported that the mortality rate is decreased; however, it is also reported that severe injuries during pregnancy and disabled babies still exist. From this point of view, the basic concepts related to development, developmental areas, developmental principles, factors affecting development and Prenatal of development (embryo, fetal, placenta) were explained in detail in the antenatal development

The ninth chapter is entitled "Childbirth and Development in Infancy." Birth is a transition process in which human offspring meets life. After a healthy birth, a new period begins for the baby and the mother. The first years following birth, infancy is a process in which the baby tries to adapt to life and gains basic skills. In this section, childbirth and the birth process were examined. In the postnatal period, it continued with the characteristics of the mother and newborn and the evaluation of the newborn. Development in infancy; physical development, cognitive development, language development and social and emotional development were analyzed.

The name of the tenth chapter is "Early Child Development." First childhood is a period that is linked with the development process of children. It was determined that physical, motor, cognitive, language, emotional, and social development, which was created in accordance with the characteristics of the developmental period of children in the first childhood, did not change much direction in later ages, the improvement in the same direction was higher. In the first childhood, children begin to learn the rules of society limited to their own environment. They understand language and show great development. Socialization and gender awareness begin to form. The researches revealed that behaviors and attitudes gained during this period shape the habits, attitudes, beliefs, value judgments and personality traits of the individual when they are adults. In this section, the changes in physical, motor, cognitive, language, emotional-social development of the first childhood were examined.

The name of the eleventh chapter is "Middle Childhood Development." The middle childhood period involves the ages between 6 and 11 years. One of the milestones of this period is the schooling. Middle childhood children start learning about the world; their roles, responsibilities and how to participate in this world by communicating with others besides the family members. These abilities are gained mostly at school with peer relations. While children in this period have a lot to crown their social-emotional development and to master the cognitive skills that demonstrate the concrete stage of Piaget's cognitive development theory, most of the physical abilities have been performed. In this chapter, the developmental areas of middle childhood children were discussed with examples from recent studies.

The name of the twelfth chapter is "Development in Late Childhood." In the classification about the development periods of people, late childhood period is between the ages of eleven until eighteen, and it is also expressed as puberty. In this period the process of transition to adulthood is observed. Further, the teenager starts to become independent from his/her family and acquire personal values and purposes in this period. In this study, the late childhood period was explained in detail.

In the third section entitled "Current Issues in Child Development" the following subjects are discussed in detail; Bilingualism in Early Childhood, Early Literacy Skills, Early intervention in Child Development, Yoga for Babies and Children: Introduction, Definition and Importance of Yoga, Benefits of Yoga, Yoga in Education Contribution of Outdoor Play Activities to Infants and Toddlers Developments, Interactive Story Reading and Language Development in Early Years, Perinatal and Early Childhood Development: Education, Innovation and Sustainable Development Goals Problems, Adaptation of Cognitive Abilities Scale into Turkish A Validity and Reliability Study: Cognitive Abilities Scale Preschool Form. The chapters within this section are described below.

The name of the thirteenth chapter is "Bilingualism in Early Childhood." Bilingualism that arises from different reasons (such as wars, migrations, geographical proximity, and trade, tourism, marriages, living to better standards, work and education) influences children, so the children have to grow up in environments where more than one language is spoken. Therefore, the bilingualism subject is regarded important. There are various arguments about whether being a bilingual for a child is advantageous or disadvantageous. The studies carried out in this field have shown that the disadvantage can be converted into advantageous situations on condition that the process about the acquisition or learning of the second language has been managed well. From this point of view, the importance of concepts such as language development in early childhood period, views on language development, bilingualism, language acquisition and teaching in early years were explained in detail in the bilingualism in the early childhood section.

The fourteenth chapter is entitled "Early Literacy Skills." Early literacy skills are related to many developmental areas of children such as language development and motor development. The development of children's early literacy skills influence their following development about literacy. Therefore, teachers should promote children's early literacy skills. In addition, the activities of conscious parents with their children positively affect their early literacy development. Based on these, it is important to address the issue of early literacy. For this reason, the definition of early literacy and concepts related to early literacy, the importance of early literacy, the factors affecting early literacy skills, the development process of early literacy, educational approaches related to early literacy, promotion of early literacy skills were included in this book chapter.

The name of the fifteen chapter is "Early intervention in Child Development." Early intervention indicates to programs developed for infants and young children with developmental delays or inadequacies or who are at risk group, and their families. The aim of the early intervention is providing services for children with disabilities, minimizing their developmental delay and disability levels and guiding parents for positive development of children. Early intervention includes extremely important studies for children in need of developmental support and their families. The effective early intervention treatment requires a gradual progress, teamwork with a multidisciplinary approach, family should be considered as the most important element of the treatment and participate in the decision-making at every phase, effective use of the Individualized Family Service Plan and updating the plan when it is necessary. It should be noted that early intervention is an ongoing process. In this process, the family and the child should be considered as a whole and supported in a holistic way. Accordingly, Early Intervention was explained in detail in this section.

The name of the sixteenth chapter is "Family Support Programs." Family support programs are defined as the programs which are organized to provide parents to take part in all the stages of their children's development and education periods. Through the family support programs, it is aimed for parents to gain their self-confidence about their children's development and education and increase their knowledge and skill they need about their children's development. Therefore, it is very important to consider social and cultural views of the parents, prioritize the voluntariness and privacy, and establish a clear communication within the frame of respect while organizing programs aimed at this field. Many family support programs are being applied both in our country and abroad. Based on this thoughts family support programs were discussed in this chapter.

The seventeenth chapter is entitled "Contribution of Outdoor Play Activities to Infants and Toddlers Development." The natural environment has generally a structure to support all areas of development of children. It can be remarked that the interior spaces are deprived of naturalness and have less stimulant compared to the outdoor spaces. Playgrounds outdoors, forms, textures, heights, and mobility in a natural or artificial landscape have the energy to encourage children to play and move. The first years of early childhood period are of a vital importance in terms of all areas of development. Educators or parent should provide the activities and environments that will enable the infants and toddlers to get pleasure out of life and that can create a learning skill and give new skills to the children at same time. Therefore, in this study, outdoor and outdoor play activities that support the development areas of 0-36-month-old children were discussed.

The name of the eighteenth chapter is "Perinatal and Early Childhood Development: Education, Innovation and Sustainable Development Goals Problems." This chapter looks at the systems approach in addressing the dilemma of investing in neonatal health in resource constraints environment. Is investing in neonatal care the key to attaining sustainable development goals? Is it just or even just to invest in Neonatal care in the midst of competing interests? It applies causal loop diagrams as part of systems dynamics approach in seeking to identify the intended and intended consequences of investing in Neonatal Health. A look into countries have done well in Neonatal health and their investment early childhood education was done in order to give insight into the relationship between the investments.

The name of the nineteenth chapter is "Interactive Story Reading and Language Development in Early Years." Interactive story reading is one of the most studied formats for increasing language learning in children. Many research studies have shown that children make significant gains in various areas of development through interactive story reading experiences. Parents can support their children's language development and thereby their children's literacy development by interactive story reading. Interactive story reading is developed by Whitehurst et.al (1988) and it is a process that leads the children improve their analytical thinking, expressing themselves, problem solving skills and inquiring. Moving from this, in this chapter, the effects of interactive story reading on language development of children was discussed through research done in the area.

The name of the twentieth chapter is "Adaptation of Cognitive Abilities Scale into Turkish: A Validity and Reliability Study: Cognitive Abilities Scale Preschool Form." This study aimed to adapt Cognitive

Abilities Scale-2 (CAS-2) Preschool (24-47 months) Form into Turkish by focusing on 24-36-month children and to conduct validity and reliability studies. The study utilized survey model and the study group consisted of 60 socio-culturally disadvantaged 24-36-month children in Bolu province. CAS-2 Preschool Form includes five subscales: oral language, reading, mathematics, handwriting and enabling behaviors. Translation and reverse translation procedures were used to adapt the scale developed in English into Turkish. Expert opinion was sought for content validity. Test-retest correlation and item-total score correlation consistency measures were used to examine the psychometric properties of the Turkish form. Results demonstrated that the scale is a valid and reliable instrument to determine cognitive abilities in 24-36-month Turkish children. Infant and pre-school forms are recommended to be used for 3-47 month at every socio-cultural level.

The name of the twenty first chapter is "Yoga for Babies and Children: Introduction, Definition and Importance of Yoga, Benefits of Yoga, Yoga in Education." Yoga is defined as merging or keeping under control in Sanskritic. Yoga; means relaxing in the body, slowing down in breathing, and mentally calming down. Yoga in general includes; physical postures / poses, breathing practices and movement meditation to improve self-regulation skills within the body and mind. Yoga; means relaxing in the body, slowing down in breathing, and mentally calming down. Yoga; is a holistic application system that includes various techniques such as physical postures / exercises, breathing exercises, deep relaxation techniques and meditation / awareness practices. In this chapter yoga for babies and children were dwelt on.

Section 1
Introduction to Development

It is regarded very important that individuals, who have an effect on children's development and education, personnel and researchers about the children and particularly the parents should have information about the basic subjects related to child development. This section serves as an introduction to child development. Subjects such as; foundations of development, factors affecting development, developmental tasks, children at risk, exceptional children, children in hospital and children with adaptations and behavior problems are included in this section.

Chapter 1
Foundations of Development/ Introduction to Development

Beyhan Özge Yersel
https://orcid.org/0000-0003-4736-1391
Ankara University, Turkey

Ender Durualp
Ankara University, Turkey

ABSTRACT

Development is a dynamic process starting from prenatal period and lasting for life. In order to understand development, some basic concepts such as development, growth, progress, change, maturation, learning, readiness, period, critical period, time, and historical time should be mentioned. To understand development, it is necessary to know the principle of development. Development is affected by biological and many environmental factors. It is accepted that development is a product of heredity and environment. The periods of development, which take part in the life of the individual and defined with certain characteristics, are classified as prenatal, infancy, early childhood, middle childhood, adolescence, adulthood, and old age. Development is examined by branching such as physical, motor, cognitive, language, social and emotional, personality, moral, perceptual development, and sexual development. Families and educators play an important role in supporting the development of children. Ethics is a very sensitive issue in child development.

INTRODUCTION

Development is considered as a dynamic process that starts from the prenatal period and continues throughout life. In this lifelong process, systematic and regular changes are observed in certain periods. During the life period starting from fertilization to death, the changes that occur in the organism following each other are defined as "development" (Gürsoy & Özaslan, 2016).

Human beings are regarded as "children" in the period from fertilization to adolescence. Considering that the changes and continuities in the child's life constitute the concept of "development", it is

DOI: 10.4018/978-1-7998-2952-2.ch001

inevitable that there is a field that examines the child's development. Good knowledge of development and development processes is of paramount importance for parents and all professionals working with children to raise healthy and happy children. Knowing the developmental stages of children in full detail and taking into account the individual differences ensure that the expectations of parents and professionals working with the child are realistic and provide opportunities for effective learning environments for the child (Baran, 2011; Erden & Akman, 2006; Şahin, 2018). Especially, since development in infancy and early childhood periods is a rapid progress, it is important to know the development in order to identify the developmental problems that will occur in the future, to monitor the development and to support the development (Rauh, 2005).

The aim of this section is to give information about the basic concepts of development such as development, growth, progress, change, maturation, learning, readiness, period, critical period, time, historical time; to provide understanding of development, and to explain the elements and principles of universal development. At the same time, development periods and development areas are included, biological and environmental factors affecting development are emphasized, heredity and environmental impact are discussed. Perspectives on development are summarized and suggestions are provided to families and educators to support development. Ethics in child development is explained briefly and its importance is mentioned.

BASIC CONCEPTS OF DEVELOPMENT

There are some basic concepts that must be known to understand development. Many concepts related to development are confused and used interchangeably. In this chapter; basic concepts related to development such as development, growth, progress, change, maturation, learning, readiness, period, critical period, time and historical time are explained.

Development

Development is stated as changes and continuity in human behaviors over time depending on structure, thinking, or biological and environmental factors. In other words, development is an expression of the functional changes of the individual (Bayhan & Artan, 2011; Özer & Özer, 2009).

The general course of development is that it is a "lifelong" process, and as a result, a "lifelong development" approach has come to the fore. The lifelong development approach is an approach that has been developed to examine the development of an individual from birth to death and to explain human behavior in this process. The most important difference of this approach is that it does not terminate development with adulthood and develops an inquiring perspective on the causes of these changes by examining life-long changes (Eryılmaz, 2011; Papalia & Ods, 2004).

Growth

Growth is a concept often confused with the concept of development and used interchangeably from time to time. While the increase in body structure as a result of the proliferation and expansion of cells is indicated as "physical growth", the concept of growth includes all physical growth and is used more extensively in the field of development (Gallahue et al., 2014). Growth is the quantitative expression of development. For example, when the height increases from 160 cm to 165 cm, or when a baby's weight increases from five to 10 kilos, it is growth. The child does not only grow physically, but there are also changes in the structure and size of the internal organs. Growth occurs at various speeds in various organs of the body (Aral et al., 2000).

Progress

The concept of progress is confused with growth. The difference that distinguishes progress from growth is the emergence of qualified and quantitative changes in development in terms of organization and functioning. A child's progress in the psychomotor field is a sign of growth, while a child's progress from a primitive stage to a more skillful stage is an expression of progress (Ataman, 2003).

Change

Changing from one state to another is defined as change. Change can be positive or negative. When a child does not know how to write, learning to write means change. That a child who does not use slang words starts using slang words is an example of the negative state of change (Aslan, 2017; Özyurt, 2012).

Maturation

Maturation is the biological change undergone with the effect of heredity, regardless of learning experiences, in order for the organs of the body be able to perform the functions expected of them. The success of many psychomotor behaviors depends on maturation. For example, if the child's bone and muscle structure does not reach enough maturation, no matter how much effort is made, the child cannot learn to walk (Senemoğlu, 2007). A child cannot be expected to functionally hold the pencil and write without developing the hand muscles and bone structure (Şahin, 2018). It is a known fact that forcing children to learn before reaching sufficient maturation does not have any benefit for the development of the child, on the contrary, it is extremely wrong (Baran, 2011). It is seen that the child who reaches a sufficient maturity level can learn the necessary knowledge, skills and attitudes much better (Başaran, 1996).

Learning

One of the most important features that distinguishes human beings from other living things is their ability to learn. The relatively permanent behavioral change that occurs as a result of the interaction of the individual with his/her environment is defined as "learning". In order for a child to hold a pencil properly, it is necessary to try to hold a pencil as much as the bones reach maturity, and to be reinforced when holding it properly (Elibol, 2016; Senemoğlu, 2013).

Readiness

Readiness can be defined as the state of cognitive, motor, physical, social and affective readiness to perform a certain learning activity (Bacanlı, 2016; Senemoğlu, 2012). For example, if a child who has reached a sufficient maturation to acquire toilet training is inadequate to use the toilet, if s/he has no experience in using the toilet and the mother has a negative attitude, the child does not have sufficient readiness for toilet training (Aral et al., 2000; Atay, 2009).

Readiness refers to both quantitative and qualitative aspects of development. Factors affecting readiness include; the individual's age, development level, maturity level, attitude, motivation and health status (Aslan, 2017).

Period and Critical Period

It is known that certain features come to the fore within certain time periods. These stages of development, in which some features come to the fore, are called "periods". New-born, infancy, childhood, adolescence, adulthood and old age are examples of periods (Selçuk, 2010). The stages that are suitable for the individual to gain certain behaviors are called "critical period" (Gündoğdu, 2013). Encountering effective stimuli in critical periods will facilitate the child's learning and ensure that their knowledge is permanent (Wilson, 1996). Providing the necessary opportunities by taking into consideration the critical period of the children will positively affect learning and enable children to gain important experiences in their lives (Şahin, 2018). For instance, early childhood is considered as the critical period of development. The first year of life is a critical period for the child to gain the basic sense of trust. The individual is at risk of not being able to gain a behavior that he/she cannot gain in the critical period (Bayhan & Artan, 2011; Elibol, 2016; Kandır & Alpan, 2008).

Time

Time is effective on learning and maturation. For example, fertility is considered a risky period for women's health. If a woman is not immune to rubella, hearing loss may occur in the fetus in the second month of pregnancy and a risk of miscarriage may occur. However, if the same woman suffers from this disease in the sixth month of her pregnancy, the baby may be born healthy. The first two to eight weeks of pregnancy are important for fetus and new-born health (Bayhan & Artan, 2011; Durualp, 2019).

Historical Time

Historical time is the concept that expresses the factors (event/phenomenon) that affect the development of the individual in the social and physical environment in which s/he lives. To be a child in today's internet and technology world, to be a child during the Gulf Wars, to experience the İzmit earthquake in 1999 can be given as an example of historical time (Gündoğdu, 2013).

BASIC PRINCIPLES OF DEVELOPMENT

Human development takes place within the framework of some universal principles, and these principles need to be known in order to understand and interpret the development correctly. The basic principles of development are described below:

- ***Development occurs as a result of the interaction of heredity and the environment:*** Development occurs as a result of both the interaction of heredity, which is the most important biological factor, and the environment. It is accepted that the impacts of heredity and environment on development are equal.

- ***Development is continuous and occurs in stages:*** Development is continuous, and it is an expression of a progressive, cumulative process. In development process, each phase is based on the previous one and initiates the next one.
- ***There are individual differences in development:*** Since each individual has different inheritance characteristics and environmental experiences, it is possible to talk about individual differences in development. Although the developmental stages follow the same order in all children with normal development, it should be noted that some differences may occur. One baby may begin to walk when s/he is 10 months old, while another baby may begin to walk when s/he is 12 months old. Some babies may teethe when they are six months old, but some babies may teethe their first tooth when they are 10 months old. Some children learn their native language early, while some children learn later. All these examples highlight the importance of individual difference in development (Anlar et al., 2008; Aslan, 2017; Gürsoy & Özaslan, 2016).
- ***Development follows a regular sequence:*** Development follows a certain order.
    - *Development is from head to toe (Cephalocaudal).* In the womb, the baby's head develops first, then the arms and legs. After birth, the baby first raises his/her head, then sits, stands up and walks at the last stage.
    - *Development is internal to external (Proximodistal).* In the prenatal period, after the inner parts of the body are formed, the outer parts form; first the body, then the arms and legs. After birth, the shoulders, then the arms and hands develop.
    - *Development is from general to specific.* The baby first gains control of the big muscle and then the control of the small muscles.
- ***There are critical periods in development.*** There are periods in which the organism is more open and more likely to learn in some areas of development. The critical period varies according to age and skills to be acquired. The absence of the mother or caregiver during infancy adversely affects the psychosocial development of the child negatively, while the first three months in the prenatal period are crucial for brain development.
- ***Development is a whole:*** Physical, motor, cognitive, language, social and emotional development is a whole. A failure that occurs in any of the development areas causes the other development areas to be affected negatively.
- ***Development occurs at different speeds in different areas.*** For example, while physical development is faster in the first years of life, the development of language and personality is faster in the following periods.
- ***The speed of development is not the same at all ages.*** Growth and development according to age do not occur at the same rate and speed. The head grows rapidly before birth, but after birth, growth slows down and decreases to one-seventh of the body in adulthood (Aral et al., 2000; Aslan, 2017; Başal, 2012; Bayhan & Artan, 2011; Elibol, 2016).

PRINCIPLES IN CHILD DEVELOPMENT

12 principles suitable for development have been declared in the "Developmentally Appropriate Practice in Early Childhood Programs Serving Children from Birth through Age 8" statement of The National

Association for the Education of Young Children of United States of America. Considering these principles, the principles to be considered regarding development are explained below:

1. Development is important with all its areas and each development is interrelated. Therefore, it should not be forgotten that activities with children will contribute to more than one development area of the child.
2. It is extremely important to know the order of learning of certain concepts and skills. Because some concepts and skills can be taught based on previously learned concepts and skills. This shows that the readiness of the child should not be ignored.
3. Since children's development and learning speeds are different, individual differences should be taken into consideration in the programs prepared for children.
4. Development is the product of the interaction of heredity and environment. Therefore, besides the characteristics that run in the child's family, environmental factors that may affect him/her should be taken into consideration.
5. It is important to be aware of the critical period of children. Early experiences of children have a significant impact on development and learning, which can be positive or negative.
6. Language, cognitive, motor, social and emotional development of children becomes more complex over time. Providing the child with a "development support program from simple to complex" will make a significant contribution to the child's learning by doing and experience.
7. The safer the relationships children establish with their environment, the more effectively the development will progress. Safe relationships with parents will contribute positively to the relationships with peers.
8. In the development and learning process of the child, especially the families, the school environment and the society in which they live have significant effects. Therefore, the sociocultural dynamics in which the child lives should be taken into consideration in the programs to be prepared for children.
9. Since each child's learning speed and learning style are different, development support appropriate to each child's learning style is required.
10. Play is the most important tool to improve children's cognitive, language and social competences and self-regulation. For this reason, it is very important that the programs to be offered to children are given through games.
11. Development and learning take place when the existing knowledge and skills of children are suitable for their upper stage. They should also have many opportunities to apply newly acquired skills. Therefore, taking into account the child's readiness and giving the child the opportunity to experience something is an important opportunity for the child's development.
12. Children's experiences are one of the most important factors that shape their motivation and approach towards learning. Children's home and school experiences determine their approach to learning. Therefore, building strong relationships with children and choosing the most effective working and teaching methods for families and children together are of great importance (NAEYC, 2009).

FACTORS AFFECTING DEVELOPMENT

Development is affected by many factors. Genetic (biological) factors stand out in determining the potential of the child, while environmental factors include the use of this potential (Şahin, 2018). Knowing the factors that affect the development of children is important in order to ensure that the necessary precautions are taken for the children that carry the risk of development and that the child's existing potential can be fully utilized (Gökçay et al., 2000).

- ***Biological Factors*:** Biological factors treat development according to a genetic plan. When discussing biological factors, the effects of "heredity" and "endocrine glands" are mentioned (Bayhan & Artan, 2011). Inheritance of all genetic traits transferred to organism from one generation to another with fertilization is explained by heredity. Genes are contained in chromosomes. Gender, eye color etc. are formed through genes. Whether a feature will appear is explained through **genotype** and **phenotype** concepts. **Genotype** is all the innate features of the organism. **Phenotype** is its effect that appears with the effect of the environment and can be observed. For example; a child that has the innate taller height *genotype* may have a shorter height than his/her potential, in other words, *phenotype* due to unfavorable environmental factors or nutritional conditions (Morgan, 2013).

Another biological factor is the hormones secreted by the endocrine glands. Hormones affect development and behavior. Endocrine glands are pituitary, thyroid, parathyroid, pancreas, adrenal and sex glands (Gürsoy & Özaslan, 2016).

- ***Environmental Factors:*** The environment is all the external stimuli that affect human development in all stages of development starting from fertilization (Baran, 2011). The effects of the environment are examined in three sections as "prenatal, during birth and postnatal".
- ***Prenatal Factors:*** Although the organism does not interact with the outside world in the prenatal period, it is a period in which the organism is most affected by the external environment and may face life-long dangers. Some factors that affect development in prenatal period have been classified as follows (Durualp, 2019).
- ***Demographic Factors:***
    - Pregnancy in adolescence (aged 16 and below)
    - Pregnancy in aged 35 and above
    - The education level of parents is less than basic education
    - Being underweight during pregnancy (gaining less than seven kilos)
    - Being overweight during pregnancy (gaining more than 12 kilos)
- ***Socioeconomic Factors:***
    - Low socioeconomic level
    - Professional conditions of parents
    - Pregnancy being out of wedlock or unwanted pregnancy
    - A history of consanguineous marriage or genetic disease
    - Domestic violence or incompatibility
    - Minority or immigration status
    - Living in the high-pressure zone
- ***Influence of Mother's Lifestyle on Baby;***
    - Mother's smoking
    - Use of drugs and alcohol by the mother
    - Teratogenic substances

  - Undernutrition
  - Stressful life
- ***Birth Factors:*** Even if a healthy pregnancy is completed, the problems to be experienced during labor cause the development to be negatively affected. Some of the problems that may occur during labor are listed below:
  - Birth in unhealthy environments
  - Very low birth weight
  - Premature birth
  - The baby's lack of oxygen
  - Cord entanglement
  - The baby's getting a blow (Gürsoy & Özaslan, 2016).
- ***Postnatal Factors:*** There are many factors that affect the development of the child after the birth.
  - Socioeconomic status of the family
  - Sociocultural level of the family
  - Behaviors and attitudes towards the child
  - Birth order
  - Neglect or abuse of child
  - Diseases, accidents and traumas experienced by the child
  - Nutrition of the child (Şahin, 2018).

Inheritance and Environment Interaction

Changes in children over time occur due to internal processes or environmental factors. When literature is examined, the effects of heredity and environment are frequently discussed; while some models attribute the changes in children to heredity, some others argue that these changes occur through environment which includes factors such as life experiences, parents, materials, television, school etc. (Slee & Shue, 2014; Tuna et al., 2018).

DEVELOPMENT PERIODS

Developmental period is a period of time that takes place in the life of the individual, and defined by certain characteristics (Aslan, 2017). The purpose of classification of developmental periods is to better understand the developmental characteristics of the individual and to facilitate the understanding of the skills that will emerge according to age periods. There are some differences in the classification of development areas. While some theorists classify development based on "age", some theorists make a classification based on development characteristics (Eylen Özyurt, 2011; Gürsoy & Özaslan, 2016).

- ***Prenatal Period:*** It covers the period from the time the egg is fertilized to the moment of birth. There is a rapid change in this period. It is the phase where physical growth is fastest and sensitivity to environmental factors is most intense (Bayhan & Artan, 2011; Gürsoy & Özaslan, 2016).
- ***Infancy (0-24 Months):*** The period from birth to 18-24 months is considered as "infancy". Infancy is examined in two parts as "infancy (first year)" and "walking childhood (second year)",

and autonomy begins when the child takes the first steps in this process (Aslan, 2017; Berk, 2013; Santrock, 2011).

- ***Early Childhood (3-6 Years):*** This period in which basic habits are acquired covers the age of three to six years. It plays an important role in ensuring that the child will sustain a healthy development and shaping the next life. The focus of attention for children during this period is still the family, with fine and gross motor skills and strength develop. Independence, personal control and self-care are increasing. There are improvements in play, creativity and imagination skills. Even though the behavior is egocentric, the ability of understanding the perspectives of other people develops (Bayhan & Artan, 2011; Gürsoy & Özaslan, 2016).
- ***Middle Childhood (7-12 Years):*** This is a period in which the child emerges from the family home and opens up to the outside world and fuses with the social environment. Since the first years of this period is a period of adaptation to school and the last year is the period of transition to adolescence, it is a very important period. In this period, children's peers become more important, self-centeredness decreases, and logical thinking begins. There is a rapid increase in children's reading, writing and numerical processing skills (Bayhan & Artan, 2011; Elibol, 2016; Yazgan İnanç et al., 2007).
- ***Adolescence (13-18 Years):*** This is the period of transition from childhood to adulthood. It starts approximately at the age of 12-13 and ends at the age of 18-21. Features such as rapid increase in height and weight, changes in body lines and rapid development of secondary sex characteristics (prominent breasts, genital organ and facial hair growth etc.) are observed. In adolescence, thoughts become more logical, abstract and idealistic. In this period, it is expected from the adolescent to accept the developing body, to gain emotional independence from the family and to establish friendships with the opposite sex as well as friends of the same sex (Berk, 2013; Santrock, 2011). The concepts of early and late maturation appear in adolescence. Early maturation means the advantage of having better muscle co-ordination in boys but may lead to negative consequences for girls. The fact that girls are taller and physically more developed than their peers may cause them to experience a decrease in their attractiveness within the group (Elibol, 2016).
- ***Adulthood:*** Adulthood is divided into first adulthood and middle adulthood. During the first adulthood, which begins in the 20s and continues until the 30s, personal and economic independence, career, spousal selection, starting a family and being with the family are at the forefront. The middle adulthood covers the period from 40 to 60 years. In this period, features such as taking personal and social responsibility, transferring their knowledge and experience to the next generation, reaching and maintaining satisfaction in a profession gain importance (Santrock, 2011).
- ***Old Age:*** Old age is the last period of life. Physical strength of the individual decreases and health problems emerge. The individual continues to adapt to these changes during this period and the individual is expected to pay attention to his/her health and adapt to the physical and social changes (Başal, 2012; Çelikkaleli, 2014).

DEVELOPMENT AREAS

Development of the individual consists of areas such as physical, motor, language, cognitive, social-emotional and moral. Areas of development are interrelated. Delay in one development area may affect other development areas. For example, delays in cognitive development may lead to delays in language development. An individual with limited language development cannot have a healthy social development. Therefore, knowing the developmental areas of children and the characteristics of these areas plays an important role in determining any problems (Berk, 2013; Forcada Guex et al., 2006).

Physical Development

It is an area of development that includes changes in body and physical appearance. Increased body weight, increase in height, the development of all organs and bones, teeth, etc. changes are related to physical development (Senemoğlu, 2013; Şahin, 2018).

Motor Development

Motor growth is defined as voluntary mobilization of the organism in parallel with physical growth and development of the central nervous system. Motor development is divided into "fine motor (small muscle motor development) and gross motor (big muscle motor development)".

- ***Fine Motor (Small Muscle Motor) Development:*** Motor skills of the child using small limbs such as hands, fingers and toes. Skills such as installing legos, holding pencils, using scissors are fine motor skills.
- ***Gross Motor (Big Muscle Motor) Development:*** Movements such as running, leaping, jumping, climbing, in which the child uses the arms, legs, feet and the whole body, are called gross motor development (Şahin, 2018).

Cognitive Development

The development of mental activities that enable the individual to understand and learn the world around him/her is called cognitive development. Cognitive development process has an important role in acquiring new skills (Anlar et al., 2008). Cognitive development is a process that starts in the womb and develops rapidly (Doğan Keskin & Aral, 2016). Children understand the environment they live in and develop their thinking skills during the cognitive development process (Şahin, 2018).

In the field of cognitive development, there are basically four rules. These rules are;

1. Periods follow a certain order and it is not possible to change the order of the periods
2. There is a hierarchy followed by each period and each period includes the gains of the previous period
3. The development of each individual must be unique and therefore individual differences in development must be considered
4. It can be explained that there are typical developmental characteristics of each period (Bacanlı, 2002).

It is possible to mention some theorists in the field of cognitive development. According to Jean Piaget, cognitive development is the adaptation of the organism to the environment. Lev Vygotsky developed the "Sociocultural Cognitive Development Theory" about cognitive development. Jerome Bruner explains cognitive development as "to become independent of responses from stimuli". Robert Gagne, on the other hand, argues that the child is born ready to learn from birth and that cognitive development arises as a result of the experiences gained through the equipment that is brought from birth (Doğan Keskin & Aral, 2016). It is a part of cognitive development as it involves processes such as thinking, memory, reasoning, problem solving. Again, the child sees that the voices s/he makes lead him/her to a conclusion, and by repeating these voices and by the reinforcement of the others for these sounds, s/he begins to learn the language by establishing a stimulating response match (Senemoğlu, 2007; Şahin, 2018).

Language Development

Language is a versatile and highly developed symbolic system that enables the transfer of thoughts, emotions and wishes to other people by making use of common elements and rules in terms of sound and meaning in a society (Aslan, 2017). Language is a feature that distinguishes human beings from other living things and is one of the most important means of communication. The basis of the child's language development lies in the need to communicate, to attract the attention of other people, and to communicate their wishes, feelings and thoughts (Küçükkaragöz, 2013). Genetic, parental interaction and social environment, gender, family structure and bilingualism, sociocultural and socioeconomic factors and media tools affect language development (Anlar et al., 2008).

Social and Emotional Development

Social-emotional development is defined as the ability of the child to express himself/herself, to control his/her emotions and to be at peace and harmony with himself/herself and his/her environment. Studies conducted on the emotional characteristics of children have revealed that emotional development occurs as a result of both maturation and learning, and none of them is effective alone (Saarni, 2001; Yıldız Bıçakçı & Durualp, 2016). Socialization is a process of learning and teaching that results in the child moving into adulthood, where s/he acts as an independent adult from infancy, where he is dependent on others. During this period, the child is in an interaction with the family, teachers, peers and close environment, and it is of great importance that the people around the child become the right role models (Şahin, 2018).

Attachment has an important role in emotional development. Attachment theory, the child's attachment to mother or any person/object etc. that comforts the child can be explained as the defense of the attachment of the figure has an important function in the life of the child (Fonagy, 2001). The realization of secure attachment is considered as an indicator of a strong relationship between mother and baby. The mother's depression, violence, etc. factors cause a negative impact on the bond between mother and child. It also causes delays and negative effects on the development of the child (Şahin, 2018).

Children show different social-emotional developmental characteristics at different age periods. While the first social behavior of the baby after the birth is to hug and hold on to his/her mother, s/he is impersonated to adopt an egocentric attitude around age one. From age three, children who enjoy spending time with adults develop empathy. From age six onwards, school and friends are gaining importance alongside their parents, while adolescents are entering a period of emotional fluctuation. As the individual enters the adulthood period, adolescent mental tensions are replaced by a calmer process (Yıldız Bıçakçı & Durualp, 2016).

Personality Development

Cüceloğlu (1996) defined personality as "a distinctive, consistent and structured way of relationship with the other individuals established by the individual with his/her inner and outer environment". Personality development is also a joint product of heredity and the environment. The reasons underlying the different behaviors of one child from the other may be explained as genetic and environmental reasons. Parents present their children with their own genes as well as the opportunities of the socioeconomic and socio-cultural environment in which they live (Şahin, 2018).

Perceptual Development

Perception occurs when the stimulus one or more sensory organs transmit to the brain is interpreted in the brain. Perceptual development, on the other hand, refers to the development in the ability to understand and interpret the stimuli coming to our sensory organs (Aral et al., 2000; Aslan, 2017).

Moral Development

Moral development is the change that occurs in an individual's perception of justice, right and wrong, and his/her behavior in the face of moral issues (McDevitt & Ormrod, 2002). Issues that some theorists discuss in moral development are the close examination of the socialization process of children and the role of family and society in socialization; points some other theorists focus on are how the children interpret the moral and how do they sense ethical issues (Özgün, 2011).

Sexual Development

The individual's growing to maturity sexually, determining one's sexual identity and behaving in accordance with it, controlling the sexual impulses, saturating and getting pleasure from the sexual impulses, process of having the necessary knowledge and skills to solve the problems faced by sexuality are sexual development (Aslan, 2017; Yılmaz, 2008).

THEORIES AND THEORISTS OF DEVELOPMENT

There are many theorists working in the field of child development, and each of them discuss different topics.

Cognitive Development Theories

Piaget conducted studies in the field of cognitive development. He evaluates development as a result of heredity and environment and he examines cognitive development in four parts; sensory motor, pre-processing, concrete operations and abstract operations (Santrock, 2011; Şahin, 2018).

Vygotsky defined the child's development in social and cultural activities. Vygotsky argued that the child's cognitive development cannot be separated from his/her social interaction with his/her more capable peers and adults (Gauvain & Parke, 2010; Holzman, 2009).

Bruner stated that children learn by researching and discovering and argued that learning is also affected by sociocultural structure (Şahin, 2018).

Theories of Language Development

- ***Behavioral Theory:*** Behavioral theory, under the leadership of BF Skinner, argues that language and speech are acquired through operant conditioning (reinforcement and punishment) (Dunlap, 2009).
- ***Social Learning Theory:*** Social learning theory is a theory that is synonymous with Albert Bandura. Bandura argues that the social context of language is an essential part of language development and that children acquire language through observation and imitation (Kürkçüoğlu, 2014).
- ***Interactive Theory:*** The first name that comes to mind when it comes to interactive theory is Piaget. According to Piaget, language plays an important role in shaping thought and language reflects the individual's cognitive level (Dunlap, 2009).

Psychoanalytic Theories

The psychoanalytic theories mainly connect the development to unconscious mind and assume that it is emotion based. When it comes to psychoanalytic theories, Freud and Erikson come into prominence.

Freud; he argues that individuals go through five psychosexual periods. These periods;

1. The period in which the baby's pleasure center is "mouth", and includes the age range of 0 to 1.5 years: *"Oral Stage"*
2. The period in which the child's pleasure center is "anus", and includes the age range of 1.5 to 3 years: *"Anal Stage"*
3. The period in which the child's pleasure center is "genitals", and includes the age range of 3 to 6 years: *"Phallic Stage"*
4. The period in which the child suppresses the focus of pleasure and develops his/her social-mental skills, and includes the age range of seven to until puberty: *"Latent Stage"*
5. The period in which the child's pleasure center is again "genitals", and includes the age range of puberty and later: *"Genital* Stage" (Değirmencioğlu, 2018).

Like Freud, Erikson also argues that individuals go through eight different periods. These periods are;

1. Trust vs. Mistrust: (0 to 1 Years)
2. Autonomy vs. Shame/Doubt: (2 to 3 Years)
3. Initiative vs. Guilt: (4 to 5 Years)
4. Industry vs. Inferiority: (6 to 12 Years)
5. Identity vs. Role Confusion: (13 to 18 Years)
6. Intimacy vs. Isolation: (19 to 25 Years)
7. Generativity vs. Stagnation: (26 to 40 Years)
8. Ego Integrity vs. Despair: (41 Years and Older) periods (Değirmencioğlu, 2018).

Behavioral and Social Cognitive Theories

When it comes to Behavioral and Social Cognitive Theory, the names that come to mind are "Skinner" and "Bandura".

Skinner drew attention to the operant conditioning. In operant conditioning, the results of a behavior change the likelihood that that behavior will reappear.

Bandura addressed the cognitive dimension of development with the environment and behavior (Güngör Aytar & Kurtoğlu Karataş, 2018; Santrock, 2011).

Ethological Theories

Lorenz and ***Tinberger*** emphasized that behavior is affected by biology, and its connection with evolution, and formed the concept of ethology in development (Güngör Aytar & Kurtoğlu Karataş, 2018). One of the contributions of the theories is that they focus on the universal and biological foundations of development and have used careful observation in the natural environment. Criticism, on the other hand, emphasizes biological basics and that critical and sensitive periods can be very strict (Santrock, 2011).

Ecological Theory

Bronfenbrenner's ecological theory consists of five environmental systems. These systems are microsystem, mesosystem, ecosystem, macrosystem and chronosystem. The microsystem includes the factors such as family, peers, and the school in which the individual lives, and the mesosystem includes the relationship between the microsystems. The ecosystem consists of the connections between a social environment in which the person has no active connection and the close context of the person. Macrosystem is the culture in which one lives. Chronosystem is how socio-historical conditions, lifelong transitions and environmental events take place in a pattern. While the lack of emphasis on cognitive factors and biological elements is the criticized aspect of the theory, that it draws attention to the connections between environmental systems and examines environmental elements systematically is thought among the contributions of the theory (Gauvin & Parke, 2010; Santrock, 2011).

Theories of Moral Development

Kohlberg was the first theoretician to come to mind in the field of moral development. In his theory, Kohlberg examined moral development in three sections: pre-traditional, traditional and post-traditional (Şahin, 2018).

PERSPECTIVES IN DEVELOPMENT

When the literature on development is examined, it is seen that experts view development from different perspectives. This section will explain some aspects of development.

Heredity and Environment Discussion

It is an ongoing debate on whether development is primarily affected by heredity or the environment. Heredity refers to the biological inheritance of the organism, while environment refers to environmental experiences. Developers with a hereditary point of view advocate that biological heritage has a major share in development, while environmentalists advocate that environmental experiences have a major share in development (Santrock, 2011). From the hereditary point of view, factors such as "gene, DNA, chromosome, genotype, phenotype, etc." are claimed to be effective on development. From the environmental point of view, factors experienced before, during and after birth are accepted as important factors. However, the idea that development is affected by both heredity and environment has also become widespread today. Characteristics related to heredity and environment affect development and both determine development and play an important role in shaping personality (Aslan, 2017).

Continuity and Discontinuity

Another point of view regarding development is "whether development is continuous or discontinuous. " It seeks whether development happens gradually based on a certain period of time or suddenly (Eryılmaz, 2011). This viewpoint, which reflects on the "predictability" of development, is concerned with whether development is "continuous or ends at some point". The belief that a four-year-old child will be cheerful and smiley throughout his/her life is shown as a proof of the fact that development is a continuing process. However, the belief that the same child could will lose joy when he is 25 points to a discontinuous aspect of development (Slee & Shute, 2014; Tuna et al., 2018).

Developers who advocate the environment define development as a "continuous" process, while those who advocate heredity often refer to development as "a set of different periods". Nowadays, it is argued that some aspects of development are continuous, and some aspects are discontinuous (show universal features) (Santrock, 2011; Yazgan İnanç et al., 2007).

Invariance to Change

Invariance means that the gains achieved in previous development periods remain constant. For example; if the question "Will the shy child be a shy individual in adulthood or will s/he continue his/her life as a talkative and social individual?" is answered as "S/he will be a shy individual in adulthood.", this shows that invariance in change is defended. Another thought that emerged from this idea is that in most developmental areas, development is gained in childhood and let-up in adulthood. Invariance emphasizes the extent in which we have developed in harmony with our previous experiences. Development experts, who focus on the concept of variance, believe that experiences gained at later ages can lead to changes in individuals. In other words, it is related to what extent we are different from our previous development periods. Experts who support the view of lifelong development focus on the speed of change and how it takes place rather than change and invariance (Baltes, 1987; Eryılmaz, 2011).

Universal Principles or Cultural Influences

One of the controversial issues in development is whether the factor shaping development is "universal" or "cultural". While some of the researchers argue that universal laws are the determining factor in development, some researchers point out to "cultural effects". By stating that the societies having an unhealthy cultural structure developed an unhealthy structure and this situation would lead to the development of unhealthy personality traits in children, Cüceloğlu (2000) highlighted the importance of cultural influences. In addition to these views, another view that has become accepted today is that "universal principles and cultural effects interact with each other" (Aslan, 2017; Özyurt, 2012; Yazgan İnanç et al., 2007).

THE ROLES OF PARENTS AND EDUCATORS IN SUPPORTING DEVELOPMENT

Early childhood is accepted as the "magical" years of life and considered to be the period when development is at its highest speed. In each phase of the period, it is highly important to support the physical, psycho-motor, social-emotional and cognitive developments of the child (Oktay, 1999; Ceylan et al., 2016).

Family environment into which a child is born is one of the most important environments that affect his/her development. Starting from birth, a child usually mostly interacts with the mother, and whoever cares for the child, their attitudes leave important traces on the child. Families play an important role in teaching their children social roles and creating positive social behaviors and values (Tümkaya, 2012). This situation reveals how important it is for the family to have knowledge about development.

In early childhood, families and educators play an important role in supporting the development of the child. The fact that families know about the developmental characteristics of their children enables them to create realistic expectations about their children. Expecting one-year-old child to gain toilet control, expecting a two-year-old child to ride a bicycle is not realistic, and often disappoints the family. Parents who are aware of the issues such as developmental periods and individual differences in development will not compare their children with other children and will display a more conscious view of their development (Baran, 2011).

School environment also plays an important role in children's lives as much as family environment. In-class attitudes and behaviors of educators have an effect on learning climate and students' motivation in addition to academic, social and emotional development. The educator should have knowledge about child development and learning styles. It should also be accepted that each child comes from a different sociocultural environment, and the individual differences of children should be taken into account when preparing their education programs. It should provide a democratic environment for children and create a moderate class climate. It should cooperate with families to ensure that children are in a healthy developmental process (Aslan, 2017; Sezer, 2018).

Ethics in Child Development

In the studies conducted with children, "ethics" topic should be given importance in order to prevent physical and psychological harm to children, and efforts should be made to protect children's rights.

- If the researcher is concerned that his or her research will harm the child, he/she should review or not conduct the research in a way that does not harm the child.
- Children should be informed about the research. Age and level of development must be taken into account when explaining to children. In addition, the parents of the children or the caregivers

of the child should be informed about the research, and the children and their families should be informed about the research process and possible research results.

- The identity of children and adults must be kept confidential. Children and adults are also entitled to discontinue the investigation.
- If the study was planned as an experimental study and the experiment yielded results that could be beneficial to the experimental group, the control group should be applied after the research was completed and the same benefit should be obtained (Bayhan & Artan, 2011).

CONCLUSION

As a result, in order to understand child development well and to provide effective health and education services for children, it is necessary to know development and the basic concepts and rules of development. Knowing different perspectives regarding development is highly important in understanding development. Families and educators play a major role in supporting development. In researches about children, ethics should be given importance in order to prevent physical and psychological harm to children.

REFERENCES

Anlar, B., Serdaroğlu, A., & Yakut, A. (2008). *Gelişimsel Cocuk nörolojisi.*. Ankara: Hacettepe Üniversitesi Hastaneleri Basımevi.

Aral, N., Baran, G., Bulut, Ş., & Çimen, S. (2000). *Çocuk gelişimi 1*. İstanbul: Ya-Pa.

Aral, N., & Doğan Keskin, A. (2016). Bilişsel gelişim ve dil gelişimi. In N. Aral (Ed.), Çocuk gelişimi. (pp. 96-126). Eskişehir: AÖF Yayınları.

Aslan, D. (2017). Gelişimle ilgili temel konular. In A. Köksal Akyol (Ed.), Erken çocukluk döneminde gelişim (pp. 86-105). Ankara: Anı Yayıncılık.

Ataman, A. (2003). Gelişim, kuramlar ve kavramlar. In A. Ataman (Ed.), Gelişim ve Öğrenme. (pp. 9-50). Ankara: Gündüz Eğitim ve Yayıncılık.

Bacanlı, H. (2002). *Gelişim ve Öğrenme*. Ankara: Nobel Yayınları.

Baltes, P. B. (1987). Theoritical propositons of life-span developmental psychology on the dynamics between growth and decline. *Developmental Psychology, 23*(5), 611–626. doi:10.1037/0012-1649.23.5.611

Baran, G. (2011). Çocuk gelişimine giriş. In N. Aral & G. Baran (Eds.), *Çocuk gelişimi*. (pp. 17-52) İstanbul: Yapa Yayınları.

Başal, H. A. (2012). *Nasıl mutlu ve başarılı bir çocuk yetiştirebilirim?* Bursa: Ekin Basım Yayın ve Dağıtım.

Başaran, İ. (1996). *Eğitim psikolojisi*. Ankara: Gül Yayınevi.

Bayhan, P., & Artan, İ. (2011). *Çocuk gelişimi ve eğitimi*. İstanbul: Morpa Kültür Yayınları.

Berk, E. L. (2013). *Bebekler ve cocuklar, doğum Öncesinden orta cocukluğa. (N. Işıkoğlu Özer, Trans.). Ankara: Nobel.*

Çelikkaleli, Ö. (2014). Bilim, eğitim, psikolojisi ve gelişimle ilgili temel kavramlar. In R. Çeçen Eroğul & F. Yurtal (Eds.), Eğitim Psikolojisi El Kitabı (pp. 1-46). Ankara: Anı Yayıncılık.

Ceylan, Ş., Gözün Kahraman, Ö., & Ülker, P. (2016). Annelerin erken çocukluk dönemine bakış açısı. *İnsan ve Toplum Bilimleri Araştırmaları Dergisi, 5*(5), 1333-1356.

Cüceloğlu, D. (2000). *İnsan ve davranışı*. İstanbul: Remzi Kitabevi.

Değirmencioğlu, H. (2018). Kişilik gelişimi. In N. Aral & Z. F. Temel (Ed.), Çocuk gelişimi (pp. 292-313). Ankara: Hedef CS Basın Yayın.

Dunlap, L. D. (2009). *An introduction to early childhood special education*. Upper Saddle River, NJ: Pearson Merrill Prentice Hall.

Durualp, E. (2019). Doğum öncesi gelişim ve doğum. In A. Köksal Akyol (Ed.), Erken cocukluk döneminde gelişim-I (pp. 114-144). Ankara: Anı Yayıncılık.

Elibol, F. (2016). 0-18 yaş çocuk gelişimi ve gelişim ilkeleri ile çocuk gelişiminde tanı ve değerlendirmenin yeri, önemi ve amaçları. In Ç. Dinçer (Ed.), Gelişimsel Tanı ve Değerlendirme Yöntemleri (pp. 26-58). Eskişehir: AÖF Yayınları.

Erden, M., & Akman, Y. (2006). *Eğitim psikolojisi. Gelişim-öğrenme-öğretme. Geliştirilmiş on beşinci baskı* (15th ed.) Ankara: Arkadaş Yayınevi.

Eryılmaz, A. (2011). Yaşam boyu gelişim yaklaşımı. *Psikiyatride Güncel Yaklaşımlar-Current Approaches in Psychiatry, 3*(1), 49–66.

Fonagy, P. (2001). *Introduction to attachment theory. In attachment theory and psychoanalysis*. New York: Other Press.

Forcada-Guex, M., Pierrehumbert, B., Borghini, A., Moessinger, A., & Muller-Nix, C. (2006). Early dyadic patterns of mother-infant interactions and outcomes of prematurity at 18 months. *Pediatrics, 118*(1), 107–114. doi:10.1542/peds.2005-1145 PMID:16818525

Gallahue, D. L., Ozmun, J. C., & Goodway, J. D. (2014). Motor gelişimi anlamak, bebekler, çocuklar, ergenler ve yetişkinler, (D. Sevimay Özer & A. Aktop, Trans.). Ankara: Nobel Yayıncılık.

Gauavin, M., & Parke, R. D. (2010). Socialization. In M. H. Bornstein (Ed.), *Handbook of cultural development science (pp. 239-258)*. New York: Psychology Press.

Gökçay, G., Köklük, S., Kayadibi, F., Eraslan, E., & Çalışkan, M. (2000). Çocuklarda ilk iki yılda gelişimi etkileyen faktörler. *Istanbul Tip Fakültesi Mecmuasi, 63*(4), 395- 405.

Gündoğdu, R. (2013). Öğrenmenin doğası ve temel kavramlar. In B. Çapri & B. Gündüz (Eds.), Eğitim psikolojisi (pp. 169-192). Adana: Karahan Kitabevi.

Güngör Aytar, A., & Kurtoğlu Karataş, B. (2018). Gelişim psikolojisine giriş. In A. Güngör Aytar (Ed.), Gelişim psikolojisi (pp. 12-44). Ankara: Hedef CS BasınYayın.

Gürsoy, F., & Özaslan, H. (2016). Çocuk gelişimine giriş. In N. Aral (Ed.), Çocuk gelişimi (pp. 2-24). Eskişehir: AÖF Yayınları.

Holzman, L. (2009). *Vygotsky at work and play*. Clifton, NJ: Psychology Press.

Kandır, A., & Alpan, Y. (2008). Okul öncesi dönemde sosyal-duygusal gelişime anne-baba davranışlarının etkisi. *Sosyal Politika Çalışmaları Dergisi*, *14*(14), 33–38.

McDevitt, T. M., & Ormrod, J. E. (2002). *Child development: educating and working with children and adolescents*. Upper Saddle River, NJ: Pearson Prentice Hall.Merrill.

Morgan, C. T. (2013). *Psikolojiye Giriş. (20th ed.) Yirminci Basım*. Konya: Eğitim Kitabevi.

NAEYC. (2009). *Position Statement (PDF) Developmentally Approptiate Practice in Early Childhood Programs Serving Children from Birth through Age 8, adapted in 2009*. National Association for the Education of Young Children. Retrieved from https://www.naeyc.org/files/naeyc/file/positions/PSDAP.pdf

Özer, D. S., & Özer, K. (2009). *Çocuklarda motor gelişim*. Ankara: Nobel Yayınları.

Özgün, Ö. (2011). Ahlak gelişimi. In M. E. Deniz (Ed.), Erken cocukluk döneminde gelişim. (pp.285-304). Ankara: Ertem Basım Ltd. Şti.

Özyurt, B. E. (2012). Gelişim konularına genel bakış. In A. Kaya (Ed.). Eğitim psikolojisi (pp. 1-27). Ankara: Pegem A Yayıncılık.

Papalia, D. E., Olds, S. W., & Feldman, R. D. (2004). *Human development* (9th ed.). New York: Mc-Graw Hill.

Rauh, H. (2005). At-risk concept. In B. Hopkins, R. G. Barr, G. F. Micheland, & P. Rochat (Eds.), Encyclopedia of child development (pp. 101-106). New York: Cambridge University.

Şahin, S. (2018). Çocuk gelişimine giriş. In N. Aral & Z. F. Temel (Eds.), Çocuk gelişimi (pp. 16-31). Ankara: Hedef CS Basın Yayın.

Santrock, J. W. (2011). *Life-span development* (13th ed.). Ankara: Nobel Yayın.

Senemoğlu, N. (2007). *Gelişim öğrenme ve öğretim- kuramdan uygulamaya*. İstanbul: Gönül Yayınevi.

Senemoğlu, N. (2013). *Gelişim öğrenme ve öğretim- kuramdan uygulamaya*. Ankara: Yargı Yayınevi.

Sezer, Ş. (2018). Öğretmenlerin sınıf yönetimi tutumlarının öğrencilerin gelişimi üzerindeki ekileri: Fenomenolojik bir çözümleme. *Hacettepe Üniversitesi Eğitim Fakültesi Dergisi*, *33*(2), 534–549.

Slee, P. T., & Shute, R. (2014). *Child development: Thinking about theories texts in developmental psychology*. New York: Routledge.

Tümkaya, S. (2012). Okul öncesi eğitim döneminde ailenin rolü ve önemi. In S. Tümkaya & F. Gülaçtı (Eds.), *Erken çocukluk eğitimi* (pp. 336–367). Ankara: Pegem.

Tuna, M. D., Sümer, H. M., & Diken, İ. H. (2018). Erken müdahale modelleri ve temel kuramlar. In P. Bayhan (Ed.), Erken müdahale (pp. 64-81). Ankara: Hedef CS Basın Yayın.

Wilson, R. A. (1996). Starting Early: Environmental Education During the Early Childhood Years. ERIC Clearing house for Science, Mathematics and Environmental Education. (ED402147)

Yazgan İnanç, B., Bilgin, M., & Kılıç Atıcı, M. (2007). *Gelişim Psikolojisi. Çocuk ve ergen gelişimi.* Ankara: Pegem A Yayıncılık.

Yıldız Bıçakçı, M., & Durualp, E. (2016). Çocuklarda sosyal ve duygusal gelişim. In N. Aral (Ed.), Çocuk gelişimi (pp. 128-156). Eskişehir: AÖF Yayınları.

Yılmaz, M. T. (2008). Cinsel gelişim. In M. E. Deniz (Ed.), Erken cocukluk döneminde gelişim (pp. 291-323). Ankara: Maya Akademi.

Chapter 2
Factors Affecting Development

Neriman Aral
Ankara University, Turkey

Gül Kadan
Sağlık Bilimleri Fakültesi, Çankırı Karatekin Üniversitesi, Turkey

ABSTRACT

Development can be defined as a concept expressing biological growth and maturation starting from prenatal period and continuing until death. Development, which involves the transformations human beings go through from the zygote period to death, proceeds quite rapidly in some periods, and somehow still in some other periods. Development, which is considered as a whole together with the physical, cognitive, linguistic, social, and emotional growth areas, can be affected by many factors. These factors are expressed as heredity, hormones, and environment. This chapter, which moves from the understanding that a healthy society comprises of healthy individuals and determination of the factors affecting development, is of crucial importance in this regard. It is aimed to address and examine the factors affecting development.

INTRODUCTION

Development is a process of regular, sequential and progressive alterations from the prenatal period to death. Development is an extraordinary process that begins with the cleavage of the fertilized egg of human being and progresses through physical and motor developments until birth, with the addition of cognitive, social-emotional developments to the prenatal developments after the birth, as well as language development in later stages. Development takes place based on a number of principles. These principles include inheritance and environment interaction, continuity, sequentiality, alternate progression, critical development periods, wholeness and individual differences (Deniz, 2017; Şahin, 2018).

Development, which is of great importance, consists of several factors and can affect one's life immensely. These factors may be chromosomal abnormalities and associated genetic diseases due to more or fewer chromosomes than the usual 46 passing through from the parents to the fetus in the prenatal

DOI: 10.4018/978-1-7998-2952-2.ch002

period (Aral & Gürsoy, 2011) or may be other conditions due to the chemical structure of dominant or recessive chromosomes carried by the mother and the father (Çifci-Tekinarslan, 2012).

In addition to chromosomes, deficiency or excessive secretion of hormones that regulate body functions cause adverse impacts on development. Excessive secretion of growth hormone by the pituitary glands leads to gigantism, whereas insufficient secretion leads to dwarfism (Deniz, 2017); deficiency of thyroid hormones leads to cretinism (Metin, 2018), whereas excessive secretion leads to behavioral and social development problems; excessive secretion of the parathyroid hormone leads to physical development and personality problems (Deniz,2017); disfunctional pancreatic hormone leads to galactosemia and phenylketonuria (Aral & Gürsoy, 2011); overactive or underactive adrenal glands leads to respiratory, circulatory problems; and deficiency or excessive secretion of the sexual glands leads to physical, sexual and psychosocial development problems (Deniz, 2017). Phenylketonuria, Tay Sachs disease, Cystic Fibrosis, Sickle Cell Anemia, Neurofibromatosis, Hemophilia, Vitamin D Resistant Rickets, Tuberous Sclerosis are non-chromosomal conditions that are not carried by the chromosomes but seen in fetus (Aral & Gürsoy, 2011; Durualp, 2017; Feldman, 2017; Metin, 2018), whereas Down Syndrome, Kleinfelter Syndrome, Turner Syndrome, Fragile X Syndrome, Williams Syndrome, Prader Willi Syndrome, Edwards Syndrome, Cri-Du-Cat Syndrome, Paton Syndrome and Trisomy 22 Syndrome are among the chromosomal conditions (Ataman et al., 2016; Aral & Gürsoy, 2011; Çifci-Tekinarslan, 2012; Durulap, 2017; Metin, 2018; Reçber &Özen, 2005).

These conditions, which are observed in the prenatal period and have negative effects on the development of the fetus, can be detected by genetic counseling upon which necessary precautions can be taken. Amniocentesis, Cordocentesis, Ultrasonography, MRI, Cord Villus Sampling Method and Fetoscopy are among the methods used to determine the condition of the baby, upon which the parents can decide on whether to continue or terminate pregnancy (Kılıç, 2018).

Environmental factors also have significant impacts on development in the prenatal period. The environmental factors in the prenatal period include maternal age, blood incompatibility and maternal health status due to the physiological state of the mother. Developmental problems may be encountered in mothers older than 35 years and younger than 20 years of age (Aral & Gürsoy, 2011; Özsoy, 2014; Yücel, 2015), whereas a Rh (-) mother and a Rh (+) combined with a Rh (+) fetus may pose a great risk for the baby (Howard, Williams & Lepper, 2011 Santrock, 2016). The mother's chronic diseases may adversely affect the development of the baby. Maternal infections (Talim, 2015), along with perinatal infections can also be quite dangerous for the fetus (Langford, 2002; Madazi & Atasu, 2000). In addition, abnormal placental location and its separation before due time affect the development of the fetus negatively (Howard et.al., 2011). Radiation received by the mother during pregnancy causes physical developmental retardations (Taşkın, 2016), whereas unprescribed medications she receives pass to the fetus through blood, causing growth and developmental retardation (Talim, 2015).

Development is significantly affected by stress, resulting from the psychological state of the mother, and the use of cigarettes, alcohol or drugs to overcome such stress (Çetin & Özözen, 2015). Imbalanced maternal nutrition during pregnancy and inability to receive health check-ups are other factors that similarly affect development negatively (Sigelman & Rider, 2017).

The infant may experience mental disabilities due to lack of oxygen as a result of delivery by inappropriate or incompetent persons, anatomical structure of the mother, cord around the neck syndrome or meconium aspiration (Aral & Kadan, 2018a).

Poverty, which affects development and is one of the postnatal factors, hinders the child's growth and development and leads to exposure to accidents, neglect and abuse in addition to restricted access to health services (Hess, Lannerdal, Hatz, Rivers & Brown, 2009; Özenoğlu & Ünal, 2015; Sak, Öz-kul, Evşen, Sak & Eviroğlu, 2009; Şener & Ocakcı, 2012). Low level of education, in addition to the socioeconomic status of the family, may also result in inability to respond to the child's developmental needs (Deniz, 2017). On the other hand, healthy inter-family communication and child's position in the family play an important role in the social emotional development (Yavuzer, 2011). Negative parental attitudes such rejectionist, indifferent, overprotective, over-tolerant, unstable and indecisive, authoritarian and perfectionist attitudes have negative effects on children's cognitive, language, motor, social and emotional development, whereas democratic attitudes contribute to all areas of development of the child and help them build their potential at the highest level. Culture, which plays a role in the family's approach to the child, also affects development (Aral & Kadan, 2018b).

A qualified school education, which is one of the environmental factors in the postnatal period, and the physical and psychological environment of the school contribute to child development (Gander & Gardiner, 2015). In particular, proper use of mass media that surround children today might be useful for all areas of child development, (Dalgalı, 2014; Yazıcı & Gencer, 2016), whereas prolonged exposure to inappropriate programs affect development negatively (Bencik- Kangal & Özkızıklı, 2015; Yavuzer, 2012). In the light of all these, the concepts related to development and factors affecting development will be explained in detail where prenatal, childbirth and postnatal period will be discussed separately.

BACKGROUND

Development is very important in the construction of a society. Healthy society is composed of healthy individuals. When these points are taken into consideration, it is very important to know the normal development and the factors affecting the development. Because, knowing the factors that affect the development and making early interventions to these factors will enable the future to be built on solid foundations.

The development that can be described as regular, progressive developments starts in the prenatal period and continues until death (Deniz, 2017; Şahin, 2018). However, an unexpected situation in these processes may affect the development.

While development is affected by hereditary factors, hormones and environment in the prenatal period, it is affected by environmental factors in the delivery process, mother and baby, and environmental factors in the postpartum period (Aral & Gürsoy, 2011). It is important that these factors do not adversely affect the development and that any measures that can be taken are known and intervened accordingly.

WHAT IS DEVELOPMENT?

Development can be considered as an extraordinary process in a living being. Development is described as regular, sequential and progressive alterations the from prenatal period to death (Deniz, 2017). It is important to be familiar with some concepts for a better comprehension of development. These concepts are explained below (Deniz, 2017; Şahin, 2018):

- Growth; It is described as increases in the human body which are expressed numerically. For example, neck growth or weight gain.

- Maturation; It is the process where organs gain the ability to perform their functions.
- Learning; It is a permanent change in behavior that occurs as a result of communication and interactions with others.
- Readiness; It is the process which is determined by taking into consideration the previous learning, interests and motivations of the child before acquiring any information, which is, in other words, to be ready for learning.
- Critical period; It is the period which is of significance in child's adoption of certain behaviors.
- Development is a product of maturation and learning interactions and is divided into different areas. These areas are briefly described below.
- Physical development; It is defined as the ability of the organs to increase in length, weight and volume, as well as the ability to perform expected functions (Alakoç-Pirpir & Yıldız-Çiçekler, 2018).
- Motor development; It is defined as the achievement of functional mobility of organs beginning from the prenatal period (Kaytez, 2018).
- Cognitive development; It includes changes in mental processes as the age increases (Bütün-Ayhan, 2017).
- Social-emotional development; Social development (Aral & Kadan, 2018b) is defined as the acquisition of socially accepted behaviors and communication and interactions with others, whereas emotional development is defined as emotional regulation and expression in a desirable manner (Çakmak & Pektaş, 2018).
- Language development; It is an area of development which involves proper communication with others, self-expression and speaking in accordance with the grammatical rules of the native language (Yıldız-Bıçakcı & Aral, 2017).

Although the areas of development are divided into different categories, they are actually a whole. The positive or negative circumstances occurring in each area reciprocally affect the other development areas (Senemoğlu, 2013).

- Development is discussed and examined under a number of basic principles. These principles (Deniz, 2017; Şahin, 2018)
- Development is shaped by inheritance and environment interactions.
- Development is a continuous process and takes place sequentially.
- Development progresses alternately. While some areas of development exhibit rapid progress, others may be more stagnant during the same process.
- There are critical periods in development. It is highly important not to miss these critical periods.
- Development progresses as a whole. Areas of development are interrelated.
- There are individual differences in development.
- Development takes place from head to foot, from inside to outside and from the general to the specific.

FACTORS AFFECTING DEVELOPMENT

Development starts in the prenatal period in just the same way as the factors which affect development. The factors affecting development are discussed under three main headings. In the following, inheritance, hormones and environmental factors, which affect development, and their impacts on development will be explained.

Impacts of Inheritance on Development

There are 46 chromosomes in the human body, each of which consists of 30 thousand to 100 thousand genes. During fertilization, one of each pair of 23 chromosomes is passed on from the mother and the father. XX chromosome in females and XY chromosome in males come together to form zygote. The mother's X chromosome combined with the father's Y chromosome forms a male zygote, whereas the mother's X chromosome combined with the father's X chromosome forms a female zygote (Deniz, 2017; Şahin, 2018).

Defects in the chromosome structure inherited from the parents, or errors during the transition may result in genetical problems. These problems, on the other hand, can harm both the fetus and development in the future. In addition to chromosomal disorders, there are non-chromosomal genetic factors affecting the development as a result of changes in the gene. The chromosomal and non-chromosomal factors that play a role in development are given below.

Impacts of Chromosomal Abnormalities on Development

Chromosomal abnormalities are genetic factors that cause various problems due to excessive or deficient number of chromosomes. The most common ones are given below.

Down Syndrome

It is the most common chromosomal disorder, which was identified by Langdon Down in 1866. Individuals with Down syndrome have 47 chromosomes instead of the usual 46. Down syndrome is caused by a random error in cell division that results in the presence of an extra chromosome or chromosomal deformations. (Aral & Gürsoy, 2011).

Children with Down syndrome have typical distinctive features. These include small skull, slanted eyes and flat head. Children with Down syndrome have a typical facial appearance, with short fingers and a single line on the palm. Their weight is generally higher. Children with Down syndrome have mild to severe mental disabilities, high social adaptation and imitation skills. Language development takes place two years after they start walking (Aral & Gürsoy, 2011; Çifci- Tekinarslan, 2012; Metin, 2018).

There are three types of Down syndrome, Mosaic Mongolism, Translocation and Trisomy 21.

Mosaic mongolism results from the differences in the number of chromosomes in cells. Some cells have 46, some have 47 and some have 45 chromosomes (Aral & Gürsoy, 2011).

Translocation

It occurs when the translocation gene in either parent passes to the zygote and the extra chromosome on chromosome 21 breaks off and attaches to another chromosome. (Aral & Gürsoy, 2011; Ceylan et al., 2009).

Trisomy 21

It occurs when an embryo has three copies of chromosome 21 instead of the usual two. These children have typical distinctive features such as small mouths, flat nose bridge and delayed motor development (Ceylan et al., 2009).

Kleinfelter Syndrome

Children with Kleinfelter syndrome, another chromosome disorder, has 47 chromosomes. This genetic problem seen in males presents with the XXY chromosome arrangement. It is typically characterized by long neck, enlarged breasts, decreased testicular volume or incompetencies (Pehlivan, Çefle, Öztürk, Akbulut & Palandüz, 2013; Peynirci & Ertürk, 2013).

Turner Syndrome

Turner syndrome, a condition that affects only females, results from a structural X chromosome abnormality. Turner syndrome, also known as X0, has some typical features. These features include short neck, lymph-edema, flat chest, high palate, congenital heart problems, delayed puberty, and low eyelids. In addition, learning disabilities and attention deficits are frequently encountered (Akcan, 2013).

Fragile X Syndrome

Fragile X syndrome, one of the most common chromosomal anomalies after Down syndrome, presents with structural X chromosome abnormalities (Gündüz et al., 2000). Mental disability, developmental delay and learning problems are seen in Fragile X syndrome. Individuals with Fragile X syndrome also have distinctive features such as a long and thin face, long and soft ears and hands, protruding foreheads and large skull. Speech intelligibility, repetition of phrases, motor development, and social and emotional problems are encountered in these children (Çifci-Tekinarslan, 2012).

Williams Syndrome

Williams syndrome, which occurs as a result of deletion of chromosome 7, result in mental disability, emotional problems and social problems in later periods (Çifci-Tekinarslan, 2012; Ng, Jarvinen & Bellugi, 2014; Tekin- Ersan & Kizir, 2016). However, it is also stated that individuals with Williams syndrome are very sensitive to music and sound, and the sentence structures they construct are quite complex (Bloomberg, Rosander & Andersson, 2006; Çifci-Tekinarslan, 2012).

Prader Willi Syndrome

It is a condition caused by both hormonal and chromosomal abnormalities. It is caused by a defect on chromosome 15 inherited from the father. The syndrome is divided into two distinct clinical stages. The first stage is characterized by sleep disorders, respiratory problems, hypogenitalism, increased appetite, growth hormone deficiency and weak sound, whereas the second stage is characterized by mental, motor and language development disabilities (Çifci-Tekinarslan, 2012).

Edwards Syndrome

Edwards syndrome, a genetic disorder caused by the presence of an extra 18th chromosome, is more common in females. Most babies lose their lives in the womb, and those who are born present with mental disability, heart failure, renal abnormalities, delayed development, digestive system problems, cleft palate, cleft lip, microcephaly, small jaw and deformities in fingers and toes (Durualp, 2017).

Cri-Du-Chat Syndrome

It is a chromosomal condition that results from a deletion or missing piece of chromosome 5. Cat-like cry, microcephaly, downward slanting eyes, simian line in hand and severe mental disabilities are seen in these children in the neonatal period (Metin, 2018).

Trisomy 13 (Patou Syndrome)

It is caused by a defect on chromosome 13. Having trisomy 13 results in miscarriage. Babies that are born with trisomy 13 present with skull and facial abnormalities, heart, kidney, stomach malformations and physical anomalies (Reçber & Özen, 2005).

Trisomy 22

It is a chromosomal problem that occurs as a result of deformation of chromosome 2 in the first trimester of pregnancy and usually results in miscarriage. The incidence is low in the second and last three months. It causes growth retardation. Infants born with trisomy 22 present with growth retardation, microcephaly, cleft lip and cleft palate, cardiac, urogenital and skeletal system problems (Ataman et al., 2016).

Impacts of Non-Chromosomal Genetic Factors on Development

These are the problems that arise in a gene and are seen as a result of chemical changes in the chemical structure of the gene. Activity of the enzyme associated with the gene decreases as a result of chemical changes, resulting in metabolic problems (Aral & Gürsoy, 2011).

These metabolic diseases and their impacts on development are given below.

Phenylketonuria

It is an autosomal recessive disorder. Related gene must be present in both the mother and the father for the formation of phenylketonuria. It is caused by a disorder of phenylalanine metabolism. Phenylketonuria, on the other hand, is a metabolic condition that results in brain damage due to non-conversion of phenylalanine amino acid to tyrosine leading to accumulation in blood and other tissues (Çifci-Tekinarslan, 2012; Ergül, 2014). The first symptoms appear after three to four months although the baby appear normal at birth. Symptoms of phenylketonuria include vomiting, bad urine odor and poor weight gain. Failure to notice the symptoms in the early period and absence of early diagnosis and treatment may result in mental retardation, developmental delay, orthopedic problems, hyperactivity and autistic behaviors (Berk, 2013; Feldman, 2017).

Tay Sachs Disease

It is an autosomal recessive genetic disorder. This disorder causes brain damage in the womb in infants. Although babies do not present with any problems until three to six months, motor retardation, inability to sit without support, losing head control, overreacting to sounds, visual and hearing impairment appear after two years of age as their muscles gradually weaken. Tay Sachs disease progresses very rapidly and children die when they reach the age of four or five (Bornstein, Arterberry & Lamb, 2014).

Cystic Fibrosis

These babies have problems in secretion glands such as sweat glands, airways and bile ducts. Coughing, wheezing, pneumonia, asthma symptoms and diseases are frequently seen as a result of blockages in the lungs (Filburn, Lahiri & Ren, 2016).

Sickle Cell Anemia

Red blood cells that are normally round present with a sickle shape. Sickle-shaped red blood cells inhibit blood flow and formation of red blood cells, which are essential for the body. It may result in anemia, severe pain, pneumonia, liver expansion, gallbladder stone, hip bone dissolution, skeletal system disorder and paralysis. Individuals with this disorder need exchange transfusion (Durualp, 2017).

Neurofibromatosis

It is an autosomal dominant disorder in which the nervous system cells are impaired. These individuals may also present with visual, hearing, and mental impairments and verbal and learning disabilities (Dalgar & Kaytez, 2018).

Hemophilia

Hemophilia is a condition in which the blood does not clot properly, and people with this disease experience prolonged bleeding in the event of injury or surgery. Intraarticular bleeding, pain, stiffness, joint damage may occur (Taşkın, 2016).

Vitamin D Resistant Rickets

Phosphorus levels are very low in these individuals. Vitamin D deficiency occurs regardless of the amount received. Growth and developmental retardation, skeletal deformity and dental problems are seen in these children (Durualp, 2017).

Tuberous Sclerosis

It is an autosomal dominant disorder. It presents with off-white spots on the skin and orange-like spots on the lower extremities. While it causes benign tumors in organs such as brain, kidney and heart, tumors in the brain cause convulsions or mental disability (Metin, 2018).

IMPACTS OF HORMONES ON DEVELOPMENT

Hormones in the human body have functions such as contributing to growth and development, regulating cell metabolism, and determining male or female characteristics (Deniz, 2017). The internal secretory glands and their functions that play an important role in development and the problems that may occur as a result of insufficient or excessive secretion of these hormones are given below.

Pituitary Gland

The pituitary gland, which controls all hormones, is located just below the hypothalamus in the anterior part of the brain in the skull. In addition to controlling all other hormones, the pituitary gland secretes growth hormone and affects the entire body. Excessive secretion of growth hormone by the pituitary gland results in gigantism, whereas insufficient secretion may result in dwarfism (Deniz, 2017).

Thyroid Gland

The thyroid gland is located in the anterior part of the brain and affects and regulates all metabolic activities of the body. Cretinism resulting from thyroid hormone deficiency causes movement restriction, growth and developmental retardation and mental disability (Metin, 2018). Hyperactivity, irritability, personality and psychosocial problems may be seen as a result of excessive secretion of the thyroid gland (Deniz, 2017).

Parathyroid Gland

The parathyroid hormone located on the back side of the thyroid gland plays an important role in maintaining the calcium and phosphorus balance of the body. Bone, muscle and skeletal system develop properly when calcium and phosphorus balance is maintained. However, as a result of excessive secretion of the parathyroid hormone, personality changes, severe psychosis, osteoporosis and easy fractures can be observed (Deniz, 2017).

Pancreatic Gland

It has important roles in digestion of lipids with secretions sent to the digestive system and regulates carbohydrate metabolism with secretions released into the blood. Galactosemia and phenylketonuria result from pancreatic insufficiency. Galactosemia occurs as a result of an inability to convert galactose, a combination of lactose also known as milk sugar, into glucose and affects mental ability, school adaptation and social relationships of the child. Concurrently, developmental delays, liver expansion and gastrointestinal problems are frequently seen as the baby refuses to eat (Aral & Gursoy, 2011).

Adrenal Glands

These glands, which are effective on respiratory, circulatory and excretory systems and sexual glands, are located above each kidney (Deniz, 2017).

Sexual Glands

The sexual glands, which determine the anatomical structure and characteristics of each gender, are called ovaries in women and testes in men. The estrogen hormone secreted from the ovaries regulates the functions of menstruation, breast growth and armpit hair, whereas the testosterone hormone secreted from the testes in men is responsible for sperm production, armpit hair, shoulder circumference and voice pitch. The absence, deficiency or excess of these hormones cause physical, sexual and psychosocial developmental problems (Deniz, 2015).

Early detection of these conditions, which have an impact on development both in the prenatal and postnatal period, and taking necessary precautions is very important. Hereditary anomalies are detected by prenatal tests. These tests include

Amniocentesis

It is performed to detect all chromosomal abnormalities and metabolic problems in week 4, upon which parents can decide for termination of pregnancy in case of any problems (Çetin & Özözen, 2015; Gander & Gardiner, 2015).

Cardiosentesis

It is performed to determine the presence of anomalies by taking blood from the umbilical cord of the baby around week 20 (Santrock, 2016).

Ultrasonography

Organ structure or anomalies of the fetus are detected with the help of sound waves (Berk, 2013; Gander & Gardiner, 2015).

MRI (Magnetic Resonance Imaging)

It is used to perform fetal anomaly screening beginning from week 17 (Özgen-Mocan, 2010).

Cord Villus Sampling Method

It is performed between 10 and 14 weeks of pregnancy in the event of detection of any problems by blood tests and amniocentesis in the first trimester, given that a hereditary condition runs in the family. Biopsy is taken from the placenta and examined for fetal anomalies (Feldman, 2017).

Fetoscopy

A camera is placed in the uterus in order to monitor the growth and development of the fetus directly, which enables early intervention (Kılıç, 2018).

IMPACTS OF ENVIRONMENT ON DEVELOPMENT

Environmental factors have significant impacts on development. These factors are examined under different headings as prenatal, childbirth and postnatal periods. The impacts of environmental factors on development are explained below.

Impacts of Environmental Factors in Prenatal Period

Environmental factors in the prenatal period include factors that arise from the physiological, psychological and social status of the mother.

Factors Caused by Mother's Physiological Status

Maternal Age

The age of the mother is one of the factors affecting development. Pregnancies after 35 years of age have a higher risk of Trisomy 21, Down syndrome and developmental delay (Aral & Gürsoy, 2011; Dede, Kandemir, Yalvaç & Altay, 2017; Yılmaz et al., 2016). Pregnancy under the age of 20 also poses a risk as in pregnancy over the age of 35. Mothers below 20 years of age, in other words, adolescent mothers may give birth to babies with translocation, which results in mental disability, developmental delay, Down syndrome, miscarriage, stillbirth, difficult and premature births (Özsoy, 2014; Yücel, 2015).

Blood Incompatibility

It occurs when a Rh (-) negative mother and a Rh (+) positive father has a baby with Rh (+) positive blood group, which leads the mother's body to perceive the baby as a foreign body and produce antibodies. Production of maternal antibodies against the baby may cause mental deficiency (Howard et al., 2011), excessive anemia, heart failure, edema and death (Santrock, 2016).

Maternal Health Status

Presence of maternal health problems may have a direct effect on the fetus. Especially, presence of certain maternal diseases in the first four weeks of pregnancy, such as hypertension and diabetes, may result in preterm birth, growth and development problems and congenital malformations (Berk, 2013). Babies of mothers with hypertension may present with multiple organ failures which may subsequently result in death. Children of diabetic mothers may experience long-term ICU stays, stillbirths, obesity and diabetes and neurological developmental problems in the future.

In addition to these diseases, heart diseases in the mother may result in neurological diseases in the baby, thyroid gland disorders, prenatal development retardation, low birth weight and infant deaths. Maternal infections during pregnancy may also have an impact on the baby's development. Infectious diseases such as rubella and chickenpox during the pregnancy may cause fetal death, premature birth, low birth weight and congenital problems (Talim, 2015).

Perinatal infections during pregnancy also have significant impacts on development. Rubella and Cytomegalovirus (CMV), which appears in the first 12 weeks of pregnancy, may cause mental disability (Kadayıfçı, Ürünsak, Ürünsak & Atay, 2001; Langford, 2002), whereas Varicella Zoster may cause microcephaly and Toxoplasma may cause hydrocephalus (Langford, 2002; Madazi & Atasu, 2000).

The structure of the placenta also affects development. Misplacement of the placenta in the lower part of the uterus instead of the upper part, premature separation of the placenta from the uterine wall, poisoning of the mother with chemical or lead, insufficient blood supply to the uterus and placenta, entanglement or sagging are among the factors that have negative effects on development (Bügü, 2001; Howard et al., 2011).

Exposure of the mother to radiation during pregnancy also causes adverse effects on the child. Such exposure to radiation during pregnancy may cause problems in the fetus such as malformations, gene mutations, miscarriage, microcephaly, head bone deformations, loss of vision, cleft palate, cleft lip and spina bifida (Feldman, 2017; Taşkın, 2016).

Unprescribed medications taken by the mother during pregnancy pass through the fetus by blood transport and result in growth and developmental retardation, miscarriage and premature birth (Talim, 2015), whereas hormonal drugs may shrink the fetal head circumference and anticoagulant drugs may cause microcephaly (Atasü and Gezer, 2000).

Factors Caused by Mother's Psychological Status

Psychological status of the mother during pregnancy and her reactions to this psychological state may also affect development.

It is very important to keep the psychological and emotional state of the mother in the best condition during pregnancy while avoiding stress and negative living conditions. Mother's presence in stressful and tense environments throughout her pregnancy results in stress hormones to get into the blood, which then reach the fetus. These stress hormones cause an increase in the heart rate, blood pressure and activity level of the fetus. Exposure of the fetus to stress hormones may cause cleft palate, cleft lip, heart deformities, low birth weight, respiratory and digestive problems. It is also stated that these children may have behavioral problems in the future (Berk, 2013; Çetin & Özözen, 2015; Yesilçiçek-Çalık & Aktas, 2011).

Smoking, alcohol and substance use during pregnancy also have negative effects on development. The toxins in these substances pass through the placenta to the fetus during pregnancy and cause negative effects on the nervous system and brain development (Ergül, 2014; Kaya & Şahin, 2013).

Nicotine in cigarettes can cause low birth weight infants, stillbirth, childhood cancers, asthma, growth retardation and behavioral disorders (Berk, 2013).

Alcohol also passes through the placenta to the fetus. Fetal Alcohol Syndrome is frequently seen in mothers who use alcohol regularly. Alcohol use may result in growth and developmental delay, behavioral disorders, mental disability and unusual facial appearance (Durualp, 2017).

Factors Caused by Mother's Social Status

Inability to get the necessary nutrients for the mother and the baby, as a result of low socioeconomic status, may result in conditions such as low birth weight and central nervous system problems in infants (İrge, Timur, Zincir, Oltuoğlu & Dursun, 2005; Sigelman & Rider, 2017). Failure to receive prenatal follow-ups regularly due to the socioeconomic status of the family may also lead to developmental disabilities and mental retardation in the fetus (Bodur, Durduran, Küçükkendirci & Doğan 2009).

Skull anomalies seen in prenatal fetus due to environmental factors also have significant effects on development. These skull anomalies are described below.

Microcephaly

Microcephaly, which means very small head, can develop both as primary and secondary. Primary microcephaly usually develops as a result of the mother's exposure to radiation, drugs, alcohol, chemicals and infections during the first or second trimester of pregnancy. Secondary microcephaly develops in the last trimester of pregnancy due to traumas at birth or postpartum accidents, metabolic diseases, and lack of oxygen (Howard et al., 2011).

Hydrocephalus

It is a condition characterized by an excessively large head as a result of excessive amount of cerebrospinal fluid accumulation in the brain ventricle (Aral & Gursoy, 2011; Howard et al., 2011). Hydrocephalus may be caused by tumor, head trauma, or excessive alcohol consumption during pregnancy (Tanyeri-Bayraktar, Bayraktar, Hepokur & Kardaş, 2015).

Environmental Factors at Childbirth

Some problems which arise during childbirth also have negative effects on development.

At birth, the baby's skull is soft, not fully-grown or fully-formed. This may result in a slightly extended head during delivery, which has an adverse effect. In addition, the baby's head is exposed to too much pressure in some cases where the baby is pulled out rapidly to shorten the labor, the mother's pelvic bones are too narrow to allow the head to pass through properly and the cervix does not expand sufficiently, which may lead to intracranial hemorrhage and subsequent brain damage (Gander & Gardiner, 2015). In this context, it is very dangerous to force the baby to come out using forceps and vacuum (Aral & Gürsoy, 2011; Çifci-Tekinarslan, 2012; Howard et al., 2011).

The lack of oxygen during childbirth also causes developmental problems. Lack of oxygen at birth is caused by several factors such as premature separation of the placenta from the uterine wall, entanglement of the umbilical cord around the baby's neck, birth crowning, and abnormal and prolonged labor. These factors may result in cerebral palsy, epilepsy or mental deficiencies due to the destruction of brain cells. Aspiration of the greenish substance called meconium by the infant may cause abnormal heartbeat changes and excessive fetus mobility, which may result in the baby to experience oxygen shortage and consequently developmental problems (Gander & Gardiner, 2015).

Low baby weight or premature birth can also lead to learning difficulties, attention deficit and hyperactivity disorder, asthma and respiratory problems (Santrock, 2016).

Postnatal Environmental Factors

Postnatal environmental factors also have significant effects on development. These factors are examined below under the headings of culture, socioeconomic status and education level of the family, inter-family communication, child's position in the family, parental attitudes, education and mass media.

Culture

Culture is the preservation of values, traditions and customs from the past and transfer to the new generations and has significant impacts on development. Desired personality traits accepted by the society are transferred to children and children are expected to behave according to these norms. In other words, it is aimed to determine social development (Ahmetoğlu, 2009). This has a significant impact on development. For example, Turkish culture places greater importance on solidarity, interdependence and family and children are expected to be respectful to their elders, whereas the values expressed in western societies remain in the background (San-Bayhan & Artan, 2009). Non-compliance with the values accepted by the society may have a negative impact on other areas of development, particularly in the area of social-emotional development (Aral & Kadan, 2018b).

Familial Factors

The socioeconomic status and educational levels of the family have an impact on development. Low socioeconomic status of the family may lead to many consequences that affect the development of children such as malnutrition, illness, accidents, abuse and neglect in different dimensions (Aral et al., 2016; Öncü, Kurt, Esenay & Özer, 2012; Özer & Çolak, 2015).

It is stated that mothers who are not fed enough during pregnancy are more likely to have low birth weight babies or stillbirths. Insufficient and imbalanced maternal nutrition leads to a higher risk of newborn infections, and more frequent and severe neurological defects, growth and developmental retardation, chronic diseases, mental disability, immune system suppression, iron deficiency anemia, pica, upper and lower respiratory tract infections and intestinal infections (Bülbül, Okan, Şahin & Nuhoğlu, 2006; Katar & Devecioğlu, 2006; Sak et al., 2009).

Breastfeeding in the neonatal period and, especially in the first six months, has significant effects on development. Exclusive breastfeeding with sufficient breast milk in the first six months boosts the infant's immune system and decreases the risk of diarrhea and respiratory infections (Yıldız, Baran, Akdur, Ocaktan & Kanyılmaz, 2008). Regular intake of minerals, vitamins, mineral salts, carbohydrates and fats every day is very important for babies and children to develop in a healthy way. Insufficient intake of these nutritients due to the socioeconomic status of families causes growth-developmental delays in children and results in more frequent and severe immune system diseases (Apodoca, 2008; Condon et al., 2009; Hess et al., 2009; Keyou, Jianbin & Hang, 2007; Walker, Chrisitan & Black, 2007).

Children in low socioeconomic status families may experience preventable diseases more severely or even lose their lives. Inability to access health services, get vaccines on time, receive appropriate health services and economic status of the family may severely hinder development (Durgun, 2011; Şener & Ocakçı, 2012). It is also reported that children in low-income families may develop psychological problems such as hyperactivity, restlessness, aggression, substance abuse and low self-esteem, which affect the child's social and emotional development (Özenoğlu & Ünal, 2015; Toran, 2010; Tümkaya, Aybek & Çelik, 2010). In addition, low economic status may seriously impair inter-family relations and cause emotional, physical and psychological abuse and neglect of the child (Altıparmak et al., 2010; Bilgin, 2009) and interrupt their education (Bilgin, 2012; Oktik et al., 2010). Children in families with a good socioeconomic status are more advantageous in terms of nutrition, prevention from accidents, continuing education and healthy living. (Deniz, 2017).

Lack of adequate education, in addition to poor socioeconomic status, results in families to adopt certain attitudes, beginning from pregnancy, based on some incorrect beliefs and ideas, which may also cause many problems. These problems result from the absence of parents as positive role-models, inappropriate communication and interaction patterns, and parental attitudes (Şahin, 2018).

Parents with higher level of education exhibit more conscious behaviour in the field of child development, provide their children with the stimuli they need, and take the necessary precautions in a timely manner as they monitor the development of their children more closely (Deniz, 2017).

In addition to the socioeconomic and educational status of the family, inter-family communication, child's position in the family and parental attitudes also play an important role in development.

Positive inter-family communication and interactions not only teach children how to solve problems or adopt healthy communication patterns, but are also the first key step for social development. As a result of positive communication and interaction between parent-parent and parent-child, the child makes significant progress in the areas of emotional and social development. However, an opposite inter-family communication and interaction may lead to aggressive behaviors and loneliness, in which child internalizes such behaviors and reflects them upon others (Çağdaş & Şahin- Seçer, 2015; Sağlam, 2017). Birth order is another factor affecting child development. Social-emotional development may be affected adversely in first born children in the even that parents exhibit more tolerant behaviors towards the subsequent siblings while, in general, they were overly harsh upon their first born children (Yavuzer, 2011).

Parental attitudes also have significant impacts. Rejectionist parental attitudes fail to meet developmental needs of the child, in which the child is ignored, abused and neglected emotionally and physically (Gültekin, 2006; Sharma, 2004). Such attitude causes the child to develop problematic behaviors in the areas of social and emotional development (Çağdaş, 2012; Önder, 2003). The overprotective attitude, which is another parental attitude, means all the needs of the child are met by the parents and the child is not allowed to be an independent individual (Gültekin, 2006; Yavuzer, 2011). Such circumstances, on the other hand, negatively affect all developmental areas of the child (Dokuyan, 2016; Sağlam, 2017; Yavuzer, 2011). Over-tolerant parental attitudes include acceptance of all kinds of child behavior (Sümer, Gündoğdu- Aktürk & Helvacı, 2010) and cause problems in the social-emotional development areas of the child (Çağdaş & Şahin- Seçer, 2015). On the other hand, parents who have perfectionist and authoritarian attitudes have higher expectations which far exceed the age and development stage of their children (Gander & Gardiner, 2015; Yavuzer, 2011). Such an expectation causes problems in all areas of development as well as exposure to emotional and physical abuse and neglect (Çağdaş & Şahin- Seçer, 2015; Sağlam, 2017).

Consistency is the most important concept in the upbringing and development of the child. However, it is known that some parents may exhibit inconsistent and unstable attitudes towards both their spouses and their children (McKinney, 2008; Sağlam, 2017; Weilburger, 2008). This attitude can cause serious harm to the social, emotional and cognitive development of children (Yavuzer, 2011). It is stated that all development areas are positively affected (Sağlam, 2017) when parents exhibit behaviour in accordance with the age and developmental stage of their children, make rules and explain the reason behind these rules and raise their children in a democratic attitude, which is an attitude that supports development (Sağlam, 2017; Şahin & Aral, 2012).

Impacts of School Education on Development

School education also has significant impacts on the child's development. School is the next institution in a child's life which contributes to development after the family. In particular, a qualified pre-school education and school environment provided for the child contribute to the development of the child in the future (Aryanti, Warsini & Haryanti, 2018; Atabey, 2018; Blair, McKinnon & Daneri, 2018; Gültekin-Akduman, Günindi & Türkoğlu, 2015; Özyürek, Bedge & Yavuz, 2014; Santrock, 2016; Sapsağlam & Ömeroğlu, 2016; Saracho, 2017; Yaşar-Ekici, 2015). It is reported that the development of the children in socioeconomically disadvantaged regions is positively affected if they receive school education and they start life on equal terms with their peers who are more advantageous in terms of socioeconomic status (Gander & Gardiner, 2015; Saracho, 2017; Tamayo, Rebolledo & Berain- Saldona, 2017). The impact of education on child development also plays a vital role in adolescence, where positive communication and interaction with teachers help them get through self-discovery process, prevent problematic behaviors in the first place and maximize their developmental potential (Austin & Sciarra, 2015).

Impact of Mass Media on Development

Mass media such as television and technological devices exhibit both positive and negative effects on development. It is stated that TV programs with appropriate language accelerate the language development of children (Dalgalı, 2014) while educational programs support cognitive development (Dağlıoğlu, 2006) and have positive contributions to social development (Dalgalı, 2014). However, it is stated that prolonged exposure to TV programs that are unsuitable for the age and development of the child will cause problems in the areas of cognitive, language and motor development in their social and emotional development (Can, 2006; Dağ et al., 2007; Işık, 2007; Uluç, 2003; Yavuzer, 2012).

Technological devices surrounding children have both positive and negative effects on development. It has been demonstrated by researches that educational programs suitable for the age and development stages of children support child development and increase their academic success (Bencik-Kangal & Özkızıklı, 2015; Lou, Higgins, Gelfer, Hang & Miller, 2005; Muis, Ranelucci, Trevors & Duffy, 2015; Plak, Merkelbach, Kegel, Von-Ijzendoorn & Bus, 2016; Schmid, Miodrag & DiFrancesco, 2008; Walter-Looger et al., 2017; Yazıcı & Gencer, 2016). However, improper use of technological devices affects developmental areas negatively, leads to academic failures and causes addiction (Bencik-Kangal & Özkızıklı, 2015; Kılınç & Kılınç, 2014; Sayan, 2016; Tarı- Cömert & Kayıran, 2010).

CONCLUSION

Development is a process that begins in the prenatal period and progresses sequentally and alternately until death. Although development is examined under different headings as cognitive, language, physical, motor, social and emotional development areas, it is a whole, each of which affects the other interrelatedly.

The prenatal period, where development progresses rapidly, is of great significance. Although physical and motor developments are at the forefront during this period, any problem that may arise has the potential to affect all development areas.

Factors affecting development in the prenatal period are inheritance, hormones and environment. Inheritance involves only the changes in the chemical structure of chromosomes, which may be chromosomal or non-chromosomal. Environmental factors include maternal age, health status, blood incompatibility, psychological and social status of the mother. Especially problems related to inheritance can be detected in the early period by genetic counseling and necessary precautions can be taken using different methods.

The birth environment as well as the state of the mother and the baby have significant impacts on development. The factors affecting development in the postnatal period include socioeconomic status and education level of the family, inter-family communication and interactions, child's position in the family, parental attitudes, education, culture and mass media.

REFERENCES

Ahmetoğlu, E. (2009). Sosyal gelişim. In Y. Fazlıoğlu (Ed.), *Erken çocukluk gelişimi ve eğitimi* (pp. 39–62). İstanbul: Kriter.

Akçan, A. B. (2013). Turner sendromu. *Konuralp Tip Dergisi*, *5*(2), 53–61.

Alakoç-Pirpir, D. & Yıldız-Çiçekler, C. (2018). Fiziksel gelişim. In N. Aral & Z. F. Temel (Ed.), Çocuk gelişimi (pp. 112-139). Ankara: Hedef.

Altıparmak, S., Yıldırım, G., Yardımcı, F., & Ergin, D. (2010). Annelerden alınan bilgilerle çocuk istismarı ve etkileyen faktörler. *Anadolu Psikiyatri Dergisi*, *14*(4), 354–361. doi:10.5455/apd.39819

Apadoca, C. (2008). Preventing child malnutrition: Health and agriculture as determinants of child malnutrition. *Journal of Children & Poverty*, *14*(1), 21–40. doi:10.1080/10796120701871264

Aral, N., Özdoğan-Özbal, E., Gürsoy, F., Çetin-Sultanoğlu, S., Aydos, S., Tosun, S., Karaaslan, T., & Kadan, G. (2016). Poverty definitions through children's eyes. *Journal of Current Researches on Health Sector*, *6*(1), 39–50.

Aral, N., & Gürsoy, F. (2011). *Özel eğitim gerektiren çocuklar ve özel eğitime Giriş*. İstanbul: Ya- Pa.

Aral, N., & Kadan, G. (2018a). Zihinsel yetersizliği olan çocuklar. In N. Aral & F. Gürsoy (Eds.), *Özel gereksinimli çocuklar ve eğitimleri: Özel gereksinimli çocuklar* (pp. 14–43). Ankara: Hedef.

Aral, N., & Kadan, G. (2018b). Sosyal gelişim. In N. Aral & Z. F. Temel (Eds.), *Çocuk gelişimi* (pp. 236–262). Ankara: Hedef.

Aryanti, A., Warsini, S., & Haryanti, F. (2018). Relationship between quality of care of young mothers and socio-emotional development in preschool children. *Belitung Nursing Journal, 4*(1), 31–38. doi:10.33546/bnj.344

Atabey, D. (2018). A study into the effective communication and social skills of preschool children. *Inonu University Journal of the Faculty of Education, 19*(1), 185–199. doi:10.679/inuefd.323598.186

Ataman, E., Yılmaz-Güleç, E., Hozan, F., Parıltay, E., Acar, D., Gedikbaşı, A., & Aslan, H. (2016). İkinci trimestersonografik taramasında multipl yapısal anomaliler gösteren Trizomi 22 olgusu ve literatür derlemesi. *DEÜ Tıp Fakültesi Dergisi, 30*(1), 39-43.

Atasü, T., & Gezer, A. (2000). Gebelikte ilaç kullanımı. In T. Atasü (Ed.), *Gebelikte fetüse ve yeni doğana etkiler* (pp. 373–444). Ankara: Nobel.

Austin, V. I. V., & Sciarra, D. T. (2015). *Çocuk ve ergenlerde duygusal ve davranışsal bozukluklar.* Ankara: Nobel.

Bencik-Kangal, S., & Özkızıklı, S. (2015). Teknoloji ve eğitim. In P. Bayhan (Ed.), Okul öncesi eğitimde teknolojinin rolü (pp. 10-31). Ankara: Hedef.

Berk, E. L. (2013). *Bebekler ve çocuklar: Doğum öncesinden orta çocukluğa.* Ankara: Nobel.

Bilgin, R. (2009). Diyarbakır'da sokakta çalışan çocuklar üzerine sosyolojik bir araştırma. *Elektronik Sosyal Bilimler Dergisi, 8*(27), 232–244.

Bilgin, R. (2012). Sokakta çalışan çocukları bekleyen risk ve tehlikeler: Diyarbakır örneği. *ZKÜ Sosyal Bilimler Dergisi, 8*(15), 79–96.

Blair, C., McKinnon, R. D., & Daneri, M. P. (2018). Effect of the tools of the mind kindergarten program on children's social and emotional development. *Early Childhood Research Quarterly, 43*, 52–61. doi:10.1016/j.ecresq.2018.01.002

Bloomberg, S., Rosander, M., & Andersson, G. (2006). Fears, hyperacusis and musicality in Williams Syndrome. *Research in Developmental Disabilities, 27*(6), 668–680. doi:10.1016/j.ridd.2005.09.002 PMID:16269236

Bodur, S., Durduran, Y., Küçükkendirci, H., & Doğan, C. (2009). Bebek ölümlerinin prenatal- natal sağlık hizmet kullanımı ve demografik özelliklerle ilişkisi: Vaka kontrol çalışması. *Dicle Tıp Dergisi, 36*(4), 288–293. doi:10.17517/ksutfd.393180

Bornstein, M. H., Arterberry, M. E., & Lamb, M. E. (2014). *Development in infancy: A contemporary introduction.* Taylor & Francis Group.

Bügü, B. Ö. (2001). *Gebelik ve laktasyon döneminde ilaç kullanımı.* Ankara: Nobel.

Bülbül, A., Okan, F., Şahin, S., & Nuhoğlu, A. (2006). Düşük doğum ağırlıklı erken doğmuş bebeklerde erken dönem hastalık ve ölüm oranı sonuçları. *Türk Pediatri Arşivi, 43*, 94–98.

Bütün-Ayhan, A. (2017). Bilişsel gelişim. In N. Aral & T. Duman (Ed.), Eğitim psikolojisi (pp. 104-128). Ankara: Pegem.

Çağdaş, A. (2012). *Anne-baba- çocuk iletişimi*. Ankara: Eğiten.

Çağdaş, A., & Şahin-Seçer, Z. (2015). *Anne- baba eğitimi*. Ankara: Eğiten.

Çakmak, A., & Pektaş, D. (2018). Duygusal gelişim. In N. Aral & T. Duman (Eds.), *Çocuk gelişimi* (pp. 264–290). Ankara: Hedef.

Can, A. (2006). *Çocuk ve çizgi film*. Konya: Öz Eğitim.

Çetin, S., & Özözen, M. (2015). Doğum öncesi gelişim. In N. Baysal-Metin (Ed.), Doğum öncesinden ergenliğe çocuk gelişimi (pp. 23-54). Ankara: Pegem.

Ceylan, G., Erol, D., & Yüce, H. (2009). DeNovo 21/21 Translokasyonu olan Down sendromlu iki olgu: Literatür araştırması. *Firat Tip Dergisi*, *14*(4), 280–282.

Çifci-Tekinarslan, İ. (2012). Zihinsel yetersizliği olan öğrenciler. In İ. H. Diken (Ed.), Özel eğitime gereksinimi olan çocuklar ve özel eğitim (pp. 137- 163). Ankara: Pegem.

Condon, E., Crepinek, M. K., & Fox, M. K. (2009). School meals: Types of foods offered to and consumed by children at lunch and breakfast. *Journal of the American Dietetic Association*, *109*(2), 67–78. doi:10.1016/j.jada.2008.10.062 PMID:19166674

Dağ, İ., Öktem, F., Yazıcı, K., Güvenç, G., Rezaki, M., Özer, Ö., & Tunçel, M. (2007). Televizyon programlarındaki şiddet içeriğinin, müstehcenliğin ve mahremiyet ihlallerinin izleyicilerin ruh sağlığı üzerindeki olumsuz etkileri. Ankara: RTUK.

Dağlıoğlu, H. E. (2006). *Üstün yetenekli çocuklar ve televizyon. M.Ü. Atatürk Eğitim Fakültesi 1. Uluslararası Okul Öncesi Eğitim Kongresi Bildiriler Kitabı*. İstanbul: Ya- Pa.

Dalgalı, H. (2014). *Televizyonda yayınlanan çocuk programlarının okul öncesi eğitim becerilerinin gelişimine olan etkisinin ebeveyn görüşlerine göre incelenmesi*. Kayseri: Erciyes Üniversitesi /Eğitim Bilimleri Enstitüsü.

Dalgar, G., & Kaytez, N. (2018). Doğum öncesi gelişim ve doğum. In P. Ünüvar (Ed.), Bebeklik döneminde gelişim ve eğitim (pp. 24-68). Ankara: Anı.

Dede, H., Kandemir, Ö., Yalvaç, S. & Altay, M. (2017). İleri anne yaşı nedeniyle yapılan ikinci trimester amniyosentez çalışmalarımız: Üç yıllık deneyim. *The Journal of Gynecology- Obstetrics and Neonatology*, *10*(38), 1586-1588.

Deniz, Ü. (2015). Üreme sistemi anatomisi ve fizyolojisi. In Ü. Deniz & Ö. R. Önder (Eds.), *Okul öncesi eğitimi öğretmen adayları ve öğretmenler için anne çocuk sağlığı ve ilk yardım* (pp. 11–22). Ankara: Nobel.

Deniz, Ü. (2017). Gelişim. In N. Aral & T. Duman (Eds.), *Eğitim psikolojisi* (pp. 32–53). Ankara: Pegem. doi:10.14527/9786052410240.02

Dokuyan, M. (2016). 12. Sınıf öğrencilerinde algılanan anne- baba tutumları ile benlik saygısı arasındaki ilişkinin incelenmesi. *Ahi Evran Üniversitesi Sosyal Bilimler Enstitüsü Dergisi*, *2*(2), 1–21.

Durgun, Ö. (2011). Türkiye'de yoksulluk ve çocuk yoksulluğu üzerine bir inceleme. *Bilgi Ekonomisi ve Yönetimi Dergisi*, *6*(1), 143–154.

Durualp, E. (2017). Doğum öncesi gelişim ve doğum. In A. Köksal-Akyol (Ed.), Erken çocukluk döneminde gelişim (pp. 114-149). Ankara: Anı.

Ergül, C. (2014). Sık rastlanan yetersizlikler. In İ. H. Diken (Ed.), *Erken çocukluk eğitimi* (pp. 258–262). Ankara: Pegem.

Feldman, R. S. (2017). *Development across the life span* (8th ed.). Pearson.

Filburn, A. G., Lahiri, T., & Ren, C. L. (2016). *Handbook o cystic fibrozis*. Springer. doi:10.1007/978-3-319-32504-0

Gander, M. J., & Gardiner, H. W. (2015). *Çocuk ve ergen gelişimi*. Ankara: İmge.

Gültekin, M. (2006). *Okul öncesinde duygu ve davranış sorunları*. İstanbul: Nesil.

Gültekin-Akduman, G., Günindi, Y., & Türkoğlu, D. (2015). Okul öncesi dönem çocuklarının sosyal beceri düzeyleri ile davranış problemleri arasındaki ilişkinin incelenmesi. *Uluslararası Sosyal Araştırmalar Dergisi, 8*(37), 673–683.

Gündüz, C., Çoğulu, Ö., Özkınay, C., Tütüncüoğlu, S., Sapmaz, G., Kösem, M., & Özkınay, F. (2000). İdiyopatik mental retardasyonda Fragil X sendromu araştırması. *Ege Tıp Dergisi, 39*(2), 97–104.

Hess, Y. S., Lonnerdal, B., Hatz, C., Rivers, A. J., & Brown, H. K. (2009). Recent advances in knowledge of zinc nutritional. *Human Health Food Bulletin, 30*(1), 5–11.

Howard, V. F., Williams, B., & Lepper, C. E. (2011). *Özel gereksinimi olan küçük çocuklar*. Ankara: Nobel.

İrge, E., Timur, S., Zincir, H., Oltuoğlu, H., & Dursun, S. (2005). Gebelikte beslenmenin değerlendirilmesi. *Sted, 14*(7), 157–160.

Işık, M. (2007). *Televizyon ve çocuk*. Konya: Eğitim.

Kadayıfçı, O., Ürünsak, İ. F., Ürünsak, M., & Atay, Y. (2001). Gebelikte enfeksiyon hastalıkları. In S. M. Beksaç, N. Demir, A. Koç, & A. Yüksel (Eds.), *Obstetrik maternal fetal tıp ve perinatoloji* (pp. 878–939). Ankara: Nobel.

Katar, S., & Devecioğlu, C. (2006). Dicle üniversitesi yeni doğan bakım ünitesinde izlenen çok düşük doğum ağırlıklı bebeklerin değerlendirilmesi. *Dicle Tıp Dergisi, 33*(4), 248–251.

Kaya, Y., & Şahin, N. (2013). Kadınlarda madde kullanımı ve hemşirenin rolü. *Hemşirelikte Eğitim ve Araştırma Dergisi, 10*(1), 3–7.

Kaytez, N. (2018). Motor gelişim. In N. Aral & Z. F. Temel (Eds.), *Çocuk gelişimi* (pp. 142–170). Ankara: Hedef.

Keyou, G., Jianbin, J., & Hang, L. (2007). Food-based dietary guidelines in China. *Annals of Nutrition & Metabolism, 51*(2), 26–31. doi:10.1159/000103564

Kılıç, K. M. (2018). Gelişimin genetik temelleri. In A. Güngör-Aytar (Ed.), Gelişim psikolojisi (pp. 46-71). Ankara: Hedef.

Kılınç, B., & Kılınç, E. P. (2014). Yeni medya ortamında çocuk birey: Yeni iletişim teknolojileri ve medya pedagojisinin önemi. *Akdeniz İletişim Dergisi*, *22*, 9–23. doi:10.31123/akil.441825

Langford, K. S. (2002). Infectious disease and pregnancy. *Current Obstetrics & Gynaecology*, *12*(3), 125–130. doi:10.1054/cuog.2001.0247

Lou, C., Higgins, K., Gelfer, J., Hong, E., & Miller, S. (2005). The effects of teacher facilitation on the social interactions of young children during computer activities. *Topics in Early Childhood Special Education*, *25*(4), 208–217. doi:10.1177/02711214050250040201

Madazi, R., & Atasü, T. (2000). Fetüs ve yeni doğana etkili enfeksiyon hastalıkları. In T. Atasü (Ed.), *Gebelikte fetüs ve yeni doğana zararlı etkiler* (pp. 319–372). Ankara: Nobel.

McKinney, S. S. (2008). *Yanlış çocuk: Çocuk yetiştirmede yapılan hatalar.* İstanbul: Ekinoks.

Metin, E. N. (2018). Zihinsel engelli çocuklar. In N. Metin (Ed.), Özel gereksinimli çocuklar (pp. 74-115). Ankara: Anı.

Muis, K. R., Ranelluci, J., Trevors, G., & Duffy, M. C. (2015). The effects of technology mediated immediate feedback on kindergarten students' attitudes, emotions, engagement and learning outcomes during literacy skills Development. *Learning and Instruction*, *38*, 1–13. doi:10.1016/j.learninstruc.2015.02.001

Ng, K., Jarvinen, A., & Bellugi, U. (2014). Toward a deeper characterization of the social phenotype of Williams Syndrome. *The Association Between Personality and Social Drive Research in Developmental Disabilities*, *35*(8), 1838–1849. PMID:24794322

Oktik, N., Çukur, C. Ş., Ünal, H., Gökalp-Yılmaz, G., Özbek, Ç., Doğan, V., & Hasırcı, D. (2010). Sosyal risk analizi: Mardin proje raporu örneği. Muğla: Muğla Üniversitesi.

Öncü, E., Kurt, Ö.A., & Esenay, F.I., & Özer, F. (2012). Çalışan çocukların ailede istismarı. *Turkish Journal of Public Health*, *10*(3), 128–140.

Önder, A. (2003). *Ailede iletişim: Konuşarak ve dinleyerek anlaşalım.* İstanbul: Morpa.

Özenoğlu, A., & Ünal, G. (2015). Açık ve yoksulluğun çocuklarda saldırganlık ve şiddet davranışları ve ilişkisi. *Gümüşhane Üniversitesi Sağlık Bilimleri Dergisi*, *4*(1), 162–179.

Özer, Y.E., & Çolak, M. (2015). Yoksulluk, dezavantajlı grup ve suç ilişkisine yönelik bir alan araştırması: İzmir kadın sığınma evleri örneği. *İktisadi ve İdari Bilimler Dergisi, 29*(3), 479-491.

Özgen-Macon, B. (2010). Fötal santral sinir sisteminin manyetik rezonans görüntülemesi. *Hacettepe Tıp Dergisi*, *41*, 155–161.

Özsoy, S. (2014). Adölesan annelerin emzirmeye yönelik düşünce ve uygulamaları. *F.N. Hemşirelik Dergisi*, *22*(2), 84–93.

Özyürek, A., Bedge, Z., & Yavuz, N. F. (2014). Okul öncesi çocuklarının sosyal becerileri ile yakın çevredeki yetişkin etkileşimleri arasındaki ilişki. *Sosyal Bilimler Dergisi*, *16*(2), 115–134.

Pehlivan, D., Çefle, K., Öztürk, Ş., Akbulut, F.M., & Palandüz, Ş. (2013). 47 Xİ(Xq)Y karyotipi saptanan bir Kleinfelter sendromu olgusu. *İstanbul Tıp Fakültesi Dergisi, 71,* 91-93.

Peynirci, H., & Ertürk, E. (2012). Kleinfelter syndrome. *Turk Jern, 17*, 63–70.

Plak, R. D., Merkelbach, I., Kegel, C. A. T., Von-Ijzendoorn, M. H., & Bus, A. G. (2016). Brief computer interventions emergent academic skills is susceptible children: A gene by environment experiment. *Learning and Instruction, 45*, 1–8. doi:10.1016/j.learninstruc.2016.06.002

Reçber, D., & Özen, S. (2005). Trizomi 13, Patou syndrome: Bir olgu sunumu. *Van Tıp Dergisi, 12*(1), 29–31.

Sağlam, M. (2017). Aile içi iletişim. In N. Aral (Ed.), *Çocuk ve iletişim* (pp. 308–324). Ankara: Vize.

Şahin, S. (2018). Çocuk gelişimine giriş. In N. Aral & Z. F. Temel (Eds.), *Çocuk gelişimi* (pp. 16–31). Ankara: Hedef.

Şahin, S., & Aral, N. (2012). Aile içi iletişim. *Ankara Sağlık Bilimleri Dergisi, 1*(3), 55–66. doi:10.1501/Asbd_0000000029

Sak, E. M., Özkul, Ö., Evsen, S. M., Sak, S., & Eviroğlu, O. (2009). Gebelik anemisinin perinatal sonuçlara etkisi. *Dicle Tıp Dergisi, 36*(1), 39–42.

San-Bayhan, P., & Artan, İ. (2009). Çocuk gelişimi ve eğitimi. İstanbul: Morpa.

Santrock, J. W. (2016). *Yaşam boyu gelişim*. Ankara: Nobel.

Sapsağlam, Ö., & Ömeroğlu, E. (2016). Examining the effect of social values education program being applied to nursery school students upon acquiring skills. *Educational Research Review, 11*(3), 1261–1271.

Saracho, O. N. (2017). Bullying young children's roles, social status and prevention programmes. *Early Child Development, 187*(1), 68–79. doi:10.1080/03004430.2016.1150275 PMID:28042899

Sayan, H. (2016). Okul öncesi eğitimde teknoloji kullanımı. *Yüzyılda Eğitim ve Toplum, 5*(3), 67–83.

Schmid, R. F., Miodrag, N., & DiFrancesco, N. (2008). A human computer technology analysis and recommendations. *The Future of Children, 10*(2), 4–30.

Senemoğlu, N. (2013). *Gelişim, öğrenme ve öğretim: Kuramdan uygulamaya*. Ankara: Yargı.

Şener, D. K., & Ocakçı, A. F. (2012). Yoksulluğun çocuk sağlığı üzerine çok boyutlu etkileri. *Ankara Sağlık Hizmetleri Dergisi, 13*(1), 57–68. doi:10.1501/Ashd_0000000098

Sharma, H. D. (2004). *Dengeli çocuk yetiştirme*. Eskişehir: Platform.

Sigelman, C. K., & Rider, E. A. (2017). *Human development across the life span* (9th ed.). Cengage Learning.

Sümer, N., Gündoğdu-Aktürk, E., & Helvacı, F. (2010). Anne-baba tutum ve davranışlarının psikolojik etkileri: Türkiye'de yapılan çalışmalara toplu bakış. *Türk Psikoloji Yazilari, 13*(25), 42–59.

Talim, B. (2015). Doğum öncesi gelişim, doğum ve yeni doğan. In F. Turan & A. İ. Yükselen (Eds.), *Çocuk gelişimi 1: Bebeklik döneminde gelişim* (pp. 99–112). Ankara: Hedef.

Tamayo, M., Rebolledo, J., & Besoaín-Saldaña, A. (2017). Monitoring inclusive education in Chile: Differences between urban and rural areas. *International Journal of Educational Development, 53*, 110–116. doi:10.1016/j.ijedudev.2017.01.002

Tanyeri-Bayraktar, B., Bayraktar, S., Hepokur, M., & Kardaş, M. (2015). Hidrosefali ile birliktelik gösteren fetal alkol sendromu. *Bezmialem Science, 3*, 48–50.

Tarı-Cömert, I., & Kayıran, S.M. (2010). Çocuk ve ergenlerde internet kullanımı. *Çocuk Dergisi, 10*(4), 166-170 doi:10.5222/j.child.2010.166

Taşkın, L. (2016). *Doğum ve kadın sağlığı hemşireliği*. Ankara: Özyurt.

Tekin-Ersan, D., & Kizir, M. (2016). Williams sendromu olan bireylerin ailelerinin deneyimlerinin ve karşılaştıkları güçlüklerin belirlenmesi. *Ankara Üniversitesi Eğitim Bilimleri Fakültesi Özel Eğitim Dergisi, 17*(3), 299–316. doi:10.21565/ozeleegitinmdergisi.268557

Toran, M. (2010). *Yoksul çocukların kaderi çalışmak: Türkiye'de çocuk emeği*. Ankara: Ütopya.

Tümkaya, S., Aybek, B., & Çelik, M. (2010). Yoksul ailelerden gelen ergenlerde psikososyal bir olgu olarak umutsuzluk ve boyun eğici davranışların incelenmesi. *Uluslararası İnsan Bilimleri Dergisi. 7*(1), 970–984.

Uluç, G. (2003). *Küreselleşen medya*. İstanbul: Anahtar Kitaplar.

Walker, F., Christian, L., & Black, E. R. (2007). Functional indicator for assessing zinc deficiency. *Food and Bulletin, 28*(3), 454–479. doi:10.1177/15648265070283S305

Walter-Laager, C., Brandenberg, K., Tinguely, L., Schwarz, J., Pfiffner, M. R., & Moschner, B. (2017). Media assisted language learning for young children: Effects of a world- learning app on the vocabulary acquisition of two year olds. *British Journal of Educational Technology, 48*(4), 1062–1072. doi:10.1111/bjet.12472

Weilburger, S. L. (2008). *Baba sakın bize vurma: Çocuk ve aile içi şiddet*. İstanbul: Ekinoks.

Yaşar-Ekici, F. (2015). Okul öncesi eğitime devam eden çocukların sosyal beceri ile aile özellikleri arasındaki ilişkinin incelenmesi. *The Black Sea Journal of Social Sciences, 7*(1), 223–259.

Yavuzer, H. (2011). *Ana-baba ve çocuk*. İstanbul: Remzi.

Yavuzer, H. (2012). *Çocuk psikolojisi*. İstanbul: Remzi.

Yazıcı, E., & Gençer, E. (2016). Okul öncesi çocuklarının bilgi ve iletişim teknolojileriyle etkileşimlerinin bazı değişkenler yönünden incelenmesi. *Kastamonu Eğitim Dergisi, 24*(5), 2235–2252.

Yeşilçiçek-Çalık, K., & Aktaş, S. (2011). Gebelikte depresyon, sıklık ve risk faktörleri ve tedavisi. *Psikiyatride Güncel Yaklasimlar, 3*(1), 142–162.

Yıldız, A., Baran, E., Akdur, R., Ocaktan, E., & Kanyılmaz, O. (2008). Bir sağlık ocağı bölgesinde 0- 11 aylık bebekleri olan annelerin emzirme durumları ve etkileyen faktörler. *Ankara Üniversitesi Tıp Fakültesi Mecmuası, 61*(2), 61–67.

Yıldız-Bıçakçı, M., & Aral, N. (2017). Dil gelişimi. In N. Aral & T. Duman (Eds.), Eğitim psikolojisi (pp. 132-150). Ankara: Pegem.

Yılmaz, E., Aydın-Tosun, Ö., Tarhan, N., Ayaz, R., Turgut, A., & Karateke, A. (2016). İleri anne yaşı gebeliklerinde perinatal sonuçlar. *Journal of Clinical and Experimental Investigations*, *7*(2), 157–162.

Yücel, M. K. (2015). *Doğum ve kadın sağlığı*. Ankara: Palme.

KEY TERMS AND DEFINITIONS

Chromosome: They are the factors that determine the characteristics that pass from parents to children.

Development: It is a progressive process starting from prenatal and continuing until death.

Development Areas: Physical, motor, language, social-emotional, and cognitive areas are divided and studied. Development areas affect human development both separately and as a whole.

Development Periods: It is classified as infancy, preschool, school, adolescence, adulthood and old age with distinctive characteristics.

Environment: It can be defined as the physical and psychological environment in which people live and affect their behavior.

Heredity: It is the transfer of dominant or recessive characteristics in the parent to fetus.

Hormone: They are the structures that regulate the body's work and affect.

Chapter 3
Developmental Tasks

Figen Gürsoy
Ankara University, Turkey

Burçin Aysu
Ankara Yıldırım Beyazıt University, Turkey

ABSTRACT

Recognizing this period and the features in this period is regarded very important in terms of evaluating, following, and supporting child development. The concept of a developmental task should be approached in each period in order to understand the development. An individual should be biologically ready to carry out the developmental tasks belonging to the period he is in. However, in addition to biological readiness, the environmental conditions like the education environment should be appropriate in order to carry out the developmental tasks. Especially in the early years, schools have crucial roles in children's fulfillment of the developmental tasks and developing. Development periods and development duties should be known in order to observe and understand the development process better. Based on this view, the development periods and the development duties are to be explained in this section.

INTRODUCTION

Recognizing this period and its features is regarded very important in terms of evaluating, following and supporting child development. The concept of developmental task should be approached in each period in order to understand the development (Bacanlı, 2002), because there are developmental tasks to be fulfilled in each period of development. An individual should be biologically ready to carry out the developmental tasks belonging to the period he is in. However, in addition to biological readiness, the environmental conditions such as the educational environment should be appropriate in order to carry out the developmental tasks. Especially in the early years, schools have crucial roles in children's fulfillment of the developmental tasks and developing healthily (Erden and Akman, 1998). Development periods and developmental tasks should be known in order to observe and understand the development process better. Based on this view, the development periods and the development tasks are to be explained in this section.

DOI: 10.4018/978-1-7998-2952-2.ch003

BACKGROUND

Development is the growth, maturation and the continuous progression of the organism with the interaction of learning. When the development is considered as a product of maturation and learning processes, the direction of the process is the development. Starting from the fertilization, the change of the organism which makes continuous progress from many aspects to the end is called development (Senemoğlu, 2007). According to another definition; the development can be defined as the characteristic changes happening in man's structure from the fertilization until the death. Various changes occur within the framework of the concept of development. The development can be examined as consisting of three dimensions; these are biological, cognitive and social developments. Such dimensions of the development are too complex a process to be evaluated separately. From those dimensions; the biological process involves the changes which may occur in parallel with the genes an individual receives from his parents hereditarily. These include the weight, the increase in height, hormonal changes and so on. The cognitive process involves the changes in the individual's thinking and intelligence capacity. The functions such as using imagination, problem solving and memorizing the lyrics are all evaluated within the cognitive development. The social process is the cluster of changes that involves the individual's relations with the other people and his personality. All the conditions such as the reactions to parents, the general characteristics of friend relationships and the passion for sports are the features which involve the social processes of the teenager's development. All the above-mentioned processes are closely related to each other. Any changes in one of them also affect the other. Rapid biological development and puberty influence an individual's social process, namely his self-concept, and such a condition can affect his cognitive, that is to say, his intelligence performance. Considering the development as consisting of three dimensions and a resultant of complex processes is a valid phenomenon for all the life cycles. The biological, cognitive and social processes continuing in man's life become effective within the framework of role and responsibilities attributed to the individual in every age period (Yazgan İnanç et. al., 2010).

Critical period of time is another important factor in the development of organism. There are times for each development to appear universally in a most intensive way. Those times are described as the critical period for that development. If the organism has the necessary hereditary potential in that period of time, when it is met with the sufficient stimulus, the emergence of the top-level development is possible in terms of some behaviors or some organs and their functions (Deniz, 2018). The environmental and hereditary factors, which provide the development, continue to show their influences starting from the fertilization until the end of the life. The biological characteristics of the baby to be born are determined by the genes in the 46 chromosomes which are in the first cell formed with the fertilization. However, the revelation of the genetic features with full potential on the organism is largely related to the experiences the organism has had, namely the environmental conditions (Erden and Akman, 1998).

The development consists of many changes that follow each other in some specific periods of life. The importance of the early years of time and the determinative influence of the interaction between the heredity and environment require the development, which starts in the prenatal period, to be examined by dividing it into some periods together with the birth (Erden and Akman, 1997). Time frame, period or stage concepts in the development are used to mention the progress that appear in specific ages and have their distinctive features. The stages in which some features are clearly seen are called periods (Deniz, 2018).

Babies, who make progress in the babyhood period from various aspects such as motor skills and coordination, language skills and emotional skills, develop a sense of attachment to family members and other caretakers in the related period. They also learn to trust or not to trust, to show their love and other emotions or not to show them. Further, the babies' learning to express their basic emotions develops the senses of self and independence. The babies in that period show distinctive differences in terms of personality and character. The physical, mental and language development carry on rapidly during the early childhood period. In this period, the children are in better position in terms of self-care skills. The child, whose self-concept develops, starts to be aware of the gender roles, too. The child in the pre-school period is willing to play with other children. The characteristic of the mother, father, and child relationship is very important in terms of the socialization process that comes true in that period. During the middle childhood period, some important progresses are observed in children's literacy, their numerical calculation skills, the way they understand the world and think logically. In addition to the harmonious relationship established with the mother and father, school success has also vital importance. The psychosocial and moral development progress rapidly, and the characteristics of the family relationships continue to be influential in the emotional and social adaptation process. Puberty, on the other hand, is a transition period between the childhood and adulthood. During the early puberty the sexual maturation occurs and the abstract thinking begins. As the teenagers gain their independence from their parents, they wish to increase their relationship with their peers and to develop the sense of belonging more with others. The development of a positive identity is an important psycho-social task. In that period, the teenagers start to make their career choices and try to enter the business world by completing their education. The ability of an individual in the puberty for establishing close and friendly relationships with the others develop; accordingly, his relationships with the opposite sex develops as well (Prechtl, 2004; Yazgan İnanç et. al., 2010). Recognizing this period and its features is regarded very important in terms of evaluating, following and supporting the child development. The concept of developmental task should be approached in each period in order to understand the development (Bacanlı, 2002) because there are developmental tasks to be fulfilled in each period of development. An individual should be biologically ready to carry out the developmental tasks belonging to the period he is in. However, in addition to biological readiness, environmental conditions such as education environment should be appropriate in order to carry out the developmental tasks. Especially in the early years, schools have crucial roles in children's fulfillment of the developmental tasks and developing healthily (Erden and Akman, 1998).

The developmental task involves the tasks man should succeed, the behaviors he should gain and the features he should improve during his life. The concept of developmental tasks was developed by Havinghurst. Havinghurst defines the developmental tasks as such: "They appear in specific period of an individual's life, when they are achieved, they lead to happiness and success for the next tasks and when they are not achieved, they lead to unhappiness and intolerance of the society and to have difficulty in the next tasks". There are some rules in the developmental tasks. These are;

- Individual growth and development are continuous.
- Individual growth can be explained with periods or stages.
- Individuals can be described with the certain general features of the life cycle they are in.
- Culture has an important effect on people's passing from similar developmental stages.
- Societies have some demands from their individuals.
- The demand expected from the individuals by the societies largely resemble to each other for all the members of the society.
- When people proceed from a developmental stage to the next one, the demands of the society also change. In other words, the demands vary according to the stages.

- When people are faced with the demand for changing their behaviors and start to obtain information about that demand, the developmental crises arise.
- People proceed from one developmental stage of maturity to other during the developmental crises and while solving those crises.
- Task emerges in a stage by itself, namely it emerges as being special to a period.
- The preparation for solving the developmental crises or the developmental tasks appear in the earlier stage than it is supposed to.
- When an individual is obliged to review the decisions he makes, the developmental task or crisis may reappear in the next stage a little differently
- The developmental task or crisis should be solved before an individual proceeds to the next stage
- While proceeding from a developmental stage to the next one, learning the developmental task and being able to cope with this crisis provide social approval, give happiness and lead to success in the next crises and in the related tasks.
- While encountering the developmental crisis or developmental task, the situation of being unsuccessful cause to the intolerance of the society.

A specific developmental task should have a meaning peculiar to the each individual. Each individual attributes a meaning to the each of the developmental tasks. At the same time, each individual differentiates in terms of his approaches to the developmental tasks. In other words, the individuals may not undertake the developmental tasks by dealing with them in the same way. Another subject about the developmental tasks is that they form a pattern by gathering. The meaning man attributes to the developmental task come to forefront at this point. Although the developmental tasks have some common elements for the individuals in a society and even for the all societies, essentially the meaning man attributes to the developmental task is important (Bacanlı, 2002).

The Developmental Tasks According to the Life Cycles

Havinghurst indicated the developmental tasks for various life cycles. The developmental tasks determined in accordance with the life cycles were explained in detail below (Bacanlı, 2002).

Babyhood and Early Childhood Period (0-6 Years)

In the babyhood period, the babies obtain basic feelings about whether they can trust the world around them or not. In this period, the characteristic of the child's relation with those around him influence his basic sense of trust. Satisfying the child's need in the first year of his life mostly depends on the mother or the adult who looks after him. The mother or the adult, who replaces the mother and looks after the child, is very important in child's world. While satisfying the child's needs, the mother's love to her child, her fondling, making her child understand her warmth and her care for him lay a foundation for child's sense of trust both for himself and for his environment. Otherwise, the distrust for himself and for his environment appears in the child, who is refused by his mother or by the adult replacing his mother, who is discountenanced and whose needs are not satisfied with love appropriately and in time. Children, as they grow up, do not want to be completely dependent to others. They want to control their environment and show their power. They determine their parents and others around them about what they can do and cannot do. The child, who carried out the developmental tasks in the previous stage, should feel his freedom in order to obtain his own control. Therefore, flexible environments, in which the children can discover freely, should be provided at home and in the schools. The children's efforts such as eating by themselves, tidying up his belongings, dressing, undressing, choosing their clothes, solving some of the problems they face should be supported and encouraged. Thus, the senses of independence are founded in the child. The essence of the self-control and respect constitutes in this period. Otherwise, the senses of suspicion of their abilities and being ashamed of their needs and bodies can emerge in children who are continuously restricted, excessively protected and strictly controlled. Together with the sociability rise in the child, the problem-creating behaviors can also increase. However, both parents and the teachers in the pre-school education institutions should allow children to run, move, jump, play and go down the slide so that the sense of sociability can develop in the children. Also, the opportunity should be provided for child to gain necessary experiences in order to discover himself. The sense of guilt can emerge in the child who is frequently scolded and prevented due to his curiosity. The child, who is punished by his parents or teachers due to his sociability, thinks what he does is wrong and can feel guilty both in this period and in the future periods of his life. But, approving whatever the child does can influence his moral development negatively. Therefore, children's sociability should be supported by balancing what they should do and should not do (Senemoğlu, 2007).

The developmental tasks in the babyhood and early childhood period can be sorted as such; learning to walk, learning to eat solid food, learning to speak, learning to control the removal of body wastes, learning the gender differences, learning to form concepts to define social and physical reality and learning language, getting ready to read, separating the correct from incorrect and starting to conscience development (Bacanlı, 2002; Senemoğlu, 2007).

Learning to Walk

The individuals in the babyhood and early childhood period should learn to walk and move by themselves. Directly proportional with the increase in weight and height and muscle development, the development of brain, spinal cord and nervous system plays an important role in the obtainment of motion abilities (Törüner and Büyükgönenç, 2011; Ziegler Delahunt, 2002).

In the first month, while lying face down, the baby can lift its head for a couple of seconds and grasps the objects reflexively. The baby can also follow the objects for a short time when they move slowly. In the second month, while lying face down, the baby can lift its chest and head with the help of its arms and while lying on its back, the baby can raise its arms. In the third month, the reflexes begin to decrease, the baby can sit with support and starts to hit the objects and follow its hands. In the fourth month, the baby sits on the lap with the support. In the fifth month, the baby turns deliberately from one side to other, reaches to the objects and grasps and can hold its feeding bottle. In the sixth month, the baby can sit on the feeding chair and can catch the objects, which are passed to it, with its palm. The baby, who can sit without support in the seventh month, can stand when held from its armpits in the eighth month. In the ninth month, the baby can remain standing by holding somewhere, starts to hold with its fingertips and sits by itself. In the tenth month, the baby crawls and can lift the cap of a box. In the eleventh month, the baby walks when held from its hands. It also squats and leans, holds a pen, can throw a toy intentionally and scribbles. In the twelfth month, the baby stands up by holding somewhere. About the thirteenth month, the baby can stand without help and starts to walk (Yazgan İnanç et. al., 2010; Santrock, 2014).

Moving is the root of the abilities of recognizing the world, gaining information and understanding for children. The children not only learn themselves and their environment, but also, they gain the coordination ability and their self-reliance develops. Moving is one of the best communication ways for the child (Şahin, 2015).

The baby, who learnt to walk, gets rid of his dependence to his mother. The individuals, who cannot learn to walk in this period, may have problems in the latter periods of their developments (Bacanlı, 2002). For example, the interest of the child, who cannot learn to walk, in his environment would remain limited accordingly.

Learning to Eat Solid Food

When the baby is born, it has no teeth, so it is breastfed. Many of the children teethe for the first time in the sixth or the seventh month. However, there are newborns who seldom teethe in the earlier period, but reason of this situation should be examined. The delay in eruption of teeth can cause to think the hypothyroidism. Lack teeth or extra teeth can emerge as a seldom genetic feature. Between the six and twelve months, middle and side teeth erupt as temporary teeth (Davies, 2010; Törüner and Büyükgönenç, 2011). First molar teeth erupt in twelve-fifteen months, the canine teeth erupt in fifteen to twenty-four months and second molar teeth erupt in twenty-four and thirty months (Törüner and Büyükgönenç, 2011). Together with the teeth eruption, child should learn to eat solid food. Therefore, the breast milk is slowly replaced by solid food (Bacanlı, 2002). If mothers continue to give the food as puree just because it is easier, this situation prevents the child from learning to eat solid food. At this point, the opportunity for eating solid food and enough time for feeding himself should be given to the child.

Learning to Speak

Another behavior, which baby should acquire, is the speaking behavior. Along with starting to perceive the aural warning, babies begin to react to those warnings when they are a few weeks old. The crying, which starts as a necessity for breathing, is in the way of vocalizing and in the characteristic of communication tool. Before the seventh week, the most important reason for crying is the hunger; noise and light are of the secondary importance. The crying differs in terms of rhythm and tone as of the second month. It is observed that together with the language development the crying decreases (Yavuzer, 1999). In the two-four months, the baby unconsciously makes simple sounds which are universal. In the production of sound, which happens reflexively, the baby randomly makes the sounds like "a, u, o" and adding "h" at the end of those sounds, it produces sounds such as "ah, uh". The baby's control in the sound mechanism increases in the four-six months; and shows the ability of rounding and sticking out its tongue and can make voices to show its satisfaction. The voices produced by the baby transform into purposeful voices in this period. The baby repeats the sound such as "ma-ma-ma" and "ba-ba-ba" while making labial sounds like "b, m, p". The murmuring is observed between the sixth and ninth months. The baby combines the voice it produces with the hearing and starts to acquire the characteristics of the language in the environment by repeating the sounds it hears. Between the ninth and twelfth months, the baby deliberately imitates the human voice (Yavuzer, 1999; Yazgan İnanç et. al., 2010). The child's control over the speaking mechanism increases progressively. Besides the child's biological capacity, the effects of the environment and education gain importance in the child's speaking. Therefore, the supportive role of education should be benefitted during the stages that the child passes through his language development. The child needs to be supported, and needs the creation of appropriate environments and stimuli so that he can speak in the early period. In the coming years, benefitting from the activities such as; drama, game, theatre, poetry, tongue twister, finger games, storytelling, repetitions and commenting both the accurate use of language can be taught and the child has fun while learning. If the child has fun, he does such activities more; therefore, he passes from the process of language development in a very good way (BağcıAyrancı, 2018).

Obtaining the Toilet Control - Learning to Control the Removal of Body Wastes

Another behavior, which the baby should acquire, is the stool control with the potty training. Four preparatory processes, three of them are related to the child and one of them is related to the parents, are needed for the potty training. In Physiological preparation; it is required to have anal and urethral stricture control, abilities of walking, sitting, squat and taking off clothes. In Mental preparation; it is required for the child to state that he needs toilet and imitate the appropriate behavior. In Psychological preparation it is required for the child to want to be liked by the parents and be able to sit in the toilet for five-ten minutes without getting bored. The preparatory process of the parents involves being willing to spend time for potty training, not having stress and being aware of the fact that the child is ready for the potty training (Törüner and Büyükgönenç, 2011).

Babies' easy defecation in the first year is regarded as the indicator of mental health. In parallel with the child's biological development and his control over his muscles, he is expected to control his stool. The child should be able to hold his defecation or take a dump whenever he wants. If he cannot hold his defecation or perform the defecation, then it means there is a problem (Bacanlı, 2002).

Learning the Gender Differences

A Child notices his/her gender when s/he is about three years old. The child also understands what boy and girl words mean. This understanding depends on the physical differences such as "The girls have long hair, and "the men have moustache". The child who noticed his/her gender is expected to comprehend what s/he means when s/he says "I am a girl/ I am a boy," and not to do an act about the gender in front of everybody by saying "I am a girl/ I am a boy." For example, the child is expected not to expose his/her sexual organ (Bacanlı, 2002).

Forming Concepts to Define Social and Physical Reality and Language Learning

The concept development, which can be linked to the language development, is another developmental task. The child is expected to interact with the outer world. In order to do it, firstly he should acquire the definitions and concepts about the outer world and form a design about them. The individuals, who cannot form that design, have difficulty in interacting with the outer world (Bacanlı, 2002).

Getting Ready to Read and Write

Child's getting ready to learn reading, writing and calculating, before he starts to the primary education, is regarded as important. At this point, the child's family and the school education he receives are important for his getting ready to the primary education.

Separating the Correct From the Incorrect and Starting to Moral Development

While the child is obtaining the daily experiences in the family and immediate environment, the senses gain different meanings and the child begins to look for a place in the social environment he is in. The child's world starts to be activity focus in order to research the components of human behaviors and his immediate environment. Nevertheless, the child needs his mother's emotional support. The emotional control begins to be formed. The child learns to organize and manage his emotions which are important for both the emotional and the social development. Thus, the child's social skills and peer relations develop synchronously. But the child does not hesitate to show his emotions. Thanks to the language development and cognitive competences, the child finds the ways so that his abreaction is approved by the society. The behaviors learned and the abilities acquired in this period underlie the social development (GüngörAytar, 2015; Yaşar, 2015).

The appropriate models have important roles in children's learning such abilities. The children can obtain new information and abilities by observing those models and imitating them. The child is alone and is in the center of life during the early years of his life. The child is in egocentrism in those years. He starts to become distant from the egocentrism as time goes by. The children need the adults as reliable role models. The adults not only teach the moral values with formal and informal training activities, but also they share their own beliefs with their children. The parents are the ones who are together with their children the most so they provide opportunities for their children to observe and take their parents as an example. The children accept the daily ways of behavior exhibited by their parents. The children try to resemble and adapt the one whose superiority they accept. They also accept the ideas and behaviors of the one whom they internalize, and they are influenced by the moral attitudes and behaviors of their parents who are taken as an example. Therefore, their decisions about the correct and incorrect along with their first moral ideas are formed. In this sense, it is very important to be a correct role model for the child (Şahin, 2015).

It is observed in this process that the children can distinguish the bad behaviors and understand when they perform an incorrect behavior. A child, who can separate the correct from the incorrect and have a conscious development, starts to learn the social rules and can cope with the new situations appeared when school starts. Otherwise, the child may face serious problems (Bacanlı, 2002; Erden and Akman, 1998).

Middle Childhood Period

Together with the children's start to primary education, a great enlargement happens in his social world in this period. While the influence of friends and teacher increases on the child, his parents' influence decreases gradually. Planning a work, carrying on the work he started, cooperation, learning and achievement have a special importance for the child. The achievement brings with the sense of hard-working and help children develop a positive attitude towards himself and his abilities. The academic self-reliance, which underlies the success in the future, develops. Otherwise, the failures cause the child to develop negative attitudes and sense of inadequacy towards himself and prevents his learning in the future. The teachers should ensure the students success by giving them responsibilities according to their power. The children, in this period, try to use the tools used by the adults and make an effort to produce something. When the children's efforts are supported, their behaviors of studying and achieving develop. The children, who are continuously criticized for whatever they do, who are not supported and are not liked, believe the worthlessness of what they do and may develop the inferiority complex. In the middle childhood period, the children are expected to achieve the followings; learning the necessary physical skills for daily games, as a growing organism developing a beneficial attitude towards himself, learning to get on well with their peers, developing basic academic skills such as reading, writing and calculating, developing the concepts necessary for daily life, conscience, moral and system of values, obtaining personal independency and developing an attitude for social groups and institutions (Senemoğlu, 2007; Bacanlı, 2002).

Learning the Necessary Physical Skills for Daily Games

The changes in the bone and tissue development from the beginning of this period have had an important effect on obtaining high level motor skills. The differences between the growing of boys and girls are in the lowest level. The protraction in the arms and legs is faster than the body and the boys may have longer arms and legs than the girls during the childhood period. On the other hand, the girls' hip width is more than the boys'. Until the preadolescence, there are small differences between the girls and boys in terms of physical features. Therefore, it is suggested not to distinguish the boys and girls according to their gender in the activities they participate in (Özer and Özer, 2000). The child should be allowed to trial so that his motion skills can mature. During the normal maturation process, the child develops his motor skills by developing his perceptive processes and by trialing. The measures should be taken so that the child does not fail in that process (Özer and Özer, 2000). Thus, the children learn to use their body effectively in the games.

As a Growing Organism Developing a Beneficial Attitude Towards Himself

Along with the differentiations in body biochemistry, the rapid increase in height, especially in girls who are about to be ten, and the secondary sex characteristics begin to appear (Erden and Akman, 1998). On one side the growing child starts to be aware of himself and his body, on the other side, he faces the developmental task of accepting himself and his body, developing a positive attitude and internalizing it. Teachers should help children recognize themselves from various aspects and enrich their sense of self (Erden and Akman, 1998).

Learning to Get on Well With Their Peers

Child socializes in this period. He also comes together with his peers within the framework of some rules. The child performs common activities with his classmates at school and with his peers in the street. This situation, depending on the child's social development, requires him to develop the skill of getting on well with his peers. Such skills are to play important roles in forming a basis for the social relationships in the future (Erden and Akman, 1998).

Developing Basic Academic Skills

The primary education comes to forefront in this period, also obtaining the academic knowledge the child needs throughout his life such as reading, writing and calculating stand out. The child becomes able to solve more complex problems in his future life with the academic skills he obtains in this period. The child should form basic concepts in order to understand those events (Bacanlı, 2002).

Developing Necessary Concepts for Daily Life

Although the child acquires the basic concepts in the pre-school period, those concepts, which are mostly related to the objects, are replaced by the concepts which should be developed so that he can understand the things happening in the daily life (Bacanlı, 2002).

Developing Conscience, Moral and System of Values

Together with the conscience development, of which foundation was laid in the previous period, the values, preferences and attitudes start to become clear. Those developmental tasks do not end in this period like the other tasks; they continue by changing their shapes (Bacanlı, 2002).

Obtaining Personal Independency

Even though child has the freedom of movement, he may not achieve to act by himself. In this period, he can attempt on his own. The best example of this situation is to be able to go to and come from the school by himself. From now on, child is expected to give decisions about starting his own behaviors and directing them. If opportunities are provided for the child to give his own decisions and take responsibility, he begins to obtain his personal independency (Erden and Akman, 1998).

Developing an Attitude Towards the Social Groups and Institutions

Towards the end of the period, along with the influence of the mental development, the child attempts to judge and comment. The mass communication influences this situation to a large extent. Together with the influence of mass communication the child can share his views about the things happening in his country and in the world. The child's mental developments start to get ready for the abstract period and beyond the concrete situations he starts the attempts for developing the judgment and comment (Bacanlı, 2002).

Puberty Period

A man is expected to acquire the types of special behaviors towards the late adulthood so that he can cope with the problems of today and tomorrow successfully. Puberty is the period of physical, social and cognitive maturation (Gander and Gardiner, 1993). The start of puberty is determined by a range of genetic and environmental factors, which are in interaction with each other, therefore, it can vary from person to person and according to the gender. For girls 11-13 and for boys 13-15 are regarded as the ages for reaching the puberty. The physical changes in the girls start about two years earlier than the boys, but both of them reach the physical maturity in terms of the sexuality approximately at the similar ages (Kulaksızoğlu, 1998; Özbay and Öztürk, 1992). The teenager is perceived by his parents and other adults neither as a complete adult nor as a child. The parents and the other adults assert different approaches from each other about what the teenagers can do and cannot do. Also, the incoherency between how the teenager perceives himself and how he is perceived by the others leads him to have identity crisis (Kulaksızoğlu, 1998). In this period the teenager is expected to have new and more mature relationships with his peers from both of the sexes, gain masculine or feminine social role, accept his body and use it effectively, gain his emotional independence from his parents and other adults, get ready for marriage, family life and acquiring a profession, form a system of values and moral system and demand and acquire the responsible behavior in terms of the society (Bacanlı, 2002).

Developing New and More Mature Relationships With His/Her Peers From Both Sexes

In this period, depending on the development of social behaviors, the teenagers' interpersonal relationships develop, rise and their characteristics change. The personal friendships come to the forefront and romantic relationships start for the individual. Especially, child develops skills for establishing social relationships by getting in touch with his peers. The skills for starting, continuing and ending a relationship are acquired in this period (Santrock, 2014). Teachers can help this development by giving students group works on which they can study together (Erden and Akman, 1998).

Gaining Masculine or Feminine Social Role

In this period, the gain of gender role, which continues since the childhood, turns into the teenagers' learning of their own gender roles for the relations with the opposite sex within the relationship complexity. Learning such roles requires the individual to make friendship with the ones from the same sex and the opposite sex (Bacanlı, 2002).

Accepting His/Her Body and Using It Effectively

The physiological and hormonal changes occur in the puberty, rapid increase in the height and body occur (Erden and Akman, 1997). In this period, the child organism, which does not have reproduction capacity, turns into a reproductive individual. This change observed in the puberty occurs with the change in the hormone releasing of hypothalamus-hypophysis-gonad axis. The age and duration of the puberty vary from one child to other. The stature which is 80% of the mature height in the beginning of the puberty, reaches to 99% of the mature height in two-four years with the influence of gonad hormones. As the puberty starts earlier in the girls than the boys, the rate of growth also emerges earlier. The peak for the rate of growth is the period when the growth is the fastest and on average it is 9cm/year for the girls and 10,5cm/year for the boys. The weight of most of the girls and boys increases due to the fat storage in their body during the preadolescence when they are eight-ten years old. The reason for the increase in the weight in the puberty is muscle growth and the increase in the skeleton mass for the boys and the fat storage in the girls. The thickness of the subcutaneous fat layer diminishes in the boys, but it keeps on increasing in the girls (Özer and Özer, 2000).

Gaining the Emotional Independence From the Parents and Other Adults

Gaining more autonomy and independence is a basic developmental task of puberty. The teenagers have big conflicts about the independence and control when they are 15-17 years old. There is a decrease in the relationship between the parents and the child. There is a non-return emotional separation from the parents. In those ages, the standards for the behaviors are formed by the peer group (Santrock, 2014). The individual, who can act on his own in the previous period, can attempt by himself in the society. The girls and boys are in the effort of evaluating their bodies. The individual, who is obliged to be supported by the adults around him in the earlier periods, should prove that he can live without them in this period (Bacanlı, 2002). Accidents and injuries increase as a result of their risk-taking behaviors. The training should be given about smoking, drinking alcohol, drug and substance use and the consultancy should be given about normal growing and development (Törüner and Büyükgönenç, 2011). It is indicated that mothers and fathers are very influential in achieving this task. However, the parents act as if their children's steps to the maturity would come true in the distant future. While the parents, in parallel with their justifications, try to extend the teenager's dependency to family, the teenager want to be independent as soon as possible. The conflicts in the families usually occur due to that reason. The studies carried out about this issue reveal that the healthy teenagers do not have crucial disconnections in their relations with their families, but they are in a struggle for revealing their own individualism while carrying on their family relationships and such a situation is only possible with the changes in the relationship types (Çuhadaroğlu et. al., 2004). As is known, mothers and fathers have a strong effect on the teenager's developing personality. The intensity of this effect and whether it is healthy or not largely depend on the attitudes of the parents by whom the teenager is brought up (Senemoğlu, 2007).

Getting Ready for the Marriage and Family Life

In the periods after this period, the individual takes place in the community life. The individual takes part in the society with his work and family life. The individual in the puberty begins to determine how a mate and marriage he foresees for himself. It is also an extension of the self-concept. He continuously asks himself who he is, what kind of person he will share his life, and how this unity will be (Bacanlı, 2002).

Getting Ready to Acquire a Profession

Choice of profession is one of the most important decisions an individual makes in his life. It is because of the fact that the choice of profession is an important decision which involves both what kind of work he wants to work and what kind of standards he wants to live in. When the profession is chosen appropriately according to the individual's interests, skills, values, expectations and personality features, it affects the individual's mental health and professional and private life positively; also, it contributes to the life quality and standards positively (Çoban, 2010). While the teenager is getting ready for his future, he should determine his job in the future and do some research about it. The determination of the job he would do throughout his life also directs his educational future. The teenager should also determine the school he is going to study (Bacanlı, 2002). It is understood from the above-mentioned that the professional consultancy activities are very important for the teenager's fulfillment of this developmental task, also they help the teenager to give the right decision as a result of the overlapping of the individual's personality features and the characteristics of the profession (Çoban, 2010).

Forming a System of Values and Moral System

Moral development aims the individual's adaptation to the environment he is in by acquiring the social value judgments and his forming own principles and value judgments (Ahmetoğlu, 2018). Together with the moral development in the puberty, the conscience development occurs as a result of the development of some value judgments. The intensification of some ideological thoughts is an image of this developmental task (Bacanlı, 2002).

Demanding and Acquiring the Responsible Behavior in Terms of the Society

The teenager wants to enter the adults' social order and take responsibility in this period. This developmental task is also related to the social development and system of values (Bacanlı, 2002).

The puberty is defined as a stressful and difficult process. In this process, the teenagers have both physical difficulties and problems in their relationships with their families and peers. Further, they experience some mind changing about the difficulties with the opposite sex, the change of his peers with whom they make friends, the demands and expectations about the education, making choices about the school and career. Therefore, the teenagers have to cope with more problems compared to many other periods of their lives. The views, expectations, attitudes of the social and cultural environment about the puberty, the opportunities provided to the teenager about the choice of profession and improving oneself can also influence the density or diversity of the problems faced in this period. In short, the social environment and the teenagers are always in interaction with each other. In such a condition, the perceptions, expectations and stereotypes of the related society should be understood in the phase of determining the teenagers' mental health problems (Şahin et. al., 2003; cited: Dinçel, 2006). It is doubtful whether the factors creating storm and stress or happiness and harmony are the same in every society in all the ages. The teenager's developmental tasks are mostly determined by the social environment. There can be social expectations in all societies about the puberty and youth. It is known that the view of the society about the teenager changes according to the social needs. The stereotypes may have effect on the teenager's interaction with the family, friend group, school or work environment. The views and expectations of the social and cultural environment about the puberty, the attitudes they internalize for their interactions with the teenagers, the opportunities and options provided to the teenager for the choice of profession, improving themselves and planning the future may affect the density and diversity of the problems faced during the puberty. It would be more appropriate to regard the relationship between the social environment and the teenager as a mutual interaction (Özbay et.al.,1991).

Such developmental tasks are mostly determined by the social environment and there can be various expectations about the puberty in every period. For example, the subject of "becoming independence" is a phenomenon of which difference from one culture to another can be observed most clearly. In the individualistic societies it is expected from the young to give their own decisions and gain his independence at early ages, but in the collective societies as the family decisions are at the forefront, the expectation of the society from the young is to gain his independence at later ages (Şahin et. al., 2003).

Young Adulthood

In this period, an individual can form close friendships with the others. Friendship and love are predominant in the relationships with the opposite sex. The friendships are built on solid basis. Instead of an exchange, a feeling lies in the relationships. The issues about the marriage and getting married have an important place in the individual's life. The person, who successfully passes through the developmental tasks in this period, has the power of giving and taking the love confidently. Otherwise, the young individual, who has difficulty in having friendly relationships with the others, can be pushed to an unwanted and unhealthy psychological loneliness. All the people around the young have responsibilities mutually so that the teenager can complete the developmental tasks positively in this period. It is observed that such conflicts can be successfully solved in a structure of society based on love and respect to human (Senemoğlu, 2007). In this context, the developmental tasks of the young adulthood period can be sorted as such; mate selection, learning to live his mate, forming a family, raising childrearing, managing the house, starting a job, taking the citizenship responsibilities, finding a social group compatible with the society (Bacanlı, 2002).

Mate Selection

After puberty, the young adult decides with whom he is going to spend the life together. The tendency of mate selection involves the social, cultural, economic, physical, spiritual variables, age and personality characteristics which are taken into consideration with the aim of evaluating potential mate candidates in terms of their suitability for the expectations (Efe, 2013). In the mate selection the factors affecting the developmental tasks are seen most clearly. The individual's biological maturity does not influence the developmental task of mate selection. Those maturating late can postpone this selection, and those maturating early wish to get marry as soon as possible. The demands of the society about this issue is also important, but together with the fact that the postgraduate education level become widespread and the change in the structure of the society the marriage and mate selection can be postponed just like the choice of profession. Moreover, some people do not undertake this developmental task (Bacanlı, 2002).

Learning to Live With His/Her Mate, Forming A Family

The individual, who fulfills the developmental task of mate selection, should learn to live with his mate. This task means to put the mate selection task into practice and foresee the formation of family as a social unit. Despite selecting his mate, if an individual is not able to live with his mate, it means that he is not able to fulfill this developmental task (Bacanlı, 2002).

Childrearing

The developmental task of the individual, who selects his mate and forms his family, is the childrearing. Therefore, a married couple is usually asked when they will make a baby. The individuals fulfill this developmental task within the bound of their lifestyles (Bacanlı, 2002).

Managing the House

Another developmental task is managing the house. In the past the men had a function as a manager and the women had a function of doing the housework. However, today those functions are replaced by a new structure depending on sharing more evenly. At this point, it is expected from the mates to fulfill their responsibilities successfully.

Starting a Job

Besides his responsibilities about the marriage, an individual is also expected to fulfill his responsibilities about the work and job. This developmental task is an individual's start and finding a job. However, he cannot carry on the work or he can change it (Bacanlı, 2002).

Undertaking the Citizenship Responsibilities

An individual, who is a part of the society, is expected to undertake the responsibilities such as vote and stand for election and pay taxes. Therefore, the criminal liability and right for political engagement is obtained in those ages (Bacanlı, 2002).

Finding a Social Group Compatible With the Society

This developmental task is the individual's finding and joining a legitimate social group of which positiveness is approved by the society. This task along with mate, work and citizenship constitutes another step for entering the society of adults (Bacanlı, 2002).

Adulthood

If an individual fulfills the earlier developmental tasks successfully, he becomes productive, fruitful and creative in this period. Continuing his generation through his children is important. An individual is also productive as long as he can do useful works for the society in addition to his house and guide the next generations. The individuals, who are deprived of those things, may have the feeling of being inefficient on the contrary of productiveness, and enter the period of inactivity. They also develop behaviors indifferent to the environment, establish fake and superficial relationships and can be unhappy by thinking that they make no progress. In the individual's getting over the developmental tasks positively in this period, those who share his house and work with him, namely the individuals with whom he has intensive interactions in his environment have important roles. Those in his environment should make him feel that he works out and is necessary for the others. The individual's all activities outside his house should be awarded (Senemoğlu, 2007).

Within this scope the developmental tasks in the adulthood period are sorted as such; obtaining adult citizenship and social responsibility, forming an economic life standard and maintaining it, helping children become a responsible adult, developing adult free time activities, as an adult associating himself with his mate and accepting the physiological changes of the middle age (Bacanlı, 2002).

Obtaining Adult Citizenship and Social Responsibility

This developmental task is the continuation of the social participation and responsibilities, which started in the previous period, by getting harder. For example, it is expressed as gaining the right to vote in the previous period, and gaining the right to stand for election in this period (Bacanlı, 2002).

Forming an Economic Life Standard and Maintaining It

It is expected from the individual, who chose his profession and started a job, to form a life standard in the level of maintaining his life in this developmental task. The individual should know his income and live accordingly (Bacanlı, 2002).

Helping Children Become a Responsible Adult

This developmental task is the adult's help to children or teenagers in his environment so that they pass to the society of adults. It is performed sometimes by leading them and sometimes by being model for them (Bacanlı, 2002).

Developing Adult Free-Time Activities

The recreation involves the activities that the participants voluntarily select and make use of their spare times. Such activities can be grouped generally under these topics; musical activities, sportive activities, games artistic activities, activities requiring ability, nature activities, social and cultural activities. The term recreation means the vitalization, the thing recreated, entertainment, resting and ways to have a good time during the free times (Kurar and Baltacı, 2015). In this context, besides the work from which they earn their livelihood, the adults should develop hobbies about the activities they can do in their spare times (Bacanlı, 2002).

As an Adult Associating Himself With His Mate

In this period the individual puts his work mate and life on the right track. The individual takes himself with his mate and presents. They plan and apply their activities together.

Accepting the Physiological Changes of the Middle Age

In this period the signs of aging appear physiologically. It is expected from the individual to adapt to those changes, accept them and learn to live with those changes (Bacanlı, 2002).

Old Age

The developmental tasks of this period are as such; adapting to decreasing physical power and health, adapting to the retirement and decreasing income, adapting to the death of mate, establishing an open intimacy with their age group, fulfilling social and citizenship obligations, forming an arrangement for a satisfying physical living (Bacanlı, 2002).

Adapting to Decreasing Physical Power and Health

In this period the physical power decreases and the health begins to impair. It is expected from the individual to adapt to such changes and do the work that he can easily do (Bacanlı, 2002).

Adapting to Retirement and Decreasing Income

A certain amount of decrease in the income due to the retirement is observed in this period, because the individual starts to receive pension. Despite earning more money previously, he starts to earn less money (Bacanlı, 2002).

Adapting to the Death of Mate

In a study carried out by Altay et. al., (2016) it was found that married old individuals had high level of life quality. Further, in the study, the effects of living with those who were close to the old individuals emotionally and socially in their houses and having a very good support system on having a high level of life quality were emphasized (Bacanlı, 2002). When the old individual, who accepts the physiological values of the middle age in the previous period, faces with the death of his mate, it is required for him to adapt to this situation and accept it. Aslan et. al., (2018) examined the old people's expressions about the effects created by the death of their mates and it was found that "the cognitive effects" were mentioned most frequently and "the physical effects" were mentioned at fewest. Further, it was determined that the old people most frequently had discourses about "life habits" and they had the fewest discourse about "receiving support from an expert" while coping with the death of their mates and the conditions stemmed from the death. In the same study the old people's discourses about their future expectations were examined and it was stated that the most frequently used expressions were about the subfield of "expectations, plans and targets". In a study done by Akgül and Yeşilyaprak (2018) it was found that The Psycho-education Program for Diminishing the Loneliness decreased the old people's level of loneliness.

Establishing an Open Intimacy With His/Her Age Group

It could be difficult for the old people to adapt to the next generations due to the generation gap. Together with the influences given by the retirement the old people have lots of free time and this situation eases their listening to each other and spending time together (Bacanlı, 2002). In their study Yavuz and Yavuz (2018) examined the effects of the mind game the old individuals in the senior center played on their cognitive skills, their levels of loneliness and their levels of psychological wellbeing and found that the game developed the cognitive skills, decreased the levels of loneliness and increased the levels of psychological wellbeing.

Fulfilling Social and Citizenship Obligations

Just like in the previous periods, the individual is expected to fulfill his social responsibilities (Bacanlı, 2002).

Forming an Arrangement for a Satisfying Physical Living

An individual should make environmental arrangements, which are distinctive to himself, in order to ease his life (Bacanlı, 2002). In this context, the gerontechnology deals with the ergonomic design of lots of equipment and technological support used by the old people with reference to the view that the sensorial and perceptual loss and the loss of strength and motion, which stem from the ageing, can be compensated with lots of special ancillary equipment and technological support. It is thought that that equipment would be beneficial in increasing the life quality of the old people and popularizing and producing the gerontechnological products (Kalınkara and Başıbüyük 2016). The following suggestions can be made for the old people; the objects on the old people's walk way which are dispersed and can cause to falling down and stumbling should be removed, the telephone and electricity wires should be strapped and be kept out from the walk way, lamps should be inserted in the rooms (firstly in the old individual's bedroom) or in the corridors for the lightening, the height of the most frequently used chair/armchair should be adjusted. The anti-slip tape should be put under the slipping carpets/ rugs, the threshold causing to stumbling and falling down should be removed, the sides of the carpet causing to stumbling should be glued, in the kitchen the frequently used utensils should be put on the lower shelf, inserting the telephones in the room (firstly in the old individual's bedroom), the place where the medicine are kept and the other rooms should be enlightened sufficiently, the shower cabins without the threshold should be used instead of the bathtub (Altuntaş and Kayuhan, 2015).

In a study carried out by Güler et. al., (2009) it was determined that the old people had difficulty in performing their daily activities, and they needed care in the fields of nutrition, ejection and sleep at most. In the old age period, the individuals are productive, fruitful, happy, adaptive to the society, loveable and respected or they are hopeless, disagreeable, ill-tempered and crabbed. In this period, the individuals go and return a conflict between the satisfactions provided by having spent a productive life and the hopelessness stemmed from having spent their lives meaninglessly. An individual, who could fulfill his developmental tasks, believes that he lived his life as he wished and is proud of himself. Also, he accepts himself with his superiority and limitations special to his age. When the developmental tasks are not fulfilled; an individual cannot be in peace in his agedness and believes that he spent his life in vain. Further, the emotions of the fear of death and hopelessness emerges (Senemoğlu, 2007).

The activities of the old people slow down in the old age period, accordingly their life satisfactions decreases, too. The life satisfaction should be considered as a unity of objective and subjective variables. The variables such as gender, health, income, social security, marital status, adult children, grandchildren, relationships with friends and neighbors, recreation, membership to voluntary agencies and associations and participating in the activities of those agencies and associations affect the old people's life satisfactions (Kurt et. al., 2010). The services should be planned in order to increase the life quality of old people in all age groups, their perception of health and the importance they give to this issue by taking the first, second and third protection principles into consideration, also the studies about this subjects should be focused on (Altay et. al., 2016). In order to provide effective and permanent solutions to the old people, the scientific studies carried out in this field are needed. The quality, the content and quantity of the service to be provided are in proportion with the accumulation of information obtained about the related area of problem. The more the information about the people to whom the service is to be provided is gathered, the more effective it gets to interfere. The collaborations should also be provided between the disciplines related to the agedness. The studies about the agedness can contribute to the solution of the old people's problems only through such collaborations. (Çevik, 2017; Kurt et al., 2010).

An individual comes across the developmental tasks in the periods he is in and tries to fulfill them in his own way. In this context, in the early years, teachers should prepare environments, in which the children can fulfill the developmental task, in order to ease students' healthy developments.

This research received no specific grant from any funding agency in the public, commercial, or not-for-profit sectors.

REFERENCES

Ahmetoğlu, E. (2018). Ahlak Gelişimi. In N. Aral & T. Duman (Eds.), *Eğitim Psikolojisi* (pp. 216–239). Ankara: Pegem.

Akgül, H., & Yeşilyaprak, B. (2018). Yalnızlığı Azaltma Psiko-Eğitim Programının Yaşlıların Yalnızlık Düzeyine Etkisi. *OPUS Uluslararası Toplum Araştırmaları Dergisi*, *8*(14), 11–52. doi:10.26466/opus.401331

Altay, B., Çavuşoğlu, F., & Çal, A. (2016). Yaşlıların sağlık algısı, yaşam kalitesi ve sağlıkla ilgili yaşam kalitesini etkileyen faktörler. *Turk Silahli Kuvvetleri Koruyucu Hekimlik Bulteni*, *15*(3), 181–189. doi:10.5455/pmb.1-1438765038

Altuntaş, O., & Kayıhan, H. (2015). Ev Düzenlemelerinin Yaşlıların Yaşam Kalitesine Etkisi. *Fizyoterapi Rehabilitasyon*, *26*(1), 1–13.

Aslan, E., Ergün, N., Duman, B., Bozdağ, F., Karataş, S., & Fakirullahoğlu, A. M. (2018). Yaşlılarda eş kaybı, kayıp sonrası başa çıkma stratejileri ve gelecek beklentileri üzerine nitel bir araştırma. Retrieved from [REMOVED HYPERLINK FIELD]http://acikerisim.artuklu.edu.tr/xmlui/handle/20.500.12514/180#sthash.Vnjj8kT2.dpbs

Bacanlı, H. (2002). *Gelişim ve Öğrenme*. Ankara: Nobel.

Bağcı Ayrancı, B. (2018). 0-12 yaş dil gelişimi uygulamaları ve yapılması gerekenler. *Kırıkkale Üniversitesi Sosyal Bilimler Dergisi*, *8*(1), 13–34.

Çevik, C. (2017). Yaşlıların Sağlık Hizmetine Erişimi. *Turkiye Klinikleri J Public Health Nurs-Special Topics*, *3*(3), 165–171.

Çoban, A. E. (2010). Mesleki rehberlik. In M. Güven (Ed.), Psikolojik Danışma ve Rehberlik (pp. 132-159). Ankara: Anı.

Çuhadaroğlu, F., Canat, S., Kılıç, E., Şenol, S., Ruganci, N., Öncü, B., ... Avcı, A. (2004). *Ergen ve Ruhsal Sorunlar. Durum Saptama Çalışması. Türkiye Bilimler Akademisi (TÜBA)*. Ankara: Raporları.

Davies, D. (2010). *Child Development*. New York, NY: TheGuilford Pres.

Deniz, Ü. (2018). Gelişim. In N. Aral & T. Duman (Eds.), *Eğitim Psikolojisi* (pp. 32–53). Ankara: Pegem. doi:10.14527/9786052410486

Dinçel, E. (2006). *Ergenlik dönemi gelişimsel ödevleri ve psikolojik problemler*. Ankara Üniversitesi Sosyal Bilimler Enstitüsü Yüksek Lisans Tezi.

Efe, E. (2013). *Üniversite Öğrencilerinin Eş Seçme Eğilimleri*. Hacettepe Üniversitesi Sosyal Bilimler Enstitüsü Yüksek Lisans Tezi.

Erden, M., & Akman, Y. (1998). *Gelişim-öğrenme-öğretme: eğitim psikolojisi*. Ankara: Arkadaş.

Gander, M. J., & Gardiner, H. W. (1993). *Çocuk ve Ergen Gelişimi* (B. Onur, Ed.). Ankara: İmge Yayınevi.

Güler, G., Güler, N., Kocataş, S., & Akgül, N. (2009). Yaşlıların sağlık bakım gereksinimleri. *Cumhuriyet Medical Journal, 31*(4), 367–373.

Güngör Aytar, A. (2015). Toplumsal ve Duygusal Gelişim. In A. Ulusoy (Ed.), Gelişim ve Öğrenme Psikolojisi (pp. 93-126). Ankara: Anı.

Kalınkara, V., & Başıbüyük, G., & Ay, F. (2016). Yaşlıların geronteknolojik ürünleri kabule yönelik tutumları. *Yaşlı Sorunları Araştırma Dergisi, 9*(2), 1–19.

Kulaksızoğlu, A. (1998). *Ergenlik psikolojisi*. İstanbul: Remzi Yayınevi.

Kurar, İ., & Baltacı, F. (2015). Halkın Boş Zaman Değerlendirme Alışkanlıkları: Alanya Örneği. *International Journal of SportCultureandScience, 2*(2), 39–52.

Kurt, G., Yücel Beyaztaş, F., & Erkol, Z. (2010). Yaşlıların sorunları ve yaşam memnuniyeti. *TurkishJournal of ForensicMedicine, 24*(2), 32–39.

Özbay, H., & Öztürk, E. (1992). *Gençlik*. İstanbul: İletişim Yayıncılık A.Ş.

Özbay, H, Şahin, N, Hınçal, G, Güngör, S., Öztürk, E, Mavili-Aktaş, A., Aybaş, M., Göka, E. (1991). Ergenlikte Benlik İmajı: Çalışan ve öğrenci ergenler arasında karşılaştırmalı bir çalışma. *3P Dergisi, 2*(2), 82-95.

Özer, D. S. & Özer, K. (2000). *Çocuklarda Motor Gelişim*. İstanbul: Kazancı

Prechtl, H. F. R. (2004). Principles Of Early Motor Development In The Human. In A.F. Kalverboer, B. Hopkıns, & R. Geuze (Eds.), Motor Development in Early And Later Childhood: Longitudinal Approaches (pp. 35-51). New York: Cambridge.

Şahin, S. (2015). Okul Öncesi Dönemde Motor Gelişim. In F. Turan, A.İ. Yükselen (Eds.), Çocuk Gelişimi 2 Okul Öncesi Döneminde Gelişim (pp. 24-35). Ankara:Hedef

Santrock, J. W. (2014). *Child Development*. New York, NY: McGrawHill.

Senemoğlu, N. (2007). *Gelişim Öğrenme ve Öğretim*. Ankara: Gönül.

Törüner, E. K., & Büyükgönenç, L. (2011). *Çocuk Sağlığı*. Ankara: Gökçe.

Yaşar, M. (2015). Okul Öncesi Dönemde Sosyal-Duygusal Gelişim. Ankara: Hedef.

Yavuz, O., & Yavuz, Y. (2018). Huzurevindeki Yaşlı Bireylere Oynatılan Zekâ Oyununun Yaşlıların Bilişsel Becerilerine, Yalnızlık ve Psikolojik İyi Oluş Düzeylerine Etkisi. *Yaşam Becerileri Psikoloji Dergisi, 2*(3), 127–141. doi:10.31461/ybpd.427039

Yavuzer, H. (1999). *Çocuk Psikolojisi*. İstanbul: Remzi.

Yazgan İnanç, B., Bilgin, M., & Kılıç Atıcı, M. (2010). *Gelişim Psikolojisi*. Ankara: Pegem.

Ziegler Delahunt, J. (2002). Motor Development Child Development. USA: Macmillan.

Chapter 4
Children at Risk

Canan Yildiz Çiçekler
Necmettin Erbakan University, Turkey

Devlet Alakoç Pirpir
Selçuk University, Turkey

ABSTRACT

Children's exposure to many risk factors such as; need for protection, living on the streets, working, abuse and neglected, pushed into crime, exposed to violence, obliged to immigrate due to war, living under socio-economic disadvantageous conditions, having chronic diseases, being a disabled child and living in divorced families can arise from both their families and from the social structure. Throughout the world, many children live at risk due to various reasons. Irrespective of the reasons, which risk group the children enter and the factors causing this situation should be examined. According to the obtained data, the factors causing to such situations should be determined and necessary precautions should be taken. Thus, the negative conditions, under which the children are, can be improved and the children can be reintegrated into society.

INTRODUCTION

Children at Risk and Their Developments

The opportunities should be provided for every individual in the world so that they can use their rights of being born healthily, living healthily, using the skills they have and the self-realization. Both the families and the institutions related to the children are responsible for providing such opportunities. Therefore, the precautions should be taken against the child neglect and abuse and it should be turned into a governmental policy in order to raise physically and mentally healthy individuals in the society (Koçtürk and Yılmaz, 2018). Today, the families and children are influenced due to some problems encountered. Especially, such a situation influences the development of children negatively (Nalbant and Babaoğlan, 2016). The problems the individuals have when they are children can cause the emergence of problems in the family lives of those children in the latter periods of their life. Therefore, the problems

DOI: 10.4018/978-1-7998-2952-2.ch004

the children encounter turn into not only an individual but also a social problem (Karakaş and Çevik, 2016). In a study carried out by Nalbant and Babaoğlan (2016) it was determined that the children at risk had problems in many different topics. As a result of the study, the children whose family unity was impaired and who had problems in a circle of friends, and had problems with the school board and the lessons in the school environment were regarded as the children at risk. It was also found that the determined families were not sufficient in supporting their children and controlling them in the school environment. In the study, it was also detected that the children generally shared the problems they had with their mothers. Further, it was determined in the study that in case the children could not share their problems with their mothers, they shared them with their siblings, friends or teachers, but it was also found that some children did not share their problems with anybody (Nalbant and Babaoğlan, 2016).

It is observed that the children at risk firstly find themselves in the street, then in the prisons. Firstly, providing the continuity of the children's attendance to school is very important for the prevention of such a situation. As a result of children's being in the same environment with a child who is pushed into crime, spending time with him and sharing something together increase the possibility of the children's being pushed into crime, as well. It should not be ignored that the children, who move away from the school environment and meet cigarette and alcohol at early ages, can turn into crime easily (Öztürk Çopur, Ulutaşdemir, and Balsak, 2015). The prevention of those negativities is carried out via giving the necessary trainings. Therefore, the children will grow up as those who are aware of their rights and responsibilities, can determine the borders of their limitations, stand on their own feet and believe in themselves even if they do not experience any negative situation. In addition, in order to prevent the negative situations, it is very important to give a place to the educational items defining the negativity of those situations via the frequently used visual media or communication channels (Kır, 2013).

Children's spending their lives healthily and in a good way forms the basis and the future of the society they live in. Therefore, it crucial to determine the problems the children encounter and take the necessary measures for those problems. The starting process of the child protection in the modern sense falls with the early times of the industrial revolution. In this period, as a result of the children's employment for heavy works, the necessity of the child protection appeared and the first social policy measures were taken. In this process some regulations were made such as a decrease in the children's working hours. In the latter periods, it was considered that the children should have the rights just like the adults and Declaration of the Rights of the Child was interiorized and then this declaration was regulated internationally (Karakaş and Çevik, 2016). When the human rights are considered, it is clear that the child rights are the most basic of all. Along with the child rights, the necessary sensitivity should be displayed in order to protect the children's both physical and mental health and raise them with the conscious that they are the future of their countries (Yalçın, Bay and Ekmekçioğlu, 2014). The children at risk were explained below by dividing them into the following categories; children in need of protection, children living in the streets and working, abused and neglected children, children committing a crime and being pushed into crime, children exposing to violence, children immigrated due to war, disadvantaged children in terms of socio-economic status, children with chronic diseases, disabled children and children of divorced families.

Children in Need of Protection

In the world all the parents wish to raise their children in the best manner. However, the environments and the opportunities, under which the parents live, cannot always be suitable to carry out it. The parents become insufficient for their children as a result of mother-father divorce and some pecuniary and non-pecuniary reasons (Yolcuoğlu, 2009). Such a situation causes the children to be unprotected and at risk. When the children in need of protection and at risk are examined, it is observed that the physical, psychological and personal securities of those children are in danger. Those children, being left by their parents or caretakers, acquire bad habits such as prostitution, mendacity, alcohol and substance use and they are unprotected against those negative situations (Çoban, 2015). The children, who need to be protected, are in integration with those around them. The first people with whom the children communicate are the family members and this situation enlarges to the society the children live in and the law and policies of that society. The individuals are influenced by the cultural values of the society they live in and by the family relationships. Being influenced of the individuals also means being influenced of the whole society. It is seen that the children are influenced by the changes occurring as a result of the death of the mother or father in the family, their job loss or their divorce (Yolcuoğlu, 2009).

It is also observed that children are negatively influenced more than anyone else by the socioeconomic problems occurring worldwide. As the children's most basic needs are not met, they cannot grow health (Yolcuoğlu, 2009). The experienced economic problems have a big effect on the impairment of unity and solidarity in the families. Some parents regard nursery schools and dormitories as boarding schools and they are more inclined to give their children to such institutions in order to evade their responsibilities (Türkmen, 2005). As a result of the decrease in the number of members in the families and being unable to benefit from extended family support, some parents cannot take care of their children sufficiently and cannot meet their needs. The protection of the children by a social system emerges as a result of this situation (Yolcuoğlu, 2009). The children are in the risk group as they have not completed their developments yet. As the children cannot protect themselves against different abuse situations, they should be protected from those negative situations both physically and mentally (Karakaş and Çevik, 2016). The protection of the children is a social responsibility (Yazıcı, 2012).

Children Living in the Streets and Working

The childhood stage forms the fastest period of the development in the individuals' lives. A healthy environment in the childhood period leaves important traces in the latter life of the individual. Not all the children have the chance of living in a healthy environment. Some children are obliged to work in the street due to some economic reasons. The children working in the street expose to many risky situations such as being subject to violence, being abused, being kidnapped and getting involved in crime. Apart from those dangers, the problems occur in the education of the children working in the street. It is observed that those children's attendance to school decreases in time and as the time spent in the street increases, they move away from the school. The street becomes attractive for the children and they spend most of their time in the street (Yıldız and Duy, 2015). After United Nations Year of the Child in 1979, the term street children was used for the first time. The children who start to work in the street as a result of escaping from the house they live in, become homeless as they left their families and houses or are obliged to live outside their houses left by their families are defined as the street children (Çoban, 2015). Contributing to the family budget is generally the most basic reason for the children's working in the street. It is realized that the children working in the streets are expressed in different categories according to their ways of working and life. It is observed that some children earn money by working except for the time when they are at school. Some other children undertake the responsibility of maintaining the family. It is seen that among the street children, there are those who are forced to work or those who are sold to be worked by the others. Some of the children go back to their houses except for their working hours. However, some other children are prevented from going back to their houses by the one who make them work just because the child could not earn enough money. Therefore, the children are drifted apart from their houses and expose to many dangers (Okumuş, 2009).

The children are called with different names according to the situation of living in the street. The children, who cannot get the necessary financial support from their families or wish to contribute to the family economy, start to work. The children, who go to their houses regularly, but spend most of their times far from their houses due to problems with the family members, abuse and economic difficulties, are defined as "the children in the street". The children, who have problems in the family environment, does not have family, escape from house, are not wanted, are exposed to violence and struggle for life without receiving support from the families are expressed as "the children of the street". The children, who live with their families in the street as they do not have a house due to the conditions of war, poverty and disasters, are called "children who are a part of the street family" (Çoban, 2015; Güngör, 2013).

The reasons for the children's living in the street are indicated by Okumuş (2009) as such; uneasiness experienced in the home environment, breaking of the family and divorce, unemployment, problems encountered in the adaptation process as a result of the immigration to city, problems related to adaptations to school in the urban environment, insufficient love, abuse, being exposed to violence, lack of parents, children's being neglected, being exposed to sexual harassment, children's desire for going on an adventure, children's failures in the education lives, unable to meet the children's physical needs in the house environment where they live, children's being encouraged to work in the street, the influence of family group, children's desire to prove themselves and the negative influences of the media. When the reason pushing the children into the street were examined by Kasatura (2002) it was stated that the children who start to live with their relatives as they do not have parents but leave from them due to financial insufficiency, leave from home due to being exposed to violence by their step father or step mother, have behavioral disorders, who leave from their home due to the bad habits of their family members, leave from their home in order to have parents who steal, use alcohol and live in the streets, are deceived by an illegal organization or make to work for panhandling, are exposed to physical, sexual, economic and emotional abuse, are obliged to live in the streets. As a result of the study carried out by Göngör (2009), it was determined that the children aged 5-16 started to work in the streets. Also, it was found in the study that the children's reasons for starting to work were their desire to contribute to the family economy and meet their school expenses. In addition, it was detected that the family members, who were supposed to prevent the children from working in the street, did not have a level of consciousness to prevent the children from working in the street. As a result of the study it was found that the children were registered to school, but they did not go to school and even they dropped out the school. The reasons for children's stopping to go to school were expressed as such; the economical insufficiency of the children's families and their dislike of school due to being excluded. When the results of the study were examined, it was found that the education levels of the families were low and they were disadvantageous in terms of their socio-economic structures. In the study, it was revealed that the low-income levels of the families made them inadequate in meeting their children's basic needs. Last but not the least, it was stated in the study that as the social structure and type, in which the children grew, directly influenced the development of the children, the family structures of the children was one of the most important factors in their turning into the crime.

Children's start to work at very early ages is a problem of many countries. The reasons for children's working vary according to the conditions of each country. No matter which county the children are in, they are usually damaged socially, psychologically and physically. It is observed that when children cannot solve the problems they have, they feel the sense of inadequacy. As a result of this situation, the children become pessimistic and find themselves in depressive situations by abandoning themselves to despair. The economic insufficiency in the children's families and the level of consciousness in the socio-cultural environment are effective on the children's working at early ages (Bilgin, 2009; Yıldız and Duy, 2015). The immigration and poverty can be taken as the two important reasons in the child labor. The precautions should be taken in order to remove the problems stemming from the immigration and economic reasons. Thus, the children's working in the street can be prevented. Also, in case the families, which come to urban areas due to the immigration, wish to go back, the necessary support should be prepared and given to them. Further, the need-oriented social projects should be formed for the migrant families that do not want to go back. In short, all kinds of support such as material, non-material and the support for education should be provided to migrant families, the link between the child and school should be strengthened, the vocational training programs should be prepared and shared with the families and the individuals, who need socio-psychological support due to the immigration, should be supported (Bilgin, 2009).

The problem of child labor is to continue in the world due to the poverty and the economic changes stemming from the globalization and economic crises (Efe and Uluoğlu, 2015). The street children appear as a global problem (Okumuş, 2009). The situations about the children living in the street should be evaluated in detail, not only in terms of the illegal employment or staying out of the social structure, but also in terms of what kind of job they do, under which conditions they live in the streets and their being exposed to abuse or neglect. Revealing those situations, which cause children to stay or live in the streets, will come into view with the determination of economic, political, social and cultural reasons leading to the emergence of such situations (Güngör, 2013). Efe and Uluoğlu (2015) indicated that enlarging the scope of social protection, increasing the support oriented to social security, determining the number of child labors all around the world and internalizing the necessary legislation are necessary in order to remove the problem of child labor in the countries.

Abused and Neglected Children

People's childhood stages shape their puberty and adulthood stages of their lives. Therefore, the situation of the individuals' being physically and mentally healthy and being able to be beneficial both for themselves and for the society they live in is directly linked to their previous childhood experiences (Acehan, Bilen, Oğuzhan Ay, Gülen, Avcı and İçme, 2013). When the issues of neglect and abuse are examined it detail, it is found that such issues involve the children's being damaged in terms of physical, psychological and sexual aspects (Çoban, 2015). The situation of parents or the other individuals', who take care of the children, not meeting the children's basic needs which should be met is "child neglect". Further, the situation of child's being damaged in terms of physical, sexual, emotional and economic aspects by the act of someone older than him/her is "child abuse" (Koçtürk, 2018).

The child neglect is the children's being unable to be fed sufficiently, the child's not having clean clothes, not providing suitable clothes to the child, the child's being obliged to live under inappropriate conditions, not taking the adequate measurements both at home and at the outside against the child's being exposed to accidents (Acehan et al., 2013). In other words, the person's, who is in charge of taking care of the child, not fulfilling his duty and neglecting the child emotionally or physically are called the child neglect. Not taking the necessary care to the nutrition, one of the primary needs, the needs related to health, wearing, emotional needs or the most appropriate conditions for life is also regarded as the neglect. The children, who show maladaptive behaviors in terms of psychosocial aspect, have growth delay, are not benefitted from protective health services and whose educational needs are not met, are exposed to the neglect (Polat, 2016). In the child neglect, if the children are given physical or emotional damage deliberatively, it means that active child neglect is being performed. Further, if there is neglect due to impossibility, indifference and not having information about what should be done, then it means a passive neglect is being performed (Çoban, 2015).

In the children's being exposed to abuse and neglect, both the children as an individual and the families of the children are affected by this situation. Frequently, such a situation can cause to the impairment of family unity and a social deterioration in the latter periods (Dilsiz and Mağden, 2015). Recently, the common use of technological devices in the studies about the child neglect and abuse makes it easier to prove the experienced situations, share them and inform many people about those situations in a short time. As a result of this situation it is stated that there is an increase in the child neglect and abuse. At this point the important thing is to use the media effectively in order to create awareness for the parents about the care and protection of their children, and it is also indicated that it would be appropriate to carry out more studies with the families and increase the parents' training (Bakır and Kapucu, 2017).

According to the effect created on the child and the characteristic of the behavior, the child abuse is dealt with and explained within four types as; physical, emotional, sexual and economic.

Although these abuse types can be observed separately, they are generally seen altogether (Koçtürk, 2018). Among the abuses the easiest one to be determined is the physical abuse (Pelendecioglu and Bulut, 2009). In its broadest sense, the physical abuse can be defined as "children's injury except for due to an accident". Beating the child and exposing violence to him is the most commonly seen type of physical abuse. The physical abuse is a type of abuse which cause scald, fracture or other similar injuries of all kinds leading to damage in the individual's body. In this abuse, the children's bodies are bruised or injured except for due to an accident (Polat, 2016). It can be observed that in the bodies of the children, who are exposed to abuse, there are scratches, scald and fractures. The children are not taken to the hospitals by their parents in order to hide the situation of the children who were exposed to physical abuse. Further, when those children are taken to the hospitals, it is expressed that those situations have happened as a result of an accident, the child himself has fallen down or this situation has been performed by his/her sibling (Pelendecioglu and Bulut, 2009). It is observed, especially in the physical abuse, that the children, who are exposed to physical abuse by the family members, regard this situation as a normal behavior by internalizing it and in the future they behave towards their children in a similar way (Dilsiz and Mağden, 2015).

The process of an adult's giving damage to the child by committing violence through hitting, burning or biting to any part of the child's body and leaving a trace on it with the aim of making the child obey him, punishing the child or easing his own anger can be defined as the physical abuse (Kurtay, 2018). The children's exposure to the physical abuse can be caused by the parents, the children or the social condition (Pelendecioglu and Bulut, 2009). When the reasons stemmed from the children are examined the followings can be mentioned; the child with premature birth, low birth weight of the baby, being a physically or mentally handicapped child, being hyperactive or bad-tempered or baby's leaving apart from his/her mother after the birth for a long time due to various reason. Bathea (1999), Şahin (2001), Şahiner, Yurdakök, Kavak and Çetin (2001) tell the reasons stemmed from the families as such; parents' having a physical or sexual abuse history in their childhood, parents' getting married at early ages, being an only parent or stepparents, parents' having emotional inadequacy and low communication skill, parents' having low self-respect, parents' being addicted to alcohol or drugs, parent's having the lack of social support, having violence in the family, inadequacy in the parental functions, the lack of preparation for the stress caused by having a newborn baby, having too many children in the family, parents' being uneducated, spiritual problem in the mother or father, unwanted pregnancy and unaccepted child. Last but not the least, the social reason are the high levels of crime in the society, the situation of having a few social services or not having them at all, the high levels of poverty and unemployment in the society (cited: Deveci and Açık, 2003).

It is observed that as a result of being exposed to physical abuse, the children generally become very nervous, have difficulty in sleeping, experience a sudden outburst of anger, behave aggressively to those around them, do not value to themselves, object to the rules and have high risk of being addicted to alcohol and drug by moving away from the school and home (Kasatura, 2002). In a study carried out by Güler, Uzun, Boztaş, and Aydoğan (2002) it was found that the girls were disturbed at most and as the age levels of the children increased the disturbance decreased. It was also obtained in the study that, as the parents' level of education increased, the disturbance decreased, the younger parents disturbed their children more than the older ones. As a result of this study it was found that 87.4% of the mother physically abused/neglected their children, 93% of them emotionally abused/neglected their children. The mothers, who were beaten by their husband and had lots of children, showed the behaviors of abuse/neglect on the high levels.

Although the physical abuse is the most commonly determined type of abuse, it is also very difficult to determine the emotional and sexual abuse cases along with the physical abuse (Özyürek, Çetin, and Yıldırım, 2018). Another type of abuse is the sexual abuse. Being used of the children, who are underage and whose psychosocial development have not been completed yet, by the adults in order to reach their sexual targets is called as the sexual abuse. An adult's use of a child to reach his/her own sexual sensation is also the sexual abuse (Polat, 2016). The sexual abuse can be carried out by the individuals in the family or outside the family. The sexual abuse is the occurrence of sexual interaction and in this situation the child being exposed to the abuse is forced or convinced by someone, who is at least four-five years older than the child, with the aim of obtaining the sexual pleasure. Also, exposing the children to inappropriate behavior such as touching the children's private parts, the children's being exposed to ugly words improper to their age and making the children watch pornography are dealt within the sexual abuse. While the sexual abuse affects the child negatively, the duration of the sexual abuse, the child's age group, the degree of familiarity of the abuser to the child, the age difference between the abused and the child and the abuser's committing violence determine the degree that the child is affected by this situation (Dilsiz and Mağden, 2015; Kurtay, 2018).

Emotional abuse involves the child being left on his/her own, being deceived, frightened, threatened, humiliated, and treated insensitively towards their needs (Acehan et al., 2013). Not meeting the love, interest, and care needed by the children and being damaged psychologically according to the scientific and social norms are defined as emotional abuse. The behaviors of the children who are exposed to emotional abuse are performed by the people or person who have influence on the related young and children (Polat, 2016). Also, giving the adult responsibilities to the children at a very early age, preventing the child's relations in the social environment are approached as emotional abuse. It is very difficult to measure the negative traces on the child left by the emotional abuse. The economic abuse, on the other hand, is to make children, who are under eighteen, work for a long in improper environments and jobs affecting their developments negatively (Kasatura, 2002).

The school principals, teachers and psychological counselors at schools have a critical importance in preventing and determining the child neglect and abuse and going into action in this problem (Koçtürk, 2018). In this issue, both the teachers and the other public officials have legal responsibilities and obligations to inform (Bakır and Kapucu, 2017; Dilsiz and Mağden, 2015; Pelendecioglu and Bulut, 2009). In a study carried out by Dilsiz and Mağden (2015), it was found that approximately more than 1/3 of the teachers came across with the child abuse and neglect cases throughout their professional lives. More than half of the teachers stated that they did not know the required legal responsibilities and obligations when they came across a child who was abused or neglected and more than 1/3 of the teachers expressed that they did not know how to approach a child who was abused or neglected. In a study carried out by Özyürek, Çetin and Yıldırım (2018) the views of the medical and education staff about the child abuse and neglect were examined. In this study it was indicated that majority of the participant family doctors and teachers did not receive pre-service training about the abuse/neglect during their undergraduate study or the training they received was insufficient, also in the work life they did not receive in-service training. Further, it was found that the medical and education staff did not have adequate information about the child neglect and abuse. It was also revealed that the teachers, as education staff, were more sensitive to the sexual, emotional abuse and neglect more than the family doctors. It was also observed in the study that as the awareness of family doctors and teachers about the child neglect and abuse increased, they tended to show a more sensitive attitude for such situations.

Although children can feel themselves in secure at most in their families, the riskiest place for the abuse is also in their families. The mothers generally abuse/neglect their children physically or emotionally in the process of raising their children or problem solving. It is possible that an individual, who is exposed to behaviors of the physical or emotional abuse/neglect and violence, uses the behaviors of violence as a problem-solving technique when s/he come across with a similar situation in the future. Today, it is seen that the number of people exhibiting such behaviors is increasing day by day (Güler et al., 2002).

Children Committing Crime and Being Pushed into Crime

Crime is defined as the activities and actions that are clearly expressed and not approved by the legislator due giving harm to society. The juvenile delinquency is defined as the individuals', who are under eighteen years old, being pushed into crime due to various reasons. The crime has appeared as a social problem since the ancient times. The crime which was present in the past, is extant, and is believed to be in the latter periods, is regarded as a universal situation When the reasons of children' being pushed into crime are examined, the insufficiency the children have about the education and giving insufficient education to the child in the family can be articulated. It was determined with the studies that committing crime causes psychological problems in the children, they also have social problems and have difficulties in the school lives (Erkan and Erdoğdu, 2006; Öztürk Çopur et al., 2015). As a result of the study carried out by Güngör (2013) it was stated that as the children being pushed into crime are popular and reputable in a circle of friends, this situation become attractive for the other children, and they even do not feel discomfort due to being pushed into crime.

It is observed that the children who both go to school and keep on working in the street join the gangs in the neighborhood or in the school. Among the children, there are those who are taken into custody. Also, when the families of the children being pushed into the crime were examined, it was revealed that majority of them had committed crime earlier and even they had been in prison. The environment of the children, who live in the street and are pushed into crime, their family structures and economic conditions, are effective on their committing crime.

The important link between the social problems the children, who are involved in crime, have and the juvenile delinquency cannot be ignored (Çoban, 2015). Today, the increase in the number of children being pushed into crime due to the changes in the social order and family structure has been observed. Therefore, taking the necessary precautions by determining the reasons, which cause children to be pushed into crime, appears as a social duty. The necessary precautions should be taken in order to prevent the wars, immigration, poverty and unemployment all over the world. Also, the followings should be done on the basis of the country to prevent children from being pushed into crime; increasing the support for education, decreasing the proportion of broken families by determining the related reasons, organizing the supportive trainings to the married or divorced parents about raising children and preparing projects to prevent the children being pushed into crime from become drug addicted. A system, which provides necessary consultancy service to reintroduce those children to the society and keeps them away from the criminal environments with social and cultural activities, should be organized (Bülbül and Doğan, 2016). Further, it is required to determine the reasons why the children are pushed into crime confirm how this process develops and develop preventive programs and policies in the light of the obtained findings (Topçuoğlu, 2014).

Children Exposed to Violence

The concept of violence, which appears together with the rise of humanity, forms a complex structure by combining many social and individual elements. Therefore, it is difficult to understand and reveal what the violence is. The phenomenon of violence, which we frequently come across in our individual and social lives, may appear in lots of different ways. The types of violence, which are seen step by step but continuously in all societies, are; badgering, putting pressure, punishing, murder, suppression and revolting. The structure of a society, its cultural characteristics and its value judgments play an active role in determining the behaviors that can be regarded as violence in that society (Kocacık, 2001).

Violence and acts of violence are defined as facts which are common in all societies, continue by increasing day by day and interest the nations. The concepts of violence and acts of violence which were developed to protect the life, but unfortunately became the main reasons of the pain the societies and people exposed to, show themselves in every part of life There are many political, economic, socio-cultural, medical and psychological reasons that create the violence and maintain it. It is still impossible to comprehend what the reasons lying behind the types of violence created by such factors are. The acts of violence which are frequently witnessed in the daily lives of healthy societies are very important in that they have the power to predict the characteristics of an individual and show inside the spiritual dynamics. The deepening of the studies which were carried out to understand the violence performed on the healthy communities plays an effective role in understanding what the violence is and preventing the acts of violence (Güneri Yöyen, 2001).

Being exposed to the lack of primary needs such as hunger and thirst and being deprived of spiritual needs such as care, love and touching can cause aggressive behaviors in the people. Evaluating the violence from different aspects and its different definitions stem from the reasons of aggressive behaviors in people, the way they occur or their size, their characteristic and their direction. In the social and economic areas, the type of violence to which the children are exposed at the largest extent is the violence in the family (Ayan, 2007).

The most commonly seen type of violence is the physical violence. The physical violence, which starts disturbing or beating an individual, can result in the death of people (Polat, 2016). The physical violence is generally understood with scratches, injuries, fractures and scald on the body. Accordingly, the injuries generally occur on the face, arms, chest, back, lower parts of the legs and genital areas (Tezel, 2002).

Another type of violence the children are exposed to is sexual violence. The children being exposed to sexual violence and their families generally hide this situation from other people (Tezel, 2002). A violent crime, which is committed to an individual with destructive and compeller sexual behaviors, is called the sexual violence. The attacker, who commits the violence, not only benefits from the victim's sexuality but also aims to capture her by controlling her behaviors, force her, embarrass her, give damage to her and make her submit to himself. Regardless of the age group and gender, each of the committed sexual violence is a crime. The individuals, who are exposed to this crime, are women, men, the old people, the children and the handicapped people, in short everybody in the society. However, the most victimized group consists of women and children. The behaviors which are called as the sexual harassment are the most difficult behaviors to prove. The cynical, violent or nonviolent verbal, physical and visual behaviors for the personality, gender, religious affiliation, ethnic origin and preferred sexual features are included in the scope of harassment. The most important factor determining whether there is harassment or not is the trace it left on the victim (Polat, 2016).

Another type of violence is emotional violence. It is observed that the children experiencing this violence are wasted away emotionally. Also, this type of violence leaves the effects, which last throughout their lives, on the children (Tezel, 2002). The emotional violence differs from the other types of violence in that no tangible physical marks are found in the emotional violence and it can be detected both by itself and along with the other types of violence. The individuals, who are exposed to physical or sexual violence, also experience the emotional violence, as well. The verbal violence is usually seen in the emotional violence cases. This type of emotional violence is the most commonly seen violence for the women and old people. Some behaviors (rejection, leaving someone alone, discouraging, making someone turn into crime, rejecting to give emotional reactions, humiliation, exploitation, giving an adult role earlier) are used in exposing the emotional violence. In the rejection the followings are seen; not meeting the child or adult's needs, not accepting the fact that s/he is a different individual, disregarding him, doing nothing in order to support an individual so that his positive features appear, rejecting the requests for help, make the individual feel that he is useless, holding him responsible for all the negative things, making him a scapegoat, avoiding from his personal touching in order not to reveal the closeness.

Another behavior is leaving someone alone. In this behavior, the individual keeps his wife or child away from himself, isolates himself from the social relations, the right of the women and children in the social relations are taken away or the emergence of such rights are prevented, and the individual makes his partner or his child believe that they are alone. In discouraging, the child or an individual are frightened by being menaced, they are being threatened and they are exposed to an environment in which they live in fear. Making someone turn into crime is being encouraged to unsocial behaviors, being directed to them and other ways which are regarded as bad by the society and being a bad example. In the basis of the rejecting to give emotional reactions lies not giving the reactions which support an individual's healthy social and emotional development. Humiliation is performing actions which cause a child or an individual to feel humiliated and hurt his pride, calling them with nicknames which make them feel bad, and applying those behaviors systematically. The exploitation is defined as the individuals' use of someone else for their own interests. Giving an adult role earlier is having meaningless expectations from the individual and also expecting a success as a result of it, pressurizing them to carry out the works which they cannot overcome and giving them some duties inappropriate for their developmental features (Polat, 2016).

Another type of violence is economic violence and takes away the economic freedom of women, as an individual, is explained as the economic violence. Using the money and other economic resources as a threat, a sanction or a control tool for the women are the facts which are regarded as the economic violence. Some of the behaviors that involve economic violence are; preventing the women's work life, not meeting the needs of the family, giving too little money at home and asking for impossible expenses with those money, depriving of women from the opportunities which carry them onward in the work life, preventing the women's work life and taking the money she earns from her, leaving the home frequently and being indifferent to the necessary spending. In the economic violence there are two contrasting behaviors. On one side, the women's freedom to work is taken away which forms the one leg of the economic violence, and on the other side, the women and children are forced to work and their income is taken from them, this forms the second leg of the economic violence (Polat, 2016).

The women, the children and the old people take part in the group which exposes to violence at most in the society. The individual violence issue is studied under the topic of violence in the family. Besides, the individual violence is a factor that gives momentum to the collective violence. While evaluating what the violence is, the groups that experience and expose to the violence are classified as such; a person's violence against himself, violence against the woman, violence against the child, violence against the old people, violence among the peers, violence among the siblings, violence against the refugees, and violence against the disabled person. In other classification which is carried out according to the form of violence, the followings are determined; physical violence, emotional violence sexual violence, economic violence and cyber violence (Polat, 2016). In his study Ayan (2007) revealed that the children who were exposed to violence without receiving positive behaviors from their mothers showed more aggressive behaviors compared to the children who were exposed to the violence and received positive behaviors, even if just a bit, from their mothers. The study group consisted of the secondary grade students of the elementary school in Sivas central districts. In the study it was aimed to measure the tendency to aggressiveness of the children exposed to violence and determine the factors increasing the aggressiveness. As a result of the study it was found that the children exposed to the violence were more inclined to the aggressiveness and the children's socio-demographic features had an important factor in their showing more aggressive behaviors. Also, when the literature review was carried out, the insufficiency of the studies about the violence in the family was noticed and what made this study more important was that, it revealed the negativities in the social structure caused by the violence. The concepts of violence and aggressiveness, as revealed in the intergenerational mobility theory, are learned behaviors. Not taking the necessary precautions and not intervening, when they are needed, may create results that are unlikely to correct. Children's not becoming a potential attacker criminal individuals, namely, the children's having a healthy mentality in the future and their becoming the individuals with consistent behaviors, the protection of the family life both legally and economically and its improvement, preventing to regard the violence as a legal behavior, decreasing the perception of a culturally approved behavior about the violence are only possible with changing the social point of view that tolerate the violence. The non-governmental organizations and public institutions have big responsibilities about this issue. The children should be educated in order not to be exposed to violence in the family, they should be aware of the situations stem from as a result of the violence and they should be taken under protection when it is necessary.

War and Child: The Immigrating Children

Immigration is a fact that has directed the humanity socially, culturally and economically since the ancient times. The new order in the globalizing world cause to experience the immigration more. The lives of the nations in the world have undergone great changes due to the effects of immigration that sometimes change its form and direction. With the influence of industry, the immigration, which is carried out from rural areas to urban areas due to economic reasons, is recorded in the history by being repeated periodically. The chaos in some areas, First and Second World Wars and some conflicts caused international mass immigration. The civil wars in the countries, from which people immigrate in mass, are the most important reason of mass immigrations. People immigrate in order to become distant to some situations such as material and nonmaterial damage caused by the wars and the death. While choosing a country for immigration, its population, its health and education services are of great importance. The changes in the social structure, local unemployment, rise in the unemployment rate, differentiation in the forms of employment are seen in the immigrant-receiving country. As people's living spaces change with the immigration, new human affairs appear. The social problems can be experienced due to intercultural conflict or mixture. The number of children who sell handkerchief and panhandle in the street, cannot go to school due to economic reasons, increased. In addition to social issues, the immigrating people bring the diseases and labor force with them. After World War II, many western countries received immigrations in order to meet their labor force need. The works, which were not wanted to be done by the people living in the western countries, were carried out by the immigrants, thus the lack of labor force was met. People, who were obliged to leave their countries as a result of political pressures and wars in their countries, constitute the forced immigrant community named the refugee. As it is accepted in the world, working of the immigrant children constitute problems in terms of many aspects such as social security, health and education (Harunoğulları, 2016). The movements for the change of location are examined within the existence of the immigration fact lasting for long years. The immigration fact has gained different senses by changing in time and developing, also has provided the basis for the production of different concepts. The immigration fact causing by the social, cultural and economic reasons affect both the receiving and the sending countries significantly due to its results. As a result of integration, the immigration has become a fact that maintains its presence all the time and is present in all the countries. Although it is a social movement in its essence, the immigration has turned into a means of change influencing every aspect of the life. The lexical meaning of the immigration is "a movement for change of location", but it is regarded as facts that affect the cultural, economic, social and political structures of the societies deeply (Sayın, Usanmaz and Aslangiri, 2016).

Urbanization has gathered speed with the experienced immigrations. The insufficient urban investment has led to crucial structural and social problems. Those immigrating to the cities generally live near the cities and in the slums which are far from the city standards. Those immigrating like this and living in the slums cannot reach to the health services and have difficulties in adapting the conditions of the environment and the city. At this point, the determination of the present risks in the environmental conditions, making plans for the service and doing studies for the improvement of the mother and child health are of great importance (Tezcan and Eryurt, 2006). The attitudes exhibited by the families during the immigration and the way they cope with this situation determine the degree they are affected by the immigration (Polat, 2007). The permanent change of place carried out in specific periods from somewhere to somewhere within borders of a country is called migration. It is not merely a change of place, at the same time, the change in the social environment causes negativity in terms of the mental health. As it is a stressful life, the immigration has become the subject of many studies due to its mental damages given to individuals, families and even societies. The school changes caused by the immigrations have a critical importance in terms of children's adaptation to new and large environment. It is predicted that the children, who adapted to school change, a negative influence of the immigration, can cope with the other negative influences caused by the immigration easier (Polat Uluocak, 2009). The numeric data about which markets children, who do not continue to school, joined is unknown. But, as it is predicted, it is obvious that the children are employed without security. The income of those working at early ages is lower in proportion to the job they do and their peers. Due to the families' need for income, the money coming from their children is needed despite being low. The works carried out by the immigrant children are generally manufacturing in the small places and serving in the small businesses, further differentiations according to the gender are seen in these jobs. Sometimes, the similarities between a family member's work and the child's work can be observed. It is also noticed that the children join the work life regardless of their gender (Lordoglu and Aslan, 2018).

In his study Polat Uluocak (2009) examined how the domestic migration affected the children's adaptations to school. It was concluded in the study that the children who were exposed to the migration had more risk of mental health disorder than those who were not exposed to the migration. It was also observed in the study that the adaptive problems of the migrated children in their home and school environment changed, they exhibited more introverted behaviors in the school environment. It is believed that the teachers who give education to the migrated children are an important factor for improving the children's behaviors and attitudes according to their mental problems. In addition to change of location, when the change of the social environment, cultural values, and even the language, and the emotional and behavioral problems, possibly seen in the children, are known and when the necessary precautions are taken earlier, such problems can be prevented. Contrary to this, when the problems, which are seen at the beginnings, are ignored, they can be more serious. As a result of this situation, some negative problems such as dropping out the school, getting used to the streets, being a drug addicted and turning into crime can be observed. Also, a disagreement between the teacher and mother's views about the children's mental adaptations was detected in the study. It is believed that, in order to prevent such a situation, the collaboration between the school and parents would be effective on improving the mental conditions of the children and increasing their school successes. While the psychosocial adaptation problems are being developed, child should be evaluated with his/her family and the social environment s/he is in. Thus, the adaptation process is applied successfully. It is thought that forming such adaptation programs by blending with health, social and education fields would be beneficial.

The migration causes huge problems both for the migrants and for those who are already in the immigrated cities. After the immigration, a situation called culture shock can also be experienced. Important traumas can occur in the individuals who were born and grew up in the village. The migrant could freely wander in their villages earlier, but when they come to city, some limitations can be experienced in their behaviors. When the children come to city, they suffer from being a stranger for the people outside. However, the opportunities provided by the city can be attractive for the children. The children try to have those opportunities by looking for different ways, working and living in the streets. When the problems and difficulties in the families encountered due to the migration is added on this situation, the children begin to live in the streets. They become distant from being a child, they children living in the streets can have unhealthy social relations. The migrated children, who get used to the street language and mentality, move away from their houses both physically and mentally. The children escape from the problems they have at home with the sense of freedom they experience in the streets. As a result of it, some unwanted cases such as robbery, tendency to violence, snatching and drug addiction emerge in the children (Okumuş, 2009).

In their study Tezcan and Eryurt (2006) found that in Turkey there is a significant difference between the migrant and non-migrant in the child death, just like the difference between rural and urban about the same topic. When the survival possibility of the children, whose mothers migrated from rural areas to urban areas, was compared with the survival possibility of the children, whose mothers lived in the rural areas, it was found with a significant difference that the migrants' children are better, but it was also indicated that the migrants' children had no advantages compared to those living in the city. It was also determined in the study that the migrant women benefitted from the health services during the birth less than the women living in the city, they were mostly deprived of the postnatal vaccinations and their children were inadequate in terms of nutrition compared to the children living in the cities. Due to the inadequacy in the educational status of the women, who lived in the rural areas or migrated to the city, their information about the child diseases could be insufficient. There is no doubt that the migrant women's interaction with their environments, their being exposed to the communication instruments such as television and radio, make them more conscious, compared to the women living in the rural areas, about what the health services are and how they can benefit from them. In the study it was also observed that the children's duration of survival varied according to their duration of living in the city. The migrant families' environmental conditions under which they live should be improved in order to fasten this process.

The opportunity and educational conditions of the migrant families' children should be equalized. Thus, the problems can be defined and appropriate solution ways can be offered. The children, who have difficulties about the adaptation, should be supported so that they receive psychological support. Providing the control of contagious diseases and make the migrants benefit from health services are very important for the public health. Immunizing the children is one of the most crucial practices for their health. It is inevitable to follow the babies since they are born and organize the necessary controls periodically. The child labor which is one of the big problems, should be prevented by periodically controlling the places the children work and not allowing the children to be employed illegally (Baş, Molu, Tuna, and Baş, 2017). According to Gencer (2017) regular studies should be carried out with the related organizations and institutions about the schooling of children and education rights, child-focused reintegrative programs should be organized against the situations which can cause the exclusion in the peer groups and discrimination. Further, in some problematic issues such as children's being deprived of school, children's getting married at early ages, dropping out the school, poverty, child labor, non-adaptation and language problem the families, the schools, the society and the related institutions should prepare the necessary services by acting together. The children should be helped for their adaptation to new society, new environment and new life.

Disadvantaged Children in Terms of Socio-economic Status

Children being disadvantaged creates many problems both for the children and for the society they are in. These problems increase the children's not reaching their fundamental rights, which they have since their birth, such as right for having education, health and housing. The problems caused by the adults' not getting rid of their disadvantageous situations underlie in many other problems encountered in our country and in the world. Also, in the development of the children, who are exposed to disadvantageous situations, some negative things occur. Further, the situation of disadvantageousness are transferred to next generations. As this cycle turns out to be continuity, it prevents the integration of healthy individual into the society. Growing of the children in such an environment along with the poverty create many negative effects on the society they live in. The children should benefit from their fundamental rights such as having education, housing and health in a best way in order not to be in those disadvantageous situations, the children should also the title characters in the created policies (Kahraman and Kahraman, 2017). According to Bilgin (2012) the children' becoming child labor at early ages and their being deprived of their education should be prevented. Working of child at early ages can expose him to the same bad conditions when grows up and can cause him to live under the necessary standard for their life and in a financially bad position. In such a case, as no differences occur for the child between his adulthood period and childhood period, this bad situation can turn out to be a vicious cycle.

Children With Chronic Diseases

Together with the decreases in a child's life quality, some influences occur in the development of the child. It is stated that these influences vary according to the variables such as whether it is a chronic disease or contracted and the age of the child. Such variables have developed by shaping since the baby period (Er, 2006). The children, who have a chronic disease and are at risk, are being employed in the works bringing low income and having heavy conditions, also those children are about to become drug addicted and turn into crime, they live in these dangers (Çoban, 2015). It is remarked by the psychologists that; the wrong sexual perceptions in the children, the effects of mass communication, mostly negative and illegal content coming from the social media, exposure of the worldwide known people, who have effects on children, as sexual objects are influential in forming the basis for chronic diseases. Further, the negative effects of those things on the children and their negative results are also emphasized. What is more, the environments such as their houses and the places they receive education can lead to the occurrence of wrong perceptions in the children, at the same time, the lack of adults, whom the children can trust without criticizing and take as a role model, pushes the children into loneliness and makes it easier for them to be affected by the negative environments faster, and forms a basis for becoming a criminal individual (Kır, 2013).

Of the families with children, who have a chronic disease, those with financial difficulty suffer at most because the psycho-social problems are among the most common problems caused by the poverty. Moreover, the poverty influences the mental development negatively. It is observed that even the brain, which develops chronically at the beginning, is affected negatively. Also, it is more possible for those poor families' children to be exposed to hazardous substances (Hatun, 2002). In order to remove this problem, the financial opportunities should be provided to families living in the slums; therefore, the problems stemmed from financial reasons can be decreased to lowest level and children's moving away from the family environment can be prevented. Taking actions only for the improvement of the financial opportunities would be beneficial to solve many problems (Türkmen, 2005).

As a result of the study carried out by Yıldız Akkuş and Bütün Ayhan (2018) it was determined that the family type and the disease type were influential on both the behavior and the quality of life. Also, it was detected that continuous drug use and the number of hospital stay had a significant influence only on the quality of life. Therefore, it was emphasized in the study that taking into consideration the developmental periods of the children with chronic disease, starting from the period when they are diagnosed, the strategies for supportive trainings, which are preparative for the disease and hospital, should be developed during their treatment and follow-ups. For Hatun (2002), the policies to decrease the poverty should be prepared, they should be put into practice and the spending for the health and education should be increased. Non-repayable financial support should be provided by the government to all families with children starting from those whose financial status is not sufficient and regular. The nutritive values of the food given in the school for the nutrition can be increased. Free kindergartens and nursery schools should be built especially in the areas in which mostly people with financial difficulties live. The mothers should be informed about the health and training for lactation should be given to all mothers with children.

Disabled Children

Getting harm of the physical, mental, psychological, emotional and social competences in different proportions innately or due to a disease or an accident is defined with the concept of "disability" (Kavaklı and Özkara, 2012). The concept of disabled have individual and social results and psychological, physical and medical extensity (Camkurt, 2013). Damage occurring in the physical and mental functions of an individual and the difficulty and lack, which are caused by this damage, in motion skills, manner and attitudes makes the individual different than the other members of the society. This difference is the basic reason for the discrimination and uncertainty a disabled person encounters in his life. The word discrimination involves the idea of "being outside the usual". Due to the influences created by the disorders of physical and mental functions on attitudes and manner, an individual keep out of the social environment. The disabled person cannot benefit from social environment evenly as a result of the lack of social support and unconsciousness (Subaşıoğlu, 2008). The negative conditions from which the child cannot be isolated, risk factors and the difficulties he may have in taking a model can prevent him from using his cognitive and behavioral skills, which are effective on acquiring the social competences, according to his needs. In order to minimize such negative conditions, the early respond programs for the children in early childhood period, who consist of the risky group, should be applied (Çetingöz and Yıldırım Doğru, 2016).

The children share socially, emotionally and economically as long as they are linked to family biologically and spiritually and the concept of family is very crucial for all the individuals. The child, who has just joined the family, should make an effort for adaptation whether he is disabled or not. As long as the development level of the family improve in parallel with the development level of the child, the family members can adapt to this development, but when a mentally disabled person stays behind this development it gets difficult for family member to adapt. The development level of a mentally disabled person progresses very slowly compared to a child showing normal intelligence development. Therefore, being the parents of a mentally disabled child requires being more sensitive and careful in the development periods (Arıcıoğlu and Gültekin, 2017). The support provided to the disabled children and the education given them are very important. Through the trainings presented to the children, who are described as the disabled, their isolation from the society can be prevented and the deprivation caused by the disability status can be minimized. At schools the special training services for the disabled children, who stay out of the inclusive education are very important (Orhan and Genç, 2015).

Children of Divorced Families

History of marriage goes back to four thousand year earlier and it is an artificial cultural fact formed by the humankind. Like all other cultural facts, it can undergo changes and can turn into new shapes. This structure, which comes true with the togetherness of man and woman, is described as the smallest unit of social system. This unit needs some intervenes in order to be repaired in the course of time, just like other structures formed by the humankind. Despite all the difficulties encountered, the marriage is a basic social unit, of which continuance is regarded as important for the health of the society, and it is accepted that it can be renewed and changed in accordance to the changes of the century (Türkaslan, 2007).

Among the families' different functions, which are determined with laws and traditions, raising the children physically and psychologically healthy and building their characters on sure ground are of great importance (Aral and Başar, 1998). The concept of marriage, which is formed with the togetherness of two different people, naturally faces with conflicts and disharmonies. It cannot be possible for two people, who are brought up in different cultures, in different environments and have different personalities, to carry on a lifelong happy marriage life without conflicts. At this point, the divorce of both people by deciding to end their marriage has gradually reached the increasing numbers in our age (Türkaslan, 2007). It is an accepted reality that the case of family fragmentation, which is frequently seen in this age, shakes the children. The healthy development of the society, its success in the next years and the transfer of the cultural inheritance can be carried out by the children of today who are well-educated, self-confident and self-respecting and whose personalities are built on sure grounds; therefore, the parents have great responsibilities about this subject (Aral and Başar, 1998).

The divorce is one of the most important factors differentiating the present family concept. As the case of divorce is a more different situation than the usual, it influences the each family member significantly. There is no doubt that the divorce decision of the parents influences their children at most. In the separation period the parents are focused on their own stress and problems very much so they notice their children's emotions late. While keeping pace with the unusual family order, the children also have some difficulties such as; guilt, fear, anger, sadness, problems about eating, sleeping problems, loneliness, rejection, comedown, school problems and physical problem, and they try to cope with them (Türkaslan, 2007). The children, whose parents separated, are under the risk of the sense of self, depression, low social responsibilities, low self-respect, academic failure, pregnancy at early ages and various health problems in the rest of their lives (Emery, 2009).

The child needs someone guiding to share what s/he learns, understand the subject well and solve the problems s/he encounters in the life. This call for help can be met best by the child's own parents. Parents helping their child when he has a problem and caring about his problems contribute to the psychological and mental development of the child positively. The presence of the parents, their good communication with the child and the spirit of togetherness are important so that the child can reach his targets and his creativeness can improve. But, parents' being separated and child's feeling the lack of one of his parents affects the child's academic success negatively (Şentürk, 2012). Parents' being together and maintaining a peaceful and happy unity is as important and effective as their being alive for children's acquiring a steady personality, being in peace with his environment and having a good academic success. The family members' way of communication with each other and their attitudes influence the child physically and psychologically. The possible togetherness, along with the love and respect between the mother and father, has an important place in the children's lives. In divorce cases the family unity is endangered and the child can find himself in a blank by harboring into his emotional world too much. The children, who are academically and emotionally successful in the development period, are those who grow up in a good a unified family. The child, who faces the communication conflicts and some other problems in the family, are deprived of the warm care and love of his parents, meets the situations that remain unsolved against the problems and lack of love. This lack of love and solutionlessness can cause to various disorders of attitude and behavior (Yavuzer, 2016).

A child's feeling of being curios and willing about learning should be stimulated with parents' support and encouragement. The desire for learning of the child, who is deprived of the parents' care and love, is not motivated and the difference between the child's potential and his performance enlarges. Parents' conflict, their indifferent attitudes towards the child, the gap that they make the child feel, and negative family togetherness prevent the child from exhibiting his skills. The child cannot find a positive family environment to discover his skills about the different topics and exhibit what he discovers, due to the problems and difficulties he has (Şentürk, 2012). It is known that the parents' relationship with the children is very effective on the children's attitudes and manner. Therefore, it is important to inform the child about with whom he is going to live after the divorce and ensure that the child is damaged as low as possible so that the children can acquire a correct and reliable personality. Giving the necessary information to the parents and leading them can be provided with specific programs through the mass communication along with the child counselling bureaus, centers for maternal and infant health, dispensary for mental health, child psychiatry clinics and some associations (Aral and Başar, 1998).

This research received no specific grant from any funding agency in the public, commercial, or not-for-profit sectors.

REFERENCES

Acehan, S., Bilen, A., Oğuzhan Ay, M., Gülen, M., Avcı, A., & İçme, F. (2013). Çocuk istismarı ve ihmalinin değerlendirilmesi. *Arşiv Kaynak Tarama Dergisi, 22*(4), 591–614.

Aral, N., & Başar, F. (1998). Çocukların kaygı düzeylerinin yaş, cinsiyet, sosyoekonomik düzey ve ailenin parçalanma durumuna göre incelenmesi. *Eğitim ve Bilim Dergisi, 22*(110), 7–11.

Arıcıoğlu, A., & Gültekin, F. (2017). Zihinsel engelli çocuk annelerinin sosyal destek ve gelecek algıları: Psikolojik danışmanların rolleri. *Mehmet Akif Ersoy Üniversitesi Eğitim Fakültesi Dergisi, 42*, 1–26.

Ayan, S. (2007). Aile içinde şiddete uğrayan çocukların saldırganlık eğilimleri. *AnatolianJournal of Psychiatry, 8*, 206–214.

Bakır, E., & Kapucu, S. (2017). Çocuk ihmali ve istismarının Türkiye'de yapılan araştırmalara yansıması: Bir literatür incelemesi. *Hacettepe Üniversitesi Hemşirelik Fakültesi Dergisi, 4*(2), 13–24.

Baş, M., Molu, B., Tuna, H. İ. & Baş, İ. (2017). Göç eden ailelerin sosyo-kültürel ve ekonomik değişiminin kadın ve çocuk yaşamına etkisi. *İnsan ve Toplum Bilimleri Araştırmaları Dergisi, 6*(3), 1680-1693.

Bathea, L. (1999). Primaryprevention of childabuse. *American Family Physician, 59*, 1577–1591. PMID:10193598

Bilgin, R. (2009). Diyarbakır'da sokakta çalışan çocuklar üzerine sosyolojik bir araştırma. *Elektronik Sosyal Bilimler Dergisi, 8*(27), 232–244.

Bilgin, R. (2012). Sokakta çalışan çocukları bekleyen risk ve tehlikeler: Diyarbakır örneği. *ZKÜ Sosyal Bilimler Dergisi, 8*(15).

Bülbül, S., & Doğan, S. (2016). Suça sürüklenen çocukların durumu ve çözüm önerileri. *Çocuk Sagligi ve Hastaliklari Dergisi, 59*(1), 31–36.

Camkurt, M. Z. (2013). 2022 Sayılı kanuna göre engelliliğin tespiti ve engelliliğe dayalı aylıkların bağlanması işlemleri. *Mali Çözüm Dergisi, 23*, 79–118.

Çetingöz, D., & Yıldırım Doğru, S. (2016). Gelişimi risk altında olan okul öncesi dönem çocuklarının sosyal beceri düzeylerinin değerlendirilmesi. *MediterraneanJournal of Humanities., VI*(2), 151–172. doi:10.13114/MJH.2016.291

Çoban, S. (2015). Türkiye'de risk altındaki çocuklar ve çocuk suçluluğu üzerine bir değerlendirme. *Sosyoloji Konferansları, 52*(2), 791–810.

Deveci, E., & Açık, Y. (2003). Çocuk istismarı nedenlerinin incelenmesi. *Arşiv, 12*, 396–405.

Dilsiz, H., & Mağden, D. (2015). Öğretmenlerin çocuk istismar ve ihmali konusunda bilgi ve risk tanıma düzeylerinin tespit edilmesi. *Hacettepe Üniversitesi Sağlık Bilimleri Dergisi, 1*(2), 678–694.

Efe, H., & Uluoğlu, S. A. (2015). Dünyada çocuk işçiliğiyle mücadelede gelinen nokta ve geleceğe dair bazı öngörüler. *Eğitim Bilim Toplum Dergisi, 13*(51), 46–72.

Emery, R. C. (2009). Stay for the children? Husband violence, marital stability, and children's behavior problems. *Journal of Marriage and the Family, 71*(4), 905–916. doi:10.1111/j.1741-3737.2009.00643.x

Er, M. (2006). Çocuk, hastalık, anne-babalar ve kardeşler. *Çocuk Sagligi ve Hastaliklari Dergisi, 49*, 155–168.

Erkan, R. & Erdoğdu, M. Y. (2006). Göç ve çocuk suçluluğu. *Aile ve Toplum, 3*(9), 79-90.

Gencer, T. E. (2017). Göç ve eğitim ilişkisi üzerine bir değerlendirme: Suriyeli çocukların eğitim gereksinimi ve okullaşma süreçlerinde karşılaştıkları güçlükler. *Uluslararası Sosyal Araştırmalar, 10*(54), 839–851.

Güler, N., Uzun, S., Boztas, Z., & Aydoğan, S. (2002). Anneleri tarafından çocuklara uygulanan duygusal ve fiziksel istismar/ihmal davranışı ve bunu etkileyen faktörler. *Cumhuriyet Üniversitesi Tıp Fakültesi Dergisi, 24*(3), 128–134.

Güneri Yöyen, E. (2001). Şiddet türleri ve kişilik özellikleri. *Yaşam Becerileri Psikoloji Dergisi, 1*(1), 35–50. doi:10.31461/ybpd.316724

Güngör, M. (2009). Eğitim hakkı ve sokakta yaşayan/çalışan çocuklar: Mersin ili örneği. *Mersin Üniversitesi Eğitim Fakültesi Dergisi, 5*(1), 28–42.

Güngör, M. (2013). Risk altındaki çocukların aile yapıları ve suça yönelimleri (Mersin İli Örneği). *Mersin Üniversitesi Eğitim Fakültesi Dergisi, 9*(2), 421–434.

Harunoğulları, M. (2016). Suriyeli sığınmacı çocuk işçiler ve sorunları: Kilis örneği. *Göç Dergisi, 3*(1), 29–63.

Hatun, Ş. (2002). Çocuk hakları sözleşmesinin 13. yüzyılında yoksulluk ve çocuklar üzerine etkileri. *Türk Tabipleri Birliği Merkez Konseyi:5-37*. Retrieved from http://www.ttb.org.tr/eweb/yoksulluk_ve_cocuklar.pdf

Kahraman, F., & Kahraman, A. B. (2017). Dezavantajlılığın çocuk halleri: Dünyada ve Türkiye'de dezavantajlı gruplarda yer alan çocukların durumu. *Gümüşhane Üniversitesi Sosyal Bilimler Enstitüsü Elektronik Dergisi*, *8*(20), 255–268.

Karakaş, B., & Çevik, Ö. C. (2016). Çocuk refahı: Çocuk hakları perspektifinden bir değerlendirme. *Gazi Üniversitesi İktisadi ve İdari Bilimler Fakültesi Dergisi*, *18*(3), 887–906.

Kasatura, İ. (2002). Sokaktaki çocuklar sokakta çalışan çocuklar. *İstanbul Ticaret Üniversitesi Dergisi*, 211-227.

Kavaklı, U., & Özkara, E. (2012). Kişisel, sosyal ve iş hayatında özürlü bireylere tanınan haklar ve özürlü olguların medikolegal değerlendirmesi. *DEÜ Tıp Fakültesi Dergisi*, *26*(1), 65–74.

Kır, E. (2013). Çocuklara yönelik cinsel taciz ve istismara karşı önleyici eğitim çalışmaları. *İÜHFM C. LXXI*, *1*(786), 785-800.

Kocacık, F. (2001). Şiddet olgusu üzerine. C.Ü. *İktisadi ve İdari Bilimler Dergisi*, 2(1), 3-7.

Koçtürk, N. (2018). Çocuk ihmalini ve istismarını önlemede okul çalışanlarının sorumlulukları. *MSKU Eğitim Fakültesi Dergisi*, *5*(1), 38–47.

Koçtürk, N., & Yılmaz, N. (2018). Çocuk istismarı ve ihmali için risk altındaki çocukları belirlemeye ve müdahale etmeye yönelik model/veri tabanı önerisi. *Kastamonu EducationJournal*, *26*(6), 1919–1927.

Kurtay, D. (2018). Çocuk ihmal ve istismarının önlenmesi. Retrieved from http://ahto.meb.k12.tr/meb_iys_dosyalar/06/25/733770/dosyalar/2018_01/03144225_cocukihmalveistismarınınönlenmesi.doc

Lordoglu, K. & Aslan, M. (2018). Görünmeyen göçmen çocukların işçiliği: Türkiye'deki Suriye'li çocuklar. *Çalışma ve Toplum*, *2*, 715-732.

Nalbant, A., & Babaoğlan, E. (2016). Risk altındaki öğrenciler: Yozgat örneği. *E-Uluslararası Eğitim Araştırmaları Dergisi*, *7*(3), 11–30.

Okumuş, E. (2009). Sokak çocuklarının sosyolojisi-Diyarbakır örneği. *Dinbilimleri Akademik Araştırma Dergisi*, *IX*(1), 9–37.

Orhan, S., & Genç, K. G. (2015). Engellilere yönelik ülkemizdeki özel eğitim hizmet uygulamaları ve örnek ülke karşılaştırması. *Sosyal Politika Çalışmaları Dergisi*, *35*(2), 115–146.

Öztürk Çopur, E., Ulutaşdemir, N., & Balsak, H. (2015). Çocuk ve suç. *Hacettepe Üniversitesi Sağlık Bilimleri Dergisi*, *1*(2), 120–124.

Özyürek, A., Çetin, A. & Yıldırım, R. (2018). Aile hekimi ve öğretmenlerin çocuk ihmal ve istismarı konusundaki görüş ve tutumları. *İnsan ve Toplum Bilimleri Araştırmaları Dergisi*, *7*(1), 436-453.

Pelendecioğlu, B., & Bulut, S. (2009). Çocuğa yönelik aile içi fiziksel istismar. *Abant İzzet Baysal Üniversitesi Dergisi*, *9*(1), 49–62.

Polat, G. (2007). İç göçün çocuk ruh sağlığına etkisi ve sosyal hizmet müdahalesi. *Toplum ve Sosyal Hizmet*, *18*(1), 89–106.

Polat, O. (2016). Şiddet. *MÜHF - HAD*, *22*(1), 15-34.

Polat Uluocak, G. (2009). İç göç yaşamış ve yaşamamış çocukların okulda uyumu. *Dokuz Eylül Üniversitesi Buca Eğitim Fakültesi Dergisi*, *26*, 35–44.

Şahin, F. (2001). Çocuk istismarı ve ihmalinin önlenmesinde hekimin rolü. *Sürekli Tıp Eğitimi Dergisi*, *10*, 246.

Şahiner, Ü. M., Yurdakök, K., Kavak, U. S., & Çetin, İ. (2001). Tıbbi açıdan çocuk istismarı. *Katkı Pediatri Dergisi*, *22*, 276.

Sayın, Y., Usanmaz, A., & Aslangiri, F. (2016). Uluslararası göç olgusu ve yol açtığı etkiler: Suriye göçü örneği. *KMÜ Sosyal ve Ekonomik Araştırmalar Dergisi*, *18*(31), 1–13.

Şentürk, Ü. (2012). Parçalanmış aile çocuklarının eğitimdeki başarı/başarısızlık durumu (Malatya örneği 2006). *Sosyal Politika Çalışmaları*, *7*(29), 105–126.

Subaşıoğlu, F. (2008). Üniversitelerin bilgi ve belge yönetimi bölümlerinin "engellilik farkındalığı" üzerine bir araştırma. *Bilgi Dünyası*, *9*(2), 399–430.

Tezcan, S., & Eryurt, M. A. (2006). Türkiye'de iç göç ve çocuk sağlığı. *Nufusbilim Dergisi*, *28-29*, 15–28.

Tezel, A. (2002). Çocuğa yönelik şiddet. *Atatürk Üniversitesi Hemşirelik Yüksekokulu Dergisi*, *5*(1), 93–100.

Topçuoğlu, T. (2014). Çocuk suçluluğu ve gelişimsel (risk-odaklı) suç önleme. *KMÜ Sosyal ve Ekonomik Araştırmalar Dergisi*, *16*, 217-226.

Türkaslan, N. (2007). Boşanmanın çocuklar üzerine olumsuz etkileri ve bunlarla başetme yolları. *Aile ve Toplum*, *3*(11), 99–108.

Türkmen, B. (2005). Sosyal risk altındaki çocuklar, gençler ve yaşlılar. *Gazi Üniversitesi Kastamonu Eğitim Dergisi.*, *13*(1), 11–22.

Yalçın, H., Bay, M., & Ekmekçioğlu, E. (2014). Türkiye ve Türk dünyasında çocuk hakları ile ilgili çalışmaların yararları. Retrieved from http://www.bilgesam.org/Images/Dokumanlar/0-279-2014070118sosbilkongre33.pdf

Yavuzer, H. (2016). *Çocuk psikolojisi*. İstanbul: Remzi Kitabevi.

Yazıcı, E. (2012). Korunmaya muhtaç çocuklar ve çocuk evleri. *Mustafa Kemal Üniversitesi Sosyal Bilimler Enstitüsü Dergisi*, *9*(18), 499–525.

Yıldız, M. A. & Duy, B. (2015). Sokakta çalışmak ve gelir düzeyi erinlerin benlik saygısı, yaşam doyumu ve depresyon düzeylerine etki eder mi? (Diyarbakır ili örneği). *İlköğretim Online*, *14*(2), 522-537.

Yıldız Akkuş, S. & Bütün Ayhan, A. (2018). Kronik hastalığı olan çocukların davranışlarının ve yaşam kalitelerinin incelenmesi. *Türkiye Çocuk Hastalıkları Dergisi*, 1-7.

Yolcuoğlu, G. İ. (2009). Türkiye'de çocuk koruma sisteminin genel olarak değerlendirilmesi. *Aile ve Toplum*, *5*(18), 43–57.

Chapter 5
Exceptional Children

Ridvan Karabulut
Kırşehir Ahi Evran University, Turkey

ABSTRACT

In addition to the children who show typical development, there are also children who show atypical development. Individuals whose development varies compared to their peers and whose differences are identified by experts are called individuals with special needs. Individuals with special needs can be children with intellectual disabilities, sight, hearing impairment, language and speech disorders, physical disability and chronic illness, specific learning difficulties, children with autism spectrum disorder and gifted children. The causes of special needs are generally prenatal causes, which can often be hereditary and can be classified as natal and postnatal. The identification of individuals with special needs from early ages and the use of appropriate intervention methods are vital for these children. Children who are happy and a future producing part of society due to early recognition and accurate education methods are directly related to the quality of the intervention performed.

INTRODUCTION

Among the people involved in social life, there are individuals who show different development. This difference can be accepted by the people around them. These differences among people can be expressed as personal abilities, interests, vital values and physical characteristics. There are individual differences that prevent individuals from fulfilling their own personal responsibility and basic skills. These differences can be seen as individual differences that are not accepted by other people in social life (Özkan, 2009).

In the 1960s, the understanding that individuals with special needs receive education in special education classes has become a common way of thinking in the world. However, although individuals with special needs continued their education in separate classes, criticisms emerged and lawsuits were filed. In the 1970s, some countries enabled individuals with special needs to receive education with their peers in general educational settings. Countries that accept to take courses in the same educational environment are the USA, Norway, England, Italy, and France (Kuz, 2001).

DOI: 10.4018/978-1-7998-2952-2.ch005

Individuals in need of special education are classified as individuals with physical disorders, communication disorders, visual disturbances, gifted and gifted, severely disabled, behavioral problems, hearing disorders, learning difficulties, mental disability, and other health problems (Sucuoğlu, 2006).

The fact that individuals in the society are different from each other ensures that there are differences in educational needs (Dikici Sığırtmaç, 2013). It takes place in children who differ in terms of developmental characteristics other than children who develop normally during the education process. It is no problem that children with some different developmental characteristics receive education in normal educational settings. However, it is an obstacle for children with disabilities in areas such as vision, hearing and learning to benefit from normal educational environments (Eripek, 1992). For individuals who cannot benefit from the opportunities of normal education environments, it is necessary to establish different education environments (Özkan, 2009).

Special Required Individual

Different definitions and nomenclatures have been made from past to present periods. Definitions such as "disabled," "disabled," those with special needs, those with developmental differences, "those with different development," "those with learning differences" and individuals requiring special education kullanılmış were used in these definitions. Definitions such as children with special needs, "children with developmental differences," children with different development," children with learning differences," "individuals with special needs" are used today without labeling them (Baykoç Dönmez, 2010).

Various classifications are made considering the disability types of children with special needs. The reclassifications made in Turkey, children with special needs "language and individuals with speech difficulties," "individuals with hearing impairment," "individuals with physical disabilities," "individuals with mental disabilities," "individuals with specific learning difficulties," "attention deficit and individuals with hyperactivity disorder," individual with cerebral palsy," individual with autism, " individual with visual impairment," individual with chronic illness "and "gifted individual (Eripek and Vuran, 2008).

While there are accepting positive attitudes towards individuals with special needs in society, there is also an attitude formed with rejection and prejudiced feelings (Aker, 2014). When the studies conducted in the last thirty years are examined, it was determined that there are negative attitudes towards individuals (Özyürek, 2006).

Education needs of individuals with special needs are different than other normal children. Children with special needs need to benefit from special education and services related to special education in order to be able to manifest themselves educationally (Solidtaş, 2019). There are certain objectives in educating individuals with special needs. At the beginning of these goals, they can be expressed as maintaining their lives without being dependent on others, being self-sufficient and providing a society (Cavkaytar, 2000 cited in Sağıroglu, 2006).

Children with special needs in society have the right to education in public or private schools, such as peer groups that develop normally. Educators and parents have a great responsibility to find a place for them in life. School administrators should follow the plan and program implementations required for the education of children and provide the education environment necessary for children (Pamukçu, 2016).

Reasons for Being an Individual With Special Needs

The reasons for being an individual with special needs are discussed in 3 groups. These; prenatal causes, causes that occur during birth and postpartum causes (Taşcı, 2017).

- Prenatal Causes: Traumas during pregnancy, drugs used by the mother during pregnancy, diseases passed during pregnancy, maternal age, drink-cigarette-drug use during pregnancy, maternal nutrition in pregnancy, consanguineous marriages, blood incompatibility, genetic diseases, radiation exposure during pregnancy factors may cause the child to have special needs (Baykoç Dönmez, 2010).
- Causes During Birth: Premature birth, risky and difficult birth, cord wrap around the neck of the baby, any trauma during birth, late birth, vacuum used during labor, delivery in unhygienic environments, problems during the use of tools such as forceps, birth of the baby immediate oxygen-free causes such as effects (Dikici Sığırtmaç and Deretarla Gül, 2010). Syphilis, tuberculosis diseases or various microorganisms transmitted to the child during childbirth can be counted as the loss of vision of the child (Taşcı, 2017).
- Postpartum Causes: Inadequate environmental conditions, infections, nutritional disorders, accidents, microbial and febrile diseases, physical traumas, poisoning, mental traumas, child abuse, natural disasters, wars can affect individuals' need for special needs (Özgür, 2008 act Tasci, 2017).

CLASSIFICATION OF CHILDREN WITH SPECIAL NEEDS

Children With Communication Disorders

It includes children with language and speech disorders. It is seen in two forms as stuttering and voice disorder. In these children, students should be given the opportunity to express themselves in different environments. In addition, the appropriate environment in their education should be adjusted to eliminate the concerns about speech.

Mentally Retarded Children

According to the 2002 directive of the American Association on Intellectual and Developmental Disabilities (AAIDD); Yetersizlik Inadequacy characterized by significant limitations in both adaptive behaviors that begin before the age of 18 and manifest in mental functioning and conceptual, social and practical adaptive skills Er (Eripek and Vuran, 2010).

Pamukçu, (2016) is a mentally handicapped person: who develops before the age of 18, differ in terms of mental functions under two standard deviations of the average, and therefore have limitations or deficiencies in social, conceptual and practical adaptation skills, as an individual in need of services.

In the 2018 Regulation on Special Education Services published by the Ministry of National Education, individuals with intellectual disabilities have made descriptions in four sections. These sections; An individual with mild mental disability, an individual with moderate mental disability, an individual with severe mental disability and individuals with very severe mental disability are as follows:

Individuals with mild mental disability; social, practical adjustment skills and mental functions and conceptual levels. Therefore, it is the individual who needs special education and support education service to a limited level.

An individual with moderate mental disability; is an individual in need of special education and support education services in daily life, gaining business skills and providing basic academic skills due to limitation of social, practical adaptation skills and mental functions and conceptual levels.

An individual with severe mental disability; social, practical adaptation and self-care skills, mental functions and conceptual skills because of their lack of life throughout their lives, is an individual who needs intensive training and support education services.

Individuals with very severe mental disabilities; Daily life with mental disability is an individual who cannot acquire basic academic and self-care skills and needs constant care and supervision throughout his / her life (MEB, 2018).

Intelligence levels of individuals are measured with the help of certain measurement tools and methods. By looking at the age of the person, it is determined whether the level of intelligence he possesses is at the same level, behind or advanced, or not, compared to the children of the same age. The harmony between intelligence age and calendar age is examined. The following formula can be used as a method.

Formula; intelligence department = (age of intelligence) / (calendar age) × 100

Sample

If the intelligence age is 10 and the calendar age is 20, when the intelligence section is calculated by looking upwards, the result is 50 ([10/20) × 100 = 50]. Individuals who are below 70 to 75 in intelligence are defined as mentally disabled (Kaya, 2016).

The behaviors that each person performs very easily can be difficult for individuals with mental disabilities. Individuals with intellectual disabilities experience problems in areas such as self-direction, work life, communication, home life, academic skills, functional social skills, being with the community, self-care, leisure time, health and safety, in social life (Sucuoğlu and Kargın, 2006)

Children With Learning Difficulties

They constitute the largest group of children with disabilities. Learning disabilities; A disorder that affects one or more of the psychological processes related to the inability to understand or understand speech, reading and writing, which affects the ability to read, write, listen, spell, speak, think, and perform mathematical operations. The majority of the children in this group are inadequate in ordering, planning and directing their lives, such as completing a job, making plans for learning. In addition, approximately 50% of these individuals have language and speech problems (Kaya, 2016).

According to Solmaz (2016), the developmental characteristics of individuals with special learning difficulties are explained below. These features:

Motor Skills: Children with learning disabilities may experience difficulties in the use of scissors, pencil-holding, button-buttoning, shoelace and tying, which are small muscle exercises. Sitting, holding his head upright, crawling, walking delays are seen in the skills. In addition, these children may suffer from inability to perform movements that require body control and rhythm, such as jumping, lifting objects, jumping, jumping from a certain height.

- Language Development: People with special learning difficulties may develop slow speech and may have problems in language expression, verbal cause-effect relationship, comprehension and verbal memory skills. - they may experience delays in learning the sound relationship, and may have difficulty in rhyming words.
- Gelişim Social Development: Self-efficacy and self-assessment of individuals with learning disabilities affect their motivation and academic development towards the school. The social development variable is directly proportional to academic achievement. A child with a learning disability is often described as withdrawn, disinterested in the environment. It is an important reality that academic success will increase due to the increase in self concept.
- Cognitive Development Characteristics:. Children with learning difficulties have limitations in understanding and understanding. They have no problems in perceiving concrete concepts, but often fail to perceive abstract concepts. Memory retention, concept development, problem solving, abstract thinking and symbolization problems arise in individuals with learning difficulties. These problems are thought to be caused by lack of thought, attention, memory and upper memory and upper language function.

Children With Behavior Disorder

Individuals with behavioral disorders are individuals who need support and special education because of their emotional behaviors and reactions that are different from social and cultural norms that are not suitable for age level (Akçamete, 2010).

The exact cause of the behavioral disorder is not known. Symptoms include excessive mobility, easy distraction, aggression, excessive anxiety, strange motor movements, anger attacks (Sucuoğlu and Kargın, 2006).

Educators with adequate knowledge and skills about individuals with behavioral disorders should ensure that these individuals have positive communication with other individuals. They should also support them to be productive individuals (Smith and Robinson, 2006 cited in Saraç, 2019).

The teacher should give clear and understandable instructions for the students with behavioral disorders in the classroom, adjust the speed of the lecture according to the students, make eye contact, constantly monitor whether the student is interested in the course and adjust the physical conditions of the class. In particular, it should be in close cooperation with parents, however, it should also give written and oral assignments (Kaya, 2016).

Visually Impaired Children

In 2018 Special Education Services Regulation issued by the Ministry of National Education, individuals with visual impairment are explained. According to this regulation; an individual with visual impairment is defined as birey an individual in need of special education and support training services due to partial or total loss of vision M (MEB, 2018).

According to Kılıçoğlu, (2007), the visually impaired people stated that the quality of education and social cohesion were negatively affected due to the total or partial insufficiency of vision.

According to Kaya (2016), visually impaired people consist of two groups. These groups; less sighted and completely blind. Even if they perceive the shadows, the light, and those who do not fully see, they often use the touch and hearing senses in the learning process.

Some arrangements are required for visually impaired students. These arrangements are as follows:

Doors must not be fully open or fully closed.

Tactile tools should be used as much as possible.

In which light the students with visual debris can see clearly, it should be determined and a suitable light environment should be provided.

In the course of the course, if there are graphics and pictures, they should also be expressed orally and plenty of descriptions should be used.

For the students to move freely, the entrances, exits, corridors and classrooms should be corrected and secured.

A suitable environment should be created for those using additional vehicles (voice recorder, etc.).

The tools used in the lessons should be in a way that does not cause danger. (Özokçu, 2013)

These individuals with visual disabilities continue their education in special education classes, boarding schools or normal classes in our country (Akçamete, 2010).

Hearing Impaired Children

In the 2018 Special Education Services Regulation issued by the Ministry of National Education, individuals with hearing impairment are defined as birey individuals in need of special education and support education services due to loss of hearing sensitivity completely or partially M (MEB, 2018).

Pamukçu (2016) stated that the person with hearing impairment is in need of special education and support education services because of difficulties in using language, acquiring speech and communication because of hearing loss is completely or partially lost.

Coarse motor and fine motor skills due to damage to the auditory canal and nerves, and difficulties in balance and coordination are among the general characteristics of hearing impaired individuals. They are unable to hear their own voices and other voices around them. however, due to their communication problems, they also face various difficulties in their social development (Akçamete, 2010).

Some arrangements should be made in the classrooms with hearing impaired students. Arrangements such as attaching anti-noise tools to the legs of the desks, arranging the classes acoustically, covering the floors with carpet, and keeping the class away from noise are important. In addition, students should be close to the teacher and the classroom environment as long as the rows should be arranged in a circle or U-shaped. It is also very important to include visual elements in the education process (Özokçu, 2013).

Children With Orthopedic Disabilities and Permanent Diseases

They are individuals in need of special education and support training services due to illness requiring long-term care or continuous treatment (Pamukçu, 2016).

Children with physical disability / chronic disease are most commonly expressed as children with disabilities due to various factors of the body's movement-related organs (Cavkaytar and İbrahim, 2007 cited in Saraç, 2019).

Some of the students with special needs have diseases that make their lives difficult, while others have physical problems. diseases such as asthma and epilepsy indicate chronic diseases, while orthopedic obstacles include muscular problems. Children with epilepsy and asthma should be observed, drug use should be provided regularly. During the seizure; a soft thing should be put under the head of the person to be hit, the course of the seizure should be followed, if the seizure lasts longer than five minutes should be called experts and seizure children should not be trying to wake up. For children with orthopedic disabilities, they should be placed close to the door in the classroom where they can easily enter and exit (Sucuoğlu and Kargın, 2006).

The problems of the physically inadequate individuals are usually related to movement. They will have the opportunity to receive education with peer groups in normal classes in general education institutions in line with the adaptations and practices made considering the needs of these students (Akçamete, 2010). Due to the high level of regulations and adaptations for individuals with multiple or very severe disabilities, they are provided with special education classes. However, separate educational environments for individuals with many physical disabilities are not seen as an appropriate educational environment (Özyürek, 1985).

Autistic Children

Autism spectrum disorder is defined as gelişim a developmental disorder that occurs before the age of three in children, which adversely affects verbal and non-verbal communication, social interaction and children's educational performance S (Sucuoğlu, 2006).

Pamukçu (2016). Individuals who need special education and support education services due to their limitations in verbal and nonverbal communication, social interaction, interests and activities that occur in early childhood are called as autism spectrum disorder.

It is usually recognized in the first three years of life and is more common in boys than in girls. It is divided into five categories as autism, rett, asperger syndrome, atypical autism, and childhood disintegrative disorder syndrome. (Akçamete, 2010).

These children are based on idioms, abstract words, metaphor

they have difficulty understanding jokes. They also pay attention to only one of many stimuli. This makes it difficult for them to grasp the whole. Children with autism can memorize a lot of information, but they cannot use it because they have difficulty in remembering what they memorize (Kaya, 2016).

Autistic children have some prominent features. We can briefly list these features: Repetition, insensitivity to danger, attachment to objects unnecessarily, hypersensitivity or hypersensitivity to sounds, irregular motor movement development, inappropriate and unreasonable laugh or cry, excessive mobility or excessive inactivity, insensitivity to pain, change overreacting, wanting everything to be the same, having difficulty communicating with other children, disliking being touched, turning objects around, feeling sadness for no reason, stressing, playing the same games all the time, and specifying with gestures rather than speaking are the prominent features of autistic children (Karasu, 2013; Kaya, 2016).

Special Skilled Individual

Special talented individual; They can be expressed as individuals who learn faster than their peers, who are ahead in the capacity of art, creativity and leadership, who love to act independently in their fields of interest, can understand abstract ideas, have special academic skills and perform at a high level (MEB, 2018).

When the development of the concept of giftedness is examined in the historical process, it is stated that there are many different definitions of giftedness, which vary from culture to culture and from society to society. Giftedness refers to individuals with creative thinking skills, awareness of task responsibility, an ability that is above what it should be, and high potentials that put these in place. (Renzulli, 1986 cited in Ozcan, 2018). The first definitions about giftedness are the definitions based on intelligence tests aiming to distinguish the level of ability (Bozkurt, 2007).

I. Special Education Council According to the definition adopted by the Ministry of National Education (1991); “Gifted children are children who are determined by the experts of the subject who perform at a high level according to their peers according to their general or special abilities”. I. According to the Special Education Council report, there are some characteristics that distinguish gifted children from normal children:

- Superior mental ability

Özel Special talent in different areas

- Sensitivity, sensitivity and creativity

Yoğun Intense motivation at the highest level (Özcan, 2018).

In 1992, the General Directorate of Special Education Guidance and Counseling Services was established under the Ministry of National Education for gifted children and the aim was to improve and develop special education and guidance activities (Çamdeviren,2014).

Çitil and Ataman (2018) stated that it is important to develop an individualized and enriched education program for the student in line with their performance to gifted individuals. In addition, students are required to notice situations different from the class level. In this case, the boredom of the child in the classroom can be seen as a step towards gaining the child by preventing distraction.

Gifted individuals; It is seen as individuals who easily adapt to new social situations, come to the forefront in the groups they are in and produce solutions to social problems existing in society. When personality traits of gifted individuals are examined; It is stated that they have a structure that respects the thoughts of others, is determined, patient, hardworking and enjoys success (Çağlar, 1972). However, gifted children also have negative characteristics. It features; irregular and disorganized, not open to cooperation, stubbornness, irritability, resistance to forced activities, excessive sensuality, resistance to common habits and unpretentious features can be listed as (Karabulut, 2018).

Gifted children have different developmental characteristics from an early age. These children exhibit different developmental characteristics in different areas. It is necessary to know the characteristics of gifted children in different development areas. This situation is very important for families and educators (Karabulut, 2018).

there are distinctive features that distinguish gifted individuals from their peers. These features can be listed as follows;

Ilgi Being able to be interested in various fields,

- To be able to concentrate attention for a long time,

Arı Careful observation and being curious,

Geniş Having a wide vocabulary at a very young age,
Ği Ability to criticize others and themselves,
Anlay To understand complex concepts,
Me To be able to remember the past in detail,

- It is in the form of abstract thinking and establishing relationships between concepts (Ataman, 1982).

Gifted children are at a disadvantage because of their characteristics. These children to mobilize their potential, healthy and happy individuals who receive satisfaction from life is closely related to the attitudes of parents. Therefore, the family has a great relay in the upbringing and upbringing of gifted children (Karakuş, 2010).

Children grow up by taking their families as models. While families with normal development children have a heavy duty in this respect, the workload of families with mentally handicapped or gifted children increases more. Turkey's future is mentally disabled and gifted children, or on behalf of the rise of the welfare society, the entrepreneur has the collaborative skills, responsibility, and their families to be happy individuals have the greatest impact. Therefore, perceptions of quality of life, life satisfaction and social support of parents are important (Toprak, 2018).

REFERENCES

Akçamete, G. (2010). Children with special needs. In G. Akçamete (Ed.), Students with special needs in general education schools and special education (pp. 31-76). Ankara: Root Spring.

Aker, G. (2014). *Pre-service teachers' attitudes towards mainstreaming education*. Unpublished master's thesis, Trakya University, Institute of Social Sciences, Edirne.

Ataman, A. (1982). How can parents and teachers help gifted children? *Ankara University Journal of the Faculty of Educational Sciences*, *15*(1), 335–344.

Baykoç Dönmez, N. (2010). Children with special needs and special education. In N. Baykoç (Ed.), Special education for teaching programs (pp. 13-25). Ankara: Daytime.

Bozkurt, Ö. S. (2007). Examination of the developmental characteristics of children who are nominated as superior and special talented by their teachers according to their peers. Unpublished Master's Thesis, Hacettepe University Institute of Health Sciences, Ankara.

Caglar, D. (1972). Characteristics of gifted children. *Ankara University Journal of the Faculty of Educational Sciences.*, *5*(3), 95–110.

Çamdeviren, Ş. (2014). Difficulties faced by parents of gifted children attending Science and Art Center (Bilsem) (case of Sakarya province). Unpublished Master's Thesis, Sakarya University Institute of Educational Sciences, Sakarya.

Citil, M., & Ataman, A. (2018). Reflection of behavioral characteristics of gifted students in primary school age on educational environments and problems that may arise. *Gazi University Journal of the Faculty of Education, 38*(1), 185–231.

Dikici Sığırtmaç, A. (2013). Family participation in the education of children with special needs. In Y. Aktaş Arnas (Ed.), Family education and pre-school family participation (2nd ed., pp. 174-176). Adana: Visa.

Dikici Sığırtmaç, A., & Deretarla Gül, E. (2010). *Preschool special education*. Ankara: Visa.

Eripek, S. (1992). Disabled and their education. *Anadolu University Journal of Education Faculty, 3*(1-2), 79–84.

Eripek, S., & Vuran, S. (2008). *Education of children with intellectual disabilities. In G. Akçamete (Ed.), Special education* (pp. 245–278). Ankara: Kök Publishing.

Eripek, S., & Vuran, S. (2010). Education of children with intellectual disabilities. In G. Akçamete (Ed.), Students with special needs in special education schools and special education (pp. 245-282). Ankara: Kök Publishing.

Karabulut, R. (2018). Examining the effect of education on problem solving skills of gifted children attending primary school. Unpublished PhD Thesis, Gazi University Institute of Educational Sciences, Ankara.

Karakus, F. (2010). Difficulties faced by gifted parents. *Mersin University Journal of the Faculty of Education, 6*(1), 127–144.

Kaya, A. (2016). Views of Turkish teachers about mainstreaming and special needs students. Unpublished Master's Thesis, Gazi University, Institute of Educational Sciences, Ankara.

Kilicoglu, M. (2007). *Effective teacher for active learning*. Istanbul: Morpa Culture Publications.

Kuz, T. (2001). Esi Examination of Attitudes towards Inclusive Education. Unpublished Thesis, Prime Ministry Administration for Disabled People Publications, Ankara.

Ministry of Education. (2018). *Ministry of National Education Special Education Services Regulation*. Ankara: Official Gazette.

Özcan, M. (2018). Comparison of attitudes of gifted and non-gifted students towards physical education and sports lesson. Unpublished Master's Thesis, Health Sciences Institute, Ondokuz Mayıs University, Samsun.

Özkan, Y. Ö. (2009). Effect of inclusion preparatory activities of students with normal development in providing social acceptance of individuals with special needs. Unpublished Master's Thesis, Institute of Educational Sciences, Anadolu University, Eskisehir.

Ozokcu, O. (2013). Mental disability, visual impairment, hearing impairment. In I. Diken (Ed.), Inclusion in primary education (p. 56-87). Ankara: Pegem Academy.

Özyürek, M. (1985). Orthopedic defective placement in educational environments. *Ankara University Journal of the Faculty of Educational Sciences, 17*(1).

Özyürek, M. (2006). *Changing attitudes towards people with disabilities.* Ankara: Kök Publishing.

Pamukcu, M. (2016). Perceptions and competencies of primary school principals in the education of children with special needs bir araştırma A research in Maltepe district of Istanbul Province. Unpublished Master's Thesis, Institute of Social Sciences, Maltepe University, Istanbul.

Sağıroglu, N. (2006). Expectations of families with special needs from special education and rehabilitation centers where their children continue. Unpublished Master's Thesis, University of Abant İzzet Baysal.

Sarac, S. (2019). Examination of attitudes of special gifted 5th, 6th, 7th grade students towards individuals with special needs affected by disability. Unpublished Master's Thesis, Institute of Educational Sciences, Bahcesehir University, Istanbul.

Solidtaş, S. (2019). Principal, teacher and parent views on the roles of school principals in providing children with special needs. Unpublished doctoral dissertation, Ankara University, Institute of Educational Sciences, Ankara.

Sucuoğlu, B. (2006). *İlköğretimde Kaynaştırma Uygulamaları-Yaklaşımlar, Yöntemler, Teknikler.* Morpa Kültür Yayınları.

Sucuoglu, B., & Kargin, T. (2006). *Ilkögretimde kaynastirma uygulamalari: Yaklasimlar, yöntemler, teknikler.* Istanbul: Morpa Yayincilik.

Tasci, K. N. (2017). Feelings and thoughts of parents and siblings in families with special needs. Unpublished Master's Thesis, Institute of Social Sciences, Cukurova University, Adana.

Toprak, F. (2018). Comparison of life satisfaction, quality of life and social support perceptions of families with mentally retarded and gifted children. Unpublished Master's Thesis, Hacettepe University Institute of Social Sciences, Ankara.

Chapter 6
Children in Hospital

Gül Kadan
Sağlık Bilimleri Fakültesi, Çankırı Karatekin Üniversitesi, Turkey

ABSTRACT

Visiting a hospital for hospitalization or treatment is one of the foremost factors triggering stress for children. Under this stress, lie both the restriction of a child's freedom and independence due to getting into an unknown environment and the rapidly changing events around the child. The child's reactions may become even more aggressive, especially if they feel the fear and anxiety experienced by their parents. Based on this importance, this chapter focuses on the responses of children to hospital life according to their developmental stages and on what can be done regarding their adaptation in hospitals.

INTRODUCTION

Disease and consequent hospitalization or inpatient treatment can be expresses as a very challenging situation that temporarily or permanently changes the psychological/physical structure of the human being and its connections with the environment. In particular, this situation can affect children more deeply due to immature development areas (Gültekin &Baran, 2005). Although there are many factors underlying this situation, especially family structure, child attachment and emotional relationship with the family, individual characteristics and age can be seen as the cause of emotional and behavioral problems during hospitalization (Baykara, Güvenir & Miral, 2011).

The reactions of the children towards the hospital are emotional and behavioral problems such as separation anxiety, sadness, fear, anxiety, anger- irritation, aggression, regression, depression, sleep problems, enuresis and finger sucking (Baykoç, 2006). In addition to these children's reactions, families' reactions to the hospital life and the disease can also occur. Giving education to families to resolve family fears and concerns, insights approached and could increase to be answered correctly collaboration of questions and will also be useful to parents in accepting children the process of parents (Beyazıt & Bütün-Ayhan, 2015).

Children's hospital in preparation for the life of the hospital is a great mission in the first place parents and the health professionals. Parents should provide information to their children in accordance with their child's developmental level, to understand the situation of the child, to take care to be with the child, and

DOI: 10.4018/978-1-7998-2952-2.ch006

to show the necessary sensitivity in the discipline (Baykoç, 2006). In this context, health personnel are responsible for determining the parents' information about the disease and the hospital and transferring the real and correct information to the parents, communicating effectively with the parents and preparing the children physically and psychologically (Altıntaş & Yıldız-Bıçakçı, 2015). Health personnel should welcome the child who comes to the hospital with a smiling face, introduce the environment to the child, inform the hospital about the rules, how long they will stay in the hospital, what should happen in this period and explain them in ways that are appropriate for the development of the child and prepare the children and family for surgery (Baykoç, 2006). In doing all this and in all of its practices, should act in accordance with the sick child rights (Purutçuoğlu, 2015).

Different programs are implemented to ensure the adaptation of children to the highly traumatic hospital life. Environmental regulation of children according to their developmental periods (Gönener & Gürak, 2009), children's participation in game rooms and hospital classes, providing children's preparatory trainings to the hospital, music and arts activities are some of them (Baykoç, 2006). Based on these considerations, the sections aim to examine the reactions of children to hospital life according to their developmental stages, the role of families and health personnel in preparing children for hospital life, and the programs used to prepare children for hospital life.

BACKGROUND

Hospital life and going to hospital for treatment are seen as a very important life experience that can cause stress and tension in all people. Although the individual knows that he/she will go to the hospital to solve the existing health problems and reach the standard of living, he/she can also react to limiting his/her life. Moreover, given the immaturity of children's development, this has to potential to further affect children. The developmental vulnerability of children also causes them to become ill more often and to go to hospital. When the reactions of the children to the hospital were examined, it was seen that they had separation anxiety, fear, anxiety, anger, regression behaviors, depression, sleep and eating problems (Baykoç, 2006). In this context, it is very important that children can get out of hospital life which is a very difficult life experience with the least harm.

Children get used to life in the hospital, including their families and especially adapt at the hospital, the medical staff have an important role and responsibilities. In particular, health personnel are critical in helping families help their children by reducing the stress of the family. Besides all of these children's development in the world and in Turkey as a medium term arrangement made according to the hospital, preparatory programs are implemented (Baykoç, 2006; Beyazıt & Bütün-Ayhan, 2015; Gönener & Gürak, 2009).

Considering all these, it is necessary to contribute to the protection of the psychological health of children by preparing the hospital life which has a very difficult and stressful life experience. In this context, this section it is aimed to focus on the reactions of the children to the hospital life and the things to be considered in preparing the children for the hospital life.

CHILDREN'S REACTIONS TO HOSPITAL LIFE ACCORDING TO DEVELOPMENTAL PERIODS

Going to the hospital and being treated the hospital is a very difficult and compelling situation for all individuals. Moreover, this may be even more stressful in relation to the immature age and developmental level of children (Gültekin & Baran, 2005). This stress can cause psychological problems after a certain period of time. However, these psychological problems are not the same in all children. Factors that may cause psychological problems are determined by a number of factors. These factors (Baykara et al, 2011):

Individual Differences

It has an important place in individual differences. Past experiences of children or familial factors play an important role. Especially the hospital experiences used while disciplining children and the reactions of the children who encounter the hospital as a result of intimidation of the children can be even more severe.

Developmental Period of the Child

Another important factor that determines the child's perception of the hospital is the developmental period of the child.

Hospitalization is a stressful phenomenon at any age and at any period. However, this may cause different reactions in children. The reactions of children during hospitalization can be examined in three dimensions. These are (Beyazıt & Bütün-Ayhan, 2015):

Separation Protest

In the separation protest, the child experiences extreme and intense anxiety due to separation from the parents and hospitalization. As a result of this anxiety, the child can cry and scream.

Sadness

As the child begins to accept hospitalization, the child becomes silent and enclosure.

Independence

It is a process in which the child is overly active, cheerful and chatty, and communicates and interacts with people around him.

Below the response of children to hospital life according to their developmental periods will be examined.

Reactions to Hospital Life in Infancy

Infancy is a period which is very important in human life and has the property of determining the future life. Although the basic needs of the baby are seen as simple and thought during this period, it is vital to meet the basic trust needs. As a result of satisfying the feeding, sleeping and toilet habits which are within the physiological needs of the child in time and with love, basic feelings of trust can be formed. The formation of a basic sense of trust can be an important investment for the baby to trust around him, and at the same time an investment in the future (Aral & Kadan, 2018). In the formation of the basic sense of trust in infancy, it is important that person who will meet these needs is the same, besides meeting their needs regularly and timely (Baykoç, 2006). However, meeting the needs of the baby in the hospital can be regulated according to the routines in the hospital and this situation may jeopardize the consistency and reliability of the baby and prevent the development of the basic sense of trust (Karaaslan, Beyazıt, & Bütün-Ayhan, 2015).

The hospital environment is also an environment, different machines, constantly changing people and a hustle and bustle dominate (Baykoç, 2006) In this case, it may life the baby's fear and anxiety, and harm of the basic sense of trust (Karaaslan et al., 2015).

After the age of two, babies are trying to explore their surroundings and want to move constantly with their developing walking skills. At the same time, two years of age are important in terms of the period in which toilet training is started to be gained in babies (Aral & Kadan, 2018). However, the freedom of movement of children who have to stay in the hospital is restricted and toilet training they start to receive may fail (Karaaslan et al., 2015).

Generally, reactions such as restlessness, crying, sleep disturbances, loss of appetite and overreacting to separation from the mother can be seen among the reactions of children under three years to the hospital (Baykoç, 2006).

Reactions to Hospital Life in Preschool Period

The preschool period is a period in which the child learns a lot of new knowledge, skills and experiences, and at the same time is experiencing relatively stagnant compared to the previous period but rapid progress in development. In the physical development of the child, the connections between the synapses in the brain cells are enriched by the stimuli they receive (Ceylan, 2017). This enrichment will form the basis of their future learning (Mayes &Cohen, 2002; Santrock, 2016). Especially between the ages of three and six, the child constantly observes his/her environment, likes to play and move, and asks questions to learn new situations (Bütün-Ayhan, 2017; Yıldız-Bıçakçı & Aral, 2017). However, since the cognitive development of children in this period is still in the pre-procedure period, their imagination is very rich and they show developmental features such as animation and artificialism (Bütün-Ayhan, 2017). In terms of social development, peer relationships, communication and interactions with the people around them become important in the preschool period (Fazlıcğlu, 2017). The fact that the child has to be hospitalized in such a period can cause both psychosocial stress due to the restriction of the movements of the child and different fear and anxiety reactions due to the imagination. Reactions such as eating and sleeping disorders, loss of intestinal control, excessive mobility, carelessness, sensitivity to external stimuli, distrust guilt and fear of imaginary beings can be seen between the ages of three and six (Baykoç, 2006).

Reactions to Hospital Life in School Period

Although the school period is relatively stagnant in development, it is a process where academic skills and socialization are intense. During this period, the child is preparing for adolescence with physical growth and development. At the same time, it is the first time that they meet the school for children who have not attended preschool education (Kılıç, 2018). When the child starts school, teachers and peers come to the forefront and the child wants to be successful especially in academic skills, sports and friendship relations, and the failures he may encounter may bring many problems for the child (Senemoğlu, 2013). In this context, the child's going to hospital for treatment or having to stay in hospital may cause emotional and behavioral problems in the child. So stay if you go to the children's hospital and to move away from the peer group, by ridicule among their peers, exclusion be, of having to go the hospital it can lead to remote remain and failures of course, if such a situation can deeply affect children (Baykoç, 2006). In the period from school to adolescence, children can express their reactions verbally, but they can resist treatment interventions (Başbakkal, Sönmez, Şen-Celasin, & Esenay, 2010).

Reactions to Hospital Life in Adolescence

Adolescence can be considered as a very special period. This specialty stems from both the developmental characteristics of adolescents and being the first step of stepping into adulthood. In adolescence, rapid development is experienced in all areas of development and the adolescent tries to keep up with this rapid change. In this period, especially two concepts, personal tale and imaginary audience come to the fore. Based on these concepts, which are the extensions of adolescent self- centeredness, the adolescent thinks that everyone is following him, and wants to be the most beautiful, attractive, most admired and handsome. He/she likes to take risks around him and he/she involved in behaviors that he/she will take risks with. He/she likes being alone, being together with technological tools, listening to loud music and playing a very important role in his/her peer group life (Santrock, 2016). It is known that the child may develop some emotional and behavioral problems as a result of going to hospital as a treatment or having to stay in hospital. First of all, the adolescent child sees this as an attack on his/her freedom and independence and tries to keep up with the changes in his/her body. However, he/she rebels because he/she contradicts the "width" of the adolescent child and express this rebellion in difficult ways. In addition to disregarding individuality, the fact that the hospital environment cannot reach out to friends at any time and cannot meet with them can also be a source of additional stress on the child during adolescence (Baykoç, 2006).

With the maturation of cognitive development in adolescence, children can behave like suppressing their emotions and avoiding confrontation with reality because they predict more accurately what the disease is, its consequences and how it can affect their lives (Başbakkal et al., 2010).

EMOTIONAL AND BEHAVIORAL RESPONSE OF CHILDREN IN HOSPİTAL

The most common emotional and behavioral reactions in children are given below.

Fear

It is a reaction that an individual uses to protect himself in the face of visible or invisible dangers and is necessary for the survival of life. Although fear is natural at any age, given that children do not know and do not know their surroundings, all kinds of new things around them are fearful (Bakırcıoğlu, 2013).

Infants and young children have quite a lot of fear. One of them is foreign fear. The fear of strangers emerges intensely when there are familiar and unusual people and environments around the baby and the child. In this context, hospital and hospitalization is an environment that triggers foreign fear. Because this environment is unknown and is far from the home environment where the child is see and does not resemble this environment at all. Therefore, the fear of foreigners is frequently seen in hospitalized children (Aujoulat, Simoneli & Deccoche, 2006; Başbakkal et al., 2010).

At the same time, the child has pain due to the wounds and illnesses caused by the disease. With these pains, the child is afraid of losing his/her body functions, being rejected by his/her friends for his/her illness, death or having surgery. In the light of this thought, he/she tries to eliminate these feelings with magic words (Gültekin & Baran, 2005). In addition, there are fears such as doctors, nurses, taking blood, examination, drinking medicine, and having an injection in children (Maraşuna & Eroğlu, 2013).

Anxiety

Anxiety is a common problem in hospitalized children. Children are worried about needles and painful procedures during treatment, medication children who experience anxiety intensely upset and excited more quickly and are constantly nervous and anxious (Foster &Park, 2012).

Anger and Rage

Anger and rage arising as a result of this, the child's wishes is a mood that occurs when blocked, often seen in hospitalized children. Because of the pain and suffering they experienced as children may feel anger because of their parents leaving them both in the hospital and can show this behavior with furious anger (Beyazıt & Bütün-Ayhan, 2015).

Depression

Depression is an emotional and condition seen in sick and hospitalized children. Depression, characterized by loneliness, uncertainty, separation, introversion, sleep and eating disorders, is a mood disorder that affects children as well (Abrams, Hazen & Penson, 2007). Children who are depressed cannot express their anger and resentment, and they become trapped, feel distressed, exhausted and may be without energy, they have difficulty concentrating (Mash & Wolfe, 2007; Saygılı, 2013).

Sleep Disorders

Night fears and nightmares can be seen in the hospitalized children when they are in the hospital and from the hospital and return home. While night fears appear as resistance when he/she wants to sleep with the mother, wanting to sleep with parents, not wanting to go to bed at night, nightmares can be related to separation from parents, hospitalization, accidents and injuries. It is seen that they express anxiously the fear of death especially in their dreams (Gültekin & Baran, 2005).

Regression

The act of acting as the child's age and returning to the period in which he/she feels safe. Finger sucking and enuresis are frequently seen in regression reactions (Karabekiroğlu, 2011; Parke & Gauvain, 2009).

Enuresis

Although the child has reached the age of gaining control of urine, making the enuresis again based on a number of reasons, is defined as a enuresis. Urinary or fecal incontinence may occur as a reaction to stressful events (Babaroğlu, 2018). It is seen that children can enuresis due to the stress of hospitalization (Beyazıt & Bütün-Ayhan, 2015).

Finger Sucking

Finger sucking that begins in the prenatal period can also be seen as a reflex after birth. Finger sucking, which has important functions in the baby's nutrition should be discontinued after one year of age (Shaffer & Kippy, 2014). However, children can also suck fingers when they feel fear, anxiety and feel unsafe and anxious. It is seen that children can suck fingers because of the emotions they feel in hospital life.

There are a number of programs and practices that are used to prepare children and their families for the highly traumatic hospital life and to overcome this process with minimal trauma. These programs and applications are given below.

DUTIES OF FAMILIES AND HEALTH PERSONNEL IN PREPARING CHILDREN FOR HOSPITAL LIFE

In preparing children for hospital life, first of all families and then health personnel have a great role and responsibilities. In this process, the points that should be taken into consideration by the families and health personnel are given below.

Roles and Responsibilities of Family in Preparing Children for Hospital Life

The effect of the family on hospital life is very important. The reactions of the family to the disease and the hospital determine the child's approach to the situation. Parents experience fear, anxiety as their children are sick and hospitalized. Under these emotions, their concerns about their children's recovery, their encounter with medical devices that they encountered for the first time in a hospital environment and that are so scary, that they think they are inadequate in their care, that they do not have enough and correct information about the disease, that they blame themselves for being sick, other family members staying at home are worried about their thoughts and economic problems (Altıntaş & Yıldız-Bıçakçı, 2015; Aykanat & Gözen, 2014; İnal-Emiroğlu & Pekcanlar-Akay, 2008). However, regardless of the cause and experience of the parents, such anxiety and fears may cause emotional and behavioral problems and treatment may become difficult (Baykoç, 2006). The family is worried and anxious being felt and experienced by the child as soon as the increase in intensity and does of it for the response of the child. In this context, it is important for the family to approach the child with understanding, listen to it, and intervene in the situation by leaving aside their own concerns and worries (Baykoç, 2006; Beyazıt & Bütün-Ayhan, 2015).

Parents have a great role and responsibilities in preparing the child for hospital life as well as hospital personnel. Parents should be aware of the fact that the child's behaviors are caused by the situation, allow the child to express his/her feelings and thoughts, not to underestimate the fears of the child and to stay away from excessive reactions (Mash & Wolfe, 2007; Saygılı, 2013). In addition to this, parents' acceptance of the disease firstly, not being pessimistic, not giving behaviors that give hope to the child, and giving explanations according to the age and development level of the children can help to overcome the rather traumatic life with less tremor. It is very important to make honest statements against the child and to answer the child's questions (Karabekiroğlu, 2011; Peykerli, 2003).

One of the mistakes that parents make after the child's hospitalization is that the child can do whatever he or she wants and behave extremely tolerant and loose. Such a situation may lead to disregard of the discipline and the child to use the situation. It may also result in the seeds of jealously among the siblings and the symptoms of disease in other siblings. Therefore, the rules and discipline in the family should continue to do so before the children's hospitalization, not to make too much sacrifice, to allow the child to do the child himself/herself, to make sure that the child understands the rules, to make the siblings very important (Abrams et al., 2007; Baykoç, 2006; Beyazıt & Bütün-Ayhan, 2015).

The child in the hospital may insist on a number of requests. He/she doesn't want his/her parents to leave, no matter what. Parents need to act with awareness of this situation, determine a common strategy against the child's behavior and act accordingly. Both mother and father should determine in advance how the child will react in case of moodiness, stubbornness or aggressive behavior, and behave according to this reaction and should not stretch in the rules (Karabekiroğlu, 2011).

As long as the child is in the hospital, he/she needs not only his/her parents, but also his/her family members, especially his/her siblings. In order for the child to feel abandoned, family visits should be continued. Activities such as reading books and painting by siblings, especially for children who are hospitalized for a long time or who cannot get out of bed, are among the applications that increase communication and interaction which are very useful (Baykoç, 2006; Beyazıt & Bütün-Ayhan, 2015).

Roles and Responsibilities of Hospital Staff in Preparing the Child for Hospital Life

Hospital staff have a big role and responsibilities in preparing the child for hospital life. Hospital staff play a key role both in the child's first admission to the hospital and in reducing parents' fears and concerns. Child developmental, doctor and nurse working together in a team spirit can increase the success in treatment and compliance (Altıntaş &Yıldız- Bıçakçı, 2015).

First of all, when the child and his/her parents come to the hospital the hospital staff should welcome them with a smile and understanding, make brief explanations about the environment, introduce them to the hospital staff and answer any questions (Baykoç, 2006). Hospital personnel should reduce the fear and anxiety of both parents and children. In this context, first of all, it is effective for parents to determine what they know about the disease, the course of the disease and the outcome of the disease, to provide parents with the right information, to actively participate in the care of their children and to stay with their children (Gönener & Görak, 2009).

It is also very important that the health personnel provide to preparatory training to the hospital in order for the parents to understand the child's situation, reactions and help their children (Baykoç, 2006). Generally, parents explain to the children what the hospital is, why it should be visited and what the duties of the employees are before they go to the hospital. However, it is important that the correct explanations are made according to the age and development level of the children. For this reason for the use of the tools and materials used in the treatment, the visiting hours and hospital rules (Altıntaş &Yıldız-Bıçakçı, 2015). When giving education to families, it should be explained that children's developmental and individual characteristics are different and explanations should be made according to developmental characteristics (Başbakkal et al, 2010).

Parents' training can be done either in writing or orally. Generally, in books and brochures, there may be introductory information about the hospital, perceptions and reactions of the hospital life according to the developmental characteristics of children, and information about the types of diseases. However, what needs to be done is that the doctor who decides to hospitalization should be give their family short and concise information about the situation reduce their anxiety and fears and then give written brochures (Baykoç, 2006).

In the statements that will be made by the health personnel for the child the children should be informed about the procedures to be applied and all the explanations to be made should be appropriate to the age and development level of the children. Children should also be informed that they can have a good time in the hospital, make use of the playroom and visit times and reduce their anxiety and fears. The child should not be subjected to harsh treatment in any way, and avoid making any statements or making false statements with the fear that his or her anxiety may increase or make treatment difficult. The child's reactions should be patient and understanding of his/her behavior, but especially his/her behaviors such as aggression should be prevented that will harm himself/herself or others. Medical staff should also learn from parents what the children's interests, wishes and daily routines are. This is especially important or children whose parents cannot stay in hospital. Again, children and their parents with fatal diseases should be given explanations in a way that they can understand without giving much hope. Another point that health personnel should pay attention is that they approach all children equally and do not discriminate otherwise, different emotional and behavioral reactions may occur in children. It is also very important that children do not speak about their illnesses and special situations, and that the examination and treatment procedures are performed in the treatment room. It is very important to talk to children from the beginning about how long they will stay in the hospital, what they will encounter as a result of this period and what they will experience during this period, and it helps the child to trust the health personnel and adapt to the hospital life more easily (Baykoç, 2006).

Preparing the children and their families or the diseases requiring surgery is also among the duties of the hospital staff. Although the preparation of children for surgery varies according to their development and age levels, it is important that children are given explanations and encouraged to ask question (Baykoç, 2006). In cases where surgery will take place during infancy, it is necessary to make a statement to the family, to treat the baby gently and to speak with a soft tone. It may be beneficial to have the baby with which the baby is accustomed and loved before and after the operation. Bringing families together with babies who have undergone similar surgery can reduce stress in the family. The family is informed about the medical devices such as catheter and tube that may be on the baby after surgery (Durualp, 2015).

Preschool children may think that surgery and hospitalization in connection with the immaturity of their cognitive development is done to punish them. Children should be particularly emphasized that the surgery was not performed to punish them, and why the operation had to be performed and what would happen before and after the operation should be explained to the child in accordance with their level of development. It is very important to make use of the play and answer the child's questions during the narration. It is important for the child to be told three to five days in advance that the operation will be performed and to inform them that no intervention will be performed before this time, in order to develop a sense of trust in the child (Ball & Bindler, 2003; Ünver & Yıldırım, 2013).

Children in school and adolescence can think more clearly about their situation and make sense of what is happening. Children should be informed about the operation date one or two weeks in advance and they should be provided with realistic information about how the operation will take place, whether they will be awake during the operation, when they stand up after the surgery, when they can eat, and how they will look after the operation (Durualp, 2015).

In addition to these, health personnel should act in accordance with the rights of the sick child what these rights are as follows (Purutçuoğlu, 2015):

A child lying in the hospital

- Receive treatment based on clinical needs
- Respect for privacy and dignity
- To receive treatment in a pediatric hospital, unit or ward suitable for age and medical condition with play and educational materials
- To be cared for by well-trained pediatric nurses
- Participate in discussions (if they have a level of understanding or reasoning) and take into account their wishes
- Introducing the caregiver
- Performing or supervising surgeries by experienced and experienced surgeons trained in childcare
- Expected to be anesthetized by experienced anesthesiologists trained in childcare

In the preparation of children for hospital life, importance is also given to environment regulations according to their development periods.

ENVIRONMENTAL ARRANGEMENTS ACCORDING TO DEVELOPMENT PERIODS

In order to adopt to hospital life, it may be important to configure the hospital environment according to the developmental characteristics of the children and in line with the available facilities.

The physical environment arrangements made according to development periods are given below.

Environmental Arrangements Made to Ensure the Adaptation of Infants to the Hospital During Infancy

Considering that the most important need in infancy is the basic sense of trust and the necessity of the baby to get used to regular and daily routine activities of the baby (Çavuşoğlu, 2013; Törüner & Büyükgönenç, 2012). At the same time, health personnel should talk to the baby in a reassuring, loving and calm tone and make eye contact. It is important not to make sudden and compelling movements to the baby to be patient or patient to act in terms of the development of trust. Care should be taken to keep the baby with the mother in the hospital, and if this is possible, keep a toy that the baby knows or a piece of belongings by the mother and father. Increasing the number and duration of visits as much as possible, placing the baby on the mother's lap or playing with her is very important for the development of trust. In addition to this, taking care of one to three or five babies, performing the treatment procedures when the baby is awake, talking with the baby, calming the baby after the painful procedures are important behaviors in terms of preparing the babies for hospital life (Baykoç, 2006; Doğan, 2010; Gönener & Görak, 2009).

Environmental Regulations for Preschool Children to Adopt to the Hospital

In the preschool period is very important the mother or the primary caregiver. For this reason, it is important that the child is in contact with the mother or primary care giver during the period of stay in the hospital, and if this is not possible the child is able to see family members and friends and bring flexibility to visit times and durations. Preschool period is also important in terms of being a period in which the child starts to gain self-care skills. However, the physical environment in the hospital may cause problems in meeting the toilet needs of the child. For this, it is very important that there are toilets and sinks close to the room where the child sleeps and suitable for the child's height. Thus, the child will support his/her own needs and support the feelings of independence and self- confidence. Ensuring that children are together at mealtimes is another environmental arrangement. In this way, both children continue their eating behavior and will try to alleviate the family life they have longed for. Again, preschool children's favorite things, feel themselves free, easily express their feelings and thoughts play environment. For this reason, the game rooms in hospitals are very important and allow the child to get used to the environment. Children are told how to use the playroom and at what time intervals they will go, not interfering with the child during play hours will help the child relax and find an opportunity to express himself (Baykoç, 2006).

Environmental Regulations for School Children to Adopt to the Hospital

School children want to be successful in their activities, feeling inferiority when they fail, and this situation may cause problems. Therefore, taking into consideration the disease of the child it is necessary to prepare activities that can be successful. Since peer relationships are very important during school years environments should be created to allow the child to come together with his/her peers, and if this is not possible, to talk by phone or internet. Academic success is also very important in addition to those expressed in children of this is period. The fact that the child has to be hospitalized causes him/her to fail to attend the courses, to fail his/her courses and in this case to cause academic failure and consequently to develop problematic behaviors in the child. Considering all these situations, hospital classroom practices are carried out in hospitals so that the child can continue his/her education immediately after discharge from hospital (Altıntaş & Yıldız-Bıçakçı, 2015; Baykoç, 2006).

Another of the developmental characteristics of the child during school period is the desire to take responsibility and the need to take into account the feelings and thoughts. For this, the environment should be introduced after the child is hospitalized, information should be given about the procedures to be performed and explanations should be made regarding hospital rules and visiting hours. Again, the participation of the child in his/her care to take responsibility, giving the child the chance to choose in the procedures to be made, making the bed in the ward and allowing the activities to arrange the room are among the useful application (Çavuşoğlu, 2013; Er, 2006; Gültekin & Baran, 2013).

Group discussions about school children together, living and feeling are among the activities that relieved children. Also, allowing children to bring their favorite books, pictures, toys, phones and computers with them to the hospital helps them to get out of this traumatic experience easily (Altıntaş &Yıldız-Bıçakçı, 2015).

School children, also try to prove that they are not afraid by appearing brave. However, it is very useful to ensure that parents are with the child during this procedures, and, especially after the painful procedures, allowing the child to express their feelings and thoughts by telling stories or painting (Altıntaş & Yıldız- Bıçakçı, 2015).

Environmental Regulations for Adaptation of Children in Adolescence

Adolescence is a risky period due to its developmental characteristics. The child in adolescence also pays attention to privacy, wants to create his personal space and expects the adults around him to pay attention to these issues. Consequently environments can be created for children who have to be hospitalized during adolescence where they can make small changes according to their interests (Baykoç, 2006).

Peer relationships easily are vital in adolescence. For this reason, it is necessary to create quiet environment where children can communicate with their peers and also should be allowed to use materials such as telephone and internet (Altıntaş &Yıldız-Bıçakçı, 2015).

Making use of children's interest in music, creating environments in which they may be interested in music, and ensuring their attendance to hospital classes in order to continue their academic success are among the effective practices in preparing adolescent children, for hospital experiences (Baykoç, 2006).

The following are the programs implemented in the world and Turkey in order to ensure their compliance with the children hospital.

HOSPITAL PREPARATORY TRAINING PROGRAMS

Hospital preparatory education programs are the practices to reduce the fears and worries of children due to hospital life and to make the treatment process more positive in order for these applications to succeed, some principles must be taken into consideration. These principles (Altıntaş & Yıldız-Bıçakçı, 2015):

- The developmental characteristics, interests and needs of children should be taken into consideration while preparing the programs.
- Individual differences of children should be taken into consideration in the education programs to be prepared.
- Attention should be paid to the needs and limitation of the disease.
- The training programs should take into account the environment and the possibilities of the hospital.
- In education programs, issues should progress gradually, starting with what the child knows.
- The education program should be structured in such a way that the child can understand both herself/himself and those around her/him.
- Play should form the basis of programs for preschool children.
- In education programs, besides the characteristics of the environment, familial characteristics should be taken into consideration.
- Effective participation should be ensured of children and families in education.

Although education is shaped according to age and developmental characteristics of children, there should be some common issues in preparatory education for the hospital. These issues are given below (Baykoç, 2006):

Health

Growth and development, organs in the body and functions of these organs, how to protect body health, environmental health issues are addressed.

Hospital

Information can be given about the departments in the hospital, game rooms, operating room, x-ray room, examination room, laboratories, pharmacy and the things done in these places can be given.

Hospital Life

In the hospital, rules such as feeding, sleeping, meeting toilet needs, visits, companionship, communication and games are mentioned.

Hospital Staff

The duties of the doctor, nurse, child developmentalist, psychologist, dietitian, social worker, administrative staff, secretary and other assistant personnel in the hospital should be explained to the child.

Instruments Used in the Hospital

The materials used during the treatment and examination are explained to the child about the materials that the child may encounter and what works.

Medical Procedures in Hospital

Children should be informed about which medical procedures are performed in the hospital and why.

The fact that are hospitalized preparatory activities both in Turkey and in the world. These activities given below:

Game

Game plays an important role in education programs. Play is the most important activity that enables the child to learn and make sense of the environment in which the child is surrounded by supporting all development areas. In this context, play in the hospital is very important (Aral & Doğan- Keskin, 2015). Playing in the hospital will both help the child to make sense of the environment that is alien to her/him and will help her/him to provide emotional relief by carrying his/her fear and anxiety to this environment (Baykoç, 2006). The combination of play and treatment can help the child to accept and cooperate (Kyle, 2008). In this context, the fact that importance is given to the establishment of play rooms in the hospital emerges (Aral &Doğan- Keskin, 2015). In order for the play rooms in the hospital to reach the stated aims, certain features must be taken into consideration. The first one is the characteristics of the game room and the other is the selection of the materials used. When arranging playrooms, physical and psychological environmental arrangements should be made to help children feel at home. Attention should be paid to colors and lighting, there must be heroes that children love on the walls, and the furniture should be suitable for children. The tables and chairs in the play room should be chosen according to the height of the children, the floors should be wooden, the play room should have a door opening to the garden and children should be directed to the garden games when necessary and to the extent that they allow diseases (Baykoç, 2006; Karalp, Dinç, & Teke, 2015).

Allowing children to play with both natural and real materials can help children express themselves. While the child exhibits his emotions, thoughts, fears, worries and frustrations in the play he established, he uses the play materials during this time. Therefore, care should be taken to ensure that play materials chosen to help children express themselves. The game materials to be used may be active game materials, dramatic game materials, medical game materials, game materials that bringing out art and creativity, and table game materials. Miniature houses should be allowed to have miniature garden toys. Hospital game room should be available crayons, paints, plastron, salt ceramics, paper, water and sand, house toys, various kitchen utensils, sitting and bedding, telephone, dolls, clothes, cover, washing rope, pegs, carpenter tool, car, truck, dominoes, matching playing cards, Legos, nicknames, unloading-stuffing toys, touch books, colored picture, story books, dictionaries, atlas, children's novels, newspapers and magazines should be available (Baykoç, 2006).

During the play, children are also allowed to express their feelings, thoughts and dreams through projective methods such as drawing pictures and completing sentences (Çavuşoğlu, 2013, Şen-Beytut, Balışık, Solak, & Seyfioğlu, 2009; Törüner & Büyükgönenç, 2017).

Language Activities, Reading Books, Communication Letters

Another activity that is as important as the game is reading. Picture books play an important role in preschool children learning information. According to the age and developmental characteristics of children can be kept in the preparatory education program story books, books novels and biographies for the hospital and the disease. In addition, there are communication boxes that children can report to the health personnel by writing their feelings, thoughts, wishes and requests or drawing their pictures for preschool children (Baykoç, 2006; Cihangir-Altay, 2008).

Music Events

In the hospital, music is used effectively to relieve pain, provide relief and relieve depression. Music contributes to the motor, cognitive and social development of children and reduces muscle tension. Considering all these effects, musical instruments should be available in the playrooms, music broadcasts in the examination room or patient rooms, and the child should be allowed to perform, rhytm exercises and dance. At the same time reading poetry, as can be utilized in children's voices singing, music can also be done for children in adolescence (Baykoç, 2006; Hendon & Bohan, 2008).

Art Activities

Art activities are another activity that helps children to relax by revealing their emotions, thoughts and dreams. In the hospital, activities such as painting, sculpture and ceramics are used to support both the quality of life of children and help the child to express himself/herself easily and to free up energy and relax (Baykoç, 2006; Teksöz & Ocakçı, 2014).

Theater and Clown Activities

In addition to performing theater plays at the hospital, the arrival of clowns to the hospital not only increases the humerous elements, but also helps the child to relax and helps to strengthen the communication and interaction between the child, child's parents and hospital staff. At the same time, these types of activities are very important in coping with pain and suffering (Baykoç, 2006; Koç, 2011; Teksöz & Ocakçı, 2014).

Computerized Activities

In today's technological world, computers have an important place in children's lives. For this purpose, computer-related activities should be included in education programs and children should be allowed to use the computer and the internet easily (Altıntaş & Yıldız-Bıçakçı, 2015).

Training and Discussion Group Activities

In this activity which especially for school and adolescent children, children feel that they are understood, belong to a group and get rid of their feelings of loneliness. In this activity, the questions, thoughts, feelings, and wishes of the children about the hospital life and illness are shared with the health personnel. Thus, children develop their coping mechanisms by seeing their peers with the same problem (Baykoç, 2006).

Meeting Activities

Meeting activities are also very important. When the child and his/her parents come to the hospital, introducing them to the environment, introducing themselves and other health care personnel, learning the child feel comfortable. Feeling comfortable, the child feels confident and affects the treatment process positively (Er, 2006).

Hospital Classes

Hospital schools or hospital classes are opened in order to continue the education of pre-school, primary and high school children who have inpatient treatment, who have chronic illness and who cannot attend school for a long time. Teachers can take part directly in the hospital classrooms or continue their education in the form of distance education depending on the developing technology and technological development (Yılmazer, Yıldız, & Şahinöz, 2015). Hospital classes are of great importance for the continuation of academic success, which is especially important for children in school. In a hospital, at least ten students and sufficient space are required to open a hospital class (Baykoç, 2006). At the same time, parents' and physicians' opinions are taken in order to attend the hospital class, education services are provided in their rooms for children who cannot come to the class, and pre-school, classroom teacher, field teacher and special education teachers are given training by the Directorate of National Education. Duration of the lesson is decided by considering the child's health status (Baykoç, 2011).

In addition, supporting families is very important in preventing depression. Meetings on children's situations help families with similar experiences to develop a coping mechanism, and through the experts attending these meetings, families are provided with psychological support and help to find appropriate solutions for their situations (Yüzer, Yiğit, & Taşdelen, 2006).

In addition, preliminary admission programs are applied in the preparatory training programs for hospitals around the world. In these programs families are given an appointment one week before the child is admitted to the hospital, the hospital environment is introduced to children and their parents, books about the hospital life are read and movies are watched. Family trainings are organized on the parent's approach to the child. Children are closely monitored and examined by the medical devices in the hospital, and the process of introducing the child to the hospital staff is performed. After the child is admitted to hospital, flexibility is provided during the visiting hours. Help the child relax in the large garden allocated to the child. In the large and private rooms in the hospitals, there are big screen televisions for children and their families, materials for arts and crafts, computer and board games connected to the internet, sketches, game shows, theater, music and dance performances are provided and music and game therapy services are provided (Altıntaş &Yıldız- Bıçakçı, 2015).

CONCLUSION

Although the disease and hospital life affect every human being deeply it can affect children differently due to the characteristics of their developmental areas. Under these different considerations, the characteristics of the child's developmental period may be found, as well as the family environment in which the child is raised and individual differences in children. Children may show some common reactions, regardless of their developmental period. These reactions are fear, anxiety, sadness, anger, aggression and depression. In addition to these children's reactions, their parents can also experience feelings of fear, anxiety and regret because of their condition and cannot help their children.

The healthcare staff has a great responsibility to prepare children and parents for hospital life. Health personnel should be able to identify the parents' knowledge of the situation, communicate the correct information to the parents, and be able to approach parents and children with understanding and patience in their emotional and behavioral responses. At the same time, they should welcome the children and introduce the environment and provide realistic information appropriate to the age and development level of the children. Health personnel should be able to make the principle of acting in accordance with the rights of sick children in all practices and initiatives.

Sick children practiced in the world and in Turkey between programs delivering training for hospital and illness to prepare the hospital experience, game rooms, environmental regulations, reading, theater music, photos, discussion, situated arts events and hospital classes.

REFERENCES

Abrams, A., Hazen, E., & Penson, R. (2007). Psychosocial issues in adolescents with cancer. *Cancer Treatment Reviews*, *33*(7), 622–630. doi:10.1016/j.ctrv.2006.12.006 PMID:17434265

Altıntaş, T. T., & Yıldız-Bıçakçı, M. (2015). Çocukların hastane yaşantısına hazırlanması ve hastaneye hazırlayıcı eğitim. In A. Bütün-Ayhan (Ed.), Hasta çocukların gelişimi ve eğitimi (pp. 80-109). Eskişehir: Anadolu Üniversitesi.

Aral, N. & Doğan- Keskin, A. (2015). Hastanede oyun ve oyun odaları. In A. Bütün-Ayhan (Ed.), *Hasta çocukların gelişimi ve eğitimi* (pp. 138-159). Eskişehir: Anadolu Üniversitesi.

Aral, N., & Kadan, G. (2018). Sosyal gelişim. In N. Aral & Z. F. Temel (Eds.), *Çocuk gelişimi* (pp. 236–262). Ankara: Hedef.

Aujoulat, I., Simanelli, F. & Deccache, A. (2006). Health promotion needs of children and adolescents in hospitals: A review. *Patient Education and Counseling, 61,* 23-32 doş:. doi:10.1016/j.pec.2005.01.015

Aykanat, B., & Gözen, D. (2014). Çocuk sağlığı hemşireliğinde aile merkezli bakım yaklaşımı. *Gümüşhane Üniversitesi Sağlık Bilimleri Dergisi, 3*(1), 683–695.

Babaroğlu, A. (2018). *Çocuk psikolojisi ve ruh sağlığı. (5. Baskı).* Ankara: Vize.

Bakırcıoğlu, R. (2013). *Çocuk ve ergenlerde ruh sağlığı.* (5. Baskı). Ankara: Anı.

Ball, J. W., & Bindler, R. C. (2003). *Pediatric nursing caring for children.* Upper Saddle River, N.J: Pearson Education.

Başbakkal, Z., Sönmez, S., Şen, N., & Esenay, F. I. (2010). 3-6 yaş grubu çocukların hastaneye yatışa karşı davranışsal tepkileri konusunda verilen eğitimin annelerin anksiyete düzeyi üzerine etkisinin incelenmesi. *Atatürk Üniversitesi Hemşirelik Yüksekokulu Dergisi, 12*(4), 59–65.

Baykara, A., Güvenir, T., & Miral, S. (2011). Hastalık ve hastaneye yatışın çocuk üzerinde etkisi. In A. Ekşi (Ed.), *Ben hasta değilim* (pp. 398–403). Ankara: Nobel.

Baykoç, N. (2006). *Hastanede çocuk ve genç.* Ankara: Gazi.

Baykoç, N. (2011). Uzun süreli (süreğen) hastalığı olan çocuklar ve eğitimleri. In N. Baykoç (Ed.), *Öğretmenlik programları için özel eğitim* (pp. 276–300). Ankara: Gündüz.

Beyazıt, U. & Bütün- Ayhan, A. (2015). Hasta çocukların ruhsal özellikleri ve hasta çocuğa yaklaşım. In. A. Bütün- Ayhan (Ed.). *Hasta çocukların gelişimi ve eğitimi* (pp. 58-78). Eskişehir: Anadolu Üniversitesi.

Bütün-Ayhan, A. (2017). Bilişsel gelişim. In N. Aral & T. Duman (Eds.), *Eğitim psikolojisi* (pp. 104–128). Ankara: Pegem.

Ceylan, R. (2017). Fiziksel gelişim. In N. Aral & T. Duman (Eds.), *Eğitim psikolojisi* (pp. 56–73). Ankara: Pegem.

Cihangir-Altay, N. (2008). Çocuklarda ameliyat öncesi hazırlık. *Sağlık Bilimleri Fakültesi Hemşirelik Dergisi,* 68-76. doi:10.14527/9786052410240.03

Çavuşoğlu, H. (2013). *Çocuk sağlığı hemşireliği 1.* Ankara: Sistem.

Doğan, B. (2010). *Hastanede yatan çocuğun bakımına ebeveynlerin katılımı konusunda hemşirelerin, doktorların ve ebeveynlerin görüşleri ve beklentileri.* Ankara: Başkent Üniversitesi/Sağlık Bilimleri Enstitüsü.

Durualp, E. (2015). Çocukların ve ailelerin tanı, tedavi ve ameliyat sürecine hazırlanması. In A. Bütün-Ayhan (Ed.), *Hasta çocukların gelişimi ve eğitimi* (pp. 112–136). Eskişehir: Anadolu Üniversitesi.

Er, M. (2006). Çocuk, hastalık, anne-babalar ve kardeşler. *Çocuk Sagligi ve Hastaliklari Dergisi, 49*, 155–168.

Fazlıoğlu, Y. (2017). Sosyal gelişim. In N. Aral & T. Duman (Eds.), *Eğitim psikolojisi* (pp. 154–176). Ankara: Pegem. doi:10.14527/9786052410240.07

Foster, R., & Park, J. (2012). An integrative review of literature examining psychometric properties of instruments measuring anxiety or fear in hospitalized children. *Pain Management Nursing, 13*(2), 94–106. doi:10.1016/j.pmn.2011.06.006 PMID:22652282

Gönener, D., & Görak, G. (2009). Okul yaş grubu çocukların hastane ve hastalığı ile ilgili Bilgilendirme Durumlarının Endişe Kaynakları ile Etkileşimi. *Gaziantep Tıp Dergisi, 15*(1), 41–48.

Gültekin, G., & Baran, G. (2005). Hastalık ve çocuk. *Aile ve Toplum, 7*(2), 61–68.

Gültekin, G., & Baran, G. (2013). 9-14 yaş grubu akut ve kronik hastalığı olan çocukların denetim odağı düzeylerinin incelenmesi. *Türk Pediatri Arşivi, 40*, 211–220.

Hendon, C., & Bohan, L. M. (2008). Hospitalized children's mood differences during play and music therapy. *Child: Care, Health and Development, 34*(2), 141–144. doi:10.1111/j.1365-2214.2007.00746.x PMID:18257785

İnal-Emiroğlu, F.N. & Pekcanlar- Akay, A. (2008). Kronik hastalıklar, hastaneye yatış ve çocuk. *DEU Tıp Fakültesi Dergisi, 22*(2), 99-105.

Karaalp, T., Dinç, P., & Teke, A. (2015). Sağlık yapılarında iyileştirme önerilerinin geliştirilmesi: Çocuk psikiyatrisi kliniği örneği. *Anatolian Journal of Clinical Investigation, 9*(3), 121–138.

Karabekiroğlu, K. (2011). *Çocuğuma nasıl davranmalıyım?* İstanbul: Say.

Karaaslan, T., Beyazıt, U., & Bütün-Ayhan, A. (2015). Gelişim dönemler açısından hasta çocuklar ve hastane ortamının düzenlenmesi. In A. Bütün-Ayhan (Ed.), *Hasta çocukların gelişimi ve eğitimi* (pp. 26–56). Eskişehir: Anadolu Üniversitesi.

Kılıç, Ş. (2018). Orta çocukluk dönemi. In. F. A. Güngör-Aytar (Ed.), Gelişim psikolojisi (pp. 118-158). Ankara: Hedef.

Koç, S. (2011). Hastane palyaçolarının ameliyat öncesi dönemdeki çocuklar ve ebeveynlerinin anksiyete düzeylerine etkisinin belirlenmesi. *Hemşirelikte Eğitim ve Araştırma Dergisi, 8*(3), 26–31.

Kyle, T. (2008). *Essentials of pediatric nursing*. Philadelphia, PA: LWW.

Maraşuna, O., & Eroğlu, O. (2013). Ortaokul öğrencilerinin tıbbi işlem korkuları ve etkileyen faktörler. *Güncel Pediatri, 11*(1), 13–22. doi:10.4274/Jcp.11.03

Mash, E., & Wolfe, D. (2007). *Abnormal child psychology* (3rd ed.). USA: Wodsworth Learning.

Mayes, L. C., & Cohen, D. J. (2002). *Çocuğunuzu anlama rehberi*. İstanbul: Beyaz Balina.

Parke, R., & Gauvain, M. (2009). *Child psychology: A contemporary viewpoint* (8th ed.). USA: McGraw Hill.

Peykerli, G. (2003). Ölümcül hastalıklara psikolojik yaklaşım. *C.Ü. Tıp Fakültesi Dergisi, 25*(4), 62–65.

Purutçuoğlu, E. (2015). Hasta çocuk hakları: Hasta çocuğa ve ailesine yönelik sosyal destek. In A. Bütün-Ayhan (Ed.), Hasta çocukların gelişimi ve eğitimi (pp. 184-213). Eskişehir: Anadolu Üniversitesi.

Santrock, J. W. (2016). *Yaşam boyu gelişim (13. Baskı)*. Ankara: Nobel.

Saygılı, S. (2013). *Çocuklarda davranış bozuklukları*. İstanbul: Elit.

Senemoğlu, N. (2013). *Gelişim, öğrenme ve öğretim: Kuramdan uygulamaya*. Ankara: Yargı.

Shaffer, D., & Kipp, K. (2014). *Developmental psychology: Childhood and adolescence*. USA: Wodsworth Cengage Learning.

Şen-Beytut, D., Bolışık, B., Solak, U. & Seyfioğlu, U. (2009). Çocuklarda hastaneye yatma etkilerinin projektif yöntem olan resim çizdirme yoluyla incelenmesi. *Maltepe Üniversitesi Hemşirelik Bilim ve Sanatı Dergisi, 2*(3), 35–44.

Teksöz, E., & Ocakçı, F. A. (2014). Çocuk hemşireliğinde sanat uygulamaları. *Dokuz Eylül Üniversitesi Hemşirelik Yüksekokulu Dergisi, 7*(2), 119–123.

Törüner, E. K., & Büyükgönenç, L. (2012). *Çocuk sağlığı temel hemşirelik yaklaşımları*. Ankara: Göktuğ.

Ünver, S., & Yıldırım, M. (2013). Ameliyat sürecinde çocuk hastaya yaklaşım. *Güncel Pediatri, 11*, 128–133. doi:10.4274/Jcp.11.35229

Yıldız-Bıçakçı, M. & Aral, N. (2017). Dil gelişimi. In. N. Aral & T. Duman (Eds.), Eğitim psikolojisi (pp. 132-150). Ankara: Pegem.

Yılmazer, Y., Yıldız, S., & Şahin-Öz, A. (2015). Hasta çocuğa yönelik eğitim düzenlemeleri. In A. Bütün-Ayhan (Ed.), *Hasta çocukların gelişimi ve eğitimi* (pp. 162–183). Eskişehir: Anadolu Üniversitesi.

Yüzer, S., Yiğit, R., & Taşdelen, B. (2006). Çocuğu hastanede yatan annelerin aldığı sosyal destek ile depresyon düzeyleri arasındaki ilişki. *Atatürk Üniversitesi Hemşirelik Yüksekokulu Dergisi, 9*(4), 54–62.

KEY TERMS AND DEFINITIONS

Game in the Hospital: It is an activity that is used especially for preschool children to adapt to the hospital.

Hospital Classes: They are necessary units for the continuation of academic success of children who are sick during school adolescence and cannot attend school for long time

Chapter 7
Children With Adaptation and Behavior Problems

Ebru Hasibe Tanju Aslişen
https://orcid.org/0000-0003-3306-2438
Ankara Hacı Bayram Veli University, Turkey

ABSTRACT

In this chapter, the adaptation and behavior problem is explained conceptually, and detailed information is given about causal factors related to biology, family, and school. Particularly in the preschool period, adaptation and behavioral problems such as thumb-sucking, nail-biting, enuresis, encopresis, masturbation, lying, aggression, stubbornness, jealousy, fear, anxiety, stealing, sleep disorders, and eating disorders are explained in this chapter, and precautions to reduce/prevent these problems are mentioned.

INTRODUCTION

Adaptation is the ability of individuals to meet their needs; to realize their wishes; to establish balanced relationships with other individuals in their society and to maintain these relationships in a healthy way (Yavuzer, 2012a; Bakırcıoğlu, 2011). Accordingly, it is expected that adaptive children will exhibit some mental, physical, social, emotional, and spiritual behaviors required by their age and characteristics (Yılmaz Ünal and Deniz, 2018).

Children may encounter many problems including some adaptation and behavioral problems while gaining various skills and adapting to their environment during their developmental period. Adaptation and behavioral problems are disorders that are mostly observed as emotional reactions caused by the adverse effects of the environment together with the inherent difficulties of developmental stages (Yılmaz Ünal and Deniz, 2018; Yörükoğlu, 2011). The child with rapid development and learning process may exhibit some unwanted behavior patterns. It is essential to note whether the observed behaviors are indeed indicative of a problem or a part of the child's development. Therefore, when defining and intervening in the behavior of preschool children as problem behavior or guiding children for self-regulation and cooperation, it is necessary to consider the developmental level of children as well as their needs (Tanju, 2011; Campbell, 1995).

DOI: 10.4018/978-1-7998-2952-2.ch007

CAUSES OF ADAPTATION AND BEHAVIORAL PROBLEMS

The studies investigating the causes of adaptation and behavioral problems revealed the causes as age, gender, temperament, developmental level, family life, parents' educational levels, changes in the family, school environment and teacher relations etc. (Genç, 2018; Işık, 2018; Ulu, 2018; Özbey, 2015; Aydın, 2008; Weaver, Shaw, Dishion & Wilson, 2008; Fields & Boesser, 2002). They are discussed in three main categories.

Biological Factors

There are biological risk factors for behavioral problems, including the mother's state of health before pregnancy; surplus weight of the mother; her history with drug abuse; her malnutrition; her age (being very young or old during pregnancy), viral infections; lead and mercury poisoning; baby's lack of oxygen at birth because of mother's alcohol and drug abuse; birth by forceps or vacuum extractor; premature birth or low birth weight (Işık, 2018; Ulu, 2018; Drossos, 2004). Also, inadequate intake of vitamins and minerals or excessive consumption of sugar or caffeine can lead to apathy, indifference, social withdrawal and school failure (Kauffman, 2001; Dawkins, Fullilove, & Dawkins, 1995).

Temperament, another biological factor, is the biological behavioral tendency that is innate (Kaner, 2009) and affects the way parents perceive their children correctly. Cheerful, mild-mannered, harmonious children make their parents happy and confident, while angry children who are always whining, in any way dissatisfied, can cause parents to doubt their parenting skills (Genç, 2018; Çetintaş, 2015).

Familial Factors

The first years of a child's life take place at home and personality traits are shaped in this period. Children may develop appropriate behaviors when their parents are sensitive to their needs and create an atmosphere of love and trust at home. The opposite attitudes may cause adaptation problems in children. Family-related risk factors which include: indifference towards children, lack of discipline or severe punitive discipline, inadequate parent-child interaction, wrong parent attitudes, broken (abandonment / death) families, problematic marriages, family size, child's birth order, unemployment, poverty, alcohol or substance abuse, domestic violence and child neglect / abuse. These factors cause many behavioral and adaptation problems in children, especially the tendency to commit crime, aggression, and regressive behaviors (thumb-sucking, enuresis, etc.). (Genç, 2018; Özbey, 2015; Taner Derman and Başal, 2013; Yumuş, 2013; Driscoll & Pianta, 2011; Dursun, 2010; Kaner, 2009; Hughes, 2003; Eripek, 2002; Walker & Golly, 1999; Kazdin, 1998; Çağlar, 1981).

School-Related Factors

The school is an important socialization institution where children spend a great deal of time, acquire and develop many skills cognitively, emotionally and socially, and establish social relationships with their peers and teachers that shape the personality development of children significantly. Additional behavioral problems in school include: ineffective teaching methods that cause academic failure, inexplicit school rules, punitive disciplinary techniques applied inconsistently for problem behavior, teacher's inadequate reinforcement of appropriate social and academic behavior (Yumuş, 2013; Pierce, Reid & Epstein, 2004;

Gunter, Coutinho & Cade, 2002; Kauffman, 2001; Dwyer, Osher & Hoffman, 2000; Sprague & Walker, 2000; Lago-Delello, 1998; Mayer, 1995).

POSSIBLE ADAPTATION AND BEHAVIORAL PROBLEMS IN PRESCHOOL PERIOD

Thumb-Sucking

Sucking behavior is one of the most powerful innate reflexes babies learn in the womb. Babies, which can suck non-nutritional objects such as fingers, pacifiers, toys, take pleasure and satisfaction from this instinctive behavior and accelerate their motor development at the same time. Around the age of two, sucking can be associated with hunger, sleep, anger, and fatigue. Thumb-sucking is normal developmental behavior that can be seen up to three to four years of age. It usually occurs at night with behaviors such as twisting and pulling hair, rubbing the cheek, sucking blankets or covers. Daytime sucking behavior can often be accompanied by relaxation activities such as watching television or listening to stories (Seven, 2018; Aral and Gürsoy, 2007). Research has shown that thumb-sucking behavior is not harmful when it ends at the age of five or six. If it persists for a longer time, it will lead to abrasion and darkening of the finger and deterioration of the teeth and palate structure. Girls seem to do it more than boys (Öz, 2003; Brunet & Sarfati, 2000; Bulut, 2000).

Another behavior similar to thumb-sucking is tongue thrusting, which can often occur when the child is detained from thumb-sucking. Children try to comfort themselves by thumb-sucking or tongue thrusting when they cannot express their mental state by speaking and feel helpless with various problems (Aral and Gürsoy, 2007; Çağlar, 1981).

There are many opinions about the causes of long-term thumb-sucking behavior. Some claim that sucking is a learned habit, while others claim that it is a form of expression driven by infantile impulses. However, the most common view is that it is a self-relieving behavior (Seven, 2018). The possible factors causing thumb-sucking include: loss of a loved one because of separation or death; being criticized, reprimanded, mocked and compared with others; getting beaten; living under strict discipline; not feeling 'safe' or 'loved' as a result of lack of care and love from parents; being kept away from parents for a long time; being jealous of the siblings; witnessing or experiencing traumatic incidents like an earthquake, car accident or a fight (Bakırcıoğlu, 2007; Öz, 2003). Moreover, thumb-sucking can develop in consequence of early/late weaning and not using a nursing bottle or pacifier as it can also occur as a sign of embarrassment or shyness (Aral and Gürsoy, 2007).

To prevent thumb-sucking behavior;

- Understand and eliminate the cause,
- Consider that it will decrease with age, it should be ignored at the beginning, and the child is expected to give up instinctively,
- Reducing the conflicts and tensions in the family; the child should easily share his thoughts and feelings with parents, and feel secure in the family,
- Avoiding sanctions such as applying peppers on the finger, tying the hands, putting on gloves; avoiding insistent and intolerant attitude towards thumb-sucking; not condemning, reprimanding, blaming or mocking the child.

- Supporting the child in his lack of self-confidence in communication with friends and adults,
- Giving responsibility to the child for the active participation and socialization by the teacher in the face of behaviors such as shame, guilt, and boredom,
- Avoiding situations that allow the model to take thumb sucking behavior from a friend or adult, telling the harms of the behavior to the oral and dental health,
- It should be ensured that the child is distracted when s/he is thumb-sucking (Seven, 2018; Aral and Gürsoy, 2007; Bakırcıoğlu, 2007; Kapçı, 2005; Brunet & Sarfati, 2000; Çağlar, 1981).

Nail-Biting

Nail-biting is used for the act of gnawing or picking the nail or the skin around the fingers. As a result of the behavior, problems such as flushing, inflammation, bleeding, and infection may be encountered, and the ingestion of an organ with a high probability of carrying germs may cause some health problems in children. As a result of the behavior, problems such as redness, inflammation, bleeding, and infection may arise, and microbial health problems may occur. It is reported that nail-biting behavior, which is usually seen after three to four years of age, continues until adulthood in some cases (Seven, 2018; Aral and Gürsoy, 2007).

Nail-biting is a learned behavior and emerges to reduce stress and tension, to get rid of intense anxiety and distress, and in some children, it is seen as a imitation of family members (Yılmaz Ünal and Deniz, 2018). Furthermore, nail-biting can be encountered in cases of excessive pressure and authoritarian attitude in the family, together with constant reprimanding and criticism, jealousy, insufficient attention, and affection.

To prevent nail-biting behavior;

- The behavior should be ignored after a while by the parents; the causal factors should be identified and eliminated at the same time as the results are explained calmly,
- Encouraging the child to share his /her thoughts and feelings easily and giving the responsibilities that he/she can achieve to increase his /her self-confidence,
- Strict disciplinary, repressive, authoritarian and intolerant attitudes and conflicts in the family should be reduced as much as possible,
- The use of pepper on the nails, wearing gloves, and reactions such as reprimand, blame and rebuke should be avoided,
- The strong and successful aspects in the family and school environment should be emphasized; participation in social activities and sports should be supported
- Children should always feel they are loved and valued in the family (Seven, 2018; Aral and Gürsoy, 2007; Kapçı, 2005; Bakırcıoğlu, 2007; Brunet & Sarfati, 2000; Çağlar, 1981).

Enuresis

Toilet training is one of the developmental tasks such as walking, running, eating, and is associated with the development of the nervous system and bladder muscles (Kapçı, 2005). Therefore, the occasional enuresis (bedwetting) should be considered normal until the age of four, and the child should not be put under pressure (Aral and Gürsoy, 2007).

Enuresis is incontinence in children, at least five years of age, in a repetitive, involuntary or purposeful manner, day and/or night in bed or clothing, and is seen at least three consecutive months and twice a week (Erermiş and Aydın, 1999a). There are two types of enuresis. If the child has never gained bladder control and the continue bedwetting after five years of age, it is called primary enuresis. If the child starts wetting again after being dry for at least a year for psychological reasons, it is called secondary enuresis. Additionally, enuresis is divided into two categories depending on the time of the wetting; if it is at night it is called nocturnal, and if it happens during daytime it is called diurnal; while total or continual enuresis occurs both day and night (Rathus, 2006; Kapçı, 2005; Ersoy and Avcı, 2001; Çağlar, 1981).

If the bedwetting behavior lasts long, it is necessary to check whether there is an organic disorder. Feverish diseases, diabetes, kidney, and urinary tract infections can lead to this condition, and if treated, the problem disappears (Aral and Gürsoy, 2007; Kapçı, 2005; Erermiş and Aydın, 1999a). Moreover, the primary physiological reasons are; the inability of the child to receive waking signals related to bladder overflow due to deep sleep, below average bladder capacity and constipation as a result of insufficient secretion of the hormone which reduces the discharge of water from the kidney (Seven, 2018, Aral and Gürsoy, 2007).

The cause of the primary enuresis may be related to the development of the nervous system due to hereditary factors. The second type of enuresis may be caused by a newborn sibling, parental loss/separation, change of teacher, deprivation of love, oppression of the child and physical traumas (Seven, 2018; Kapçı, 2005; Ersoy and Avcı, 2001). If the involuntary enuresis is seen with problems such as aggression, disobedience, uncontrolled impulses, tic disorder, stuttering, it means that the child needs treatment. Enuresis alone should not be regarded as psychological incompatibility, but rather should suggest a physiological problem (Yörükoğlu, 2011; Ersoy and Avcı, 2001).

To prevent enuresis;

- Receive medical treatment if the cause is organic,
- Give and maintain toilet training with a gentle attitude at the most appropriate time for the child,
- Collaborate with the child for the behavior to fade away and to avoid public embarrassment,
- Avoid neglecting the child when s/he has a newborn sibling and to ensure the continuity of the love and interest,
- Make children participate in social activities as much as possible and assisting them to go to the toilet whenever needed,
- Keep the feet warm and not to use diapers to give a sense of cleanliness and dryness (Seven, 2018; Aral and Gürsoy, 2007; Kapçı, 2005; Çağlar, 1981).

To cure enuresis family counseling, psychotherapy, behavioral approach, and medical treatments are used. Successful results can be achieved in the treatment of psychological enuresis caused by loss of a parent or another traumatic experience by keeping a record and rewarding, fluid restriction and wakening the child in the middle of the night, bladder exercises and the use of alarm devices (Ersoy and Avcı, 2001; Erermiş and Aydın, 1999a).

Encopresis

Encopresis happens when a child defecates in her/his pants or any other inappropriate place even though s/he is old enough to control his/her bowel movements. To be accepted as a problem: the child should be four years of age or older; the behavior should repeat itself for at least once a month for three months; the child is seen voluntarily or involuntarily defecate in inappropriate places; the behavior shouldn't occur due to the direct effect of laxatives or a medical condition (except constipation) (American Psychological Association, 2001). Inadequate and inconsistent toilet training and psychosocial stress are considered as preparatory factors for the occurrence of this disorder, and it is divided into two. The primary type of encopresis is the inability to control bowel movement due to inadequate and loose training, while the second type is the deterioration of regulated defecation, usually due to psychological factors.

Encopresis can occur for many organic, social, and emotional reasons. Constipation, various bowel problems, defecation dynamics problems are among the organic causes. Repressive attitudes in toilet training, disturbance in parent-child relations, the disintegration of the family, a newborn sibling, aggression, shyness, weakening of intestinal control because of anxiety and stress are also major social and emotional reasons (Seven, 2018; Ersoy and Avcı, 2001).

To prevent encopresis;

- Firstly, whether the child has an organic disorder should be determined by a physician, and the treatment should be carried out,
- Avoid unnecessary oppression, punishment, and violent practices, excessive rigorous attitudes; ensure the child sits in the toilet 3-4 times a day, especially after meals,
- In cases where the family relations are very bad, child mental health departments should be consulted to eliminate the problems or tensions (Seven, 2018; Yörükoğlu, 2011; Ersoy and Avcı, 2001; Erermiş and Aydın 1999).

Masturbation

Childhood masturbation, which is more common in girls, is a condition that annoys parents even though it is not a medical disorder and manifests itself as signs of discovering one's sexuality with the symptoms of redness, sweating and frequent breathing during stimulation of the child's genitalia (Özakar and Gözen, 2012). Children discover their genitals in the first four months of their lives. The transition to eager masturbation begins accidentally with the process of using diapers and cleansing, and by the end of two years, it becomes a disconnected, self-directed behavior (Öz, 2003; Ünal, 1999). From the age of three to five, the movements seen in the form of rubbing on an object or self-stimulation by hand to repeat the pleasure discovered in the sexual area can be performed alone or in crowded places. When a healthy baby is breastfed, it falls asleep satisfied and peaceful. If this is not the case, the baby looks for other ways to satisfy itself. In this case, it is known that some babies lacking the love and warmth of the mother, suck their thumbs and others stimulate their genital areas to provide satisfaction and fall asleep. There are many similarities between thumb-sucking and masturbation (Aral and Gürsoy, 2007; Ünal, 1999).

Informing the family about childhood masturbation, determining the causal factors and correcting them are the first step to eliminate masturbation. To prevent masturbation;

- First, the medical conditions such as vaginal discharge, intestinal worms, genital itching caused by parasites should be diagnosed and treated immediately,
- Have the child do intense physical activities and fall asleep as soon as s/he goes to bed; make the child get up immediately after s/he wakes up; teach him/her proper toilet habits,

- The child shouldn't be allowed to ride a bicycle, carousel, seesaw, swing, etc. for a long time; avoiding to dress the child up with underwear made from very tight and silky fabric, trousers with front pockets,
- Going on to play different games after playing pretend for a long time,
- Instead of using intimidation, punishment, and warning, try to attract the child's attention to something else,
- It is important that the child's questions about sexuality are answered age-appropriately and their curiosity is eliminated (Özakar and Gözen, 2012; Kapçı, 2005; Ersoy and Avcı, 2001; Çağlar, 1981).

Masturbation fades away eventually but takes longer time to disappear when parents overreact to the behavior since the child wants to attract attention or to relax alone. The accusations and the threats of the concerned parents make children frightened may bring additional problems (Arı, 1986).

Lying

A lie is something untrue, said and rationalized by a person deliberately to deceive, humiliate, or take advantage of something (Seven, 2018). Telling the truth is a behavior that is always accepted in society, and children need to learn the difference between real and unreal. Most adults get angry and overreact when they notice a child is lying. However, cognitive-developmental stages show that this behavior seen in young children is the effect of the developmental process (Fields & Boesser, 2002). Therefore, the self-centered thinking styles of children aged three to four, and their level of imagination creates fictional events and objects, making their conversations unreal. They believe what they want will come true when they dream because they don't mature mentally to separate the fiction from reality until they are five years old (Aral and Gürsoy, 2007; Kapçı, 2005).

The following are factors that cause lying: insecurity in the family; repressive or indifferent parental attitudes; desire for love and attention in children; inconsistent behaviors of peers and teachers; positive attitudes towards those who lie; feelings of fear and shame; and false models.

The following behaviors can reinforce lying in children: expectations beyond children's abilities; to underestimate them; comparison with other children; to lie in their presence; to embarrass, scare and punish children. Families should accept their children as they are, support them in line with their abilities, display affectionate, relevant, explanatory and supportive attitudes, be the right role model, share their children's wishes, needs and dreams with them (Seven, 2018; Weilburger, 2008; Kapçı, 2005; Öz 2003).

Aggression

Aggression is any behavior performed with the intent of psychological or physical harm. It is a primitive form of defense used by an individual by adopting anger, frustration, and self-blocking elements. It is a problem behavior, starting at an early age, and must be punished or eliminated. Causes include domestic violence, inconsistent parental behavior, wrong models, physiological disorders, reckless social media use, and poor interpersonal relationships (Seven, 2018; Yılmaz Ünal and Deniz, 2018; Akçay and Emiroğlu, 2016). The aggressive behaviors emerge especially in terrible twos period and increase with the expectation that the rules within the family should be followed. It makes social relations difficult to maintain for 3-year-olds because of their self-centered characteristics. Children who tend to take more than they give exhibit the behaviors of biting, pushing, pinching, and so on. After three years, physical aggression is replaced by verbal aggression (Seven, 2018; Özbey, 2015). An aggressive child is insecure, has no self-esteem, believes that s/he is not loved, cannot restrain his/her impulses, insatiable, exposed to inconsistent parental behaviors, raised intolerably, feels guilt secretly and therefore has a low level of social acknowledgment. Studies show that physical and verbal aggression is seen mostly in boys and girls use verbal expressions more (Yörükoğlu, 2011; Özgün, 2008; Aral and Gürsoy, 2007; Alink et al., 2006; Hughes, 2003).

Firstly, parents should be role models to reduce aggression. It is also important to pay attention to reckless media use and to provide the child with activities to discharge his / her energy. The child should be approached with love, given certain responsibilities during the day, and parents and teachers should not be insensitive towards aggressive behaviors. To cope with aggression;

- Negative behaviors should be avoided in the family; parents should not be inconsistent, punitive, or permissive, and they should avoid criticizing and scolding the child.
- The child should be encouraged to participate in various sports activities to develop a sense of self-esteem,
- The curriculum should be arranged according to the needs and interests of children, considering their developmental level,
- Stories should be created and shared with the child to criticize his/her behavior and empathize (Seven, 2018; Özgün, 2008; Aral and Gürsoy, 2007; Öz 2003; Çağlar, 1981).

Stubbornness

Stubbornness is children's persistence in acting in a certain way and resistance in changing their behavior, thought, or status. This can be seen as the result of self-centered thinking in children. Children, who are selfish, expect everything they wish to come true, and they react angrily and stubbornly when met with an obstacle. Stubborn children who show aggressiveness through passive resistance, do everything slowly, stare blankly and met with harsh responses after infuriating someone, are reluctant to learn anything in school (Seven, 2018; Yörükoğlu, 2011; Aral and Gürsoy, 2007; Çağlar, 1981).

Stubbornness can stem from the way the child was raised, children's developmental characteristics or their disposition. Stubbornness is a result of a tense mother-child relationship and temporarily observed during the autonomy period and the fact that the mother is strict and persistent on toilet training or food will lead the child to passive resistance. Inconsistent behaviors towards children can lead to a stubborn personality; postponing the child's needs, frequently punishing the child and acting in an overly controlling way can lead to the establishment of stubborn behavior (Çağdaş and Şahin Seçer, 2015; Öz, 2003). Grumpy or hypersensitive children may exhibit stubbornness due to their temperament. Although it is difficult to eliminate stubbornness, it can be kept at acceptable levels with positive behaviors (Seven, 2018).

To prevent stubbornness;

- Exhibit tolerant, sharing, flexible, determined and a consistent attitude when it comes to discipline and establishing rules,
- Meet the child's needs at the right time and to set a model including accepted ways of behavior in order for the child to learn to socialize,
- Support the child when it comes to postponing and controlling their wishes,

- Try to heal the stubborn behavior instead of blocking it; to avoid any persistent attitude and behavior,
- Prevent the emergence of disagreements in schools, to distinctly identify wishes, interests, and needs, exhibit a reconciliatory attitude when met with stubborn attitude and to cooperate with the family and experts (Seven, 2018; Aral and Gürsoy, 2007; Öz 2003; Çağlar, 1981).

Jealousy

Jealousy, which is the result of sharing a loved one or assuming that a loved one is shared with someone else, is a natural response that includes anger and fear. Jealousy, which manifests itself through denial, intolerance, and resentment, mostly emerges in close relationships. Jealousy can be observed in a child as a result of being afraid to lose the relationship between children and a person who is important to them like their parents and teachers or being unwilling to share that person with someone else. The first traces of jealousy can be encountered when a new sibling is born. Children feel insecure and defenseless when they share the attention and love of parents with their siblings. Therefore, they start to compete to win the love of their parents again. It is impossible to prevent the jealousy that children show to their sibling, but it should not be forgotten that it can be reduced (Seven, 2018; Çağdaş and Şahin Seçer, 2015; Yavuzer, 2012b; Bakırcıoğlu, 2011; Yörükoğlu, 2011).

It is accepted that jealousy is the combination of mixed feelings like; low self-respect, uncertainty, loneliness, lack of confidence, anger, hatred, thoughts of revenge, self-pity, sadness, boredom, fear, and depression (Aral and Gürsoy, 2007; Kapçı, 2005). Infantile behaviors and regression symptoms such as crawling, baby talk, bottle-feeding, enuresis, nail-biting, and thumb-sucking that may be seen in children who experience extreme jealousy greatly affect their mental health. Besides, these children can exhibit behaviors like uneasiness, being mad for various reasons, crying, screaming, damaging the surrounding objects, and can disturb their parents with reasons like insomnia, fear, feeling cold or feeling hungry. Children who feel worthless and insufficient can feel jealousy more easily while children who develop positive self-respect can overcome this issue within a shorter period (Çağdaş and Şahin Seçer, 2015; Yavuzer, 2012b; Aral and Gürsoy, 2007; Çağlar, 1981).

Jealousy is at its peak between 3-5 years of age when children need their mothers in many ways. During this period, when children need attention and love, when children are not told that they will have a new sibling, when the mother doesn't pick up her child up because it is risky for her to lift heavy things and when the parents direct all their attention to the newborn baby after they come home can make children turn their jealousy-based hostility towards the mother instead of the sibling. Thus, children may start to wet the bed, disobey and stop eating (Seven, 2018; Çağdaş and Şahin Seçer, 2015; Yörükoğlu, 2011; Çağlar, 1981).

Therefore, parents should inform their children about the new-sibling and give them responsibilities like feeding the baby, cleaning, and clothing, in a controlled manner. Children who are jealous of their newborn siblings can adapt themselves as long as the parents have balanced and consistent attitudes. Parents should avoid comparisons and should make children realize each of them has a special place within the family (Seven, 2018; Çağdaş and Şahin Seçer, 2015; Yavuzer, 2012b; Yörükoğlu, 2011; Çağlar, 1981).

In the cases of twins, aside from competition, verbal statements of family and the community on their physical similarities are frequently encountered. No matter how similar they are, they are two different individuals and making them wear the same clothes, buying them the same toys and making them sleep in the same bed will inhibit their individualization.

To reduce the jealousy between siblings;

- Incorporate children into the prenatal preparations of the newborn sibling, convey the things that will happen in a way that they would be able to understand and make them feel they will always be valuable,
- Take care of children after the birth of the new sibling, to give children small responsibilities regarding baby-care and to ensure there are active interactions,
- Be patient with and tolerant of children's changing behaviors, grumpiness, crying jags, fits of anger or introversion,
- Make children feel they are not alone by showing them love and paying them attention sufficiently; to spend quality time together,
- Consult an expert or provide them with the required help if they are physically hurting their newborn sibling (Seven, 2018; Çağdaş and Şahin Seçer, 2015; Yavuzer, 2012b; Yörükoğlu, 2011; Çağlar, 1981).

Fear

Fear is an instinctive and transducer reaction shown to visible or invisible dangers. It changes according to the environment, the degree of the stimulant, the psychological and physiological state, and past experiences. An individual, who feels s/he should take measures in case of danger, reacts by running, hiding, or fighting. Fear manifests itself with symptoms like speeding of heart and pulse rates, stomach contraction and cramps, increment of adrenalin, problems in controlling urine and stools, fasciculation, shaking, nervousness, irregularity in breathing, teeth grinding, frequent swallowing, itching, cooling of the body and turning pale, and its most important characteristics is that the stimulant that stimulates fear appears suddenly and in an unexpected situation (Seven, 2018; Çağlar, 1981).

Most of the fears stem from unknown reasons, situations that are new or of an unknown quality generally stimulate fear. In this regard, many objects or situations can be frightening to a newborn baby. For example; children of 0-1 years of age are scared when there is no physical support; their mothers are gone, scared of loud noises and strangers; children of 2-3 years of age can be scared of animals, the sound of flush, vacuum cleaners and thunders. At the age of three or four, children fear losing the support of the mother, loneliness, darkness, fires, accidents, etc.; at the age of four or five children can be afraid of dreams, noises heard during night and at the age of five or six they can be afraid of fantasy-based objects like ghosts, witches, zombies as a result of the impacts of fairy tales (Seven, 2018; Yavuzer, 2012b; Weilburger, 2008; Kapçı, 2005).

One of the most important reasons behind fears is related to the family. The studies have emphasized that children are afraid of the things their parents were afraid of and this is totally normal. Some families try to frighten their children emotionally, to change their attitude, and this makes children feel guilty, nervous and be afraid of losing their mothers (Çağdaş and Şahin Seçer, 2015; Yörükoğlu, 2011; Aral and Gürsoy, 2007). Things that are beyond the power of the family and children may also cause fear or increase the levels of already existing fears. Experiencing an accident or a robbery or a natural disaster are fears related to the childhood memories. Their impacts can last throughout the adult years. Big fights witnessed outside of the home, accidents that have caused wounds or casualties can create uneasiness that lasts for some time. Additionally, it is natural for fears to occur in children who are frequently hospitalized or undergo surgeries one after the other (Yörükoğlu, 2011).

To prevent the negative effects of fears;

- Search for the reasons and to eliminate them,
- Be patient with children, avoid blaming embarrassing them,
- Give children an opportunity to interact with different social groups, to create an environment where children can openly talk about their fears and can relax, to make them face their fears and giving them an opportunity to promote their confidence,
- Avoid horror tales, games and movies, and to talk about the reality regarding animals and other frightening events
- Avoid using sayings like "If you throw bread on the floor God will make you blind" and to know that fear is not a tool for discipline,
- Help them learn to cope with their fears and that they are not alone, help them understand that everyone has fears and to support them on this,
- Use the threshold method to overcome fears. In this method, it will be helpful to face fears in small amounts and slowly. For example; Children who are afraid of dogs should be given an opportunity to watch the dog and children playing with dog at first and then slowly approach the dog and touch it (Seven, 2018; Çağdaş and Şahin Seçer, 2015; Yörükoğlu, 2011; Aral and Gürsoy, 2007; Kapçı, 2005; Nelsen, Lott and Glenn, 2002; Ersoy and Avcı, 2001; Çağlar, 1981).

Anxiety

Anxiety is the abnormal fear that manifests itself through physical symptoms and develops without reason. It is the reaction shown to the possibility of the occurrence of the frightening situation. High levels of anxiety affect the physical and social functions of the individuals and cause them to be unhappy (Seven, 2018).

Every child can be afraid of dangerous and strange things. When frightening situations and the level of fear increase; children start to develop an anxious personality. Therefore, anxiety is an emotional reaction mixed with fear. If a mother starts to be afraid of the possibility that her child may die because s/he has tonsillitis, then this feeling is called anxiety. Nevertheless, if the child is going through a severe disease, then the reaction the mother shows is fear. In other words, the danger causing fear is visible and objective; therefore, people know the reason behind their fears. However, the danger causing anxiety is invisible and subjective. The person is not aware of the real reason behind this feeling (Çağdaş and Şahin Seçer, 2015; Yörükoğlu, 2011).

Most of the time, anxious children have problems like: uneasiness, nervous facial expressions, apparent stiffness of the muscles, hyperactivity, sweating, racing heart, difficulty in breathing, xerostomia, elevated serum glucose level, elevated blood pressure, accelerated heart rate and respiratory rate, aggression, vomiting, loss of appetite, speech disorders, overreliance, disruption in socialization and domestic relations, academic failure, frequently crying and sleep disorders (Çağdaş and Şahin Seçer, 2015; Çetintaş, 2015; Shelton, 2013; Paktuna Keskin, 2012; Cook & Cook, 2005; Çağlar, 1981).

The root causes of anxiety are biological elements, physical illnesses, and learned behaviors. The inheritance of some temperament traits to the child and disorders in the nervous system are among the biological causes of anxiety. Physical illnesses like diabetes, epilepsy and drug habit; emphasizing the horror elements in the environment and intensifying frightening situations and elements are among the factors causing anxiety (Seven, 2018; Sheehan, 1999). Additionally, restricting the wishes of the child, establishing strict rules, being overprotective while raising the child and making him/her dependant, not letting children express themselves, underestimating their talents and abilities, expecting an outstanding performance from them, having overly anxious and depressed personalities as parents are also among the triggering factors (Çağdaş and Şahin Seçer, 2015; Austin & Sciarra, 2013; Yörükoğlu, 2011).

The most encountered anxiety disorder during the preschool period is separation anxiety. Fearing that something terrible will happen to a loved one; being afraid of losing relatives or of being kidnapped can cause separation anxiety. Most of the children show their first clear reactions to separation when they are between the 9th and the 30th months old and are separated from their mothers. Most of these reactions are crying, grabbing onto their mother's clothes, and trying to run after her. Starting from when they are three years old, these reactions start to decrease, and they can be more easily calmed down. Children who are experiencing separation fear mostly avoid going to school. They may have trouble falling asleep alone at night and during the day they may refuse to go to another room even though their families are at home too. In order for this to become a problem it should be continuous, not be fitted to their age and should be affecting the daily activities of the family and the child negatively (Seven, 2018; Gardner and Shaw, 2008; Kapçı, 2005).

To prevent separation anxiety, children should be informed about separation and reunion, games like peek-a-boo, which are aimed at object permanence should be played, and they should be tried to be relieved by implying that their anxiety is acknowledged (Kapçı, 2005). To prevent anxiety in children;

- Determine the reasons and to eliminate them,
- Avoid using expressions that can cause anxiety and fear, to provide opportunities for children to openly talk about their existing fears and anxieties and to show them love and pay them attention,
- Avoid the things that would make children lose confidence by expecting achievements from them that are far above their age and development level and to avoid punishing them while educating them,
- Know that panic anxieties decrease the efficiency of learning and professional help is needed (Seven, 2018; Çağdaş and Şahin Seçer, 2015; Aral and Gürsoy, 2007).

Stealing

Stealing is the unauthorized or unacceptable possession of something that does not belong to the child. This is observed in all children as a normal developmental characteristic until a specific age. Kids don't think that taking something that belongs to someone else without their permission is wrong. Because the concept of possession is not developed in kids, they think that everything they touch or love is theirs, and they put this thought into practice by doing things like taking their friend's toy or picking up the flowers in their neighbor's garden without permission (Seven, 2018; Özbey, 2015; Çağdaş and Şahin Seçer, 2015).

During their play age, children can take another kid's toy without taking permission when they see a toy that is better than theirs. If the other kid also thinks that his/her toy is more colorful, shiny, charming, and beautiful s/he would not want to give away the toy. But this is not caused by the possession instinct of the kid; rather, it is because the sense of sharing is not developed yet (Öz, 2002).

Stealing behavior is observed in children because the sense of belonging is not yet developed in children, children's needs are not sufficiently met, children want to experience exciting things and because of the inappropriate reactions that parents show to the behavior of their child, lack of harmony in the family, cognitive inadequacy, lack of confidence, rivalry and jealousy (Seven, 2018; Yılmaz Ünal and Deniz, 2018; Çağdaş and Şahin Seçer, 2015; Özbey, 2015; Yörükoğlu, 2011).

Repetitive stealing stems from a lack of attention and love. Children, who think they are not loved enough, try to compensate for the love they don't get by stealing someone else's belongings. Therefore, abandoned children, children who are excluded from their families, and street children tend to steal more. Adopted children can also pick and steal food, collect objects or steal, making their adoptive parents worried. Stealing in children, which stems from the lack of confidence caused by maternal deprivation, disappears in time as children become sure that adoptive parents love and embrace them (Çağdaş and Şahin Seçer, 2015; Yörükoğlu, 2011). Children who don't feel love or compassion within the family start stealing to eliminate their psychological dissatisfaction. It is vital for families to observe their children well, share their emotions, pay attention to their problems, spend time with their children and to know their friends (Öz, 2002). Before accusing them of thievery children should be given an opportunity to correct their attitude and put the stolen thing back into its original place (Nelsen, Lott and Glenn, 2002).

To prevent stealing in children:

- Children should be taught how to respect individual and social rights and to restrain the urge to possess. The material needs of children should be provided in the right place at the right time. Private spaces should be provided for children for the sense of belonging to develop.
- Parents and teachers should set good examples for children.
- Reacting in a firm and an abrupt way to children who stole something should be avoided. Children should be given time to realize what they have done and to correct their mistakes.
- The situation should be among the parents, the child, and the teacher. The child should not be embarrassed next to peers. Children should not be labeled a thief, and the stealing behavior should never be ignored (Seven, 2018; Çağdaş and Şahin Seçer, 2015; Yavuzer, 2012a; Aral and Gürsoy, 2007; Senemoğlu, 2005; Fields & Boesser, 2002; Çağlar, 1981).

Sleeping Disorders

Sleep is a vital condition related to brain function and psychology. It is the temporary, partial, and periodic disappearance of the organism's communication with the environment in a reversible manner with different stimuli. Many theories suggest that sleep has an important effect on many biological functions, including physical and psychological regulations, conserving energy, emotion discharge, brain development, and the system. In this sense, sleep serves many purposes in a developmentally complex structure (Seven, 2018; Yılmaz Ünal and Deniz, 2018).

Due to the biological maturation of the sleep-wake cycle and its relation to psychological development, sleep disorders are common among children. Studies show that almost 25-50% of preschoolers, 20-30% of school-age children and teenagers have various sleep disorders (Akt; Perdahlı Fiş et al., 2010). Another study in Turkey, which analyzed children aged between 2 months and 16 years, showed that mostly observed sleep disorders are 8% snoring, 7% gaping, 5% talking in sleep, 4% teeth grinding, 3% snoring+sleep apnea (Koçak, Albayrak, Erol and Şanlı, 2012).

In oral stage, physical factors like hunger, gas pains, having a wet diaper, extremely hot environment can cause sleep disorders. Between ages one and three, losing the sense of security and trust, early potty training, lack of a mother, teething and other illnesses can affect the emergence of sleep disorders. During 3-5 years of age, nightmares are more common, and this situation may stem from a bad environment, frightening TV programs, and punishments (Yavuzer, 2012a Yörükoğlu, 2011). Sleep disorders, which mostly caused by anxiety, fear, continuous punishments, abuse, stressful memories, wrong parenthood, and inflammatory diseases, can manifest themselves as insomnia, sleepwalking, night terrors, nightmares, sleep paralysis or hypersomnia (Seven, 2018; Çetintaş, 2015; Özbey, 2015).

Insomnia is having trouble falling asleep or sleep-maintenance. The most frequent reason behind having trouble falling asleep and waking up in the night is related to starting to fall asleep. Laying down, relaxing, and falling asleep without the help of the parent is an acquired behavior. If a child needs his/her mother to fall asleep (rocking, feeding, holding) or is used to a particular activity (toys, music) s/he would want these conditions after waking up in the middle of his/her night's sleep. The best method of treatment for sleep disorders is behavioral therapy, including stimulus control and graduated reduction (Seven, 2018; Durand & Mindell, 1990).

Sleepwalking is meaningless, pointless, and complex physical actions shown during sleep. This happens during the first two or three hours of sleep when it is at its deepest point, and while sleepwalking, the eyes of children are open and dull. The things children say during sleepwalking are incomprehensible, and when they wake up, they don't remember that period of time. Along with inheritance, developmental, psychological, and physical factors may cause sleepwalking. The families of these children should put them to bed without waking them up, take the necessary precautions for them to not get physically damaged and should consult an expert about a solution (Seven, 2018; Yılmaz Ünal and Deniz, 2018; Yörükoğlu, 2011; Rathus, 2006).

Night terror, which is frequently observed in children, mostly happens during the first three hours after the child falls asleep. Children wake up by screaming or crying, and their heart and respiratory rates increase, their muscles can contract, they can sweat and redden. They calm down by themselves after a few minutes and don't remember what happened after they wake up. A night terror is not medically dangerous; however, it is a wearing situation for the parents. Because, trying to calm children down can worsen the situation. The parents try not to touch them except for preventing them from hurting themselves. When these attacks become more and more frequent, the reasons should be questioned and frightening or worrisome events that happen during the day should be reduced or removed (Seven, 2018; Çetintaş, 2015; Rathus, 2006; Ersoy and Avcı, 2001; Arı, 1986).

Nightmares manifest themselves as awakening from highly frightening dreams towards the end of the night's sleep, often remembered in detail, and they can be easily distinguished from night terrors in terms of their occurrence in the beginning/ending of sleep and their recall. While treating night terrors, it is important to calm the person down and reducing daily stress (Yılmaz, Ünal and Deniz, 2018).

Sleep paralysis is the condition in which the child dreams about the threatening things that have happened during the day. It can be observed particularly after the mother leaves the house or after being frequently hospitalized. Traumatic experiences, sights of violence, accidents, or fears can cause children to have sleep paralysis, in other words, nightmares. In this condition, children can express their fear while sleeping, and the effect of what they saw can continue when they wake up (Seven, 2018; Yörükoğlu, 2011).

To prevent sleep disorders; especially kids should be physically and emotionally relaxed, and positive behaviors should be encouraged. Punishment for discipline and threatening with leaving or banning from the family should be avoided. The place where children sleep should be organized in a way that they can feel comfortable in, the reasons and the frequency of the disorder should be researched, and an expert should be consulted (Seven, 2018; Yılmaz Ünal and Deniz, 2018).

Eating Disorders

Eating disorders are diversions of eating habits which can stem from a combination of social, emotional, and physical factors and can cause diseases and malnutrition (Yılmaz Ünal and Deniz, 2018). When these disorders become common they cause physical-mental deterioration (Raynor and Rudolf 1996; Wilensky, Ginsberg, Altman, Tulchinsky, Ben Yishay, &Auerbach, 1996; Puckering, Pickles, Skuse, Heptinstall, Dowdney, & Zur-Szpiro, 1995) and variety of attitude problems (Lindberg, Bohlin, Hagekull, & Thunström, 1994; Dahl & Sundelin, 1992; Dahl, 1987). Therefore, the diagnosis and the treatment of these disorders during early childhood is important. Lack of appetite is the first complaint of the parents, who think their children are playing with food because they cannot feed themselves or they don't want to eat during this period. Children who are forced to eat with fear of weight loss or developmental latency can develop major eating disorders in the future (Özbey, 2015).

Eating provides psychological satisfaction during babyhood and social satisfaction in the following periods, paving the way for the social and emotional patterns to form. The situations negative affecting children's appetite include: showing excessive love to the child while eating; being indifferent to the child; annoying the child; fast feeding; force-feeding the things s/he does not like, etc. (Özbey, 2015).

Earth-eating is a type of eating disorder and defined as eating things that are not food, such as dirt, chalk, paper, threads, paint, etc. It is more frequent in abandoned, neglected, emotionally deprived children, who can also bite their nails and suck their thumbs. One of the other eating disorders, observed during this period, is rumination. Rumination is re-chewing the food after being swallowed and partially digested. It is mostly observed in children aged between three and ten, children who are mentally incompetent and it is more frequent in boys (Arı, 2018; Ulu, 2018).

Obesity is a major problem that develops due to the increase of fat tissues when the energy intake is much more than energy consumption, and can create major medical, psychological and social issues (Şarbat and Demirkol, 1999). Studies in Turkey show that the prevalence of obesity varies between 25-30% among children and teenagers (Akt. Ergül and Kalkım, 2011). Obesity is caused by genetic factors, diet, and environmental factors like communication problems within the family, time spent with television and computer, land low acceptance by peers. It should be diagnosed early because it has a negative effect on children. Also, after learning the causes, energy intake should be decreased and energy consumption should be increased (Çetintaş, 2015; Ersoy and Avcı, 2001).

Anorexia nervosa is an eating disorder that mostly starts to develop during teenage years and young adulthood. It is characterized by body dysmorphia related lack of eating to the extent that physical integrity is threatened (Tahiroğlu, Fırat, Diler and Avcı, 2005). The distinctive feature of this weight loss is that it cannot be explained with a physical disorder. Psychological characteristics of the disorder are being extremely afraid of getting fat and the misperception of the body image. These people may stop eating, vomit, use medications or they may try to lose weight by exercising too much (Crisp, 1999; Şarbat and Demirkol, 1999). The life quality and the physiological/psychological health of these people can be affected to a great extent. Therefore, early diagnosis and carrying out the treatment by psychological support with a multidisciplinary approach that includes the family and is aimed at changing the dietary habits are significantly important. (Arı, 2018; Ulu, 2018; Özenoğlu et al., 2008; Crisp, 1999; Şarbat and Demirkol, 1999).

Bulimia nervosa, which is another eating disorder encountered at the end of adolescence and young adulthood, is accompanied by losing control and is characterized by gaining weight by binge eating and at the same time trying to stop gaining weight with improper ways (vomiting, using laxatives, and diuretics, extreme dieting, fasting, intensive exercise etc.) (Arı, 2018; Ulu, 2018; Çetintaş, 2015; American Psychiatry Association, 2001). These people have low self-esteem and have weak interpersonal relations. They can isolate themselves from society, hurt themselves, and can be impulsive. The aim of the treatment is to provide a regular eating habit and to avoid eating uncontrollably (Arı, 2018; Şarbat and Demirkol, 1999).

The treatment to reduce or prevent eating disorders takes time. To boost the confidence of children certain activities should be performed within the family and in school. Children should feel they are valued and loved; comments regarding their physical appearances should be avoided. Tension within the family can also cause eating disorders to develop and persist. It is crucial to note that early diagnosis eases the treatment (Seven, 2018; Yılmaz Ünal and Deniz, 2018; Çetintaş, 2015).

Most of the children who have attitude problems have difficulty in establishing positive relationships with adults. To minimize these issues, it will be helpful for parents and teachers to cooperate. Accordingly, it is important for parents to support their children in acquiring skills and strengthening their positive characteristics that will help change their attitude. Moreover, if teachers behave in a natural, sincere, meaningful, determined and an informative way it will help children establish positive fiduciary relationships. Also, being in contact with parents in a cooperative and empathetic way is one of the most important steps to solve the attitude problems of children.

REFERENCES

Akçay, D. & Emiroğlu, O. N. (2016). Medya etkileşimine bağlı okul öncesi dönemdeki çocukta saldırganlık davranışı. *Türkiye Klinikleri, 2*(1), 116–120.

Alink, L. R., Mesman, J., vanZeijl, J., Stolk, M. N., Juffer, F., Koot, H. M., ... van IJzendoorn, M. H. (2006). The early childhood aggression curve: Development of physical aggression in 10-to 50-month-old children. *Child Development, 77*(4), 954–966. doi:10.1111/j.1467-8624.2006.00912.x PMID:16942499

Amerikan Psikiyatri Birliği. (2001). Psikiyatride hastalıkların tanımlanması ve sınıflandırılması el kitabı. Ankara: Hekimler Yayın Birliği.

Aral, N. & Gürsoy, F. (2007). Özel eğitim gerektiren çocuklar ve özel eğitime giriş. İstanbul: Morpa Yayınları.

Arı, M. (1986). Çocukluk döneminde uyum bozuklukları. In Ş. Bilir, (Ed.), Özürlü çocuklar ve eğitimleri (pp. 52-59). Ankara: Ayyıldız Matbaası.

Arı, M. (2018). Çocuklarda uyum bozukluğu, duygu ve davranış sorunları. In E.N. Metin, (Ed.), Özel gereksinimli çocuklar (2. Baskı, pp. 319-356). Ankara: Anı Yayıncılık.

Austin, V. L., & Sciarra, D. T. (2013). Çocuk ve ergenlerde duygusal ve davranışsal bozukluklar (M. Özekes, Çev. Ed.). Ankara: Nobel Yayıncılık.

Aydın, A. (2008). *Sınıf yönetimi (9. Baskı)*. Ankara: Pegem Akademi.

Bakırcıoğlu, R. (2007). *Çocuk ruh sağlığı ve uyum bozuklukları*. Ankara: Anı Yayıncılık.

Bakırcıoğlu, R. (2011). *Çocuk ve ergende ruh sağlığı*. Ankara: Anı Yayıncılık.

Brunet, C. & Sarfati, A. C. (2000). 1-7 yaş arası çocuğun eğitimi (K. Bıçakçı, ed.). Ankara: Kapital Medya Hizmetleri A.Ş.

Bulut, Ş. (2000). *Altı yaş çocuklarında davranış problemleri ile anne ve öğretmenlerin uyum düzeyleri arasındaki ilişkinin incelenmesi.* Yayınlanmamış Yüksek Lisans Tezi, Ankara Üniversitesi.

Çağdaş, A. & Şahin Seçer, Z. (2015). Anne-baba eğitimi. Ankara: Eğiten Kitap.

Çağlar, D. (1981). Uyumsuz çocuklar ve eğitimleri. Ankara: Ankara Üniversitesi Eğitim Fakültesi Yayınları No: 103.

Campbell, S. B. (1995). Behavior problems in preschool children: A review of recent research. *Journal of Child Psychology and Psychiatry, and Allied Disciplines, 36*(1), 113–149. doi:10.1111/j.1469-7610.1995.tb01657.x PMID:7714027

Çetintaş, S. (2015). *5 yaş çocuklarında görülen davranış problemleri ile annelerin çocuklarına uyguladıkları disiplin yöntemleri arasındaki ilişkinin incelenmesi.* Yayınlanmamış Yüksek Lisans Tezi, Gazi Üniversitesi.

Cook, J. L., & Cook, G. (2005). *Child development principles and perspectives.* Boston, Mass: Pearson.

Crisp, A. (1999). Anoreksia nervoza: Psikanalitik bakış. In A. Ekşi (Ed.), Ben hasta değilim (pp. 223-228). İstanbul: Nobel Tıp Kitabevi Ltd. Şti.

Dahl, M. (1987). Early feeding problems in an affluent society. III. Follow-up at two years: Natural course, health, behaviour, and development. *Acta Paediatrica (Oslo, Norway), 76*(6), 872–880. doi:10.1111/j.1651-2227.1987.tb17257.x PMID:3480683

Dahl, M., & Sundelin, C. (1992). Feeding problems in an affluent society. Follow-up at four years of age in children with early refusal to eat. *Acta Paediatrica (Oslo, Norway), 81*(8), 575–579. doi:10.1111/j.1651-2227.1992.tb12303.x PMID:1392380

Dawkins, M. P., Fullilove, C., & Dawkins, M. (1995). Early assessment of problem behavior among young children in high-risk environments. *Family Therapy, 22*(3), 133–141.

Driscoll, K., & Pianta, R. C. (2011). Mothers' and fathers' perceptions of conflict and closeness in parent-child relationships during early childhood. *Journal of Early Childhood and Infant Psychology, 7,* 1–24.

Drossos, N. T. (2004). *Execute function and externalizing behavior problems in preschool children* [Doctoral Dissertation]. The Graduate College Of The Illinois Institute Of Technology.

Durand, V. M., & Mindell, J. A. (1990). Behavioral treatment of multiple childhood sleep disorders: Effects on child and family. *Behavior Modification, 14*(1), 37–49. doi:10.1177/01454455900141003 PMID:2294900

Dursun, A. (2010). *Okul öncesi dönemdeki çocukların davranış problemleriyle anne- baba tutumları arasındaki ilişkinin incelenmesi.* Yayınlanmamış Yüksek Lisans Tezi, Dokuz Eylül Üniversitesi.

Dwyer, K. P., Osher, D., & Hoffman, C. C. (2000). Creating responsive schools: Contextualizing early warning, timely response. *Exceptional Children, 66*(3), 347–365. doi:10.1177/001440290006600306

Erermiş, S. & Aydın, C. (1999). Çocuk ruh sağlığı yönünden enürezis. In A. Ekşi (Ed.), Ben hasta değilim. İstanbul: Nobel Tıp Kitabevi Ltd. Şti.

Ergül, Ş. & Kalkım, A. (. (2011). Önemli bir kronik hastalık: Çocukluk ve ergenlik döneminde obezite. *Turk Silahli Kuvvetleri Koruyucu Hekimlik Bulteni, 10*(2), 223–230. doi:10.5455/pmb.20110119104521

Eripek, S. (2002). Uyumsuz çocuklar ve eğitimleri. Y. Özsoy, M. Özyürek ve S. Eripek (Eds.), Özel eğitime muhtaç çocuklar özel eğitime giriş (pp. 199-223). Ankara: Karatepe Yayınları.

Ersoy, Ö. & Avcı, N. (2001). Özel gereksinimi olan çocuklar ve eğitimleri-özel eğitim. İstanbul: Ya-Pa Yayınları.

Fields, M.V., & Boesser, C. (2002). *Constructive guidance and discipline: Preschool and primary education*. Upper Saddle River, NJ: Merrill Prentice Hall.

Gardner, F. & Shaw, D. S. (2008). Behavioral problems of infancy and preschool children (0–5). In M. Rutter, D. Bishop, D. Pine et al. (Eds.), Rutter's child and adolescent psychiatry (5th ed., pp. 882-894). London: Blackwell Press.

Genç, A. (2018). *Okul öncesi ve ilkokul döneminde olan işitme kayıplı çocukların davranış problemlerinin incelenmesi*. Yayınlanmamış Yüksek Lisans Tezi, Anadolu Üniversitesi.

Gunter, P. L., Coutinho, M. J., & Cade, T. (2002). Classroom factors linked with academic gains among students with emotional and behavioral problems. *Preventing School Failure, 46*(3), 126–132. doi:10.1080/10459880209603357

Hughes, E. (2003). Learners with emotional or behavioral disorders. In D. R. Hallahan & J. M. Kauffman (Eds.), *Exeptional learners introduction to special education* (9th ed., pp. 221–261). Boston: Allyn & Bacon.

Işık, H. K. (2018). *Çalışan/çalışmayan ebeveynlerin çocuklarına karşı tutumları ve bu tutumların 5-6 yaş grubu çocukların davranış problemleri ile ilişkisinin incelenmesi*. Yayınlanmamış Yüksek Lisans Tezi, Maltepe Üniversitesi.

Kaner, S. (2009). Duygusal ve davranış bozuklukları olan çocuklar. In A Ataman, (Ed.), Özel gereksinimli çocuklar ve özel eğitime giriş (pp. 171-194). Ankara: Gündüz Eğitim ve Yayıncılık.

Kapçı, E. G. (2005). Erken çocukluk döneminde sıklıkla karşılaşılan duygusal davranışsal sorunlar. In Y. Uzuner (Ed.), Çocukta ruh sağlığı, uyum bozukluğu (pp. 74-92). Eskişehir: Anadolu Üniversitesi Yayınları.

Kauffman, J. M. (2001). *Characteristics of emotional and behavioral disorders of children and youth* (7th ed.). Upper Saddle River, NJ: Merrill/Prentice Hall.

Kazdin, A. E. (1998). Conduct disorder. In R. J. Morris & T. R. Kratochwill (Eds.), *The practice of child theraphy* (3rd ed., pp. 199–230). Boston: Allyn & Bacon.

Koçak, Ü., Albayrak, M., Erol, R., & Şanlı, C. (2012). Evaluation of sleep patterns and sleep disturbances in children: a preliminary study in Kırıkkale. *Türkiye Çocuk Hastalıkları Dergisi, 6*(2), 81-87.

Lago-Delello, E. (1998). Classroom dynamics and the development of serious emotional disturbance. *Exceptional Children, 64*(4), 479–492. doi:10.1177/001440299806400404

Lindberg, L., Bohlin, G., Hagekull, B., & Thunström, M. (1994). Early food refusal: Infant and family characteristics. *Infant Mental Health Journal, 15*(3), 262–277. doi:10.1002/1097-0355(199423)15:3<262::AID-IMHJ2280150303>3.0.CO;2-Q

Mayer, G. R. (1995). Preventing antisocial behavior in schools. *Journal of Applied Behavior Analysis, 28*(4), 467–478. doi:10.1901/jaba.1995.28-467 PMID:16795877

Nelsen, J., Lott, L., & Glenn, S. (2002). *A'dan Z'ye pozitif disiplin (M. Ersin, Çev.).* İstanbul: Hayat.

Öz, İ. (2002). Çocuğum neden çalıyor? *Çoluk Çocuk Anne-Baba-Eğitimci Dergisi, 21*, 20-21.

Öz, İ. (2003). *Çocukta uyum ve davranış bozuklukları*. Ankara: Kök Yayıncılık.

Özakar, S. ve Gözen, D. (2012). Erken çocukluk döneminde masturbasyon sorunu olan çocuğa hemşirelik yaklaşımı. *Ege Üniversitesi Hemşirelik Fakültesi Dergisi, 28*(1), 113–122.

Özbey, S. (2015). Çocuklarda uyum ve davranış problemleri. In Y. Aydoğan & G. Gültekin Akduman (Eds.), Çocuk ruh sağlığı (pp. 173-188). Ankara: Eğiten Kitap.

Özenoğlu, A., Özdemir, A., Tecer, Ö., Balcı, H., Uğurlu, S., & Tüzün, Ş. ve Uğur M. (2008). Anoreksiya nevroza tedavisine ekip yaklaşımı. *Gözde Tıp Dergisi, 23*(2), 66–69.

Özgün, Ö. (2008). Sınıfta istenmeyen davranışlar, nedenleri ve yönetimi. In Y. Aktaş Arnas & F. Sadık, (Eds.), Okul öncesi eğitimde sınıf yönetimi (pp. 243-276). Ankara: Kök Yayıncılık.

Paktuna Keskin, S. (2012). *Çocuk davranışlarındaki korkuyu tanımak ve başetmek*. İstanbul: Boyut Yayınları.

Perdahlı Fiş, N., Arman, A., Ay, P., Topuzoğlu, A., Güler, A. S., Gökçe İmren, S., & Ersu, R., & Berkem, M. (2010). Çocuk uyku alışkanlıkları anketinin Türkçe geçerliliği ve güvenilirliği. *Anadolu Psikiyatri Dergisi, 11*, 151–160.

Pierce, C. D., Reid, R., & Epstein, M. H. (2004). Teacher-mediated interventions for children with EBD and their academic outcomes: A review. *Remedial and Special Education, 25*(3), 175–188. doi:10.1177/07419325040250030501

Puckering, C., Pickles, A., Skuse, D., Heptinstall, E., Dowdney, L., & Zur-Szpiro, S. (1995). Mother-child interaction and the cognitive and behavioural development of four-year-old children with poor growth. *Journal of Child Psychology and Psychiatry, and Allied Disciplines, 36*(4), 573–595. doi:10.1111/j.1469-7610.1995.tb02315.x PMID:7650084

Rathus, S. A. (2006). *Childhood: voyages in development*. Belmont, CA: Thomson/Wadsworth.

Raynor, P., & Rudolf, M. C. J. (1996). What do we know about children who fail to thrive? *Child: Care, Health and Development, 22*(4), 241–250. doi:10.1111/j.1365-2214.1996.tb00427.x PMID:8818427

Şarbat, G. & Demirkol, M. (1999). Yeme bozuklukları-nutrisyonel bakış. In A. Ekşi (Ed.), Ben hasta değilim (pp. 212-222). İstanbul: Nobel Tıp Kitabevi Ltd. Şti.

Senemoğlu, N. (2005). *Gelişim öğrenme ve öğretim kuramdan uygulamaya (12. Baskı)*. Ankara: Gazi Kitabevi.

Seven, S. (2018). *Çocuk ruh sağlığı (5. Baskı)*. Ankara: Pegem Akademi.

Sheehan, E. (1999). Kaygı bozuklukları: Kaygı, fobiler ve panik ataklar (M. Sağlam, Çev.). İstanbul: Alfa Yayınları.

Shelton, R. C. (2013). *Anksiyete bozuklukları (Çev. İ. Ekmekçi Ertek). In Psikiyatri tanı ve tedavi* (pp. 351–365). Ankara: Güneş Yayınları.

Sprague, J., & Walker, H. (2000). Early identification and intervention for youth with antisocial and violent behavior. *Exceptional Children*, *66*(3), 367–379. doi:10.1177/001440290006600307

Tahiroğlu, A. Y., Fırat, S., & Diler, R. S. ve Avcı, A. (2005). Erkek çocuklarda yeme bozuklukları; bir anoreksiya nervosa vakası. *Çocuk Sagligi ve Hastaliklari Dergisi*, *48*, 151–157.

Taner Derman, M. & Başal, H. A. (2013). Okul öncesi çocuklarında gözlenen davranış problemleri ile ailelerinin anne-baba tutumları arasındaki ilişki. *Amasya Üniversitesi Eğitim Fakültesi Dergisi*, *2*(1), 115–144.

Tanju, E. H. (2011). Okul öncesi dönemde davranış problemleri ve alınacak önlemler. In G.U. Balat & H. Bilgin (Eds.), Okul öncesi eğitimde sınıf yönetimi (pp. 45-92). Ankara: Eğiten Kitap.

Ulu, Y. (2018). *Okul öncesi 3-6 yaş grubu çocuklarının davranış problemleri ve anne-baba tutumlarının incelenmesi*. Yayınlanmamış Yüksek Lisans Tezi, Çağ Üniversitesi.

Ünal, F. (1999). Çocukluk mastürbasyonu. In A. Ekşi (Ed.), Ben hasta değilim (pp. 72-84). İstanbul: Nobel Tıp Kitabevi Ltd.

Walker, H. M., & Golly, A. (1999, November/December). Developing behavioral alternatives for antisocial children at the point of school entry. *ERIC Clearing House*, *73*(2), 104–107. doi:10.1080/00098659909600159

Weaver, C., Shaw, D., Dishion, T., & Wilson, M. (2008). Parenting self-efficacy and problem behavior in children at high risk early conduct problems: The mediating role of maternal depression. *National Institutes of Health, 31*(4), 594-605.

Weilburger, L. S. (2008). *Çocuk ve disiplin: Çocuğumuza davranış yöntemleri (İ. Özbaş, Çev.)*. İstanbul: Ekinoks.

Wilensky, D. S., Ginsberg, G., Altman, M., Tulchinsky, T. H., Ben Yishay, F., & Auerbach, J. (1996). A community based study of failure to thrive in Israel. *Archives of Disease in Childhood*, *75*(2), 145–148. doi:10.1136/adc.75.2.145 PMID:8869197

Yavuzer, H. (2012a). *Çocuk psikolojisi (34. Baskı)*. İstanbul: Remzi Kitabevi.

Yavuzer, H. (2012b). *Anne-baba ve çocuk (23. Baskı)*. İstanbul: Remzi Kitabevi.

Yılmaz Ünal, S. & Deniz, Ü. (2018). Davranış ve uyum problemleri olan çocuklar. In N. Aral & F. Gürsoy (Eds.), Özel gereksinimli çocuklar ve eğitimleri-Özel gereksinimli çocuklar (pp. 85-111). Ankara: Hedef CS Basın Yayın.

Yörükoğlu, A. (2011). *Çocuk ruh sağlığı: Çocuğun kişilik gelişimi, eğitimi ve ruhsal sorunları (31. Baskı)*. İstanbul: Özgür Yayınları.

Yumuş, M. (2013). *Okul öncesi eğitimcilerinin 36-72 ay aralığındaki çocukların davranış problemleri ile ilgili görüşlerinin incelenmesi ve başa çıkma stratejilerinin belirlenmesi.* Yayınlanmamış Yüksek Lisans Tezi, Hacettepe Üniversitesi.

ADDITIONAL READING

Austin, V., & Sciarra, D. (2016). *Difficult students and disruptive behavior in the classroom: Teacher responses that work.* New York: W. W. Norton & Company.

Campbell, S. B. (1995). Behavior problems in preschool children: A review of recent research. *Journal of Child Psychology and Psychiatry, and Allied Disciplines, 36*(1), 113–149. doi:10.1111/j.1469-7610.1995.tb01657.x PMID:7714027

Campbell, S. B. (1997). Behavior problems in preschool children. In T. H. Ollendick & R. J. Prinz (Eds.), *Advances in Clinical Child Psychology. Advances in Clinical Child Psychology* (Vol. 19, pp. 1–26). Boston, MA: Springer. doi:10.1007/978-1-4757-9035-1_1

Campbell, S. B. (2006). *Behavior problems in preschool children. Clinical and developmental issues* (2nd ed.). NY: The Guilford Press.

Flook, L., Goldberg, S. B., Pinger, L., & Davidson, R. J. (2015). Promoting prosocial behavior and self-regulatory skills in preschool children through a mindfulness-based kindness curriculum. *Developmental Psychology, 51*(1), 44–51. doi:10.1037/a0038256 PMID:25383689

Hughes, E. (2003). Learners with emotional or behavioral disorders. In D. R. Hallahan & J. M. Kauffman (Eds.), *Exceptional learners introduction to special education* (9th ed., pp. 221–261). Boston: Allyn & Bacon.

Mallants, C., & Casteels, K. (2008). Practical approach to childhood masturbation a review. *European Journal of Pediatrics, 167*(10), 1111–1117. doi:10.100700431-008-0766-2 PMID:18575886

KEY TERMS AND DEFINITIONS

Aggression: Behavior that is intended to harm others physically or psychologically.
Anxiety: Abnormal and unreasonable extreme fear.
Eating Disorder: Deviations in eating habits that can cause disease or malnutrition.
Encopresis: Involuntary defecation even though the child is old enough to hold it.
Enuresis: Unintentional or intentional urinary incontinence during day and / or night.
Jealousy: A natural reaction that occurs when a loved one cannot be shared with someone else.
Masturbation: Taking pleasure by stimulating the genitals.

Section 2
Periods of Life

Development is a long process which starts from prenatal period and continues until the end of life; therefore it is important to know the characteristics of each of the periods. Accordingly, the developmental periods are included in this section. Child development is approached starting from the prenatal period; childbirth and development in infancy, early childhood development, middle childhood development, development in late childhood are explained in detail.

Chapter 8
Prenatal Development

Nezahat Hamiden Karaca
Afyon Kocatepe University, Turkey

ABSTRACT

The prenatal period is a vital process for the mother and father to become conscious about the development of the baby. Particularly in this process, in which mental and physical changes are observed in the mother, both the mother and father should be supported in order for the baby in the mother's womb to have a healthy gestation and to prevent a risky pregnancy. In this regard, particularly with the impact of recent technological innovations and with the increasing awareness of parents through increasing literacy, it is reported that the mortality rate is decreased; however, it is also reported that severe injuries during pregnancy and disabled babies still exist. From this point of view, the basic concepts related to development, developmental areas, developmental principles, factors affecting development and prenatal development will be explained in detail in the antenatal development section.

INTRODUCTION

Development includes continuous quantitative and qualitative changes in the period from fertilization of the organism, that is, from the fall of children to the womb, to death (Aral, Baran, Bulut, and Çimen, 2000; Trawick-Swith, 2013). Development and change move on continuously. Due to the fact that height growth stops, or the growth is too little by age, it is assumed that the development stops. However, both the changes in the body of the individual and the development continues. Furthermore, development should not be confused with the concept of growth. While growth is mostly used for physical properties, development is used to include both physical and other features (psychological). At the same time, according to Senemoğlu (2018), development occurs through the interaction of growth, maturation and learning. In this respect, development occurs over time as the organism continuously progresses through growth, maturation and learning interactions.

Every development means a change. But not every change is a development. Because development is the ongoing progress of the organism from prenatal period onwards in physical, mental, emotional and social aspects of maturation and the interaction of these maturation with learning. Change is the transition of the organism from one state to another. Change is also systematic, sequential and adaptive. In

DOI: 10.4018/978-1-7998-2952-2.ch008

addition, the change can be random, negative or temporary. For example, in the case of a child's learning to skate; skating is both development and change. However, the child's broken foot as a result of falling from the skate is not the development of the child; it is rather a negative change.

Prenatal development (from Latin *natalis*, meaning 'relating to birth') includes the development of the embryo and of the fetus during gestation. Prenatal development starts with fertilization, in the germinal stage of embryonic development, and continues in fetal development until birth. In human pregnancy, prenatal development is also called antenatal development. The development of the human embryo follows fertilization and continues as fetal development. By the end of the tenth week of gestational age the embryo has acquired its basic form and is referred to as a fetus. The next period is that of fetal development where many organs become fully developed. This fetal period is described both topically (by organ) and chronologically (by time) with major occurrences being listed by gestational age.

In this chapter, concepts regarding development as a whole will be given initially together with the principles and factors related to the developmental process. Next, prenatal developmental periods are explained, and prenatal care facilities are listed just before the stages of delivery.

FUNDAMENTALS OF DEVELOPMENT

In this section, developmental terminology will be given, and related concepts are explained according to the literature. These are, namely, growth, maturation, learning, readiness, critical period and education.

Growth

It is the quantitative (numerical) increase in the body mass of an individual such as weight gain, height, size and weight growth of organs, etc. (Aral et al., 2000; Rathus, 2006; Myers, 1996). Growth begins with fertilization and continues until the end of puberty. Growth ends when puberty ends. In addition, the rate of growth varies according to the age and development level of the individual. For example, while the growth rate is very high in infancy and preschool period, it slows down during school and accelerates again in adolescence.

Maturation

It is the physical and mental readiness of an individual to perform a behavior. In other words, it is the state of an individual's body limbs to reach a state that can exhibit that behavior. Maturation does not have to be learning. However, maturation is not possible without growth (Ünver, 2013; Metindoğan Wise, 2015). Maturation refers to the emergence of one's innate potential over time. In this sense, maturation occurs spontaneously, and one does not need to make any effort because maturation is determined by heredity. That is, if the parents started to speak early when they were a child, their children also tend to speak early. In short, maturation has no effect on the environment. However, unusual environmental conditions such as exposure to radiation, climatic conditions and environmental pollution affect maturation. If we take development as a product, we can consider maturation as a process that depends on development. Maturation has a certain pace as well as growth. Unless an organism has reached the level of maturity enough to achieve a particular behavior, it cannot learn or do it properly. Therefore, no matter what we do, we cannot accelerate this development with external effects. From this point on, the child should not be forced to perform some developmental tasks if he / she cannot perform at the expected age. A child who has just learned to hold a pencil cannot be expected to write properly or to write as an adult. The child cannot learn how to walk, no matter how long we practice walking, until the muscle and bone structure of the child is mature. For example, a two-year-old child goes to the sink for adults with the help of a mother and with the help of a stool; but when he reaches 3-4 years of age, he or she reaches without the need for assistance and without a stool. In this case, both growth and maturation seem to have an effect.

Learning

One of the most important concepts of development is learning. Because for the development to take place, besides the growth and maturation, the learning process should be included. Individuals gain knowledge and skills by interacting with their environment in order to do something new in their lives. If these acquired knowledge and skills are persistent changes, learning occurs. According to Senemoğlu (2018), learning is defined as a relatively permanent trace behavior change that occurs in behavior as a result of repetition and experience in the individual. Thus, in order to learn, it must have the following characteristics:

- Observable change in behavior.
- The change in behavior is relatively permanent, i.e. continuous.
- The change in behavior has been gained as a result of life experience.
- The change in behavior is not only a result of growth and is not affected by factors such as fatigue and disease.

From the moment individuals are born, they become capable of doing some skills with the effect of maturation and growth. But without learning it becomes ineffective. For example, in order for a child to eat, he or she must first grow up and reach a certain maturity level to hold the spoon. Then it is necessary to learn from the individuals around how to take it to the mouth and how to eat. When individuals are born, unlike other living things, their instinctive behaviors they bring through are very few. Therefore, they perform life-long knowledge, skills, attitudes and behaviors (e.g. speaking, eating, personality traits, etc.) through learning (Aral et al., 2000).

Readiness

It covers all the features that the individual has before, in short, the preliminary learning in order to realize a new learning situation. Readiness is a broader and more comprehensive concept than maturation. It contains knowledge, skills, attitudes, behaviors, growth and maturation. Readiness is to reach the maturity level for what an individual need to do (Aral et al., 2001; Ünver, 2013) and is a product of the interaction of maturation and learning. In addition to the necessity of reaching the maturation required for an individual to be able to do a task, he / she must have acquired the necessary knowledge, skill and attitude. Readiness, thus, includes both the concept of maturation and the pre-qualification required for a task. For example, the child who will learn to perform the diving action must reach the maturity to be able to comprehend the features of the diving tools, be ready to learn and have the basic knowledge and skills necessary for diving. Trying to teach a child who is not ready to learn a life can leave some permanent traces of negative behavior in the child.

Critical Period

There are certain times, i.e. critical times, for an individual to perform certain behaviors. If it does not exhibit behaviors in these critical periods, it is very difficult to exhibit that behavior in later periods. Therefore, it can be said that children not being exposed to the necessary stimuli during these periods may have a significant effect on their future life. These periods are called critical periods. For example, some diseases that a 3-4-month pregnant woman experiences, would cause her to give birth to a baby with anomaly.

Education

Education is the process of deliberate change in the behavior of an individual through his own life experiences. Education is a process that maintains its existence since the birth of the individual and exists at different levels in every society and continues in every society regardless of the level of civilization. For this reason, it is aimed to be able to make plans and programs in the education process, to know and understand the level of development of the child in order to support the child's development areas and to prepare the necessary educational environment in line with this information. From this point of view, education has four purposes. First, education tries to cultivate the individual. Culturing is the process of ensuring that the cultural values that an individual develops by accumulating from generation to generation are appreciated by other individuals, and that they are able to contribute to the development of these values. The second aim of education is to try to socialize the individual. It is the process of ensuring that the individual adopts and applies the written or unwritten rules set by the nation and society and reaches the competence that can contribute to the development of the individual. The third aim of education is to ensure that the individual is productive. Being productive can be defined as the process of ensuring that he / she has a job or profession that can provide for himself and his family and that is suitable for his / her abilities. The fourth and final aim of education is to ensure the individualization of the individual. Individualization can be defined as the process of developing the individual's personality and directing the latent powers of the individual with inheritance to the ability to use them for the benefit of himself, his family and nation (Dewey, 1930; Brown, 1993).

MAJOR PRINCIPLES OF DEVELOPMENT

- Regarding the development of an individual from the prenatal period onwards, there are some principles which are as follows (Aral et al., 2001; Senemoğlu, 2018; Durmuşoğlu, 2018; Aslan, 2019):
- Development occurs through the interaction of heredity and the environment. It cannot be said that only heredity or only the environment is effective on development. Inheritance refers to the physical and psychological characteristics transmitted to the individual from his / her parents by hereditary units (genes) on chromosomes. The individual resembles his / her parents in many ways. On the other hand, the baby is able to improve the characteristics it brings from its parents by interacting with and in an environment. The environment includes all the conditions in which the cells live during and after fertilization until the cells multiply and become organisms. The organism has internal and external environment. In this environment, the individual must live

from fertilization to death. Some aspects of development are under the influence of heredity and some aspects are more influenced by the environment. In short, heredity and environment are effective together. Because the effect of inheritance is mentioned in features such as eye color and hair color, temperament, temperament, and how personality will be formed by the effect of the environment.

- Development is continuous and involves certain stages. Development covers the process from prenatal period to death. Development in this process is forward, during the prenatal period and the first two years the development is rapid but slows down in later years. It is therefore continuous and sequential. Each development period is the pioneer and preparer of the next period. For example, gross motor skills (hand, arm, leg movements) precede fine motor skills (use of finger muscles). For this reason, children use their arms and hands before their fingers. Walking takes place before scissors and pencil holding skills. The child walks first, then runs. Later, it is possible to cut in detail with scissors.
- There are individual differences in development. Each individual has its own and different characteristics. One of the two children of the same age can be tall, one is short, or one has curly and the other has straight hair. For example, there are differences in reading rates when two children aged seven are compared. There are also differences in their ability to paint.
- Development follows a regular sequence. Development follows a sequence from head to toe (head development occurs faster than other organs from fertilization onwards), inward to outward (body's internal organs are formed, then skin, outer surface is formed) and from general to specific (in children first develops great muscle control and then small muscle control).
- There are critical periods in development. Children are expected to exhibit certain behaviors at certain periods. It may occur with children who are not exhibited at certain times or who have been forced to exhibit before. For example, one and a half to three years is a critical period for toilet training.
- Development is a whole: There are different areas of development in the life of the individual and all areas of development are integrated with each other. It cannot be separated. Changes or disruptions in any development area affect other development areas. These areas of development include physical development (encompassing fine and gross motor skills), cognitive development (covering a wide range of abilities such as thinking, intelligence, language, perception, understanding, attention, memory, problem solving, reading and writing), social development (interpersonal) emotional development (includes the process of expressing and controlling their own emotions), language development (covering the communication process such as speaking, listening), sexual development (sexual maturation of the individual, adopting his / her own sexual identity and acting accordingly) (Berk, 2013; Rathus, 2006; Bredekamp, 2014).
- Development continues in turns. As one development area accelerates, the other development area may slow down. For example, the language development of a child who starts to walk or run may slow down.

In the above headings, basic concepts and principles of development are explained. Knowing the factors affecting the development of the baby in the womb and in the postnatal period plays an important role in the healthy development of the process. Therefore, the factors affecting development are explained below.

FACTORS AFFECTING THE DEVELOPMENTAL PROCESS

Scientists have given different opinions on the factors affecting development at various times. Some scientists have argued that heredity is effective on development, while others have argued that environment is effective. Today, it is accepted that both heredity, environment and time are effective on development (Bee and Boyd, 2009). These factors are explained below as prenatal, delivery and postnatal factors.

Factors Affecting Prenatal Development

Heredity is the leading factor affecting development before birth. Inheritance is the traits that are passed on to the individual through genes transferred from the mother and father. The genetic material (hereditary structure) in the organism is called the genome and the genes are made up of DNA. DNA contains information about all the characteristics of living things. In this information, genes are dominant (dominant / dominant: brown eye, A, B, AB blood groups) and recessive (recessive / faint: yellow hair, straight hair, blue eye). The number of chromosomes in human is 23 pairs. These characteristics, which cannot be changed by the effect of the environment, are transferred from generation to generation by randomly bringing together 23 different chromosomes from each parent during fertilization. (Sax, 1918).

The inheritance structure from mother to father is called genotype. Genotype and intelligence and physical characteristics are largely determined. Phenotype is the form of this genetic structure that can be observed from outside, that is, it is physically manifested in the individual. From a genotype, many phenotypes can occur, and phenotypes are determined by the environment. In other words, the environment is effective in the formation of phenotypes before birth, during birth and after birth. The characteristics of the child will be girls or boys, eye-hair color, personality-body characteristics are determined during fertilization. It is argued that genes cause some prenatal problems. Because these problems can sometimes lead to physical disorders, sometimes mental disabilities and sometimes unexpected adverse health conditions. Some anomalies, diseases or disabilities that may occur in children may also be attributed to genes and chromosomes. For example, down syndrome is caused by a deficiency in the division of the twenty-first chromosome pair during meiosis. Chromosomal disorders other than Down's syndrome generally affect the development seriously and may cause miscarriage at birth. In addition, the mother's age and mother-infant blood group discrepancy can also cause negative developments. Another important factor affecting development is the environment. All the external stimuli since fertilization and many negative experiences may also affect the development of the fetus in the womb. For example, the mother's exposure to infectious disease, mental state, radiation during pregnancy, drug use, drugs, cigarette use, accident, such as nutritional disorders (Aral et al., 2000; Er-Sabuncuoğlu, 2015; Senemoğlu, 2018). In addition to heredity and environment, time is another factor that affects prenatal development. It is known that some events are more important and effective in certain time periods in development. In particular, these periods are known as critical periods in development. This time period cannot have the same effect on the development of expected behavior before or at the next stage. For example, Freud argues that the first six years of life are more important in personality development than in other years, and negative experiences in this period indicate that the individual will leave negative traces in personality development (Caspi and Roberts, 2009; Maree, 2018). Therefore, the experiences of these first six years emphasize that the child will largely determine what kind of an adult he will become in the future. In the emergence of certain behaviors, it is stated that the quality and quantity of stimuli within certain time periods can support or hinder development (Lewis and Goldberg, 1969). For example, viral infections in the first months of pregnancy may pose a risk of prenatal anomalies in the fetus (Petropoulou, Stratigos, and Katsambas, 2006).

Factors Affecting Development During Delivery

There are many factors affecting the development of the child at birth. The first of these is whether the birth is performed at home or in a fully equipped hospital. Because any complication during childbirth may cause life-long risks for the child's development. Also, delay in delivery and inability to deliver on time, failure to perform the appropriate method, cord entanglement or prolonged stay in the birth canal may cause insufficient blood supply to the brain and become oxygen-free. Such situations may cause the baby's brain to die due to the death of cells in the mental functions and developmental problems in the baby can cause. Reversal of the baby at birth, the use of instruments such as vacuum or forceps by non-specialists, and unexpected accidents at birth are among the factors (Senemoğlu, 2013; Er-Sabuncuoğlu, 2015).

Factors Affecting Postnatal Development

There are many factors affecting the development of children after birth. These factors affect each other indirectly or directly. The first is the exposure of children to the environment and chemicals. For example, drinking water and air affect many areas of development in children's environments. According to the WHO (2002) report, 40% lead is found in children's drinking water worldwide and is among the most important risks. South Asia, Bangladesh, Mexico and Taiwan due to arsenic in drinking water in settlements such as children's IQ levels were found to be lower than other children (WHO, 2002; Wasserman et al., 2006). The other factor that affects child development is poverty. Children in low- and middle-income countries worldwide are below their potential development levels due to poverty, malnutrition and care, and inability to benefit from learning opportunities (Walker et al., 2007). Moreover, developmental processes may be adversely affected by the exclusion of children who are in disability, orphaned children who have been abused, children who are prone to crime or have committed a crime.

Another important factor in the child's development process is his or her parents. Children live with their parents from the moment they are born. However, in some cases, such as death and divorce, they may have to live with a single parent. There are factors that adversely affect the development of children in both parent and single parent life. Parents' working conditions, living standards or attitudes and behaviors towards parents' children are important factors that may affect the development of the child. The relationship between parents and children varies from culture to culture, and its results significantly affect all areas of development of children, especially their social-emotional and cognitive development. For example, there is a great difference between the behavior of the parent of democratic parent attitudes and behaviors and that of the parent of an overprotective or authoritarian attitude. In democratic parental attitudes, children are respected children who can express their thoughts and opinions in a reassuring and supportive environment, and most importantly, they are recognized as having the opportunity to know their own self (Gordon, 2014; Mackenzie, 2014). In authoritarian parenting, parents try to control their children's behavior in a repressive way. As a result of these attitudes and behaviors, children may be shy, low self-confidence, unable to act on their own, not knowing themselves, and experiencing adaptation problems (Çağdaş, 2008). In overprotective parental attitudes, parents are extremely protective, considering that their children will be harmed. This environment causes the child to grow up as an addictive individual who cannot move alone, while in the permissive parental attitude, the children control their parents and the parents are overly tolerant, rules and limits are not defined in the family (Semerci, 2015; Nelsen et al., 2002; Demir and Yavuzer, 2016).

In addition to parental attitudes that have an impact on the development of children, negative factors such as exposure to violence and sexual abuse are unfortunately too many. In addition, there are children who are under institutional care due to poverty in the family, death of one or all the parents, abuse, violence. It can be said that the development of children under institutional care is below the potential for growing children alongside their parents. Because these children may be deprived of the factors that affect their development, such as inadequate care, nutrition and love, away from a warm home and family environment. In a study conducted by Yaşar (1996) in Turkey, it was found that growth retardation, anxiety and fears were high in the children in need of protection in the kindergartens and orphanages, and there were borderline or mild mental retardation.

Factors affecting development are examined under three main headings. The development of the baby in the womb is explained below.

DEVELOPMENTAL PERIODS

It was stated that individuals' development took place in periods and that each period had characteristics to be acquired. Development is reviewed in periods to better organize and understand. In development; prenatal period, infancy (0-24 months), first childhood (3-6 years), middle childhood (7-11 years), adolescence (12-18 years), adulthood and old age section. This section describes prenatal development.

Prenatal Development Periods

The individual is in continuous development and this development process is divided into two parts as prenatal and postnatal development. The prenatal period covers the period from fertilization to the moment of birth. The prenatal development process is divided into three parts as ovum period (zero to two weeks), embryo period (two to eight weeks) and fetus period (from the eighth week to birth). These periods prepare the baby for birth and after birth. These three terms are explained in detail below.

Ovum Period

The combination of female egg cell (ovum) and male egg cell (sperm) is called fertilization (Carroll, 2018). A new cell is formed with 23 chromosomes and 46 chromosomes from women and men and this new cell is defined as zygote.

The zygote moves from the fallopian tubes into the uterus. In this process, it begins to divide. In this first period, the volume of the embryo remains constant but there is a numerical increase. The divisions continued as increasing by 16 cells on day 3 and 58 cells on day 4, and the dividing cells were in the form of stacked clusters. This shape is called morula. When the resulting morula reaches the uterine cavity on the 5th day, the secretion of the glands begins to infiltrate slowly into the morula mass. The cells become a liquid filled sphere for what is called a blastocyst. The inner cell cluster (embryoblast) in the inner part of the blastocyst is called the outer cell cluster (trophoblast) in the outer part. Trophoblasts are composed of placenta, embryo of embryoblasts. In short, the structures that will feed and protect the developing organism begin to take shape (Yücel, 2015).

The outer layer (ectoderm) and inner layer (endoderm) are formed in the first two weeks and the middle layer (mesoderm) is formed in the third and fourth weeks.

- Ectoderm: The formation of brain, spinal cord organs, nervous system, hair, skin, teeth, nails, secretory glands,
- Endoderm: The formation of digestive and respiratory system, bladder and thyroid gland,
- Mesoderm: Lymph system, muscle, bone, heart, circulatory system, kidneys and reproductive organs provides the formation (Jirasek, 2001).

Blastocysts are now ready to join the uterine wall, the endometrial layer. During the first week of fertilization, blastocysts merge with the endometrial layer. This condition is defined as implantation (Moore et al., 2013).

In addition, the characteristics of the mother and father and sex is determined by moving through chromosomes. The mother has XX chromosome and the father has XY chromosome. If the X chromosome comes from the mother and the father, a girl is born, and if the X chromosome comes from the mother and Y chromosome from the father, the boy is born (Levine and Munsch, 2014). Miscarriages are most common during this period.

Embryo

The second stage of prenatal development, called the embryo period, starts from the second week and continues until the eighth week. At the end of the second week after fertilization, the implantation is completed and is usually located on the posterior wall of the uterus. When implantation is not located in the posterior wall of the uterus, i.e. in other places (for example, in the uterine cervical mouth, abdominal cavity, over the ovary), it may cause problems such as preterm labor, ectopic pregnancy or bleeding in the later stages of pregnancy. Some unhealthy pregnancies result in spontaneous abortion in this period. Although approximately half of the pregnancies terminating with spontaneous abortion are related to chromosomal anomalies, there may be many pregnancies that have completed their pregnancy abnormally and gave birth to the anomalous baby. During the ongoing pregnancy, amniotic cavity, umbilical cord and placenta begin to form at the end of the second week.

In these early weeks of pregnancy, the embryo is attached to a tiny yolk sac which provides nourishment. Placenta begins to develop in the third-fourth week and is completed in the third-fourth months. Placenta is a vital bridge between the mother and the fetus. It performs the respiratory and excretory function between mother and fetus. The weight of the placenta is about one-sixth of the fetus, that is, about five hundred grams, two-three centimeters thick, and an average diameter of twenty-twenty-two. Placenta's tasks are: Carries oxygen from mother to fetus - carbon dioxide from fetus to mother. It transports glucose, protein, fat, water, vitamins, minerals and electrolytes from the mother to the fetus. It transports the fetus's metabolic wastes to the mother's blood and ensures that it is excreted by the mother's kidney. The enzymes of the placenta help to eliminate the effects of some toxins. At the same time, the placental barrier protects the fetus from harmful substances found in maternal blood. The placenta is also responsible for the production of many hormones and hormone-like substances (MEB, 2012).

For the feeding of the fetus, the trophoblast cells fuse with the decidua and become a single membrane, forming the chorion membrane first. As the chorion membrane is formed, the amniotic membrane and amniotic sac begin to develop. There are two pieces of amniotic membrane inside the amniotic sac and the chorion outside. Between these two membranes are phospholipids. Phospholipids help to start birth. The first pouch that surrounds the embryo is amniotic fluid. The tasks of amniotic fluid are as follows: It protects the fetus from external influences and traumas. It prevents the amniotic membrane from sticking to the fetus; enables the exchange of water and substances between the fetus and the mother, prevents heat loss of the fetus, protects against infections, allows the fetus to move freely, it also helps the musculoskeletal system to develop, allowing it to move freely. Also, it helps the mechanism of labor by providing the aseptic fluid to be washed at birth at birth (MEB, 2012).

The umbilical cord begins to form in the third-fourth weeks when the placenta begins to form. From the fourth week of fertilization, the embryo develops and feeds through the umbilical cord. At first it looks like a primitive handle. However, the size reaches one to three meters. Longer or shorter may cause problems in the development of pregnancy. The umbilical cord contains two arteries that transmit nutrient-laden blood and carry waste (MEB, 2012).

In the third week, the mother understands that she may be pregnant due to the absence of menstrual bleeding. In the fourth week, the embryo continues to grow rapidly. Now the brain and spinal cord tissue of the embryo begins to form. Muscles, spine, ribs and digestive system appear. Heartbeat starts but the mother can hear the heart sound from the sixth week. Between the fourth and eighth weeks, all major internal and external organs develop. (Moore et al., 2013). At the end of the sixth week, the embryo is approximately 22-24 millimeters in length, and the ear, nose, eye, mouth, arms, legs, fingers and toes begin to appear in draft. At the end of the seventh week, the first body movements begin in the embryo. The mother can feel these movements in the sixteenth weeks. However, some mothers may feel this movement even though they are very weak. In the eighth week, the tongue, lips, teeth and palate structure of the embryo begins to form. Changing body ratios allow the embryo to become steeper (Jirasek, 2001; Moore et al., 2013).

Fetal Period

The period from the second month to the moment of birth is the period when the embryo period ends, and the period of the fetus begins. In this period, development continues rapidly, and these developments are divided into months.

- **Third month (weeks 12-16):** In the twelfth week, the fetus has all the organs. Development continues rapidly. The fetus kicks, bends his arms, turns his head. His face becomes prominent. Features are noticed in the eye sockets and nose. At the twelfth week, the external genitalia become more pronounced. However, as a result of the testis and penis of the male fetus, the sex of the baby can be differentiated. Parents can learn the sex of the baby during this period. Hair and nails begin to grow. Vocal cords are formed. The fetus is seven to nine inches tall and weighs twenty-five to twenty-eight grams (MEB, 2012; Durualp, 2019).
- **Fourth month (weeks 16-20):** The fetus is twelve inches tall and weighs one hundred and ten grams. The movements of the fetus can be felt by the mother. The first excrement of the fetus occurs in the intestines. During this period, a white cheese-like substance called vernix is formed,

which prevents the baby's skin from cracking in amniotic fluid. It also consists of a large number of nerve cells. Gender is determined correctly during this stage (MEB, 2012; Durualp, 2019).

- **Fifth month (weeks 20-24):** Twentieth week coincides with the middle of pregnancy. The eyes of the fetus begin to open. The hairs called vernix and lanugo cover the whole body of the fetus. Its hair is colored and the heart beats. The heartbeat of the fetus can be heard from outside with a stethoscope. In the twentieth week, all systems can be assessed by ultrasound to a large extent. At the end of this month, the fetus reaches twenty-five - thirty inches in length and weighs two hundred and fifty grams. Since the swallowing reflex also begins, the fetus may swallow the amniotic water surrounding itself (MEB, 2012; Durualp, 2019).
- **Sixth month (Weeks 24-28):** Babies born during this period have a chance to survive and are therefore called the "age of survival". However, a baby born in the seventh and eighth months needs oxygen support to breathe, even though the brain's respiratory center develops. Therefore, it is not ready for birth. The skin of the fetus is wrinkled, begins to store fat, eye movements and eyebrows and eyelashes movements can be recognized. Muscles and nails are developed. Gain the ability to capture the thumb. The fetus is twenty-eight to thirty-six inches tall and weighs six hundred to seven hundred grams (MEB, 2012; Durualp, 2019).
- **Seventh month (Weeks 28-32):** This month, the fetus looks old because the subcutaneous fat depot has not fully developed. The skin is red. Perceptions such as taste, sound and smell develop. As the fetus is in the uterus; the uterine growth rate slowed down. The respiratory movement of the fetus can be seen, and the fetus can rotate around itself. The fetus can live outside the uterus but survives in intensive care. In this month, the fetus is thirty-five to thirty-eight inches tall and weighs one thousand hundred thousand two hundred grams (MEB, 2012; Durualp, 2019).
- **Eighth month (weeks 32-36):** Lung development accelerates and the adaptation process to normal life begins. The skin color changes from red to light pink. The appearance begins to improve as the elderly because subcutaneous fat increases. Because the fetus grows rapidly, it stops its rotation around itself. However, the hands and feet are in motion. The amount of amniotic fluid remains constant as of the thirty-third week. If it is born at the end of the thirty-fifth week, it can adapt to life without any problems. At the end of this month, the fetus reaches a length of thirty-eighty-forty-three centimeters, weighing one thousand eight hundred and two thousand two hundred grams (Jirasek, 2001; MEB, 2012; Yücel, 2015; Durualp, 2019).
- **Ninth month (36-40 weeks):** The fetus is about to be ready for delivery. All development has been completed. It can now swallow, stretch, startle and sob. Occasionally, it sucks your thumb. Amniotic fluid is reduced and laguno hairs begin to shed. At the end of this month, the fetus is forty-two to forty-nine inches long and weighs two thousand two hundred and two thousand seven hundred grams (Jirasek, 2001; MEB, 2012; Yücel, 2015; Durualp, 2019).
- **Tenth month (Week 40):** Fetus development has completed. It is now ready for birth and life. The hair is long. The skin is covered with pink and abundant vernix. At the end of this week, the fetus is fifty centimeters long and weighs between two thousand seven hundred seventy-three thousand nine hundred grams (Jirasek, 2001; MEB, 2012; Ercan, 2013; Yücel, 2015; Durualp, 2019).

The development of the baby in the womb is explained above. In the process of development in the womb, requirements for a healthy pregnancy is explained below.

PRENATAL CARE

Pregnancy and birth are among the leading causes of female death and disability. It is estimated that approximately four hundred thousand maternal deaths occur each year in the pregnancy process worldwide. Therefore, reducing maternal mortality in our country has been among the development goals. Care before and during pregnancy is important for the health and protection of parents and babies. Healthy preparation of parents for pregnancy helps prevention of transmitting genetic diseases. However, although some mothers know that they are pregnant, they do not seek care before and during pregnancy. In particular, approximately 20% of mothers in the United States do not receive prenatal care (Klerman, 2000; Park et al., 2007; Berk, 2013). There are many reasons for this. For example, being low-income, finding a doctor or dissatisfaction with the care, pregnancy symptoms, psychological stress, family crises, lack of knowledge about the benefits of getting prenatal care can be listed as reasons. It is extremely important to educate the public about the importance of early and regular prenatal care for all pregnant women (Berk, 2013).

In the Prenatal Care Management Guide prepared by the Ministry of Health (2014); it has been stated that while the maternal mortality rate was 64 per hundred thousand live births in 2002, the National Maternal Mortality Study conducted in 2005 reached 28.5, 15.4 in 2012, and the infant mortality rate was 31.5 per thousand in 2002 and 7.4. Strengthening prenatal, delivery and postnatal care services, ensuring the delivery of births in hospitals, increasing the number of health personnel in the priority regions of development especially in the east and southeast, developments in adult and newborn intensive care services in hospitals. and 112 emergency aid and rescue services have contributed a lot to the health care improvement in Turkey.

If there is no problem in pregnancy while providing care; follow-ups are performed once a month from the beginning of the pregnancy to the seventh month, every two weeks in the eighth and ninth months, and every week from the thirty-seventh week.

Especially parents' knowledge is important in this regard. For example, during the pregnancy, the baby's brain development begins at the beginning of the embryo period in the third week. By the sixth week of pregnancy, the brain develops and grows twice as large. At the end of the seventh week, the embryo; mouth, nose, eyes, lips and ears began to become clear. During the fetal period, 250,000 brain cells are formed in the brain approximately every minute. When a baby is born, it has more brain cells that will be needed in life. At the same time, sensory organs start to develop in this period. for example, the development of the first sense of touch occurs when the baby is about 7.5 weeks old. The fetus continuously detects soft touches until they are born. After the touch sense, the taste sense develops between 7.5-15 weeks. 21 weeks of hearing and between the ages of 15-23 develops the sense of vision. The development of other organs and brain development also affect the development of the baby significantly. Therefore, parents can provide a good environment for their baby before birth and take precautions for brain development and physical development. These measures can be listed as items.

Mother;

- Should walk and exercise during pregnancy.
- Should stay as far away as possible from chemicals and radiation during pregnancy.
- Should not use medication unless necessary.
- Should avoid alcohol and cigarettes.
- Should avoid stress.

- Should use folic acid during pregnancy. In addition, foods such as green leafy vegetables and orange juice contain folate. They should take care to consume these foods frequently.
- Should take vitamin D during pregnancy. In this process, she may benefit from the sun or vitamin D should be taken.

All these issues must be under the control of the doctor.

First Follow-up

It is performed within the first 14 weeks of pregnancy. The health personnel dealing with the pregnant women at the first visit records personal information (age, marriage of relatives, education level, etc.), family history of health, habits, medical and pregnancy-related previous and current information. The personnel physically examine the pregnant woman and performs laboratory tests (urine analysis, blood tests, other tests deemed necessary).

Second Follow-up (weeks 18-24)

It is recommended that glucose screening test and ultrasonography be performed between 2 weeks of gestation. In addition to those in the first follow-up, information is given about inability to feel the fetus movements and rapid weight gain.

Third Follow-up (Weeks 28-32) and Fourth Follow-up (Weeks 36-38)

Pregnancy examination is performed. In addition to the first two follow-ups, information on pregnancy and delivery, planning where to deliver and by whom to assist delivery and breastfeeding. Unfortunately, in most developing countries, prenatal care is not sufficient and at the desired level. It was reported that most problems that may affect the mother and the baby negatively are detected during pregnancy and not enough care is provided (Coşkun, 2008). A good prenatal care service takes place before the pregnancy starts.

STAGES OF GIVING BIRTH

Birth takes place between thirty-eighth and fortieth weeks of prenatal development. About two weeks before birth, the baby's head goes down and settles there. Days and hours before the birth of the mother's body begins to contract. A sign of birth has started, the cervix opens and closes the mucus plug opens and a reddish discharge comes (Lieberman and Holt, 2005).

Birth has three stages (Berk, 2013; Trawick-Swith, 2013; Durualp, 2019):

- **Expansion and thinning of the cervix:** This is the longest stage of delivery. It may take twelve to fourteen hours for first births and four to six hours for later births. Contractions begin with a fifteen to twenty-minute break and become more frequent and severe. If the uterine contractions become more frequent and stronger, the cervix opens and becomes thinner, thus leading from the uterus to the birth canal or vagina. As the openings continue, the amniotic fluid will rupture enough to allow it to escape.

- **Fetus removal**: It is a shorter stage of labor. Fifty minutes for the first birth may occur in twenty minutes for subsequent births. At this stage, the mother feels the urge to push so that the fetus can move out involuntarily and irresistibly. Together with the mother's pushing movements, the fetus moves down the birth canal and moves from the vagina to the outside world.
- **Expelling the placenta:** When the baby first comes to the world, it breathes and usually begins to cry. Once it breathes, the dependence on the food from the placenta disappears and the umbilical cord is cut and tied. The final step is the removal of the placenta. It is the shortest process among the birth stages. With the last few contractions and strains of the mother, the placenta is separated from the uterine wall and expelled.

When approaching the completion of the baby's existence in the mother's womb, pregnant women are most curious about which birth method they should select. The methods of delivery and the problems that they may encounter in these methods are explained below.

MEDICAL INTERVENTIONS DURING DELIVERY

Above mentioned are normal or natural birth stages and have multiple advantages for mothers and babies. With the effect of hormones after birth, breast milk comes immediately, the baby can be started to breastfeed immediately, the mother begins to perform daily tasks in a short time, the mother has the chance to give normal birth later, the mother has a shorter stay in hospital, postpartum emotional bond between the baby and the mother is established more quickly and they are not subject to complications compared to the mothers and infants who are exposed to anesthesia (Yücel, 2015). However, in some cases, it may be necessary to use different methods to assist birth instead of normal birth for the health of the mother and the baby such as inducing labor, use of assistive devices or cesarean section.

Doctors sometimes must carry out medical interventions at birth. An electronic device (fetal monitor) is used to monitor the fetal heart rate at birth. Fetus monitors are mandatory in most US hospitals. This monitor is used by placing the baby directly under the baby's head during birth or along the cervix (Berk, 2013). An episiotomy is also performed just before the baby is taken out. Episiotomy is a method of preventing tears and facilitating birth by making a cut between vagina and anus during delivery (Trawick-Swith, 2013; Sigelman and Rider, 2015).

Another method is to remove the baby from the birth canal by means of a device to accelerate the delivery at birth. This is called birth clamps (forceps). If the baby cannot go out as a result of the mother's strains at birth, it is provided to be removed by using vacuum. The vacuum consists of a plastic cap placed on the head of the baby and the tube to which it is attached. However, instrumental delivery has decreased in recent years. Because it increases the risk of brain damage in the births performed with forceps and vacuum tool, and the plastic cap attached to the baby may cause bleeding under the skin of the baby and outside the skull (Trawick-Swith, 2013).

Medium doses of medication are used to reduce the pain and relaxation of the mother during childbirth. Particularly recently used epidural analgesia is applied by administering a regional painkiller to a small area of the lower part of the spinal cord to control pain during childbirth. With this painkiller, the mother feels the labor pains and can perform straining movements. Thus, the fetus can be extracted with normal delivery method (Berk, 2013). In the event that the health of the mother or baby is compromised, the induction method is used. In general, this method is performed by opening the water sac or by giving the mother the synthetic oxytocin hormone that initiates contractions. It is a natural, artificially initiated birth. Although induced labor is done according to the mother's or doctor's opinion, it can cause some problems. For example, contractions at birth may last long enough to reduce the amount of oxygen required for the baby, and control may not be in the hands of the mother at birth because she has different functions than normal births, and even if the induced labor is performed before the mother is physically ready for birth, she may fail and increase the likelihood of cesarean delivery (Berk, 2013).

TYPES OF DELIVERY

Home or Hospital?

Birth practices may vary from culture to culture. Depending on the place of residence and the birth process, the hospital environment or the home environment may or may not be preferred. For example, although home birth is common in countries such as the UK, the Netherlands and Sweden, but in America it is less preferred. It is possible for most healthy pregnant women to have their births at home without any intervention. However, it is not foreseen that no complications will occur during the delivery process. Therefore, in most countries, the delivery of all women in hospitals is recommended as the safest way. However, medical procedures and non-specialist physicians in hospitals can cause unnecessary interventions and unnecessary complications. It is possible for the pregnant woman to be supported by the midwife for the delivery of the planned home and if necessary, the pregnant woman can be referred to the hospital or doctor immediately. According to the researches, hospitalized deliveries are not safer than planned home deliveries; complications may have been reported (Olsen and Clausen, 2012).

At the same time, the most important purpose of normal birth is to avoid medication and medical intervention. It has been found that the number of medications and medical interventions decreases with birth in such houses or similar places and increases the satisfaction of mothers' pleasure and feeling from birth (Hodnett, Downe and Wlash, 2012). However, if the personnel who will assist birth at home are not well educated and do not know what to do in case of emergency, problems may occur for the mother and baby and even deaths may occur. Therefore, there are places where there are emergency personnel and trained personnel, that is, hospitals are the first places to apply for a healthy birth.

Normal or C-Section?

Caesarean delivery method is used when normal delivery is not possible. Cesarean section is a surgical method of delivery in which the doctor cuts the mother's abdomen and removes the baby from the womb. Particularly, the frequency of cesarean delivery increases due to the fear that vaginal birth is not safe, the pregnancy and fetus will be damaged, breech presentation of the fetus, multiple pregnancy, maternal diabetes, blood pressure or the absence of painless births and the mothers especially prefer. In contrast to normal / natural birth, it takes more time for the mother and the baby to heal during cesarean delivery, increases the likelihood of anesthetic substances passing to the baby, and it may take longer for the mother's milk to arrive after birth. Therefore, it is emphasized that the number of cesarean sections has increased in recent years and unnecessary cesarean deliveries have been performed (McDonagh, Osterweil, and Guise, 2005; Özçerezci Güner and Kavlak, 2016).

SOLUTIONS AND RECOMMENDATIONS

Mothers who do not receive good medical care during delivery may have problems at birth. The first is that the fetus cannot get enough oxygen. If the mother's birth canal does not widen sufficiently at birth, the fetus may stay in the birth canal longer than necessary and may have difficulty. As a result, it cannot get enough oxygen. When oxygen is not available, it can cause the formation of harmful acids in the blood of the fetus and the lack of vital substances in the blood. In the fetus, the possibility of brain damage (such as cerebral palsy) increases as a result of being deoxygenated (Clark, Ghulmiyyah and Hankins, 2008). Deoxygenation of the baby might be due to the nuchal cord (cord wrapped around the neck of the baby) or breech presentation (baby's legs come first).

Another problem that causes vital problems and results in a high rate of infant death is the early separation of the placenta from the uterus or the closing of the cervix by holding the placenta too far down the uterus. In some cases, if the baby survives the birth process in a healthy way, it may not indicate that everything is going well. For example, when the baby is born, it may not be able to breathe and may be deprived of oxygen. If the baby does not reach normal respiration in the first ten minutes, it is possible that brain damage may occur (Berk, 2013).

Babies born at least three weeks before the 38-week gestation period and babies weighing less than two thousand five hundred grams are called "premature". In particular, it was emphasized that most premature babies born less than one thousand five hundred grams carry risks and problems (Dombrowski, Noonan and Martin, 2007). For example, structural abnormalities in the baby's brain, frequent illness, lack of attention, excessive mobility; delay in speech, sensory damage, emotional and behavioral problems, such as learning difficulties may occur in the case of premature babies (Clark et al., 2008).

CONCLUSION

"What adults will give to the child is to give shape and color to all instincts, reactions and abilities that the child brings together at birth" (A. Adler)

The life of the individual begins as a result of the merging of the mother's egg with the sperm cell from the father. This period lasts approximately two hundred and eighty days and it is very important to know the developmental stages of the baby and its effects on the health of the mother in terms of both the health of the mother and the health of the baby. Pregnancy is a challenging process in which the mother experiences physical and psychological changes. The mother may have very different concerns and worries during this period both for her and her baby. Particularly, it may cause problems during pregnancy due to concerns about the baby, the inability to give birth and the fear of not being a good mother. On the contrary, a well-prepared mother will give birth to her baby more easily. Therefore, during pregnancy, the mother and the baby, to be healthy in terms of mental and physical sense, this period should be supported by many applications. for example, childbirth preparation courses are becoming widespread. In this process, both mother and father have the chance to learn what they must do to become a good parent. In addition, there are many articles, books and journals written in the field such as pregnancy process, what should be done in this process and what should not be done, baby development, factors affecting and improving their development. Parents should do research for the new individual they will bring to the world and gain the knowledge, experiences and experiences they lack. Because every new knowledge, experience, and experience gained through education in this process is a step towards the future of the child.

REFERENCES

Aral, N., Baran, G., Bulut, Ş., & Çimen, S. (2000). *Eğitimde drama*. İstanbul: YA-PA Yayınları.

Aral, N., Baran, G., Şenay, G., & Çimen, S. (2001). *Çocuk gelişimi I*. Ankara: YA-PA Yayınları.

Aslan, D. (2019). Gelişim ile ilgili temel konular. In A.K. Akyol, (Ed). Erken Çocukluk Döneminde Gelişim I (pp. 82-109). Ankara: Anı Yayıncılık.

Bee, H., & Boyd, D. (2009). *Çocuk gelişim psikolojisi (O. Gündüz Çev.)*. İstanbul: Kaknüs Yayınları.

Berk, L. (2013). Bebekler ve çocuklar, doğum öncesinden orta çocukluğa. Ankara: Nobel.

Bredekamp, S. (2014). *Effective practices in early childhood education: building a foundation*. USA: Pearson.

Brown, L. S. (1993). *The Politics of Individualism: Liberalism, Liberal Feminism, and Anarchism*. Black Rose Books.

Çağdaş, A. (2008). *Anne baba çocuk iletişimi*. Ankara: Kök.

Carroll, M. (2018). *Clinical Reproductive Science*. John Wiley & Sons. doi:10.1002/9781118977231

Caspi, A., & Roberts, B. W. (2009). Personality development across the life course: The argument for change and continuity. *Psychological Inquiry*, *12*(2), 49–66. doi:10.1207/S15327965PLI1202_01

Clark, S. M., Ghulmiyyah, L. M., & Hankins, G. D. (2008). Antenatal, antecedents and the impact of obstetric care in the etiology of cerebral palsy. *Clinical Obstetrics and Gynecology*, *51*(4), 775–786. doi:10.1097/GRF.0b013e3181870994 PMID:18981802

Coşkun, A. (2008). Yüksek riskli gebeliklerin fetüs/yeni doğana etkisi ve hemşirelik yaklaşımı. In T. Daloğlu & G. Görak (Eds.), Temel neonatoloji ve hemşirelik ilkeleri içinde (pp. 57-94). İstanbul: Nobel Tıp Kitabevi.

Demir, İ. & Yavuzer, H. (2016). Yeni kuşak anne-babalar ve çocukları. İstanbul: Remzi.

Dewey, J. (1930). *Individualism Old and New*.

Dombrowski, S. C., Noonan, K., & Martin, R. P. (2007). Low birth weight and cognitive outcomes: Evidence for a gradient relationship in an urban, poor, African American birth cohort. *School Psychology Quarterly*, *22*(1), 26–43. doi:10.1037/1045-3830.22.1.26

Durmuşoğlu, M. C. (2018). Çocukta gelişim ve öğrenme. In F. Temel & N. Aral (Eds.), Gelişimsel Destek Programları (pp. 135-176). Ankara: Hedef CS Yayıncılık ve Mühendislik.

Durualp, E. (2019). Doğum öncesi gelişim ve doğum. In A.K. Akyol (Ed.), Erken Çocukluk Döneminde Gelişim I (pp. 11-149). Ankara: Anı Yayıncılık.

Er-Sabuncuoğlu, M. (2015). Genetik, çevre ve gelişim. In F. Turan & A.İ. Yükselen (Eds.), Her Yönüyle Okul Öncesi Eğitim: Çocuk Gelişimi 1 (pp. 73-98). Ankara: Hedef CS Yayıncılık ve Mühendislik.

Erkan, S. (2010). Aile ve aile eğitimi ile ilgili temel kavramlar. In Z.F. Temel (Ed.), Aile Eğitimi ve Erken Çocukluk Eğitiminde Aile Katılımı (pp. 3-43). Ankara: Anı Yayıncılık.

Gordon, T. (2014). *Etkili anne baba eğitimi*. İstanbul: Profil.

Hodnett, E. D., Downe, S., & Walsh, D. (2012). Alternative versus conventional institutional settings for birth. *Cochrane Database of Systematic Reviews*, (8): CD000012. doi:10.1002/14651858.CD000012.pub4 PMID:22895914

Jirasek, J. E. (2001). *An atlas of the human embryo and fetus: A phofografik rewiew of human prenatal development*. NY: The Parthenon Publishing Group.

Klerman, L. (2000). Family planning: Needs and opportunities. In M. C. McCormick & J. E. Siegel (Eds.), *Prenatal care: effectiveness and implementation* (pp. 271–284). Cambridge: Cambridge University Press.

Levine, L. E., & Munsch, J. A. (2015). *Child development: An active learning approach*. Canada: Sage Publication.

Lewis, M., & Goldberg, S. (1969). Perceptual-cognitive development in infancy: A generalized expectancy model as a function of the mother-infant interaction. *Merrill-Palmer Quarterly of Behavior and Development*, *15*(1), 81–100. Retrieved from http://www.jstor.org/stable/23082954

Lieberman, A., & Holt, L. H. (2005). *Nine months and a day: A pregnancy and birth companion*. Harvard Common Press.

Mackenzie, R. J. (2014). *Çocuğunuza sınır koyma*. İstanbul: Yakamoz.

Maree, J. (2018). Promoting career development and life design in the early years of a person's life. *Early Child Development and Care*, *188*(4), 425–436. doi:10.1080/03004430.2017.1345892

McDonagh, M. S., Osterweil, P., & Guise, J. (2005). The benefits and risks of inducing labour in patient with prior cesarean delivery: A systematic review. *BJOG*, *112*(8), 1007–1015. doi:10.1111/j.1471-0528.2005.00623.x PMID:16045510

MEB / Milli Eğitim Bakanlığı. (2012). *Gebelik ve fetüs fizyolojisi*. Ankara: MEB yayınları.

Metindoğan Wise, A. (2015). Gelişim bilimi: Tanımı, kapsamı, tarihsel temelleri, araştırma konu ve yöntemleri. In F. Turan & A.İ. Yükselen (Eds.), Her Yönüyle Okul Öncesi Eğitim: Çocukluk Gelişimi 1 (pp. 18-44), Ankara: Hedef CS Yayıncılık ve Mühendislik.

Moore, K. L., Persaud, T. V. N., & Torchia, M. G. (2013). *The developing human: cilinically oriented embroyology*. Philadelphia: Elsevier Saunders.

Myers, R. (1996). *Hayatta kalan on iki*. İstanbul: Varol Matbaası.

Nelsen, J., Lott, L., & Glenn, S. (2002). *Çocuk Gelişiminde a'dan z'ye pozitif disiplin (M. Ersin Çev.)*. İstanbul: Hayay Yayıncılık.

Olsen, O., & Clausen, J. A. (2012). Planned hospital birth versus planned home birth. *Cochrane Database of Systematic Reviews*, *12*(9), 1–26. PMID:22972043

Özçerezci-Güner, Ö., & Kavlak, O. (2016). Doğum. In Ü. Sevil & G. Ertem (Eds.), Perinatoloji ve Bakım (pp. 299-372). Azmir: Ankara Nobel Tıp Kitabevi.

Park, J., Vincent, D., & Hastings-Tolsma, M. (2007). Disparity in prenatal care among women of colour in the USA. *Midwifery*, *23*(1), 28–37. doi:10.1016/j.midw.2005.08.002 PMID:16842895

Petropoulou, H., Stratigos, J., & Katsambas, A. D. (2006). Human immunodeficiency virus infection and pregnancy. *Clinics in Dermatology*, *24*(6), 536–542. doi:10.1016/j.clindermatol.2006.05.003 PMID:17113971

Rathus, S. A. (2006). *Childhood voyage in development (2nd ed.)*. New York: Thomson Wadsworth.

Sağlık Bakanlığı. (2014). *Doğum öncesi bakım yönetim rehberi*. Ankara: Sağlık Bakanlığı Yayınları.

Sax, K. (1918). The Behavior of the Chromosomes in Fertilization. *Genetics*, *3*(4), 309–327. PMID:17245907

Semerci, B. (2015). *Birlikte büyütelim*. İstanbul: Alfa.

Senemoğlu, N. (2018). *Gelişim, öğrenme ve öğretim*. Ankara: Anı Yayıncılık.

Sigelman, C.K. & Rider, E.A. (2015). *Life span human development*. USA: Cengage Learning.

Trawick-Swith, J. (2013). Erken çocukluk döneminde gelişim. Ankara: Nobel.

Ünver, G. (2007). Gelişimle ilgili temel kavramlar, gelişimin temel ilkeleri ve gelişimi etkileyen etmenler. In A. Ulusoy (Ed.), Eğitim Psikolojisi. Ankara: Anı Yayıncılık

Walker, S. P., Wachs, T. D., Meeks Gardner, J., Lozoff, B., Wasserman, G. A., Pollitt, E., & Carter, J. A. (2007). Child development: Risk factors for adverse outcomes in developing countries. *Lancet*, *369*(9556), 145–157. doi:10.1016/S0140-6736(07)60076-2 PMID:17223478

Wasserman, G. A., Liu, X., Parvez, F., Ahsan, H., Factor-Litvak, P., van Geen, A., ... Graziano, J. H. (2004). Water arsenic exposure and children's intellectual function in Araihazar, Bangladesh. *Environmental Health Perspectives*, *112*(13), 1329–1333. doi:10.1289/ehp.6964 PMID:15345348

WHO. (2002). *Health Research on Arsenic Poisoning (Scientific Debate)* http://apps.who.int/iris/bitstream/10665/127108/1/SEA-ACHR-27-10.pdf?ua=1adresindenerişildi

Yaşar, F. (1996). *Elazığ ve Malatya illerindeki çocuk yuvası ve yetiştirme yurtlarında kalan korunmaya muhtaç çocukların fiziksel ve psiko-sosyal sorunlarının araştırılması*. Elazığ: Tıpta Uzmanlık Tezi, Fırat Üniversitesi Tıp Fakültesi Çocuk Sağlığı ve Hastalıkları.

Yücel, M. K. (2015). *Doğum ve kadın sağlığı*. Ankara: Palme Yayınları.

Chapter 9
Childbirth and Development in Infancy

Şermin Metin
Hasan Kalyoncu Üniversitesi, Turkey

ABSTRACT

Living things start life with fertilization. Human offspring develop rapidly in approximately 38-42 weeks. Birth is a transition process where human offspring meets life. After a healthy birth, a new period begins for the baby and the mother. The first years following birth, infancy is a process in which the baby tries to adapt to life and gains basic skills. In this section, childbirth and the birth process are examined. In the postnatal period, it continued with the characteristics of the mother and newborn and the evaluation of the newborn. Development in infancy; physical development, cognitive development, language development and social and emotional development has analyzed.

INTRODUCTION

Development begins long before birth. Some of the most rapid and dramatic developments in the human lifespan occur in the first nine months. Gestation—the period from conception to birth—takes about 280 days, counting from the mother's last menstrual period. During this time, a single cell, smaller than the period at the end of this sentence, develops into an approximately seven-pound, twenty-inch-long baby boy or girl, a new and unique individual (Steinberg, Vandell, & Bornstein, 2010). To make the vast, interdisciplinary study of human constancy and change more orderly and convenient, development is often divided into three broad domains: physical, cognitive, and emotional and social. Yet the domains are not really distinct. Rather, they combine in an integrated, holistic fashion to yield the living, growing child. Furthermore, each domain infuences and is infuenced by the others (Berk & Meyers, 2015).

Prenatal development and infancy are also critical, as are the years of rapid development. In these years, it is important to know and support the development of children. In this section, birth, postnatal and infancy development will be discussed.

DOI: 10.4018/978-1-7998-2952-2.ch009

BACKGROUND

Bırth

Despite continuing research, the exact mechanisms that initiate labor remain unknown. Labor normally starts when the fetus is mature enough to adjust easily to extrauterine life but before it grows so large that vaginal birth in impossible. This stage (term gestation) occurs between 38 and 42 weeks after the first day of the woman's last menstrual period (Murray & McKinney, 2013). Labor is a complex and interrelated physiological process that involves harmonious signaling and communication between the fetus, the placenta, and mother's body. Intervowen stimulatory and inhibitory mechanisms are actively orchestrated to encourge uterine contractility, fetal membrane integrity, and cervical maturation (Yates, 2010).

Stage 1: Dilation of the Cervix: Labor usually starts gradually with painless "Braxton Hicks" contractions increasing in frequency and strength at the end of pregnancy until they merge into labor. The onset of the first stage of labor is conventionally defined as the presence of regular painful uterine contraction in the presence of either progressive cervical effacement or dilatation or ruptured membranes (Grant, Strevens & Thornton, 2015).

Three phases within first stage are latent, active, and transition. Each phase is characterized by changing maternal behaviors. These behaviors vary with the woman's preparation, use of coping skills, and use of medication (Murray & McKinney, 2013). In the first stage of labor;

- Contractions serve to thin/efface the cervix from 0 to 100% and to dilate/open the cervix from 0 cm to 10 cm, which is considered full dilatation.
- Latent labor is the phase of labor which takes the longest amount time and can vary most in its duritation. Generally, this phase tends to be shorter for multiparae.
- Once the active phase beging then the cervix usually dilates at around 1 cm/h or more from 3 to 10 cm.
- Transition ends the first stage of labor with the cervix dilating from 7(8) to 10 cm (Yates, 2010).

Stage 2: Birth of the baby: The second stage (expulsion) begins with complete (10 cm) dilation and full (100%) effacement of the cervix and ends the birth of the baby (Murray & McKinney, 2013). The defining physiological process during this stage is the descent of the baby's head (West, 2008). The constractions are called expulsive constractions as they are now working to push the fetus down the birth canal and out into the world (Yates, 2010).

Stage 3: Removal of the placenta: The third stage of labor begins after the delivery of the baby and ends with the delivery of the placenta and membranes. Immediately after the birth of the baby, the umbilical arteries within the umbilical cord go into spasm, reducing the flow of fetal blood toward the placenta. It is the most hazardous for the mother, because of possible compilications such as haemorrhaging. The placenta can take between 5 minutes and an hour to deliver (Grant et al., 2015; West, 2008).

Medical Interventions

Some of the more common medical interventions during labor include induction of labor, electronic fetal monitoring (EFM), episiotomy, acceleration with oxytocin, amniotomy, cesarean sections, and epidural analgesia. Physicians may recommend medical interventions due to either the baby or the mother being in distress, or to speed up the process of labor for the convenience of either the medical staff or the patient. (Hunter & Hurst, 2016).

- *Electronic Fetal Monitoring (EFM):* Electronic fetal monitoring (EFM) is a common medical intervention used during pregnancy. It is used in order to evaluate the uterine contractions of the mother and subsequent heart rate of the fetus. EFM was developed in order to provide a way to monitor the fetus for any signs of crisis and subsequent need for a quick delivery.
- *Episiotomy:* Episiotomy is a surgical incision made around the perineum. It was once thought that the proactive surgical incision would prevent more extensive tears that would have formed naturally during delivery. Episiotomies have been shown to increase the risk of pain, infection, urinary and fecal incontinence, and sexual dysfunction in the mother (American College of Obstetricians and Gynecologists ACOG, 2006).
- *Acceleration with Oxytocin:* The early use of oxytocin does accelerate labor, but does not affect the mode of birth and does not affect any other major maternal or fetal outcomes (Grant et al., 2015)
- *Amniotomy:* In normal labor, amniotic fluid retained with intact fetal membranes tends to minimize reduction in placental blood flow and prevents cord compression. Even in the absence of cord prolapse, the lack of amniotic fluid may allow the cord to be compressed by fetal parts. The rupture of membranes also allows infection to ascend into the uterus more easily. This is why traditionally obstetricians and midwives have tried to keep the membranes intact during labor (Grant et al., 2015).
- *Caesarean*: Caesarean section rates have increased substantially since the early 1970s; many women having their first babies are older and this may contribute to ineffective or difficult labour, most often because of inadequate uterine action (dystocia). The Active Management of Labour is a clinical protocol that includes early intervention with amniotomy and oxytocin to increase the frequency and intensity of uterine contractions (augmentation) when the progress of labour is delayed. Continued ineffective labour ('cervical arrest') can result in the decision to undertake a caesarean section (Wei, Wo, Qi, Xu, Luo, Roy, Fraser, 2012).
- *Epidural Analgesia*: Epidural analgesia is the most frequently used neuraxial technique to provide pain relief in labor. With local anesthetic skin infiltration, the epidural space is located using an epidural needle and a loss-of-resistance technique. A catheter is then threaded into the space, remaining there for the duration of the labor (Stirparo, 2015).

Non-Medical Interventions

The choice of one instrument over another depends on tradition, training, practitioner experience, and the clinical circumstances. Forceps, metal clamps placed around the baby's head to pull the infant from the birth canal, have been used since the sixteenth century to speed up delivery A more recent instrument, the vacuum extractor, consists of a plastic cup (placed on the baby's head) attached to a suction tube (Berk & Meyers, 2015).

Forms of Birth

Natural labor begins when forces favoring continuation of pregnancy are offset by forces favoring its end. Research is ongoing in this area because this knowledge is essential to developing effective measures to treat preterm labor. Normal labor can occur any time between 37 and 42 weeks' gestation, although unlike other mammals, human gestation is not an exact science (Stillerman, Mattison, Giudice & Woodruff, 2008). Vaginal delivery is associated with fewer complications, is less expensive, has a faster recovery, and for many women there is an important satisfaction factor. Published series indicate that about 60–80% of trials of labor after a previous cesarean result in successful vaginal births (Murray & McKinney, 2013).A cesarean is a surgical birth; the doctor makes an incision in the mother's abdomen and lifts the baby out of the uterus (Berk & Meyers, 2015). The most common indications for primary cesarean delivery included, in order of frequency, labor dystocia, abnormal or indeterminate (formerly, nonreassuring) fetal heart rate tracing, fetal malpresentation, multiple gestation, and suspected fetal macrosomia (ACOG, 2014).

POSTNATAL PERIOD

In medical terms, this period (starting as soon as the placenta is expelled) is called the puerperium, meaning "the period belonging on the child". During this time the mother's reproductive organs return to their pre-pregnant sate (West, 2008). The anatomic and physiological changes in pregnancy occur over approximately 40 weeks, while the postpartum recovery takes place in just 40 days. Immediately after childbirth, a woman's body begins the journey of returning to its postpregnant state (Couch & DeCocker-Geist, 2014).

This period becomes more difficult due to reasons such as changing family order, new roles of spouses, changing expectations, and postpartum discomfort. The birth of the baby affects the social and working lives of the spouses. After birth, each mother is very sensitive to the effects of hormones, the new condition and fatigue of pregnancy and delivery process. Anger, crying, frustration, anorexia, insomnia are symptoms of postpartum sorrow. If this condition is very intense and lasts more than one to two weeks, it should be considered as postpartum depression. As a result, the postnatal period is a stressful period and adaptation is difficult (Durualp, 2017).

THE NEWBORN

Mothers who are unhealthy, who do not receive good medical care or who have had birth problems in the past are more prone to have problems during childbirth (Berk, 2007). Nagtalon-Ramos (2014) discusses these problems; prematurity, post-maturity, meconium aspiration syndrome, respiratory distress syndrome, small for gestational age, large for gestational age.

- *Prematurity:* Infants born before the 37th week of gestation are considered preterm (premature). There is a variety of specific and nonspecific risks for prematurity, depending on gestational age. These include but are not limited to a lack of prenatal care, poor prenatal care, uterine enlargement (e.g., multiple gestation, uterine anomalies), cervical incompetence, chronic maternal diseases, tobacco smoking, anemia, substance abuse, sexually transmitted diseases, group B streptococcus infection, placenta previa, and abdominal trauma. Infants with congenital anomalies also are more at risk for a preterm birth.
- *Post-maturity*: An infant born after 42 weeks' gestation is considered postterm. The term post-mature refers to the infant born after 42 weeks of gestation who shows the effects of progressive placental insufficiency.
- *Meconium Aspiration Syndrome*: Meconium is passed in utero when the anal sphincter relaxes as a result of the hypoxic insult. Either in utero or after delivery the infant gasps and aspirates meconium into the tracheobronchial tree, resulting in airway occlusion, atelectasis, and pneumonitis. Aspiration of meconium can trigger persistent pulmonary hypertension, which can result in respiratory failure and circulatory collapse. Risk Factors: Intrauterine asphyxia, intrauterine stress.
- *Respiratory Distress Syndrome*: RDS occurs primarily in preterm infants as a result of surfactant deficiency. It is characterized by compromised lung expansion, poor gas exchange, and ventilatory failure.
- *Small for Gestational Age*: SGA describes infants with birth weights below the 10th percentile for length, weight, or head circumference for gestational age. Intrauterine growth retardation may be symmetrical, affecting all parameters of growth, or asymmetrical.
- *Large for Gestational Age:* LGA is birth weight greater than the 90th percentile for gestational age. At term, that would be 4 kg. Risk Factors; Infant of a diabetic mother, familial tendency, genetic abnormality (e.g., Marfan syndrome).

Newborn Assessment

Immediately after the baby is born and 5-10 minutes later, the general health status of the newborn is checked by the doctor, baby nurse or midwife (Sağlam, Ulutaş & Çalışkan, 2015). The most common Apgar Test, Brazelton Neonatal Behavioral Assessment and Neonatal Intensive Care Unit Network Neurobehavioral Scale are used to evaluate whether infants have developmental problems that require urgent support.

- ***Apgar* Score:** Heart rate, reflexes, skin color, and muscle tone are measured on a three-digit scale using the by American pediatrician Virginia Apgar (Saglam et al., 2015). An Apgar score of 0–3 at 5 min may correlate with neonatal mortality but alone does not predict later neurologic dysfunction (see Fig.3). The Apgar score is an expression of the infant's physiologic condition, has a limited time frame, and includes subjective components (Perrotta, 2015).

Figure 1. Apgar Scoring System.
Source: https://www.abclawcenters.com/practice-areas/diagnostic-tests/apgar-score-for-assessment-of-the-newborn/

Apgar Scoring System

	Indicator	0 Points	1 Point	2 Points
A	Activity (muscle tone)	Absent	Flexed arms and legs	Active
P	Pulse	Absent	Below 100 bpm	Over 100 bpm
G	Grimace (reflex irritability)	Floppy	Minimal response to stimulation	Prompt response to stimulation
A	Appearance (skin color)	Blue; pale	Pink body, Blue extremities	Pink
R	Respiration	Absent	Slow and irregular	Vigorous cry

- **The Brazelton Neonatal Behavioral Assessment Scale (NBAS)** is typically performed within 24 to 36 hours after birth. It is also used as a sensitive index of neurological competence up to one month after birth for typical infants and as a measure in many studies of infant development. The NBAS assesses the newborn's neurological development, reflexes, and reactions to people and objects (Santrock, 2014).
- **An "offspring" of the NBAS, the Neonatal Intensive Care Unit Network Neurobehavioral Scale (NNNS)** provides another assessment of the newborn's behavior, neurological and stress responses, and regulatory capacities. The NNNS is especially useful for evaluating preterm infants (although it may not be appropriate for those of less than 30 weeks' gestational age) and substance-exposed infants. A recent NNNS assessment (at one month of age) of preterm infants who were exposed to substance abuse prenatally revealed that the NNNS predicted certain developmental outcomes, such as neurologic difficulties, IQ, and reduced school readiness at 4,5 years of age (Santrock, 2014).

Characteristics of Newborn

The newborn period, also referred to as the neonatal period, begins the moment the infant is born up to the first 28 days of life (Nagtalon-Ramos, 2014). During this period, the newborn adjusts from intrauterine to extrauterine life. The first few hours of life, in which the newborn stabilizes respiratory and circulatory functions, are called neonatal transition. All other newborn body systems change their level of functioning and become established over a longer period of time during the neonatal period. (Ladewig, London & Davidson, 2014). The newborn has the following physical properties:

- **The weight and height:** The weight of the term newborn ranges between 2500 and 4000 grams (g) (5 pounds [lb], 8 ounces [oz] and 8 lb, 13 oz). They can be expected to lose up to 10% of their birth weight during the first week of life. Infants normally regain or exceed their birth weight by 14 days of life. Thereafter they gain approximately 30 g per day during the early months (Naglaton-Ramos, 2013). The average length of a newborn baby is between 48-53 cm. The lower limit of height is considered to be 46 cm. During the first six months, the baby's height increases by 2,5 cm (Gürsoy & Özaslan, 2017).

- **The head:** Head size in the newborn is approximately the size of or slightly larger than the chest diameter. The anterior fontanel, located at the top of the skull, is nearly diamond shaped, 3 to 4 cm long by 2 to 3 cm wide. It closes within 18 months of age. The posterior fontanel at the base of the skull is smaller and triangular. It is 0,5 by 1 cm and closes within 8 to 12 weeks after birth (Naglaton-Ramos, 2013).
- **The skin:** The newborn's skin is fragile and shows marks easily, especially in infants with fair coloring. The skin color should be pink or tan. Red, thin skin occurs in preterm infants. The discoloration may indicate that the infant was compromised at some time before birth, and it is more common in the postterm infant (Naglaton-Ramos, 2013).
- **The crying:** Crying is a major parental concern during the early weeks after birth. Crying peaks at approximately 6 weeks of age at up to 3 hours a day. It usually decreases to 1 hour or less each day by 3 months. Infants cry for many reasons, including hunger, discomfort, fatigue, overstimulation, and boredom (Naglaton-Ramos, 2013).
- **The sleep:** Newborns spend most of their time asleep, and it is normal for them to sleep 16 to 18 hours each day. Because infants' stomachs are so small, they will wake up to eat about every 3 or 4 hours throughout the day and the night. By 6 months of age, most babies will sleep through the night for 12 hours or more and continue to take daytime naps (Levine & Munsch, 2016).
- **Breastfeeding:** Human breast milk is widely considered the optimal nutrition for the newborn infant. Aside from providing the neonate with the nutritional needs for growth and development, breast milk also contains a plethora of bioactive factors that promote health and offer protection from infections. Research during the last decades has implicated functions for human milk oligosaccharides in the healthy colonization of the neonatal gut, protection from infections, maturation of the immune system and neuronal development (Jantscher-Krenn & Bode, 2013).
- **The nutrition:** The full-term newborn needs 85 to 100 kilocalories per kilogram of body weight (kcal/k) (39 to 45 kilocalories per pound of body weight [kcal/lb] each day if breastfed and 100 to 110 kcal/kg (45 to 50 kcal/lb). The infant must consume sufficient calories to meet energy needs, prevent the use of body stores, and provide for growth. Breast milk and formulas used for the normal newborn contain 20 kilocalories per ounce (kcal/oz) (Jantscher-Krenn & Bode, 2013).

Reflexes

Newborns can't move around on their own, and they don't have much control over their limbs, but from the time they are born, they have a set of involuntary, patterned motor responses called reflexes that are controlled by the lower brain centers and that help them respond to some of the stimuli in the environment. These reflexes are hardwired into the newborn's nervous system, so they are automatic and don't need to be learned. Within the first few months of life, the higher centers of the brain develop and take over from the lower centers (See Fig. 4) (Bornstein, Arterberry & Lamb 2013).

Figure 2. Primitive reflexes.
Source: (https://www.sprockettherapy.com/primitive-reflexes.html)

Primitive Reflexes

Primitive Reflex	Purpose	What the Reflex Looks Like in a Baby:	When It Appears	Should Integrate by:	Signs of Retention
FPR Fear Parlysis Reflex	protective mechanism, response to perceived threat	Freezing reaction - similar to deer in headlights	5th to 8th week in Utero	before birth	anxiety, poor self esteem, sleep/eating disorders, agression, fear of failure or embarassment, phobias
Moro	Insant arousal of survival systems	Automatic reaction to a sudden change in sensory stimuli. Startle response. Primitive fight or flight reaction	birth	2 to 4 months	hyper sensitivity, hyper reactivity, poor impulse control, sensory overload, social and emotional immaturity
Rooting Reflex	to assist baby to find food, breastfeeding	Baby automatically turns head towards touch on their cheek	birth	3 to 4 months	picky eater, thumb sucking, dribbling, speech and articulation problems
Palmar Reflex	to assist baby's grasp development	Hand closes when object is places in the palm	birth	5 to 6 months	poor fine motor skills, poor manual dexterity, poor handwriting
ATNR Asymmetrical Tonic Neck Reflex	To assist baby through birth canal and to develop cross pattern movements	Activated by turning the head to the left or right side. As the head is turned, the arm and leg on the same side will extend while the opposite limbs bend.	birth	6 months	Difficulty with: eye-hand coordination, handwriting, crossing vertical midline, visual tracking
STNR Symmetrical Tonic Neck Reflex	preparation for crawling	Arms bend and legs extend when the head is bent down. Arms straighten, legs bend when the head is bent backwards.	6 to 9 months	9 to 11 months	tendency to slump while sitting, poor muscle tone, W-sitting, poor eye-hand coordination, inability to sit still and concentrate
TLR Tonic Labrynthine Reflex	head management and postural stability	Forward TLR: Head bendt forwards causes body and limbs flexion. Backwards TLR: Head bent backwards causes body and limb extension	in utero	3 1/2 years	poor muscle tone, W-sitting, toe-walking, poor balance, motion sickness, spatial orientation issues, gravitational insecurity
Spinal Galant Reflex	assists baby with birth process, crawling and creeping	Hip rotation when back is touched on either side of the spine	birth	3 to 9 months	unilateral or bilateral posture issues, fidgeting, bedwetting, clothing issues, poor concentration, poor short term memory

DEVELOPMENT IN INFANCY

Infancy is a period in which the baby tries to adapt to life. Babies develop many skills while adapting to life. This period in which basic skills are acquired constitutes an important basis for other periods. In this section, physical development, cognitive development and socio-emotional development of infants will be explained.

Physical Development in Infancy

In this section, physical development, motor development and sensory and perception development in infancy will be examined.

Physical Development

Physical growth, which is easy to observe and quantify, tells us a lot about whether a child is developing normally. If a child suddenly stops growing, it is a sign for worry. That's one reason that a pediatric exam always includes an assessment of the child's height and weight. Physical growth also has implications for development in many other psychological domains. For example, when children start to walk, their parents change their own behavior in many ways, which in turn affect the child's cognitive and socio-emotional development (Steinberg et al., 2010).

The newborn maintains a high rate of growth for several months following birth. The fastest increase in body length occurs in the fourth prenatal month when the fetus grows about 1,5 millimeters per day. The fetus continues to grow at a slightly slower rate in the remaining prenatal weeks. The newborn provides high growth for several months after birth (Bukatko & Daehler, 2012). By the end of the first year a typical infant's height is about 32 inches— more than 50 percent greater than at birth. By 2 years, it is 75 percent greater (36 inches). Similarly, by 5 months of age, birth weight has doubled, to about 15 pounds. At 1 year it has tripled, to 22 pounds, and at 2 years it has quadrupled, to about 30 pounds (Berk & Meyers, 2015).

Brain Development

Brain development begins prenatally, and by the time a baby is born, all the basic structures of the brain are in place. Infancy is a time when there is rapid growth and change in the young brain, just as there is rapid growth throughout the rest of the body (Levin & Munsch, 2016).

Figure 3. Functional areas of the Cerebral Cortex.
https://www.shutterstock.com/tr/search/cortex

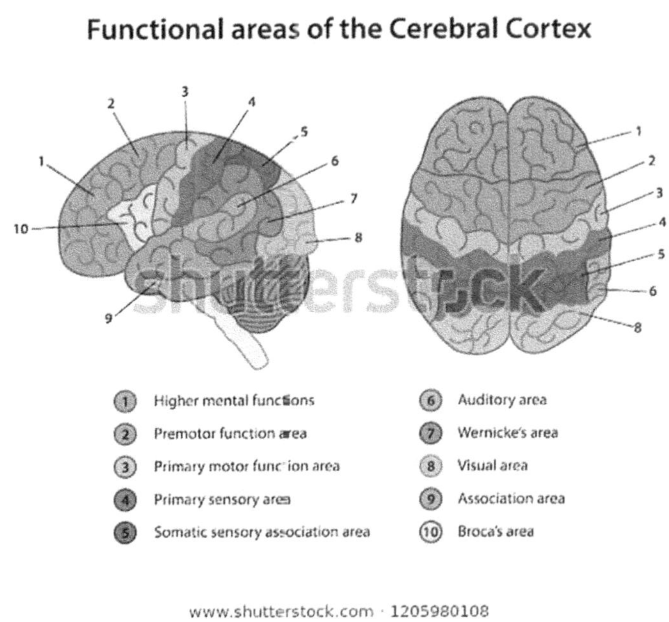

- *Structures of the Brain:* The brain is divided into two halves or hemispheres, connected by the corpus callosum. The cortex of the brain is made up of thin layers of outer tissue that cover the brain. The corpus callosum, which integrates activities of the two hemispheres by transferring information between them and the cortex does not mature fully until adolescence (Steinberg et al., 2010). Different areas of the cortex have specialized functions (See Fig. 5). Some parts of the brain are found on both sides, and some are located in only one hemisphere. For example, the motor cortex that controls the body's movements is similar on both sides, but only the right side of the brain controls the left side of the body, and vice versa. (Levin & Munsch, 2016).
- Lateralizatıon and Plasticity of the Cerebral Cortex: Specialization of the two hemispheres is called lateralization. fMRI studies reveal that the left hemisphere is better at processing information in a sequential, analytic (piece-by-piece) way, a good approach for dealing with communicative information-both verbal (language) and emotional (a joful smile). A lateralized brain is certainly adaptive. It permits a wider array of functions to be carried out effectively than if both sides processed information in the same way. Consequently, the cortex has a high capacity for learning. In addition, if a part of the brain is damaged, other parts can take over tasks that it would have handled. But once the hemispheres lateralize, damage to a specific region means that the abilities it controls cannot be recovered to the same extent or as easily as earlier. At birth, the hemispheres have already begun to specialize (Berk, 2007).
- Infant Brain Maturation: The human brain has 100 to 200 billion neurons, or nerve cells, that store and transmit information, many of which have thousands of direct connections with other neurons. Unlike other body cells, neurons are not tightly packed together. Between them are tiny gaps, or synapses, where fibers from different neurons come close together but do not touch. Neurons send messages to one another by releasing chemicals called neurotransmitters, which cross the synapse (Berk & Meyers, 2015). These neurons compose the majority of those that the individual will have

Figure 4. The Neuron and Synapse.
Source From Khan Academy. https://www.khanacademy.org/science/biology/human-biology/neuron-nervous-system/a/overview-of-neuron-structure-and-function

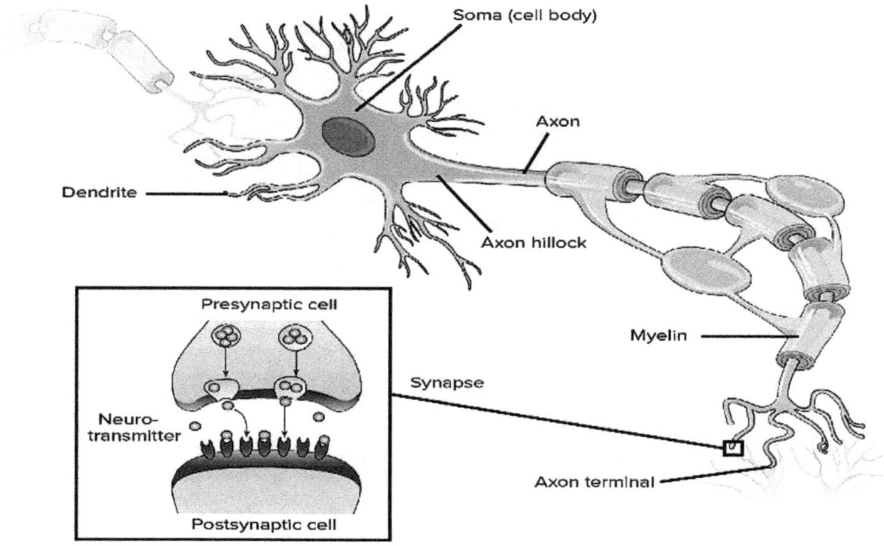

throughout life, but many of them are not yet connected in neural networks (See Fig. 6) . Only those needed for essential life processes are connected firmly at birth so the process of synaptogenesis (creating neuronal network connections) is of great importance during the first few years of life. If the brain is well nourished and adequately stimulated, each neuron can produce up to 15,000 synapses during early development (Bergen, Davis & Abbtt, 2015).

Motor Development

Like other aspects of development, motor development depends on both physical maturation and the experiences infants have. As motor systems develop, infants acquire the ability to move about and manipulate objects (Steinberg et al., 2010). The infant's ability to coordinate physical movement develops impressively over the first 2½ years. Movements seem uncontrolled for the first few months, but coordinaion develops rapidly there after. From the newborn, unable to roll over from the position in which originally placed, emerges the toddler who is increasingly deft (Bornstein, Arterberry & Lamb, 2013). Two basic forms of motor skills are gross motor and fine motor.

Gross Motor Skills

One way babies learn to use their bodies is by learning to achieve large physical tasks, or gross motor skills, such as crawling and walking (Oswalt, 2019). Gross motor skills call on the large muscle groups of the body (for example, the legs and arms) (Levine & Munsch, 2016) and it refers to control over actions that help infants get around in the environment, such as crawling, standing, and walking (Berk, 2007).

Adolph & Robinson (2013) states that gross motor skills require basic postural control. For example, to track moving objects, you must be able to control your head in order to stabilize your gaze. Newborn infants cannot voluntarily control their posture. By 2 months of age, babies can sit while supported on a lap or an infant seat, but they cannot sit independently until they are 6 or 7 months of age. Standing also develops gradually during the first year of life. By about 8 to 9 months of age, infants usually learn to pull themselves up and hold on to a chair, and they often can stand alone by about 10 to 12 months of age (Santrock, 2014). In the second year of life, toddlers continue to become more mobile and more agile. Around age 15 months, babies begin to climb stairs, high chairs, and furniture, but they will not yet be able to get back down once they reach the top. By age 18 months, toddlers' balance becomes more stable as they can move more easily on their feet around objects and begin walking backwards, sideways, in circles, and even running. Near the end of their second year, toddlers begin to develop complex gross motor skills such as throwing objects for distance and kicking. They continue to refine and to become more fluid in their movements. By age 24 months, they can jump in place and balance on one foot for a short period and may begin peddling their first tricycle. They can go up stairs easily on their own, even though they may need some help climbing back down (Oswalt, 2019).

Fine Motor Development

Fine motor skills enable us to make small movements, mostly of the hands and fingers, but also of the lips and tongue. The development of these skills is linked with the development of the brain and the entire nervous system (Levine & Munsch, 2016). Of all motor skills, reaching may play the greatest role in infant cognitive development. Reaching and grasping, like many other motor skills, start out as gross, diffuse activity and move toward mastery of fine movements (Berk, 2007).

From birth to around 2 months, babies are "pre-reaching." They will extend their arm and hand toward an object that interests them, but they will rarely be able to make hand contact with that object. Between the ages of 3 and 4 months, babies begin holding objects between their palm and their enclosed fingers in a clumsyulnar grasp. By around age 6 months, babies are refining their ability to manipulate objects as they learn by using their hands and mouths. In the second half of the first year, babies continue to mature in their ability to use their hands and can manipulate even smaller objects. By around age 10 or 11 months, they can also place smaller pellet-sized objects, like bite-sized cereal, where they want to, such as in a bowl or cup. By age 12 months, babies can now use their hands independently of one another in play. In the second year of life, by around age 15 months, toddlers begin to use tools such as cups, spoons, and crayons. By 18 months, toddlers are refining their ability to use tools such as crayons, and they can now draw lines and rough circles rather than just scribbling on a page. By age 24 months, toddlers can use their hands with more dexterity (Oswalt, 2019).

Sensory and Perceptual Development

Although various kinds of learning contribute to mastering new bahaviors and enriching each individual's skills and compotencies, these depend on other basic processes, including sensation and perception (Bukatko & Daehler, 2012). Sensation occurs when information interacts with sensory receptors—the eyes, ears, tongue, nostrils, and skin. Perception is the interpretation of what is sensed. The physical energy transmitted to the retina of the eye might be interpreted as a particular color, pattern, or shape, depending on how it is perceived (Santrock, 2018).

- *Vision and Visual Perception:* For exploring the environment, humans depend on vision more than any other sense. Infants are able to see from time they are born altough their vision much weaker than normal adult vision. Young infants' visual acuity, the ability to see things in sharp detail, is about 20/400 (Levine & Munsch 2016). Although at first a baby's visual world is fragmented, it undergoes extraordinary changes during the first 7 to 8 months of life (Berk, 2007).
- *Face Perception:* Infants are very attracted to faces, but it is not clear why (Steinberg et al., 2010). Infants show an interest in human faces soon after birth. By 3 months of age, infants match voices to faces, distinguish between male and female faces, and discriminate between faces of their own ethnic group and those of other ethnic groups (Santrock, 2014).
- *Color Vision:* The infant's color vision also improves over time. By 8 weeks, and possibly as early as 4 weeks, infants can discriminate between some colors (Santrock, 2018).
- *Depth Perception:* Although researchers (Eleanor Gibson ve Richard Walk, 1960) do not know exactly how early in life infants can perceive depth, we do know that infants develop the ability to use binocular cues to discern depth by about 3 to 4 months of age (Levine & Munsch. 2016).

Other Senses

- Hearing: Hearing becomes functional while the fetus is still in the womb, and sound fetuses hera loudly is their mother's voice. Subsequently, babies show a pereference for their mother's voice within the first of life (Bornstein et al., 2013)

- Smell: Babies know their mother's smell very early in their lives. Within the first 6 days of life, they will turn toward their mother's smell often than toward another mother's scent (Bornstein et al., 2013), and research has shown that babies who are being breastfed recognize their mother's scent in the first week of live (Levine & Munsch, 2016).
- Taste: Infant prefer sweet tastes and react negatively to salty, sour, and bitter tastes (Bornstein, Arterberry & Lamb, 2013). Mother's milk is sweet, so this draws the baby to the food and to the mother. This taste preference has been used to help infants who must undergo a painful procedure: baby boys sucking a sweet pacifier while undergoing circumcision cried less than those who an unflavored pacifier (Levine & Munsch, 2016).
- Touch and Pain: Newborns do respond to touch. A touch to the cheek produces a turning of the head; a uch to the lips produces sucking movements. Newborns can also feel pain. For many years, doctors performed operations on newborns without anesthesia. This practice was accepted because of the dangers of anesthesia and because of the supposition that newborns do not feel pain (Santrock, 2018).

Cognitive Development in Infancy

The term 'cognitive', which is derived from the Latin cognosco (to know), refers to all those psychological activities involved in the acquisition, processing, organisation and use of knowledge - in other words, all those abilities associated with thinking and knowing (Brich, 1997).

Piaget's Theory of Cognitive Development

Most psychologists would agree that Jean Piaget (1936/1952) was the most influential developmental psychologist of the twentieth century (Brich, 1997). Piaget suggested that each infant constructs an understanding of the world—including space, time, causality, and substance—on the basis of his or her own motor activity and interactions in the world. At the time, other developmental scientists viewed the infant as a well- equipped, but basically passive, recipient of information from the environment. Piaget did not think that knowledge derives from sensations or perceptions, or from information provided by others. Rather, Piaget wrote that infants actively construct what they know. This basic notion has shaped the study of infant cognition ever since. (Steinberg et al., 2010).

Piaget built the process of cognitive development on basic concepts. For Piaget, the basic element of infant knowledge is the scheme. Schemes are central nervous system structures that produce motor activities capable of processing information in the environment. In newborns, schemes are more or less automatic motor activities, although not all reflexes (sneezes and coughs) are schemes; but many can be used to acquire or process information in the environment (sucking, grasping, looking, hearing, and tasting are examples). According to Piaget, knowledge develops as an elaboration of simple reflex schemes present in the newborn (Bornstein & Lamb, 2002).

Adaptation is the process whereby knowledge is altered by experience. Adaptation involves two complementary processes, assimilation and accommodation. When new information cannot be assimilated into the baby's existing understanding, two things can happen. One is that the infant fails to assimilate and simply moves on to another activity. Alternatively, the infant changes her understanding to permit new information to be processed. The modification of existing understanding to make it apply to a new situation is termed accommodation; that is, infants actively change so that they can understand the environment better. Through the process of accommodation, the baby has developed a new way of thinking (Steinberg et al., 2010). To make sense out of their world, said Piaget, children cognitively organize their experiences. Organization in Piaget's theory is the grouping of isolated behaviors and thoughts into a higher-order system (Santrock, 2017).

Equilibration is a mechanism that Piaget proposed to explain how children shift from one stage of thought to the next. The shift occurs as children experience cognitive conflict, or disequilibrium, in trying to understand the world. Eventually, they resolve the conflict and reach a balance, or equilibrium, of thought (Santrock, 2017).

The Sensorimotor Stage

According to Piaget, infants learn through actions: looking, listening, touching, sucking, mouthing, and grasping. Within the sensorimotor period in infancy, Piaget sketched a six-stage sequence. The child's current stage defines the way the child views the world and processes information in it. (Steinberg et al., 2010). This is the stage that has been most extensively studied by Piaget. It marks a long period of transition which culminates in the emergence of operational thinking (Birch, 1997). Piaget divided the sensorimotor stage into six substages:

Stage 1. Simple reflexes, the first sensorimotor substage, corresponds to the first month after birth. In this substage, sensation and action are coordinated primarily through reflexive behaviors such as rooting and sucking. Soon the infant produces behaviors that resemble reflexes in the absence of the usual stimulus for the reflex. For example, a newborn will suck a nipple or bottle only when it is placed directly in the baby's mouth or touched to the lips. But soon the infant might suck when a bottle or nipple is only nearby. Even in the first month of life, the infant is initiating action and actively structuring experiences.

Stage 2. First habits and primary circular reactions is the second sensorimotor substage, which develops between 1 and 4 months of age. In this substage, the infant coordinates sensation and two types of schemes: habits and primary circular reactions. A habit is a scheme based on a reflex that has become completely separated from its eliciting stimulus. For example, infants in substage 1 suck when bottles are put to their lips or when they see a bottle. Infants in substage 2 might suck even when no bottle is present. A circular reaction is a repetitive action.

Stage 3. Secondary circular reaction is the third sensorimotor substage, which develops between 4 and 8 months of age. In this substage, the infant becomes more object-oriented, moving beyond preoccupation with the self. The infant's schemes are not intentional or goaldirected, but they are repeated because of their consequences. By chance, an infant might shake a rattle. The infant repeats this action for the sake of its fascination. This is a secondary circular reaction: an action repeated because of its consequences. The infant also imitates some simple actions, such as the baby talk or burbling of adults, and some physical gestures. However, the baby imitates only actions that he or she is already able to produce.

Stage 4. Coordination of secondary circular reactions is Piaget's fourth sensorimotor substage, which develops between 8 and 12 months of age. To progress into this substage the infant must coordinate vision and touch, eye and hand. Actions become more outwardly directed. Significant changes during this substage involve the coordination of schemes and intentionality. Infants readily combine and recombine previously learned schemes in a coordinated way. They might look at an object and grasp it simultaneously, or they might visually inspect a toy, such as a rattle, and finger it simultaneously, exploring it tactilely. Actions are even more outwardly directed than before. Related to this coordination is the second achievement—the presence of intentionality.

Stage 5. Tertiary circular reactions, novelty, and curiosity is Piaget's fifth sensorimotor substage, which develops between 12 and 18 months of age. In this substage, infants become intrigued by the many properties of objects and by the many things that they can make happen to objects. A block can be made to fall, spin, hit another object, and slide across the ground. Tertiary circular reactions are schemes in which the infant purposely explores new possibilities with objects, continually doing new things to them and exploring the results. Piaget says that this stage marks the starting point for human curiosity and interest in novelty.

Stage 6. Internalization of schemes is Piaget's sixth and final sensorimotor substage, which develops between 18 and 24 months of age. In this substage, the infant develops the ability to use primitive symbols. For Piaget, a symbol is an internalized sensory image or word that represents an event. Primitive symbols permit the infant to think about concrete events without directly acting them out or perceiving them. Moreover, symbols allow the infant to manipulate and transform the represented events in simple ways (Santrock, 2017).

Language Development

Language is a form of communication—whether spoken, written, or signed—that is based on a system of symbols. Language consists of the words used by a community (vocabulary) and the rules for varying and combining them (grammar and syntax). (Santrock, 2017). The basic units of a language are words and each word is made up of sounds, known as phonemes, which correspond roughly to the letters of the alphabet. Phonemes combine together to form morphemes which are the smallest units of language to have a grammatical purpose (Birch, 1997).

Language Development Process in Infancy

Language learning begins before birth. During the last trimester of prenatal development, changes in fetal heart rate and motor activity when the mother is speaking show that the fetus can hear her voice, and this affects the infant's preferences for language after birth in a number of ways (**See** Karmiloff & Karmiloff-Smith, 2001). This prenatal awareness of language sets the stage for language learning once the baby is born (Levin & Munsch, 2015). Language development, as in other areas of development, follows a certain sequence. There are three main phases of early language acquisition:

- *Babbling and Other Vocalizations:* The first sounds made are cries, which reflect the child's physical state of well-being, followed soon by gurgles, coos and chuckles, which are not strictly speaking language, but are thought to represent what Vygotsky called the 'pre-intellectual' stage. At about six or seven months, the child begins to babble, producing syllable-like sounds, for example, 'gaga', 'dada'. All babies, including those who are deaf, in all cultures produce these same speech-like sounds. This suggests that maturation, rather than learning, is responsible (Birch, 1997). Babies' sounds go through the following sequence during the first year:
- Crying: Babies cry even at birth. Crying can signal distress, but as we will discuss in different types of cries signal different things.
- Cooing: Babies first coo at about 1 to 2 months. These are gurgling sounds that are made in the back of the throat and usually express pleasure during interaction with the caregiver.

- Babbling: In the middle of the first year babies babble—that is, they produce strings of consonant-vowel combinations, such as ba, ba, ba, ba (Santrock 2014).
- Gestures: Infants start using gestures, such as showing and pointing, at about 8 to 12 months of age. They may wave bye-bye, nod to mean yes, show an empty cup to ask for more milk, and point to a dog to draw attention to it. Pointing is considered by language experts to be an important index of the social aspects of language, and it follows this developmental sequence: from pointing without checking on adult gaze to pointing while looking back and forth between an object and the adult (Santrock,2014).

One-word utterances: Infants understand words before they can produce or speak them. For example, many infants recognize their name when someone says it as early as 5 months of age. A child's first words include those that name important people (dada), familiar animals (kitty), vehicles (car), toys (ball), food (milk), body parts (eye), clothes (hat), household items (clock), and greeting terms (bye) (Santrock, 2014). At about 12 months, a child produces the first understandable words, for example, 'mama', 'dog', 'no'. Children's active vocabulary by the age of 18 months is, on average, about 30 words, though they will understand and react correctly to many more that they cannot yet utter (Birch, 1997). When young children first learn words, they sometimes apply them too narrowly, an error called underextension. As vocabulary expands, a more common error is overextension— applying a word to a wider collection of objects and events than is appropriate. As vocabulary expands and pronunciation improves, overextensions gradually decline (Berk & Meyers, 2015).

The Two- Word Utterance: Young toddlers add to their spoken vocabularies at a rate of one to three words a week. Because gains in word production between 18 and 24 months are so impressive (one or two words per day), many researchers concluded that toddlers undergo a spurt in vocabulary— a transition (Berk & Meyers 2015).

Social and Emotional Development

Infants enter the world with a basic repertoire of emotions that lets them communicate what they are feeling. They are also born with a characteristic way of responding to their experiences that we call temperament. At first the infant does not feel separate from his caregiver, but early in development he begins to distinguish the self from the other (Levine & Munsch, 2016).

Emotions

A leading expert on infant emotional development, Michael Lewis (2007, 2008, 2010, 2015, 2016) distinguishes between primary emotions and self-conscious emotions. Primary emotions are present in humans and other animals; these emotions appear in the first 6 months of the human infant's development. Primary emotions include surprise, interest, joy, anger, sadness, fear, and disgust (Santrock, 2018). The primary emotions appear well before such secondary emotions as embarrassment, pride, guilt, shame, and envy, which do not emerge until the second and third years of life. Secondary emotions depend on higherlevel mental capacities. For example, to feel guilt or shame, you have to know that you have done something wrong (Steinberg et al., 2010).

- **Happiness:** Happiness-first expressed in blissful smiles and later through exuberant laughter-contributes to many aspects of development. Happiness binds parent and baby into a warm, supportive relationship that fosters the infant's developing competences. During the early weeks, newborn babies smile when full, during REM sleep, and in response to gentle stroking of the skin and the mother's soft voice. By the end of the first month, infants smile at interesting sights that are dynamic and eyecatching. Between 6 and 10 weeks, the human face evokes a broad grin called the social smile. These changes in smiling parallel the development of infant perceptual capacities-in particular, babies' sensitivity to visual patterns, including the human face (Berk, 2007).
- **Fear:** Fear rises during the second half of the first year into the second year. Older infants, for example, look wary, hesitating before playing with a new toy. But the most frequent expression of fear is to unfamiliar adults, a response called stranger anxiety. Many infants and toddlers are quite fearful of strangers, although the reaction does not always occur. The rise in fear after 6 months of age keeps newly mobile babies' enthusiasm for exploration in check. Once wariness develops, infants use the familiar caregiver as a secure base, or point from which to explore, venturing into the environment and then returning for emotional support (Berk & Meyers, 2015).
- **Anger and Sadness:** Newborn babies respond with generalized distress to a variety of unpleasant experiences, including hunger, painful medical procedures, changes in body temperature, and too much or too little stimulation. From 4 to 6 months into the second year, angry expressions increase in frequency and intensity. And increased parental limit setting once babies crawl and walk contributes to babies' angry responses. The rise in anger is also adaptive. Independent movement enables an angry infant to defend herself or overcome an obstacle to obtain a desired object (Berk & Meyers, 2015).

Emotional Self- Regulation

Emotional self-regulation refers to the strategies we use to adjust our emotional state to a comfortable level of intensity so we can accomplish our goals. This capacity for effortful control improves rapidly in early childhood, as the result of development of the prefrontal cortex and a history of support from caregivers, who help children manage intense emotion and teach them strategies for doing so. A good start in regulating emotion during the first two years contributes greatly to autonomy and mastery of cognitive and social skills (Berk & Meyers, 2015).

Infants have little self-control and have difficulty regulating their emotional reactions. However, even young infants make some effort to exercise self-control. Early in development they begin to develop ways to soothe themselves through behaviors such as thumb sucking, holding a favorite "blankie," or avoiding a feared or frustrating object by looking away. They may also signal in a subtle way when they are being overstimulated. Being sensitive to the infant's signals—whether of hunger, tiredness, or discomfort—helps the infant learn to regulate his own emotions (Levine & Munsch, 2016). In the early months of life, infants are easily overwhelmed by intense emotion. Rapid development of the frontal lobes of the cerebral cortex increases the baby's tolerance for stimulation. At the end of the first year, crawling and walking enable infants to regulate feelings by approaching or retreating from various situations (Berk, 2007).

Temperament

Temperament is the biologically based source of individual differences in behavioral functioning. It emerges early in life and appears to be moderately stable over time. A child's underlying temperamental characteristics tend to show themselves and endure even though the specifics may change (Steinberg et al., 2010). Although Arnold Buss and Robert Plomin argued that temperament measures must show both early appearance and long-term stability, developmental changes in temperament-related processes have been found throughout childhood and beyond, and at no time are these changes as rapid as in the first year of life. Expressions of temperament differ greatly in the newborn, the 3-month-old, and the 12-month-old, and develop further into the toddler and preschool periods (Rothbart, & Gartstein, 2009).

Early Infancy (Birth–3 Months): During early infancy, smiling and laughter are emerging as part of a positive affectivity dimension, and motor activity comes to be linked to infants' positive as well as negative affect displays. In addition, infants by 2 months are demonstrating anger/frustration when their actions no longer control an interesting outcome (Rothbart, & Gartstein, 2009).

Infancy (4–12 Months): Whereas earlier distress signals may not often be easily identified as being associated with specific emotional reactions, such differentiation is achieved during this developmental period. Fear and irritability, for example, are becoming increasingly differentiated during this period, with behavioral inhibition (fear) being increasingly associated with novel and/or unpredictable experiences. Susan Calkins and colleagues have noted that high motor activity and positive affect at 4 months is associated with bold behavior in later childhood. Later, by 9–10 months, behavioral inhibition will come to exercise a good deal of control over approach. Behavioral inhibition and other aspects of fear are developing late in the first year (between 9 and 12 months of age), and coming to modulate infants' approach responses (Rothbart, & Gartstein, 2009).

Toddler/Preschool Period (1–5 Years of Age): Temperamental characteristics are not all in place at the end of the first year. Positive emotionality becomes increasingly stable later in infancy, with most of surgency/extraversion's associated characteristics identifiable by toddlerhood. Although changes in negative emotionality during infancy have been reported, considerable stability is noted by the toddler period. Development of the executive attention system during the toddler and preschool periods is particularly important for effortful control, planning, and the ability to inhibit or delay action and expression. The development of effortful control is closely linked with advances in a variety of attentional skills, which provide the basis for emerging self-regulation (Rothbart, & Gartstein, 2009).

Attachment

Bowlby defined attachment as an emotional tie that an infant constructs and elaborates with his principal caregiver(s) in the context of everyday interactions. This bond is specific in that it is directed toward a particular individual, is long lasting, and ties the two individuals together across contexts. Bowlby proposed an account of how child-mother attachment relationship develop and are organized during the first 3 years of a child's life. He argued that all infants organize an attachment behavioral control system in the context of interactions with their mother during the course of the first year. That is to say that system does not emerge automatically and it is not fully functional at birth. Its elements, behaviors such as crying, sucking, smiling, clinging, and following become integrated and coordinated into a system of behavior through learning, practice, and feedback from the mother. The behavioral system becomes functional during the second half of the first year and it is clearly seen in operation during children's implementation of the secure-base game in everyday life circumstances and especially, not exclusively, in emergency situations (Posada, 2009).

Bowlby described four main phases in the development of infant-parent attachment:

Phase 1. Preattachment (birth to 6 weeks): Indiscriminate social responsiveness (1 to 2 months)- the baby develops a repertoire of signals that help to provide comfort and security (crying) by bringing a protective, caregiving adult close. Babies then associate the presence of adults with alertness and the relief of distress (Steinberg et al., 2011).

Phase 2. Attachment in the making (6 weeks to 6–8 months): Discriminating sociability (2 to 7 months)- babies come to prefer to interact with familiar people. (Steinberg et al., 2010. Babies begin to smile at about 6 weeks of age. At first these smiles seem almost random, but by about 2 months, babies clearly have developed a social smile directed specifically at people. Very quickly, smiling becomes reserved for people the baby recognizes (Levine & Munsch, 2016).

Phase 3. Clear-cut attachment (6–8 months to 18 months–2 years) Attachments (2 to 24 months)- infants undersand the rules of reciprocity in their interactions and increasingly initiate interaction using directed social behaviors and protest when left by people to whom they attached (separation protest) (Steinberg et al., 2010). In the stage of clear-cut attachment, infants begin to move about on their own and become able to actively maintain contact with their caregiver. Infants now clearly discriminate between their attachment figures and strangers. When a parent is around and the infant can get to the parent, he is comfortable to explore, but if the parent is absent, exploration may stop (Levine & Munsch, 2016).

Phase 4. Goal-corrected partnership (also referred to as the formation of reciprocal relationships; 18 months on). Goal-corrected partnerships (year 3 onward)-children take their parent's need into account when interacting with them so their needs or wants must wait (Steinberg et al., 2010). Toddlers now are able to form symbolic representations of the particular attachment relationships they have been experiencing (See Bowlby, 1969). This concept, referred to as an internal working model of attachment, has helped psychologists understand how early attachment patterns contribute to the close relationships that children—and even adults—develop later in life (Levine & Munsch, 2016)

PERSONALITY DEVELOPMENT

Emotions and temperament form key aspects of personality, the enduring personal characteristics of individuals. The development of trust, self and independence, which are the main characteristics of personality development, occur in infancy (Santrock, 2018).

Basic Trust Versus Mistrust

According to Erik Erikson (1968), the first year of life is characterized by the trustversus-mistrust stage of development. Following a life of regularity, warmth, and protection in the mother's womb, the infant faces a world that is less secure. Erikson proposed that infants learn trust when they are cared for in a consistent, warm manner. If the infant is not well fed and kept warm on a consistent basis, a sense of mistrust is likely to develop. The issue of trust versus mistrust is not resolved once and for all in the first year of life. It arises again at each successive stage of development and can have positive or negative outcomes (Santrock, 2018)

Erikson recognized that no parent can be perfectly in tune with the baby's needs. Many factors affect parental responsiveness— feelings of personal happiness, current life conditions (for example, additional young children in the family), and culturally valued child- rearing practices. But when the balance of care is sympathetic and loving, the psychological conflict of the first year— basic trust versus mistrust— is resolved on the positive side. The trusting infant expects the world to be good and gratifying. As a result, he feels confident about venturing out and exploring it, and he emerges from this stage well- prepared for the challenges of toddlerhood. The mistrustful baby cannot count on the kindness and compassion of others, so she protects herself by withdrawing from people and things around her (Berk & Meyers, 2015).

Autonomy Versus Shame and Doubt

With the transition to toddlerhood, Freud viewed the parents' manner of toilet training as decisive for psychological health. In Erikson's view, toilet training is only one of many influential experiences. The familiar refrains of newly walking, talking toddlers- "No!" "Do it myself!"-reveal a period of budding selfhood. They want to decide for themselves not just in toileting but also in other situations. The conflict of toddlerhood, autonomy versus shame and doubt, is resolved favorably when parents provide young children with suitable guidance and reasonable choices. A self-confident, secure 2-year-old has been encouraged not only to use the toilet but also to eat with a spoon and to help pick up his toys. And parents meet his assertions of independence with patience and understanding-for example, by giving him an extra five minutes to finish his play before leaving for the grocery store (Berk, 2007).

According to Erikson, the parent who is over- or under- controlling in toileting is likely to be so in other aspects of the toddler's life as well. The outcome is a child who feels forced and shamed or who doubts his ability to control his impulses and act competently on his own. In sum, basic trust and autonomy grow out of warm, sensitive parenting and reasonable expectations for impulse control starting in the second year. If children emerge from the first few years without sufficient trust in caregivers and without a healthy sense of individuality, the seeds are sown for adjustment problems (Berk, 2007).

CONCLUSION

Birth normally starts when the fetus is mature enough to adjust easily to extrauterine life. This stage (term gestation) occurs between 38 and 42 weeks after the first day of the woman's last menstrual period (Murray & McKinney, 2013). Birth takes place in three stages: Dilation of the Cervix, Birth of the baby, and Removal of the placenta. After birth, mother and baby is the adaptation process. While the mother adapts to physical and emotional changes, the baby strives to survive. Postnatal period (starting as soon as the placenta is expelled) is called the puerperium, meaning "the period belonging on the child". During this time the mother's reproductive organs return to their pre-pregnant sate (West, 2008). Immediately after childbirth, a woman's body begins the journey of returning to its postpregnant state (Couch & DeCocker-Geist, 2014).

Infancy is the period where growth and development are the fastest. All systems and organs of the baby change rapidly during the first two years. It is very surprising that only the hungry, crying and sleeping baby turns into an individual who walks, runs, gets angry. According to Piaget, infants learn through actions: looking, listening, touching, sucking, mouthing, and grasping (Steinberg et al., 2010).

The baby, who initially tries to survive with reflexes, is able to act with physical, motor, cognitive, emotional and social development.

REFERENCES

American College of Obstetricians and Gynecologists. (2014). ACOG Practice Bulletin No. 1 Safe Prevention of the Primary Cesarean Delivery. *Obstetrics and Gynecology*, *114*(1), 2.

Bergen, D., Davis, D. R., & Abbitt, J. T. (2015). *Technology play and brain development: Infancy to adolescence and future implications*. Routledge. doi:10.4324/9781315681436

Berk, E. L. (2007). *Child development*. New York: Pearson.

Berk, L. E., & Meyers, A. B. (2015). *Infants, children, and adolescents*. Pearson.

Birch, A. (1997). *Developmental psychology: From infancy to adulthood*. Macmillan International Higher Education. doi:10.1007/978-1-349-14204-0

Bornstein, M. H., Arterberry, M. E., & Lamb, M. E. (2013). *Development in infancy: A contemporary introduction*. Psychology Press. doi:10.4324/9780203758472

Bornstein, M. H., & Lamb, M. E. (2002). *Development in infancy: An introduction*. Psychology Press. doi:10.4324/9781410602596

Bukatko, D., & Daehler, M. W. (2012). *Child development: A thematic approach*. Nelson Education.

Couch, A., & DeCocker-Geist, K. (2014). Physiological alterations during the postnatal period. In G. Robin (Ed.), *Prenatal and postnatal care a woman-centered approach* (pp. 419–422). Iowa: John Wiley & Sons, Inc.

Durualp, E. (2017). Doğum, doğum sonrası. In A. K. Akyol (Ed.), *Erken çocukluk döneminde gelişim I*. Ankara: Anı Press.

Grant, N., Strevens, H., & Thornton, J. (2015). Physiology of labor. In *Epidural Labor Analgesia* (pp. 1–10). Cham: Springer.

Gürsoy, F. & Özaslan, H. (2017). Çocuk gelişimine giriş. In N. Aral (Ed.), Çocuk gelişimi. Eskişehir: Açık Öğretim Fakültesi Yayını.

Hunter, C. A., & Hurst, A. (2016). *Understanding doulas and childbirth: Women, love, and advocacy*. Springer. doi:10.1057/978-1-137-48536-6

Jantscher-Krenn, E., Marx, C., & Bode, L. (2013). Human milk oligosaccharides are differentially metabolised in neonatal rats. *British Journal of Nutrition*, *110*(4), 640–650. doi:10.1017/S0007114512005727 PMID:23312292

Ladewig, P. A. W., London, M. L., & Davidson, M. R. (2014). *Contemporary maternal-newborn nursing care* (8th ed.). Upper Saddle River, NJ: Pearson Education, Inc.

Levine, L. E., & Munsch, J. (2016). *Child development from infancy to adolescence: An active learning approach*. Sage Publications.

Murray, S. S., & McKinney, E. S. (2013). *Foundations of maternal-newborn and women's health nursing-e-book*. Elsevier Health Sciences.

Nagtalon-Ramos, J. (2014). *Maternal- newborn nursing care: best evıdence-based practıces*. Philadelphia: F. A. Davis Company.

Oswalt, A. (2019). Infancy Physical Development: Gross Motor Skills. Grace Point Wellness. Retrieved from https://www.gracepointwellness.org/461-child-development-parenting-infants-0-2/article/10109-infancy-physical-development-gross-motor-skills

Perrotta, D. (2015). Neonatal care in labor and delivery room. In C. Capogna (Ed.). Epidural labor analgesia: Childbirth without pain (pp. 221-235). Springer. doi:10.1007/978-3-319-13890-9_17

Posada, G. (2009). Attachment. In J. B. Benson & M. M. Haith (Eds.), *Social and emotional development in infancy and early childhood* (pp. 30–41). Academic Press.

Rothbart, M. K., & Gartstein, M. A. (2009). Temperament. In J. B. Benson & M. M. Haith (Eds.), *Social and emotional development in infancy and early childhood* (pp. 480–495). Academic Press.

Sağlam, M., Ulutaş, A., & Çalışkan, Z. (2015). Children's rights in the pediod of infancy]. *Hacettepe University Faculty of Health Sciences Journal*, *1*(2), 502–510.

Santrock, J. W. (2014). *Child development* (14th ed.). New York: McGraw-Hill Education.

Santrock, J. W. (2017). *Educational psychology* (6th ed.). New York: McGraw-Hill Education.

Santrock, J. W. (2018). *Lıfe-span development* (17th ed.). New York: McGraw-Hill Education.

Steinberg, L., Bornstein, M. H., & Vandell, D. L. (2010). *Life-span development: Infancy through adulthood*. Cengage Learning.

Stillerman, K. P., Mattison, D. R., Giudice, L. C., & Woodruff, T. J. (2008). Environmental exposures and adverse pregnancy outcomes: A review of the science. *Reproductive Sciences (Thousand Oaks, Calif.)*, *15*(7), 631–650. doi:10.1177/1933719108322436 PMID:18836129

Stirparo, S. (2015). Maternal expectational, information, and satisfaction with labor analgesia. In C. Capogna (Ed.). Epidural labor analgesia: Childbirth without pain (pp. 63-69). Springer.

Wei, S., Wo, B. L., Qi, H. P., Xu, H., Luo, Z. C., Roy, C., & Fraser, W. D. (2012). Early amniotomy and early oxytocin for prevention of, or therapy for, delay in first stage spontaneous labour compared with routine care. *Cochrane Database of Systematic Reviews*, 9. PMID:22972098

West, Z. (2008). *Acupuncture in pregnancy and childbirth*. Elsevier Health Sciences.

Yates, S. (2010). *Pregnancy and Childbirth E-Book: A holistic approach to massage and bodywork*. Elsevier Health Sciences.

Chapter 10
Early Child Development

Hatice Bekir
Gazi Faculty of Education, Gazi University, Turkey

ABSTRACT

Early childhood is a period that is associated with the development process of children. It was determined that physical, motor, cognitive, language, emotional and social development, which was created in accordance with the characteristics of the developmental period of children in the early childhood, did not change much direction in later ages, the improvement in the same direction was higher. In the early childhood, children begin to learn the rules of society limited to their own environment. They understand language and show great development. Socialization and gender awareness begin to form. The researches revealed that behaviors and attitudes gained during this period shape the habits, attitudes, beliefs, value judgments and personality traits of the individual when they are adults. In this section, the changes in physical, motor, cognitive, language, emotional-social development of the first childhood are examined.

INTRODUCTION

Development is a concept that expresses the regular, harmonious, continuous progress and change that a person has gone through her life. Development is an interrelated issue that needs to be addressed as a whole. The situation in any development area affects other areas as well. Although individuals exhibit relatively different characteristics, it is assumed that each individual reaches a certain stage of development according to the age variable. In the preschool period, which is accepted as the critical period of development, development is affected by many factors. Development is a whole with physical, motor, cognitive, language, social-emotional fields. The development of the child in each field is interrelated. Without proper environment and proper interaction, obstacles occur in developmental areas. While each child is unique, all children follow a certain pattern of speed and development. If we understand the child correctly in the early childhood and offer the necessary support in this context, we can expect to see today's children as well-educated adults in the future. Therefore, it is a necessity and well-known fact that development is well known.

DOI: 10.4018/978-1-7998-2952-2.ch010

Concepts Related to Development

Clarifying the basic concepts in development will be useful for a better understanding of the issues. Below are some common concepts of development.

- Growth; refers to changes in height, weight and volume in the body. Growth occurs at different speeds in different organs.
- Maturation; It refers to the ability of the child or an organ belonging to the child to attain the ability to perform the expected movement / behavior at a certain age, so it refers to biological change. The child passes through certain maturation processes and reaches competences such as running, climbing and drawing.
- Learning; It refers to the change in behavior that occurs as a result of the child's experiences with the environment.
- Readiness; Maturation and learning, the child's ability to perform a certain skill depends on the readiness. Reaching a certain maturity is necessary for readiness. In addition, factors such as prior learning, health status, interests and attitudes, needs and motivation have an impact on readiness.
- Development Task; It refers to the skills expected to acquire according to the age and developmental period of the child and the behaviors that are supposed to be doing.
- Evolution; It can be expressed as the process of child's progress and gaining new competencies through the effect of growth, maturation and learning.
- Development; These are observable, regular and continuous changes in the child's physical, motor, cognitive, language, social-emotional, personality, morality and sexual development in terms of growth, maturation and learning and experiences. (Bağcıoğlu Ünver, 2013).

Basic principles of development

- It is a product of development, inheritance and environmental interaction.
- Development is continuous.
- The development is from head to foot, from inside to outside, from general to private.
- Development varies according to certain life periods.
- There is a certain order of ability / skill development.
- There are critical periods in development.
- The development is a whole and the areas of development are associated.
- There are individual differences in development.

In the development of the child, all areas of development are interrelated and affect each other. For example, the child's physical development (muscles, bones, hearing system, etc.) affects motor skills and language use. Motor skills and language development are effective in the process of social interaction. In summary, development is a whole and all development areas affect each other.

Factors Affecting Development

Conditions that may have positive or negative effects on development can be examined as heredity, environment, hormones and time.

Inheritance refers to the genetic characteristics of the child. The speed and rhythm of an individual's growth and development depends on the relationship between environmental and genetic factors. The effect of the environment on development is generally examined in three stages: prenatal, birth and postnatal.

Prenatal factors are related the period before birth. Mothers' diet, mother's health status, drugs used by the mother, alcohol-smoking status, the presence of blood mismatch between parents can affect development. The birth order factors indicate a number of complications during delivery. For example, oxygen deficiency due to compression of the umbilical cord during labor can cause brain damage or death. Conditions such as low birth weight and preterm labor are seen as an important cause of various developmental disabilities. Postpartum factors are related to what the child encounters after birth. For example, depression in the mother or father can cause problems in the child. Parental behavior appears to be an important factor in psychological adjustment of the child. (Berk, 2013). Factors such as nutrition, stress, sports, environmental stimuli, diseases, socioeconomic status of the family are effective on development.

Body Growth and Body Change

Growth and development are two closely linked processes. Growth is the increase in length or weight of any organ of the body and is related to every aspect of development. The development of the body covers all the changes that the organism undergoes from the womb to the death. The body does not grow as a whole at the same time and speed, different body parts and organs develop at different speeds. Growth depends on heredity and environmental characteristics and is driven by hormones. The effect of heredity on growth not only determines the speed of growth but also the limit to be achieved in growth.

In contrast to the surprising speed of physical development in infancy, the growth rate slows down during this period. Early childhood is a period of slow but steady height, weight and increase in muscle density. Significant physical development can be seen in children aged three and five. In this period, it is stated that the amount of weight gained between two and five years of age is less than the amount gained in the first year of life. The growth rates of the body parts show individual differences. In general, development is continuous and regular from head to foot and from inside to outside. Growth and development of the head first, then the upper body and arms, then the lower body and legs takes place in the form of growth. The development of human organs from birth to maturity is very rapid. The head, which is about 1/4 of the whole body at birth, is about 1/8 of the whole body although it grows twice as much in maturity. A significant physical change in the early childhood is the elongation of the arms and legs. This growth causes a shift in the center of gravity. The child's center of gravity is close to belly botton during this period. This low weight distribution allows these age children to perform actions that are impossible for them to perform a year or two years ago. (Özbar, 2007; Trawick-Swith, 2013).

Growth is more prominent primarily on the skeletal and nervous-muscular system. Ossification starts before birth and continues until late puberty. At the end of the bones, there is a bone growth center called epiphysis, which differs depending on the child's sex and bone structure. Girls are 1 year older than boys in terms of skeletal development. The ratio of bone length to tendon and muscle strength is low in children. Because the increase in bone length occurs before the development of force. In every part of the skeleton, there is a rhythm of seizures. One part of the body becomes longer and the other part thickens. Since the ossification was not completed at the age of five or six, the skull protecting the brain is still soft. Boys bones are wider than girls. (İnan, 2004; Pınar & Erkut, 2000).).

Muscles are organs that cover the bones and provide the movements of the bones and therefore the body by contracting. Muscles are innate and follow a certain sequence of development. First large muscle groups develop, then small muscle groups develop. Large muscle groups of five to six years of age are more developed than thin muscles (Mengütay, 2005).

The functions of the cardiovascular system according to age characteristics are influenced by the intensity of physical movements. Development of respiratory system; decrease in respiratory rate and increase in respiratory volume. The respiratory system is closely related to the cardiovascular system and shows a parallel development. Vital capacity and respiratory volume increase due to chest enlargement and strengthening of respiratory muscles (İnan, 2004; Özbar, 2007). Children's heart, beats and residual volume are lower in the first childhood period. Low blood pressure increases with age and reaches adulthood in youth.

Brain and Nervous System

One of the most important physical developments in the early childhood is the continuous development of the brain and nervous system. Although the brain continues to grow during this period, the growth here is not as fast as in infancy. Brain development takes place in moves rather than following a smooth and continuous path. For example, while the development is slow between the ages of two to four years, another big move takes place after infancy at the age of four. Most of these developmental moves do not cover the entire brain, but take place in one or a few sections. There is a clear relationship between these developmental moves and milestones of development. For example, the move around four years of age is accompanied by progress in both speaking and understanding. The development move, which also coincides with the age of six to eight years, is accompanied by great progress in fine motor skills and hand-eye coordination. (Bee & Boyd, 2009). By the time the child was three years old, his brain had reached the size of three quarters of an adult brain. By the age of six, the child's brain reached 95% of the adult brain size. (Lenroot & Giedd, 2006). As a result, the brain of a five-year-old child has almost reached the size of adulthood. However, the development in the brain continues throughout childhood and adolescence. (Blakemore, 2010).

One of the processes that contribute to brain development is synaptogenesis. Synaptogenesis is the process of the formation of synapses (connections between neurons). Many scientists agree that some synapses are entirely based on experience and that synaptogenesis continues as long as new skills are learned. There is a pattern of start-stop in brain development. In other words, the synaptogenesis process is followed by the pruning of synapses, that is, the synaptic pruning process. Synaptic pruning is not used at all or little used paths are pruned. For example, the most intense synapses in the parts of the brain that control language comprehension and word production are around three years of age. Intense synaptic pruning occurs at the age of four. However, the synapse density in the brain of a four-year-old child is expressed as twice that of an adult. (Bee & Boyd, 2009). The synaptogenesis and synaptic pruning processes point to the importance of the environment that contains rich stimuli in early childhood, on the other hand it demonstrates that early childhood children can easily adapt to the environment they encounter. This harmony is then processed into the brain of the child all his life. For example, a child who lives in an environment belonging to a tribe living in the forest develops the ability to distinguish certain details of the forest, while a child living in a noisy urban environment will learn to eliminate unnecessary sounds.

Some changes in the brain lead to an increase in dendritic connections such as myelination. Myelination refers to the isolation and cover of nerve cells by adipose tissue. Myelination plays a role in improving the quality and speed of information transmitted across the nervous system (Fair & Schlagar, 2008). Myelination is important in the development of many of the children's abilities. Myelination in the hand-eye coordination part of the brain has not yet been completed until the age of four. Myelination accompanies improvements in brain function. For example, neurons in the connotation area where the sensory, motor, and mental functions of the brain are interconnected are to some degree myelinized when the child starts school. At the age of six to twelve, it completely becomes myelin. This progress in the myelination process is thought to contribute to the increase in the child's capacity to process information. (Bee & Boyd, 2009; Santrock, 2012).

MOTOR DEVELOPMENT

The motor development area examines the movements of the babies from the mother's womb, the non-optional trunk movements that continue with birth, and then the change and development of controlled and voluntary movements with their own body. Walking, running, crawling, cycling motions, ability to hold and grasp objects, to thread rope to sewing needle, to hold crayons are motor skills that include rough motor skills including manipulative skills. In motor development, first develop rough motor skills, then fine motor skills. Therefore, children who are six years old can jump and jump properly, ride bicycles, climb and jump. They can do rough motor skills, but they don't have the ability to cut as shown with scissors, the ability to use crayons and pencils. In the following years, very fine improvements in fine motor skills occur, children can begin to write easily, make pictures, play music instruments. (Bee & Boyd, 2009). In the early childhood, children with varying body proportions can coordinate movements such as running, throwing, jumping, climbing, dressing on their own, buttoning the buttons in front of them, tying their shoelaces, as well as toilet habits and sleeping patterns. Sensory competences are important in motor development because infants gain insight into the world through hearing, sight, touch, smell and taste. Therefore, mobility allows them to get closer to their environment by increasing the type and proportion of sensory knowledge they have acquired, giving them the opportunity to explore the situations they encounter. (Gander & Gardiner, 2010). In the motor development that continues throughout life, firstly postural postures, primitive movements and reflexes of infancy and two-year-old children form the basis for them to acquire more voluntary movement skills in childhood by doing motion training. At the beginning of childhood, a rapid development of skills known as "basic movement skills" begins. During this period, as they move, they can discover the potentials in their bodies, develop successful control of gravity over the muscular systems (balance), and acquire the ability to manipulate and manipulate objects around them more precisely (Manipulation). (Gallahue, Ozmun & Goodway, 2014).

There is a certain order in which children acquire motor skills. This sequence is first from head to toe (cephalocaudal) and then from the center to the outside (proximodistal). First, children undergo head and trunk control and then sit and walk. To make free movements, first head, then shoulders, last buttocks can move freely. In the development from the center to the outside, the child first learns to raise his arms and reach out, then to catch and hold objects with his hands. In his studies, Gesell stated that in motor development, the maturation of the nervous system and the control of the brain shell (cortex) should be increased, while reflexes (innate learned, genre-specific behaviors) play a very important role. Motor and sensory development are completely interconnected. In the interaction of children with their environment, there is a process of neurological processes called sensory integrity which organizes the stimuli coming from the environment and the body so that they can use their body effectively. The level of cognitive development, as well as the motor development of children, as well as movements and motor activities, perceptual and cognitive development provides a positive progress (Fazlıoğlu, 2009).

The motor behavior acquired during the early childhood also forms the basis of lifelong movements. They called it the period of basic movements. Basic skills are used in all areas of life such as jumping, throwing, running, catching. These are considered in three stages as initial stage, first stage and maturity stage. The initial stage, the first stage, is the stage in which the child becomes aware of his or her abilities, which are about exploring the body. For example, it lifts an object by its weight, and learns about being able to relocate it by throwing it. If its too heavy, he can't lift or throw. Coordination may be impaired, but in the first phase, the harmony and coordination between movements and control increases. Despite this harmony, some limitations may be observed. Three-to-four-year-olds may hit the ball with their feet, but may not throw the ball exactly to the target. Environmental factors such as biological maturity, learning, experimentation and motivation are important in the maturity stage. For this reason, it is better to wait for the mature age and the age of five to seven, rather than waiting for basic skills such as jumping, jumping and running to be done like adults. (Arı, 2005). Even if they partially spill the food at the age of three, they can eat themselves, but they have to be seven to eight to cut food like meat on their plates. They can take off their clothes again at this age both they always wear and take off at the age of 4. As they cannot control their fingers, they have difficulty in tying their shoelaces and cannot wear their shoes because they do not reach the perception that can distinguish between left and right. They can take a bath on their own in a controlled manner at the age of five. If children find it difficult to do so at this age, it can be thought that it is due to the parents' lack of opportunities. 4-years-old children can run, jump and also they move in every situation. They can paint sizeable paper holding successfully brushes, pencils and scissors. (Zembat, 1996). Children of this age do not need to be kept under constant control. Because now they are on his way to become an individual who can ask questions, make his own choices and act freely on his own. The ability to use the pen by holding the pen starts around two years of age by making different signs by touching the paper with a pen for the first time. They start with vertical lines and then horizontal lines in two to three weeks. Even though they are overflowing the line at the age of four, they can paint a simple shape's inside. Because the control and detection levels have improved, they can copy the square shape. When they are five years old, they can draw the shape of a triangle by drawing shapes similar to the human figure. Make copies of rhombus or other geometric shapes when they are seven years old (İnanç, 2007). Since myelination in the early childhood is the continuation of myelination in infancy, hand-eye coordination is more successful as a result. (Santrock, 2012). Most of the basic movements are acquired in the early childhood. If the child has no disability or illness, they can run fully towards the end of the age of five.

When children reach the age of three, they can do coloring-cutting, scissors-holding skills. Before children start school, they acquire all the skills related to motor development in their early childhood. Due to muscle development in their early childhood, they are more successful in their dynamic movement skills (walking, jumping, running, etc.) than fine motor movement skills. Two years old children scribble with pencil, three-year-old, draws geometric shapes like circles, squares, 4-5 years old can paint what they see around (home, people, etc.). When they are four to five years old, they can eat on their own, at least partially, they can zip up, button up, make shapes from play dough, make shapes with jigsaws and cubes. Children 49 make a significant progress in the development of fine motor skills by coordinating bodily movements with perceptual movements. Around the age of 5, he can translate by looking at the pages of the book, but since hand-eye coordination is not fully developed, they cannot do fine motor skills like adults. Therefore, in early childhood activities, care should be taken to use large picture books and thick and large tools suitable for the hands. In these years, girls are more successful in showing fine motor skills than boys. Cultural approaches also have an impact on the development of these skills (Ergin vd, 2010).

COGNITIVE DEVELOPMENT

Cognition and Cognitive Development

When we look at the definitions of cognition and cognitive development; The word "cognition" means mental activities that involve learning and understanding our world (Ergin vd, 2010). Cognition is the form or process of obtaining information, including perception, recognition, reasoning, and judgment. (Gander & Gardiner, 2010). The most important feature of the individual is its cognitive potential. Abstract reasoning, logical thinking related to assumptions, cognitive development with the aim of organizing the rules in a complex and higher level, involves the changes in the organism of the individual through different stages in the process from birth to death.

In cognitive development, according to Piaget, environmental factors play a major role, and as a result of children's interactions with their environment, children develop techniques to understand objects and situations. To develop a method of understanding and knowing is called "cognitive development. Cognitive development includes all processes that help to acquire and use information in various ways. With its cognitive power, human beings take control over other living things, try to cope with nature, develop new cultural values and technology. (Aral, 2002).

Cognitive development process includes changes in the functions of the individual towards language, thought and intelligence (Ergin vd, 2010). The concept of cognitive development consists of cognitive processes that enable the child to obtain various information about his / her social environment. In cognitive development, there are individual differences between each child. However, it has been shown that there is no gender difference. It was also determined that socioeconomic status affects cognitive development.

All activities that occur in one's mind are examined within the field of cognitive development. Although it progresses in parallel with their physical development, it is not the same in every child. In the early stages, the child expresses his wishes and thoughts only through crying, and learns the way of passing through language as time passes. Individuals in different age groups have different abilities to understand and learn the world. This difference arises from different levels of cognitive development. In order to make progress in cognitive development, the organism must reach a biological maturity to meet this. (Fazlıoğlu, 2009).

Children take a set of information from their environment and perceive and absorb this information and store it in their memory with a cognitive system. When they think or plan to make a behavior, they remove their information from memory and use it when appropriate conditions exist. They enrich their thinking through conceptualization, reasoning and generalization of some of this information. Children can achieve change in a small group between the ages of four and six, but as the number increases, the responses to the object can be very different. They can use different ways of reasoning about whether the quantity of the object is preserved or not. (Gander & Gardiner, 2010).

Children who notice many affective events; they can express and interpret them, then save them in their memory, remember, use and repeat them when they seem necessary. In order to perform many cognitive processes such as reasoning, thinking, evaluation, problem solving and interpretation, children have cognitive abilities such as memory, perception, generalizing and evaluating information, and reasoning. The cognitive parts of these abilities are; symbols, diagrams, images, rules and concepts. Cognitive abilities form the stages of problem solving. With the accumulation of life-long knowledge and experience, problems are solved and this information is realized through the operating system. It is very important that children receive new knowledge, combine information and blend it with what they know, and start producing and developing new skills at an early stage. Early education will help children to cope with the difficulties they may face in life, to gather the necessary information in the problem solving stage, to interpret the events correctly, to make accurate evaluations and to have the courage to adapt to life. (Ömeroğlu & Kandır, 2005).

There are two important situations in perception, one of which is the sound and healthy organ of the senses (maturation), the other is experience and experience (learning). Perceptual development is examined between the ages of 3 and 6, taking into account the discrimination, selectivity, learning of object persistence, object invariance and the decrease in self-centralization. To understand objects and events, senses are primarily used in perception. All sensory organs significantly influence the way they are perceived. To sort the ways in which sensory organs that affect in different fields affect perception; it is very important in the development of perception, comprehension and understanding. It allows the child to direct his attention and supports the extension of the attention span. It makes it possible to perform its activity by perceiving in a certain order and from beginning to end, it gives the ability to perceive and listen to what is heard, and makes it easier to store the perceived ones in memory. Touch perception helps the child to recognize those around him by touch and guide them when the use of other senses is needed (Aral, 2002).

Attention has a very important place in perception development. This can be defined as the process of encoding sensory stimuli. Our sensory organs may encounter multiple stimuli in a very short period of time, but only give attention to a few. (Mağden, Şahin & Çetinkaya, 2001). Attention is also important to what one perceives as well as why. Therefore, attention includes the duration of attention and the selectivity of attention. The attention span of children whose attention span is shorter at first increases with age. Interest and attention are intertwined. The child can only focus by paying attention to the object and situation of interest.

Remembering is important in the development of thinking skills. Children in different age groups in early childhood show differences in the quality of thinking in children, some previously known concepts in children should be remembered that the storage, evaluation and re-use of stimulants can develop in the thought system. (Baloğlu, 2016). Reasoning is organizing and combining previously learned information with the aim of solving new problems. Symbols represent external stimuli, events, and objects, and are important tools for performing the thinking event. (Aral, 2002).

The pre-operational period is a process in which language and symbols are developed, not yet processed, but they are prepared to be operated (Fazlıoğlu, 2009). Piaget defines "the process" term in the form of logical thinking. In this respect, logical thinking has not yet developed in the pre-operational period. (Selçuk, 2000). Trial and error and taking as a model takes attention in the child's learning in this period. According to Piaget, the transition to the pre-operational is critical in the mental process of the child. The pre-operational period is divided into two periods as conceptual / prejudice (2-4 years) and intuitive / intuition (4-7 years). (Aydın, 2013). In the pre-operational conceptual period, children gradually begin to reason about events, even if they are not rationally based on the qualities of adults. (Baloğlu, 2016). But in this period, idea is irrational and momentary, dependent on visible conditions. (Gander & Gardiner, 2010). The child perceives only as he / she sees and is unable to see a situation from the perspective of others, to realize the feelings and needs of another person. Egocentrism is the inability of the child to distinguish his / her own point of view from others. Because of this insufficiency, they have problems communicating with both peers and adult people, because for him, those who know what they know know. (Santrock, 2013). A child looking at an object can perceive only one feature, the other features do not attract his attention, this is the state of focus. (Fazlıoğlu, 2009). Children can use words as symbols that replace objects and situations when speaking. In the same way, they can pretend to drink milk in an empty container with the toy in their games. Their way of thinking depends on concrete objects and events happening at that time and they believe in the images of the objects. According to Piaget, while associating two situations at the same time, children in this period may think that in the cause-effect relationship, one causes the other. Piaget explained this as private to private reasoning (Akman, 2017). The belief that inanimate objects are objects with realistic features and capable of movement is called anism. For example, the child may think that his doll is hungry, say it, or think that he will respond by talking to him. (Ergin vd, 2010). One of the important features of the period is symbolic play. This is imitation based symbolization. Children can use a piece of wood as a truck or baby (Selçuk, 2000). During this period, if the child constructs a picture in the mind of the object and a design about how to draw it towards himself, and uses words to describe this action, it means that he has achieved this fiction mentally on a symbolic level. Hetheringhton & Parke, on the other hand, makes children unable to understand the mutual differences between their dimensions by understanding the single property of the object and the situation cannot understand the conservation. (İnanç, 2007). Children try to imitate what they see in this period. Reasoning is not in the form of general to private (deduction) or from private to general (induction); it is private-to-private reasoning without generalization. For example, when the child who is taken to the amusement park on the day of her aunt's marriage wants to go to the same park, "will you get married? When are you getting married again?" he might ask. Develops a direct relationship between going to the park and aunt's marriage (Ergin vd. 2010). Although children show insufficiency in classifying, they have advanced more than the previous period. In this period, they can make simple addition, grouping with subdivisions of large groups. When both are requested, the large group is insufficient to make the group within itself and the small group within itself. This inability is very important for mathematics teaching (Fazlıoğlu, 2009). Children often think that if the image of the object changes during the pre-operational period, the amount may also change. (Selçuk, 2000). They classify objects according to their characteristics such as size, color and shape, and when they reach the age of five, they become aware not only as an object, but in the group it represents. Piaget found that what young children think about the problem is different from older children and that concepts such as number, volume and weight conservation have not developed. (Özer & Özer, 2016). Children can only pay attention to a detail of the situations they encounter (Ergin vd. 2010). During this period, children try to do some mental processes while classifying, characterizing, associating objects. They find it difficult to explain how and what they do. This is because they use half logical reasoning (Ömeroğlu & Kandır, 2005). Rice defines private-to-private reasoning as reasoning without generalizing from a special case to a special case (İnanç, 2007). The concepts of conservation and reversal have not yet been acquired in the Pre-Operational Period. Piaget described it as irreversibility / reversibility, saying that children may have difficulty in reversing their thoughts. In order to test fluid protection for children, when the water is drained from the short container to the long container, child will always say that the water in the long container is excessive. This indicates "the reverse conversion" feature (Aydın, 2013). Children at the age of four to five years think that there may be water when ice is melted, but they do not think that melt water can be ice if put back in the freezer. Children can classify by examining the pictures given to them, separating them as clothing and food. But the ability to classify by category is not fully developed (Ergin vd. 2010). Pre-operational children cannot mentally understand that the initial condition can be regained. In this period, animism and artificalism occur in children. Animism is to think of a lifeless object as if it were alive. Artificalism, on the other hand, is to think that one of the natural phenomena is realized or caused by one. Like a five-year-old kid thinks someone's lit the sun with a match. (Fazlıoğlu, 2009). Piaget stated that children do not believe that plants are alive, but that the sun is alive. While some of the children say that plants are alive, they can attribute this to inanimate objects. Children may know that animals and plants can grow during the early childhood and that some situations affect it. Children aged four to five, in particular, cannot realize that it is not possible for unusual and supernatural events to take place. (İnanç, 2007). Pre-operational period is an important period in which children receive pre-school education and includes many cognitive skills and thoughts.

LANGUAGE DEVELOPMENT

The language, which is often used in the same sense as speech, actually requires the use of a mental process beyond speech. Language is an effective communication tool that enables the transmission of verbal messages as well as non-verbal messages. Verbal and / or non-verbal language is used in all the actions we perform in our daily life. The multicomponent and complex nature of the language leads to the definition of the language in different ways. There are many opinions and definitions about language and acquisition of language. If we start from the common points in these definitions, language; It is a system of symbols that we use to classify, organize, and express our thoughts. Language is a traditional system in which socially shared codes or symbols are used to represent concepts based on various rules. (Otto, 2006; Owens,1984).

Language is one of the most important tools we use to define the world and get information about the world. The knowledge gained about the language spoken in the environment from early childhood helps to reach the higher levels of language development by differentiating in the developmental process. Initially only used for communication purposes through sounds and / or phonemes, language has become an indispensable element of communication with its single words, two-word combinations and its complex structure. This process, from simple to complex, follows almost the same developmental sequence, especially in individuals with single language and normal development. (Maviş, 2005).

Traditional behaviorists who claim that imitation and repetition take place at the basis of language learning; they emphasize that children repeat the models by imitating them until they reach the right language by hearing the sounds and speeches around them and make it a habit. The verbal behaviors of children are supported by those around them and shaped in a similar way to adult speech. (Ekmekçi, 1995).

Behaviorists argue that children simultaneously imitate new words they hear, reflect imitated words and structures in their daily conversations, and that children acquire language by applying and imitating idioms and phrases in the game. (Reich, 1986; Berk, 1999).

According to behavioral theory, Owens (1999) stated that language is learned through operant conditioning and behavior is shaped by external reactions stated. While positive reinforcement behaviors are more likely to be repeated, punished behaviors are less likely to be repeated. Complex behaviors are learned step by step or by shaping. In shaping, the stimuli are gradually increased or decreased in order to achieve the desired behavior. According to the behavioral approach, language is acquired in this way.

Nativist Theory assumes that human's language is a biological basis of language and that they are born with this language at birth. Researchers who adopt this view argue that children come with a certain capacity in terms of their ability to use language and that every child has some universal invariant features in language acquisition. Stating that children can produce an unlimited number of sentences and a limited number of elements in the language, the innateists have introduced the theory of productivity in language acquisition. (Başal, 2003).

Chomsky tried to explain language acquisition with two different concepts. According to Chomsky, children are able to perform their mother tongue and use features properly without any education; however, they are not consciously aware of it. Chomsky calls this unconsciously grammar of children the ability to use (linguistic), and the ability to make it available effectively. Chomsky discovered that what the child understands and says from a conversation cannot be explained solely by analogy and imitation, and how the brain comes equipped and ready to learn for all kinds of development; they are born with a special ability to understand. Claiming that children were born through language acquisition, Chomsky argues that children immediately memorize the conversations they hear around them at various levels and genres, and then use these experiences in their own sentences. (Temel, Bekir & Yazıcı, 2014; Kıran & Kıran, 2002).

Another view on language acquisition is the interactionist view. Jean Piaget, who has made important studies in the field of cognitive development and genetic epistemology, argues that thought and language development in children is not through continuity but occurs through stages and is effectively structured in individual environment relations. Piaget argues that language development is part of general cognitive development and that there is no specific mechanism for language acquisition. (Foster, 1990).

According to Piaget, the pre-contion for language acquisition is symbolic thinking. Language development is a coordination of general cognitive changes and cognitive development is not affected by language. Once the child has gained symbolic thought, he / she understands that language functions as a symbol system and can understand words, objects, actions and abstract views. The first words the child uses are not real symbols yet. Because these symbols are related to the objects and events seen. When children start speaking before they are present, they use the basic symbols. (Flavell, Millir & Millir, 1993). Vygotsky stated that language and thought emerged independently; however, he later claimed that he gained meaning by interacting with each other. Vygotsky disagrees with the idea that cognitive development is different from language, and that language education and learning affect one's ability to think mentally. According to Vygotsky, the word goes to the structure of the object and thus gains a functional meaning. At the same time, the language environment of the child affects the level of thinking. Therefore, it is emphasized that verbal thinking can be understood by examining the undeveloped, egocentric, partial and primitive stages of the child. Today, language acquisition is described as a multidimensional, continuous process that derives its origin from relationships that begin immediately after birth between the child and the closest and continues to adulthood in its most complex form. As can be understood from all these explanations, the issue of how language is acquired has been the focus of many studies and discussions. This diversity in language views stems from the fact that language acquisition is a very complex system. (Meece, 1997). Looking at these theories or researches on language acquisition, it is understood that language is a system of language and many factors are effective in the development of this system. Although a living entity, language, seems to be a complex system, it is universally a system of language and these systems also have culture-specific rules. Mutual sharing is at the forefront in the early stages of language acquisition. From birth, the child learns the rules of communicative sharing in their direct relationship with those close to them and grows up in a communicative language environment. In the first shares of this first stage, he only grasps the way he communicates about the objects that exist around him, and he can understand and produce the first words. However, these words have no representation. The second stage of language acquisition; It is defined by cognitive processes that lead the child to understand the representative dimension of words and to develop the semantic stage of language. In the cognitive process, the child learns to think and construct intangible events and objects, thus gradually moving from the concrete environment to the symbolic stage. (Kuhn, 2006; Temel, Bekir & Yazıcı, 2014).

In order to use the language effectively and in an advanced way, children must have acquired different levels of competence. Phonology, morphology, semantics, syntax and pragmatic knowledge are required for language proficiency. Phonology explores the basic sound structures of language. There are different sounds in addition to the same sounds in each language. The child first encounters sounds during the development process and gains proficiency on these sounds. These sounds are called phonemes. Morphology is the science that examines the smallest units in the language that contain meaning. These units are called morphemes. Semantics generally deals with the meaning aspect of language and makes the analysis of sentences and words in terms of meaning. Syntax is the science that examines sentences in terms of rules and structure. It is a system that examines the rules of sentences consisting of words. Pragmatic explores the rules governing different social contexts. For example, when a five-year-old tells something to a two-year-old, he uses simpler sentences, speaks slowly and repeats his words for his understanding. (Temel, Bekir & Yazıcı, 2014).

In the language development, the first year of life is defined as the prelinguistic period and the preschool years are defined as the periods in which basic language skills are acquired. The following table presents the general characteristics of language development in children aged three to six years. While this

Table 1. General language development characteristics of three-six age group

Age (month/year)	General Features
3 years old	• The expressive vocabulary includes 900-1000 words. Form sentences of three to four words. • Uses simple sentence structures, including subject and action in sentences. • Plays with words and sounds. • Follow two-step instructions. • Talk about the current situation.
4 years old	• The expressive vocabulary includes 1500-1600 words. • Ask too many questions. • Uses increasingly complex sentence structures. • He has difficulty in answering "how" and "why" questions. • Make predictions based on word order.
5 years old	• The expressive vocabulary includes 2100-2200 words. • Discuss about emotions. • Understands the words "first" and "after" regardless of word order. • Follow three-step instructions. • 90% of grammar rules have been gained.
6 years old	• The expressive vocabulary includes approximately 2600 words, and the recipient language vocabulary is 20000-24000 words. • 6 years use multiple sentences with complex structure.

Source: Adapted from Owens, (2001).

ranking in language development of three-six year old children whose general characteristics are in the table does not change, the rate of this development can be affected by factors such as physiological and genetic characteristics, gender, perceptual, cognitive and neurological development, social environment and interaction, verbal communication level between family and child, socioeconomic and sociocultural characteristics. (Lewis, 1982; Paul, 1996).

SOCIAL-EMOTIONAL DEVELOPMENT

Emotional and social development are too intertwined. The most important factor that affects emotion development is the relationship of the individual with other individuals, so social development. Emotions are the most important elements of being human. Emotions play a fundamental role in the formation of social ties. Therefore, researchers have adopted the term social-emotional to emphasize how intertwined the social emotional dimensions are. (Bayhan & Artan, 2004).

Social emotional development is the ability of the child to express themself, to control their emotions and to be at peace and harmony with them and their environment. Studies on the emotional characteristics of children have shown that emotional development is the result of both maturation and learning, none of which is effective alone. (Saarnı, 2001). Socialization is a basic social process which means the integration of the individual into the society by learning the culture and the role of the society in which the individual lives. Socialization enables the child to take part in society. This process starts after birth and continues for life. As the child socializes, the integrity of his behavior becomes increasingly complex (Aral, Baran, Bulut & Çimen, 1981; Kandır & Alpan, 2008).

Emotional competence of children three to four years of age contributes to social competence in the long term (Denham vd. 2003). Accordingly, social and emotional development are closely related. Emotional disorder in children results in social mismatch (Rubin, Coplan, Fox & Calkıns, 1995). Raver (1997) He stated that emotional and social skills are also related to children's academic achievement and that children with emotional problems have problems in their school life.

In the first childhood, the child tends to see the situations of others in an egocentric way and tries to identify the other person's point of view from his own point of view. In the early childhood, the child demonstrates social development, acquires social skills and learns pro-social behavior. Children begin to understand the values of society. They also learn how to play with his peers and make friends in this period. In this period, children's methods of achieving their goals in social situations have improved. In the beginning, children can try to get what they want, just like toddlers. For example, a three-year-old can look for an opportunity to take the toy that another child is playing. But by the age of four, verbal strategies begin to replace such physical behavior. The parent at the center of the child's life durıng the first childhood is the parent. But at the same time, the child's interest begins to focus on his peers. The child's sense of self as a social person is now strongly influenced by these friendship relationships. As a result, although a four-year-old has a warm relationship with his teachers, he still wants his friends.

During this period, with more and more acceptance among peers, the child's perception shifts towards an increasingly objective and complex worldview. As five-six-year-olds get a better grasp of other people's perspectives, they gradually move away from self-centeredness. In this way, they understand that not only their own perspectives, but also the ideas of others should be evaluated. On the other hand, children who do not have a secure attachment background or who are silent and shy may experience difficulties in making friends during this period. These children may startle from social games and as a result may become alienated from the group. This may further strengthen the child's shyness and prevent him from learning the pleasures of friendship. At the beginning of this period, parallel games of toddlers continue to be replaced by common games. For example, two three-year-olds riding a double-wheeled tricycle enjoy it. However, they do not talk much to each other and therefore do not give the impression that they share something in common. When the same children reach the age of four, they can play a more complex and interactive structure around a common idea.

REFERENCES

Aral, N. (2002). *Okul öncesi eğitim ve okul öncesi eğitim programı. Öğretmen rehber kitabı*. İstanbul: YA-PA.

Aral, N., Baran, G., Bulut, Ş., & Çimen, S. (1981). *Çocuk Gelişimi*. İstanbul: YA-PA.

Arı, R. (2005). *Gelişim ve öğrenme*. Ankara: Nobel.

Aydın, B. (2013). Eğitim Psikolojisi Gelişim-Öğrenme-Öğretim. In Gelişimin doğası. Baskı, Ankara: Pegem.

Bağcıoğlu Ünver, G. (2013). Gelişimle ilgili temel kavramlar, gelişimin temel ilkeleri ve gelişimi etkileyen etmenler. In A. Ulusoy (Ed.), Gelişim ve öğrenme psikolojisi içinde (pp. 1-17). Ankara: Anı.

Baloğlu, M. (2016). *Öğrenme psikolojisi*. Nobel.

Başal, H. A. (2003). *Nasıl mutlu çocuk yetiştirebilirim?* Ankara: Morpa.

Bayhan, S.P. & Artan, İ. (2004). Çocuk gelişimi ve eğitimi. İstanbul: Morpa.

Bee, H., & Boyd, D. (2009). *Çocuk gelişim psikolojisi*. İstanbul: Kaknüs.

Berk, L. (2013). Bebekler ve çocuklar. In N. Işıkoğlu Erdoğan (Ed.), Doğum öncesinden orta çocukluğa. Ankara: Nobel.

Berk, L. E. (1999). *Infant children and adolescents*. Boston: Ally and Bacon.

Blakemore, S.J. (2010). The developing social brain: Implications for education. *Neuron, 65*, 744-747.

Chomsky, N. (1975). *Reflections on language*. NY: Pantheon.

Denham, S., Blaır, K., Demlılder, E., Levitas, J., Sawyer, K., Auerbach, S., & Queenan, P. (2003). Preschool Emotional Competence: Pathway to Social Competence. *Child Development, 74*(1), 238–256. doi:10.1111/1467-8624.00533 PMID:12625448

Ekmekçi, Ö. (1995). *Psikodilbilim*. Çukurova Üniversitesi, Yadim.

Ergin, H., Yıldız, A., Erdoğan, G., Mercan, S.Ç., Çelikkol, K.A. ... Gazioğlu, G., E. (2010). Gelişim psikolojisi. Academic Press.

Fabes, R., & Martin, C. (2000). *Exploring Child Development* (pp. 175–195). Boston: Pearson Education, Inc.

Fair, D. & Schlaggar, B. L. (2008). Brain development. In M.M. Haith, J.B Benson (Eds.), Encyclopedia of infantvand early childhood development. Oxford, UK: Elsevier.

Fazlıoğlu, Y. (2009). *Erken çocukluk gelişim ve eğitimi. 1.* Baskı, Krite.

Flavell, J. H., Millir, P. H., & Millir, S. S. (1993). Cognitive development. New Jersey: Prentice-Hall.

Foster, S. H. (1990). *The communicative competence of young children: A modular approach.* London: Longman.

Gallahue, D. L., Ozmun, J. C., & Goodway, J. D. (2014). *Motor gelişimi anlamak. Bebekler, çocuklar, ergenler, yetişkinler.* Ankara: Nobel.

Gander, M. J., & Gardiner, H. W. (2010). *Çocuk ve ergen gelişimi.* Baskı, İmge.

İnan, M. (2004). 3-9 yaş çocukları için uygulamalı hareket eğitimi öğretmen el kitabı. İstanbul: Morpa

İnanç, B. Y., Bilgin, M., & Kılıç Atıcı, M. (2007). *Gelişim psikolojisi, çocuk ve ergen gelişimi.* Ankara: Pegem.

Kandır, A & Alpan, Y. (2008). Okul öncesi dönemde sosyal-duygusal gelişime anne-baba davranışlarının etkisi. *Aile ve Toplum Eğitim Kültür ve Araştırma Dergisi, 4*(14), 33-38

Kıran, A., & Kıran, Z. (2002). *Dilbilime giriş.* Ankara: Seçkin.

Kuhn, T. (2006). *Grammatik im englisch unterricht der primarstufe: theoretische grundlagen und praktische unterrichtsvorschläge.* Springer.

Lenroot, R. K., & Giedd, J. N. (2006). Brain development in children and adolescents: Insights fromanatomical magnetic resonance imaging. *Neuroscience and Biobehavioral Reviews, 30*(6), 718–729. doi:10.1016/j.neubiorev.2006.06.001 PMID:16887188

Lewis, G. (1982). Human migration: A geographical perspective. London: Canberra.

Mağden, D., Şahin, S., & Çetinkaya, Ş. (2001). *Üç-dört yaş grubu çocukların fotoğraf ve çizimlerdeki insanların bakış yönlerini algılamaları ve görsel dikkatin incelenmesi.* Milli Eğitim.

Maviş, A. (2005). *Anlayarak hızlı okuma ve öğrenme teknikleri.* İstanbul: Hayat Yayınları Dergisi.

Meece, J. L. (1997). *The communicative competence of youngchildren.* New York: Addison Wesley Longman.

Mengütay, S. (2005). *Çocuklarda hareket gelişimi ve spor.* İstanbul: Morpa.

Ömeroğlu, E., & Kandır, A. (2005). *Bilişsel gelişim.* İstanbul: Morpa.

Otto, B. (2006). *Language development in early childhood.* Upper Saddle River, NJ: Pearson.

Owens, E. R. (1984). *Language development: An introduction.* Ohio: Merrill Publications.

Owens, R. (1999). *Language disorders: A functional approach to assessment and intervention.* Boston: Allyn & Bacon.

Owens, R. (2001). *Language development: An introduction.* Boston: Allyn & Bacon.

Özbar N. (2007). *Hareket eğitimi programının 4-6 yaş grubu çocuklarda motor beceri ve vücut kompozisyonu üzerine etkisinin incelenmesi* [Doktora TeziM]. Ü. Sağlık Bilimleri Enstitüsü.

Özer, D. S., & Özer, M. K. (2016). *Çocuklarda motor gelişim.* Ankara: Nobel.

Paul, S. T. (1996). Search for semantic inhibition failure during sentence comprehension by younger and older adults. *Psychology and Aging, 11*(1), 10–20. doi:10.1037/0882-7974.11.1.10 PMID:8726366

Pınar, S., & Erkut, O. (2000). *Artistik Jimnastik Yaş Grupları Gelişim Programı.* Ankara: G.S.G.M. Basımevi.

Raver, C. (1997). Poor Children Gain Social Competence From Sensitive İnteraction With Parent. *The Brown University Child and Adolescent Behavior Letter, 13*(7), 1–2.

Reich, P. A. (1986). *Language development.* Englewood Cliffs, NJ: Prentice-Hall.

Rubin, K., Coplan, R., Fox, N., & Calkıns, S. (1995). Emotionaltiy, Emotion Regulation and Preschoolers Social Adaptation. Development and Psychopathology, (7), 49-62.

Saarnı, C. (2001). Cognition, Context and Goals: Significant Components in Social-Emotional Efffectiveness. *Social Development, 10*(1), 125–127. doi:10.1111/1467-9507.00152

Santrock, J. W. (2012). Yaşam boyu gelişim. Ankara: Nobel.

Santrock, J. W. (2013). Life-span development (14th ed.). McGraw-Hill.

Selçuk, Z. (2000). *Gelişim psikolojisi.* Ankara: Nobel.

Temel, Z. F., Bekir, H., & Gül Yazıcı, Z. (2014). *Erken çocuklukta dil edinimi.* Ankara: Vize.

Trawick Swith, J. (2013). Erken çocukluk döneminde gelişim (Çok kültürlü bir bakış açısı). Ankara: Nobel.

Zembat, R. (1999). *Okul öncesi eğitimde temel konular.* İstanbul: Turan Ofset.

Chapter 11
Middle Childhood Development

Seçil Yücelyiğit
https://orcid.org/0000-0002-6886-5997
Faculty of Education, TED University, Turkey

ABSTRACT

Child development is segmented into five periods and the bridge between early childhood and adolescence is named as "middle childhood." One of the milestones of this period is schooling. Middle childhood children start learning about the world; their roles, responsibilities and how to participate in this world by communicating with others besides the family members. These abilities are gained mostly at school with peer relations. In this chapter, the developmental areas of middle childhood children will be discussed with examples from recent studies.

INTRODUCTION

Research on child development is carried out under the main developmental areas of the human being. These can be listed as physical growth and motor skills, cognitive development, social-emotional development, moral development, and sexual development. Middle childhood is the period of a lifetime, which sets the stage for some crucial milestones such as developing self-identity; mastering critical thinking, reasoning and decision making; learning health literacy; gaining the habit of healthy nutrition and regular exercise, improving communication skills at school, developing peer relations, etc.

The theoretical perspective of the developmental areas concerning the middle childhood period will be presented one by one and the knowledge that forms the fundamentals of the period will be supported by the research about each developmental area.

During middle childhood, physical development is in progress. In this period, children's bones broaden and lengthen dramatically, which may result in pain in their arms or legs. At the beginning of the period, boys are taller while girls grow by leaps and become taller by the end of the period. They lose their baby teeth and replace them with permanent adult teeth during middle childhood. Both boys and girls build muscles and while their growth is similar in the early stages of the period, at the end of the period their physical growth vary in shape as they approach puberty. Girls tend to retain more fatty tissue than boys and look rounder (Berk, 2006). The children of this period improve both their fine and gross motor

DOI: 10.4018/978-1-7998-2952-2.ch011

skills. They can run faster or master the skills such as hop, skip or jump in terms of gross motor skills. As a piece of evidence in the improvement of the fine motor skills, middle childhood children develop the ability to write rigorously, draw complex pictures, use their hands to accomplish complex crafts with small-sized materials such as beads, use tools under the supervision of an adult, like a hammer or needle involving high levels of eye-hand coordination (Lindon, 2010). This is the period in which they commonly become experts of games on computers, tablets or mobile applications.

Comparing to physical development, cognitive development during middle childhood is recognized and considered more frequently. The changes in the cognitive development of the children in the middle childhood period are noticeable. They become more logical and organized in this period. Piaget's cognitive development theory suggests that children in the middle childhood period constitute "Concrete Operational" thinking stage. They are capable of operating on the "concrete" contents which implies the actual people and places or the objects they can observe. According to Piaget's cognitive theory, children in middle childhood begin to master some mental skills such as conservation, reversibility, decentration, classification, spatial reasoning or seriation (Berk, 2006; Lindon, 2010; Biehl, Park, Brindis, Pantell, & Irwin, 2002). The details of these operations will be presented in the following sections of the chapter.

Socialization is another developmental area that will be considered under middle childhood. It is defined as the process by which children learn what is expected of them, in terms of behavior and attitudes within their society, or social and cultural group (Lindon, 2010). Children in middle childhood period are experiencing the 'Industry versus Inferiority' stage according to Erikson's psychosocial development theory. While the parents and siblings keep their importance, friendships have a larger role in children's lives. During this period, they receive feedbacks concerning to peer-relation, adaptation to school rules, and etc. all of which are new to their life. If they are encouraged by their parents and teachers, they develop a feeling of competence and belief in their skills. The positive feedbacks bring the feeling of success while the negative feedbacks bring the feeling of failure on the contrary. Psychosocial development is related to a range of subjects such as emotional development, personality development, sexual development or moral development which is commonly observed interwoven. There are a number of theories concerning these developmental areas such as Vygotsky's developmental theory which highlights the contribution of social, interpersonal and linguistic factors in facilitating children's mental development, Freud's sexual development which describes how personality developed over the course of childhood, Kohlberg's moral development which identifies the moral reasoning in three levels. During middle childhood, most children accomplish recognizing and regulating their own emotions. Afterward, they start interpreting other people's emotional displays, which contributes to their perspective-taking abilities. During this period their empathy skills increase as well. Another aspect of the middle childhood period is the development of gender identity. They know the cultural definitions of feminine and masculine gender roles, attitudes or behaviors and try to adapt them (Lindon, 2010; Santrock, 2011; Damon & Lerner, 2008). Characteristic features of the middle childhood in terms of social development listed above will be addressed individually.

Knowing the developmental stages of children helps to understand and support them and this can lead us to practical ideas that will optimize a healthy development. During middle childhood, children keep growing physically and mentally. They develop self-concept and gender identity, gain self-esteem and learn empathy in this period. Promoting their development during this period will prepare children for healthy adolescence and strong adulthood. Researches on child development covering all periods of development will broaden our perspectives on meeting today's children's needs and addressing some issues that became critical. This chapter aims to foster the healthy development of middle childhood children. It is grounded in the theories of child development, enriched and updated by the current studies that catch up the new generation.

THE ROLE OF MIDDLE CHILDHOOD

Middle childhood is one of the stages of child development which takes place between early childhood and adolescence. Early in the history, children were regarded as adults by the age of 6 or 7 and they were given the responsibilities of the family life. Today in our century, this approach has been put aside. The children at this age are obliged to start their primary education. Schooling is one of the cornerstones of the middle childhood period. Another cornerstone in this period is the onset of puberty (National Research Council, 1984). These milestones give the clue for the developmental tasks of the middle childhood period. The primary tasks of development during this period cover the physical, cognitive, and social-emotional areas (Education Encyclopedia). Children acquire the skills to build healthy social relationships that will prepare them for adolescence and adulthood.

Although the theoretical attention given to the early childhood and adolescence is higher, middle childhood period also deserves to be recognized as a distinctive developmental period. Research findings indicate that middle childhood experiences are critical for a child's healthy development and lifelong learning. The responsibilities that children take in their families and their changing roles in the community help them develop new capabilities. The developmental representations of this period are directly related to the future cognitive, social, emotional, language, and physical development (National Research Council, 1984). The evidence from the research indicate that the acquisitions of middle childhood have an influence on the school readiness of the children and their later success in life (Boyden, Dawes, Dornan & Tredoux, 2019). It is also stated that some of the medical conditions of adult health such as maltreatment, chronic disease, obesity, dental-caries point to middle childhood period (Halfon, 2009; Shonkoff, Boyce & McEwen, 2009; Liew, 2012; Schore, 1999). In case of inadequate experiences in middle childhood period, the milestones of the social-emotional, language, cognitive and motor development faces the risk of delay or perish. These risks do not only cover the medical health issues, they include a range of behavioural and cognitive growth and development such as school success, peer relationships (Shonkoff & Phillips, 2002; Luby et al., 2013; Duncan, Ziol-Guest & Kalil, 2010; Duncan et al., 2007; Eccless, 1999). Understanding the role of middle childhood for a healthy adolescence will confront the limits of child's learning, growing and becoming a healthy adult.

Similar to other developmental periods the middle childhood, representing the ages 6-12, is studied by a number of disciplines ranging from psychology to education, medicine, sociology and public health. In the following part of this chapter factors affecting the development during this period will be presented.

FACTORS AFFECTING THE DEVELOPMENT DURING MIDDLE CHILDHOOD

Child development is monitored through developmental areas. The factors that affect middle childhood development are observed on physical, cognitive, social-emotional development areas. Considering the factors that affect middle childhood development, they can be categorized as school-based, family-based, medically-based, genetically-based and social-environment based.

The physical development is affected by the factors such as the financial potential of the family; the opportunities they provide for the child in terms of nutrition and health care; ethnic origin and genetics or cultural environment. The poverty is the main factor that has a wide range effect including physical, cognitive or social-emotional development. It is directly related with the opportunities presented to the child including healthy nutrition, necessary medical care, books, educational material and etc. The lack of medical resources which causes deterioration can be initiated by different reasons. The medical centre can be inaccessible due to weather conditions or the nutrition provided for the child can be limited because of the geographical reasons. The genetic factors are known to have a direct effect on the physical development along with the environment. The opportunities presented to the child is related with the socio-cultural characteristics of the society the child belongs. For example, healthy eating habits, exercising routines, physical activity options differ in cultures and these are among the factors that is reported to have effect on the physical and motor development (Weinberg et al., 2000; Yan & McCullagh, 2004).

It is reported that genes account for the most powerful effect on cognitive development yet it is not possible to deny the impact of environmental factors (Tucker-Drob, Briley & Harden, 2013). The school, the teacher or the parents offer children some opportunities related to their cognition. If the children are presented different types of stimuli to develop their memory and perception they will benefit from the given opportunity in terms of cognitive development. The type, variety and amount or frequency of these stimuli may depend on the socio-economic status of the family. It is also known that healthy nutrition of the child is also in relation with brain development which plays role in cognitive development (Prado & Dewey, 2014). The family approach depending on their cultural codes is another effect related to cognitive development (Greenfield & Bruner, 1966). Some cultures do not care about the gender equity to support their children's education which results with inequality of the opportunities. The research results reveal that play has an effect on the cognitive development during middle childhood. Children learn and practice life while playing developing their life skills such as decision making, building relationships, reasoning and etc (Bergen, 2002; Hutchison, 2015).

Children typically learn to develop relationship with their peers at school during middle childhood period. It can be said that the major factor to affect the social-emotional development is through their friends (Vitaro, Boivin, & Bukowski, 2009). Findings of the research reveal that heredity is another factor affecting emotional development since some similarities were found between the emotional development of parents and their children (Domitrovich, Durlak, Staley, & Weissberg, 2017). It is known that children's emotions are influenced by conditioning. Children can be trained under certain conditions to reflect their emotions under control which makes this an environmental factor to affect social-emotional development. When children are mature enough to control their emotions, they become more competent in socializing. Emotions can be a factor of socialization. The disabilities, disorders or impairments are the factors related to health which have influence on both emotional and social development (Smith, 1977; Yavuzer, 2011; Sørlie, Idsoe, Ogden, Olseth, & Torsheim, 2018).

MIDDLE CHILDHOOD DEVELOPMENT

Motor and Physical Development in Middle Childhood

Although it is slower than the early childhood period the physical growth continues during middle childhood. Children usually enter middle childhood with a slim body and low-fat rate. At the beginning of this stage the body shape of boys and girls are similar. Until puberty both genders grow 5-8 cm in height and gain about 3 kilograms in weight per year. Physical development in middle childhood comprises one of the milestones of the period (Jürimäe & Jürimäe, 2001). The physical growth of normally developing middle childhood children is regular and stable until the onset of puberty which is the beginning of their reproductive maturity. The normal range of onset is ages 8-14 years in females and ages 9-15 years in males. For girls, puberty begins approximately 2 years earlier than boys (Benson & Migeon, 1975). The similarity in the body shapes of boys and girls disappears after puberty. The secondary sexual characteristics of boys and girls make a distinction among the genders. These are the curves and the enlargement of breast tissue in females and deeper voice, broad shoulders in males (Smith, 1977). Since the physical growth is influenced by both nature and nurture, depending on the factors such as socioeconomic status, genetics, ethnicity or pathological symptoms, there may be rush or delay in the timing related to any of the maturities mentioned above (Sinclair, 1978). Not all of the children are growing under the same circumstances. Some of them may have challenging physical conditions, but still they can achieve their highest potential with support.

Due to the growth rate of approaching puberty, some children may experience difficulties in coordination and balance. This has a negative effect on children's emotions so it can be recommended for parents and teachers to explain children that what they experience about their balance and coordination is not lack of a skill but by the nature of growth. The same approach can be applied to early or late maturing children which may cause a short-term advantage or disadvantage among their peers. For example, a taller and stronger boy can achieve the advantage of physical activities like basketball, etc. (Harris, 2000). Children can be informed about the stages they are going through not to feel frustrated or superior to the others depending on their physical growth.

Parallel to their physical growth children achieve motor skills during middle childhood. In the beginning of the period between 5–6 years, they master the fundamentals of the period such as running, galloping, jumping, hopping, skipping, throwing, catching, striking, or kicking. It is recommended to include in their program activities that focus on having fun and developing motor skills rather than competition. In middle childhood period children do not complete their cognitive development. For this reason, simple and repetitive activities with little instruction such as swimming, throwing and catching ball will fit better to their programs. By the age of 7-9 the rules of their activities may become flexible such as tennis or ping pong and finally at age 10-12 their activities may involve entry-level complex motor and cognitive skills which require building strategy with team as in the example of basketball (Willoughby & Polatajko, 1995).

Another aspect of the physical development during middle childhood is the nervous system development. The central nervous system matures, the brain grows and reaches almost the adult size. Complex behaviours and cognitive abilities become possible for children. This increase in size especially in the frontal cortex which is responsible for personality; enables children to master their higher order thinking skills such as making plans or giving decisions. The brain and neural development are parallel with the cognitive development during this period. When the nervous system develops to a certain amount of neural connections children seem to achieve the concrete thinking (Kuther, 2016).

Social-Emotional Development in Middle Childhood

Social and emotional development is defined as the ability of expressing and regulating emotions, and managing social relationships (Carr, 2011). The most important that is accomplished in this period is the development of the self-concept. Social-emotional development in middle childhood is one of the development areas that shelters a leaping skill depending on another milestone which is transition to school. The children start school during this period and they are introduced to a set of new social contacts. They are required to develop strategies to manage the relationships in the school environment. For them, both the settings and the characters of the play turn into a challenge to cope with and consequently they are faced with a number of social tasks (Rogoff, Sellers, Pirrotta, Fox, & White, 1975). They become members of a society by developing understanding of the self, refining how the social world works and developing standards and expectations for their behaviour. First, they learn to regulate their own needs, desires, goals and expectations which is followed by the reflection of other people's perceptions. They become eligible to understand not only simple social roles but also the complex coordination of social or behavioural roles (Levine & Munsch, 2016). For example, children of the middle childhood period can understand that their elder sister can be the younger sibling of their elder brother or more complexly that she can become a wife or mother of someone else. Due to the requirement of the period, they now have friends at school, they all have the same rights and follow the same rules. When they acquire the understanding of other's perspectives this helps them to underpin the basis for the concepts such as friendship, equity or trust. They build a private world for themselves with their peers which they like to call 'best friends' and this process enables to develop self-concept. They set their own limits and learn to respect to other's limit when they meet. By the end of the period they manage to develop standards and incorporate norms of the society they are living as part of. Although it may be complicated for some of them to adjust their standards according to different components of the society such as children or adult it is reported that there is significant continuity in social emotional development from middle childhood to adolescence. (National Research Council, 1984; Aral, Baran, Bulut, & Çimen, 2000; Carr, 2011; Yavuzer, 2011).

The major titles studied in terms of social emotional development during middle childhood period include the concepts such as self and identity, self-concept, gender, self-esteem, peer relationships, empathy and attachment development.

Sigmund Freud (1856-1939) developed a theory on the development of personality. He claimed that human personality emerges from the early childhood experiences and these experiences shape the adult personality. According to his theory, there are 4 stages of psychosexual development which he named as oral stage from 0 to18 months, anal stage from 18 months to 3 years, phallic stage from 3 to 5 years, latency stage from 5 to 12 years and genital stage from 12 years to adulthood. If the individuals become unsuccessful or fail to complete any of the stages than they become fixated which leads to an unhealthy personality. Freud's theory does not say a lot about the personality in terms of middle childhood. The years between 5 and 12 corresponds to latency period in Freud's psychosexual development theory. Children have already become aware of sex differences and adopted their sexual characters in the preceding stage. The libido is dormant and there is no sign of psychosexual development during middle childhood. The sexual impulses are repressed, and the energy is directed towards school and friends (Bacanlı, 2004; McLeod, 2008).

Erik Erikson (1902–1994) studied on Freud's psychosexual development theory and modified it as a psychosocial theory. According to Erikson while mastering the skills and attitudes the ego makes positive contributions to development. He argued that personality develops through eight stages from infancy to adulthood. Each stage is finalized by a crisis with a positive or negative outcome of personality development. The way leading to a healthy adulthood is to overcome the psychological conflict successfully during each stage. The stages are, trust versus mistrust between 0 to 18 months, autonomy versus shame between 18 months to 3 years, initiative versus guilt between 3 to 5 years, industry versus inferiority between 5 to 12 years, identity versus role confusion 12 to 18 years, intimacy versus isolation between 18 to 40 years, generativity versus stagnation between 40 to 65 years and integrity versus despair after age 65 (Damon & Lerner, 2008; Lindon, 2010; Salkind, 2002).

The middle childhood period between the ages 6 to 11 corresponds to industry versus inferiority stage in Erikson's psychosocial theory. The focus of children at this age is on their peers to make comparisons and identify their own identity. After their evaluations of comparison, they result with a sense of pride and accomplishment or inferiority and inadequacy in their social activities at school, at home or at any other environment. At this stage, since the school has a very important place in the life of the children, the families share their throne with the child's teacher and the siblings share it with the classmates. These new characters in middle childhood period provide an opportunity for children to learn some specific social skills. If they can manage to get along with the new relationships and share positive experiences with them their acquisition will be the feeling of competency on their way to the next stage prior to adolescence. On the contrary, if the children fail to develop positive relationships with their peers, teacher or even with their family as a consequence of incompetency feeling they have experienced outside their home environment, this brings with it a traumatic feeling of self-esteem. It is recommended for teachers and parents to encourage children for their initiatives to boost their self-esteem. If the restrictions take place instead, the children can doubt their own abilities which avoids them reaching their potential. (Aral, Gülen, Bulut & Çimen, 2000; Boyden, Dawes, Dornan & Tredoux, 2019; Damon & Lerner, 2008; Lindon, 2010; Salkind, 2002; Santrock 2011).

Emotions are defined shortly as 'feelings' and are influenced by both genetics and experience (Kagan, 2010). Middle childhood period children become more capable of controlling their emotions unless they experience stressful moments (Cunningham, Kliewer & Garner, 2009). There are a number of studies that present the important developmental changes in emotions during middle childhood. They are listed as improved emotional understanding, understanding simultaneous emotions, awareness of the events leading to emotional reactions, ability to suppress or conceal negative emotional reactions, use of self-initiated strategies for redirecting feelings, and capacity for genuine empathy. Children in this period can understand complex and discrete feelings and take the responsibility of their emotions. They realize that they can feel the excitement with fear at the same time or find out what causes their current feelings. They can control their emotions such as anger or irritation, reflect their emotions or share the feeling with a distressed friend as if they themselves experienced (Denham, Bassett, & Wyatt, 2007; Kuebli, 1994; Thompson, 2009).

Some of the indicators of a normal social development for the middle childhood period can be listed as preferring to spend more time with friends than with their family, eager to become an adult urgently, behaving bossy, incline to play with their own gender, eager to be liked by everyone, incapacity to lose and acceptance of fair punishment. Unsuccessful completion of this stage may cause behavioural and emotional disorders such as anxiety or even depression (Bacanlı, 2004; Yavuzer, 2011; Kuebli 1994).

In middle childhood period, increasing number and expanding range of social relationships expose moral issues depending on the child's perception of right and wrong parallel to their cognitive development. Studying the cognitive development Piaget tried to explain the moral development of children following a series of stages. Later came Kohlberg to expand his earlier works and used moral dilemmas to present conflicting ideas about moral values with the aim of teaching morality. Kohlberg presented his dilemma and recorded their responses which he categorized as the stages of moral reasoning and identified the three main levels as pre-conventional, conventional and post-conventional morality. Each level encompasses two stages. According to Kohlberg, the middle childhood period corresponds to the transition from pre-conventional morality to the conventional morality before age 9. This can be interpreted as the middle childhood children is ready to complete the stages of pre-conventional level that are 'obedience and punishment orientation' and 'self-interest orientation' and move on to stages of conventional level that are 'interpersonal accord and conformity' and 'authority and social-order maintaining orientation' (Boyden, Dawes, Dornan & Tredoux, 2019; Berk, 2006; Lindon, 2010; Levine & Munsch, 2016).

Throughout the middle childhood, it is important for children to get the approval of the community they belong, so their understanding of morality is highly dependent on social relations. In the pre-conventional morality stage they accept the rules of authority figures, and when they transfer to conventional morality stage they built positive relationships and follow the societal order because they believe that it is necessary. They strictly adhere themselves to the rules without questioning the fairness or appropriateness. In interpersonal accord and conformity stage, children organize their behaviour to get the approval of others and try to avoid any disapproval. Their motivation is to do their best to be good and nice. In authority and social-order maintaining orientation stage, the focus is on the rules which they believe are the main pillars that keep society alive. In this stage rules are considered as applicable for all. They move a step further about moral reasoning in this stage. The violation of a rule makes them suspicious and alert to the necessary measures for the protection of rules and laws (National Research Council, 1984; Santrock, 2011; Damon & Lerner, 2008; Bee & Boyd, 2009).

Moral development is related to cognitive development as well as social development, as in the case of language development. The level achieved in these areas depend on the level of cognitive development. Middle childhood is a period during which the child has gained competency in language development. Grammar is mature and by the time children start school they begin to use it appropriately in their speech and written work. If the milestones of the period cannot be achieved properly; children may have difficulties socializing depending on failure of expressing thoughts and ideas verbally and in written form, and responding appropriately to questions. They may have difficulties to follow instructions, remember the events and retell them (Beck, Kumschick, Eid & Klann-Delius, 2012).

Cognitive Development in Middle Childhood

Piaget (1896–1980) and Vygotsky (1896–1934) are well-known two names who studied the cognitive development of children. Piaget observed children of the same age and noted that they make the same mistakes. This lets him land up with the idea that children follow the same patterns to develop concepts when they try to explain the world and their surroundings. He proposed four stages for the cognitive development of children. These stages reflect the increasing sophistication of children's thought (Lindon, 2010; Bee & Boyd, 2009).

The four stages of cognitive development according to Piaget are sensorimotor stage from birth to age 2, preoperational stage from age 2 to age 7, concrete operational stage from age 7 to age 11 and formal operational stage from age 11 to adolescence and adulthood. Children follow the same sequence through these stages. After their interaction with their environment they move to the next stage when they are mature enough. The timing of a new stage may differ from child to child but the one thing that will never change is the order of the stages. It is assured that none of the stages will be missed for any children. The rate of the progression depends on the individual potential of the children and their environment. Some individuals may never attain the later stages (Salkind, 2002; Santrock 2011; Aral, Baran, Bulut & Çimen, 2000, Bacanlı, 2004).

Middle childhood period corresponds to the concrete operational stage of Piaget's cognitive development theory. According to Piaget, concrete stage is a major turning point in the child's cognitive development because it is the period when the child starts to think logically and solve problems by reasoning. They become eligible to think without the necessity to touch or see what they think in the real world. This stage is distinguished from the others with the achievements of conservation, classification and irreversibility (Bacanlı, 2004).

Conservation is defined as the understanding that something stays the same in quantity even though its appearance changes. Children achieve conservation skills beginning with liquid conservation. The conservation skill is gained in number at age 6, in mass at age 7, and in weight at age 9. (Aral, Baran, Bulut & Çimen, 2000; Bee & Boyd, 2009; Bacanlı, 2004). Children who have already gained the liquid conservation can understand that when water is poured into a different shaped glass, the quantity of liquid remains the same, even though its appearance has changed. Following the liquid conservation children become ready for the number conservation which is understanding that the number of objects do not change even if they are positioned in a way that they are spread out and cover a larger area. By the time they are nine years old children become capable of drawing a conclusion about the conservation of mass and volume similar to the conservation of liquid and number (Watanabe, 2017).

Vygotsky explained the cognitive development as a continuous process rather than dividing it into stages as Piaget preferred and unlike Piaget, he put the emphasis on social context. He argued that play is not the only factor, but has an important place in the child's learning. Vygotsky introduced the term 'zone of proximal development' to the literature to explain how children's learning could be supported. Although there are no boundaries for the cognitive development specific for different ages it can be concluded that play prepares children for learning within the zone and both the family and school environment can boost their learning during the middle childhood period (Lindon, 2010).

RECOMMENDATIONS FOR PROMOTING DEVELOPMENT DURING MIDDLE CHILDHOOD

Middle childhood is the period in which the children gain their identities. Their identity is shaped by the effects of socialization, emotion control, and cognitive capacity. The skills they develop provide an opportunity for the children to learn about their own boundaries and to adapt to their environment. In this period, the support they receive from their parents and school environment, contributes to their development and sets them ready for adolescence in a healthy way (Hutchisan, 2015).

Recent studies of middle childhood show that the conducted researches cover a wide range of topics related to all developmental areas. Research titles on middle childhood include peer relationships (Mundy et al., 2017), bullying (Jansen et al., 2016), malnutrition (Van Itellie, 1979; Ammaniti, Lucarelli, Cimino, D'Olimpio & Chatoor, 2012), sedentary behaviours (Cliff et al., 2016), technology addiction (Yücelyiğit & Aral, 2018), screen time (Hale & Guan, 2015), physical activities (Rowland, 1981; Potter, Spence, Boulé, Stearns & Carson, 2017), obesity (Wijnhoven, 2015; Akdemir, Dönmez & Polat, 2017; Olaya et al. 2015), chronic health problems (Tambalis, Panagiotakos, Psarra & Sidossis, 2019), and the effects of specific tools or programs on cognitive skills such as questioning, problem solving and critical thinking (Ferrer-Mico, Prats-Fernàndezb, Redo-Sanche, 2012; Sáez-López, Román-González. & Vázquez-Cano, 2016; Demetrioua, Merrell & Tymms, 2017).

Since development is a complex process with layers, there are many factors that have influence on development commonly categorized as 'nature' and 'nurture'. Nature is used to express the biological factors while nurture depicts of the environmental components (Berk, 2006; Lindon, 2010; Santrok, 2011). As seen from the subject titles of the researches shared above, they cover the middle childhood period in a multi-faceted way targeting all developmental areas. This can be considered as an indicator that children are evaluated holistically. There are some research subjects unchanged for many years about middle childhood that causes developmental retardation. On the other hand emerging new topics about the development of middle childhood are introduced in the literature reflecting the changing life conditions. For example poverty or malnutrition has always been a concern in child development however for the last decade human being is facing a challenge because of sedentary behaviours, screen time or technology addiction and middle childhood period carries a potential risk of undesired habits settling. For the well-being of children, adults are responsible to be aware of the approaching risks and to take the necessary measures about them. Besides the fundamental requirements for a healthy development such as shelter, nutrition, health care; the needs of children should be considered as a whole in accordance with the changing conditions (Levine & Munsch, 2016).

In this context the recommendations will start with parental dimension, continue with educational dimension and finalize with the researchers to shed light on emerging issues related to the development of middle childhood. Parents have the responsibility to provide their children the environment they deserve. This includes a healthy nutrition and health care for their physical growth, a caring relationship for their social-emotional development and opportunity of education for their cognitive development (Halfon, Forrest, Lerner & Faustman, 2018). In this period children start school, so this transition period gives another responsibility to the parents to invite teachers to share their responsibility and to welcome the support they give. Since one of the milestones of middle childhood is transition to school, teachers have a significant role on the development of children during this period. The school environment, specifically the teachers help children about gaining their identity, self-esteem, self-efficacy and develop healthy relationships with their peers. Engaging children in group activities helps them to place themselves in a community and to gain the values and virtues like sharing, empathy and, solidarity (Santrock, 2011).

Promoting middle childhood period in a framework limited with parental and educational recommendations lacks to shed a light on the future studies about the concerns related to middle childhood development depending on changing lifestyles and daily circumstances. In today's world where rapid changes are taking place in every field, the most important suggestions address the researchers. It is crucial to continue the observations with children in middle childhood period and their environment; to carry out studies in order to overcome the challenges encountered, and to share their results with their recommendations with all the stakeholders. The new findings of the recent researches about emerging topics in middle childhood period such as sedentary lifestyles, technology use, unhealthy eating, cyber-bullying of children will broaden the visions of policy makers to act in concurrence with all partners for the optimum outcomes.

CONCLUSION

Middle childhood is a period which encompasses some of the milestones achieved in lifespan. It is as important as early years or adolescence in child development. In this chapter concepts about middle childhood development is shared and the importance of the period is discussed. The development process is considered under the titles of motor and physical development, socio-emotional development, and cognitive development. The characteristics of each stage is presented in details for each development area. It is important to be knowledgeable to create supportive and safe environments for children at home or at school and foster their development during this period. Evaluating the development and monitoring the possible implications make difference in prevention efforts since the measures taken can have lasting benefits on child's future development.

REFERENCES

Akdemir, D.P. (2017). The Effect of Nutritional and Physical Activity Interventions on Nutritional Status and Obesity in Primary School Children: A Cluster Randomized Controlled Study. Retrieved from https://acikerisim.cumhuriyet.edu.tr/xmlui/handle/20.500.12418/6763

Ammaniti, M., Lucarelli, L., Cimino, S., D'Olimpio, F., & Chatoor, I. (2012). Feeding disorders of infancy: A longitudinal study to middle childhood. *International Journal of Eating Disorders*, *45*(2), 272–280. doi:10.1002/eat.20925 PMID:21495054

Aral, N., Baran, G., Bulut, Ş., & Çimen, S. (2000). *Çocuk Gelişimi*. İstanbul, Turkey: Ya-Pa Yayıncılık.

Bacanlı, H. (2004). *Gelişim ve Öğrenme*. Ankara, Turkey: Nobel Yayın.

Beck, L., Kumschick, I. R., Eid, M., & Klann-Delius, G. (2012). Relationship between language competence and emotional competence in middle childhood. *Emotion*, *12*(3), 503–514. doi:10.1037/a0026320 PMID:22148995

Bee, H., & Boyd, D. (2009). *Çocuk Gelişim Psikolojisi*. İstanbul, Turkey: Kaknüs Yayınları.

Benson, R., & Migeon, C. (1975). Physiological and pathological puberty, and human behavior. In B. Eleftheriou & R. Sprott (Eds.), Hormonal Correlates of Behavior (Vol. 1). New York: Plenum Press. doi:10.1007/978-1-4615-7189-6_4

Bergen, D. (2002). The Role of Pretend Play in Children's Cognitive Development. *Early Childhood Research & Practice*, *4*(1), 1–13.

Berk, L. E. (2006). *Child development*. Boston: Pearson Education.

Biehl, M. C., Park, M. J., Brindis, C. D., Pantell, R. H., & Irwin, C. E. Jr. (2002). *The health of America's middle childhood population.* San Francisco: Public Policy Analysis and Education Center for Middle Childhood and Adolescent Health.

Boyden, J., Dawes, A., Dornan, P., & Tredoux, C. (2019). Middle childhood: A key time for healthy development and learning. In *Tracing the Consequences of Child Poverty: Evidence from the Young Lives study in Ethiopia, India, Peru and Vietnam* (pp. 73-100). Bristol, UK: Bristol University Press. Retrieved from http://www.jstor.org/stable/j.ctvkjb390.10

Carr, A. (2011). Social and emotional development in middle childhood. *Child Psychology and Psychiatry*, 56-61.

Cliff, D. P., Hesketh, K. D., Vella, S. A., Hinkley, T., Tsiros, M. D., Ridgers, N. D. ... Plotnikoff, R. C. (2016). Objectively measured sedentary behaviour and health and development in children and adolescents: systematic review and meta-analysis. *Obesity Reviews, 17*(4), 330-344. Retrieved from https://onlinelibrary.wiley.com/doi/full/10.1111/obr.12371

Cunningham, J. N., Kliewer, W., & Garner, P. W. (2009). Emotion socialization, child emotion understanding and regulation, and adjustment in urban African American families: Differential associations across child gender. *Development and Psychopathology, 21*(1), 261–283. doi:10.1017/S0954579409000157 PMID:19144233

Damon, W., & Lerner, R. M. (2008). *Child and adolescent development: An advanced course.* Canada: John Wiley & Sons, Inc.

Demetrioua, A., Merrell, C., & Tymms, P. (2017). Mapping and predicting literacy and reasoning skills from early to later primary school. *Learning and Individual Differences, 54*, 217–225. doi:10.1016/j.lindif.2017.01.023

Denham, S. A., Bassett, H. H., & Wyatt, T. (2007). The socialization of emotional competence. In J. E. Grusec & P. D. Hastings (Eds.), *Handbook of socialization. New York*: Guilford.

Domitrovich, C. E., Durlak, J. A., Staley, K. C., & Weissberg, R. P. (2017). Social-Emotional Competence: An Essential Factor for Promoting Positive Adjustment and Reducing Risk in School Children. *Child Development, 88*(2), 408–416. doi:10.1111/cdev.12739 PMID:28213889

Duncan, G. J., Dowsett, C. J., Claessens, A., Magnuson, K., Huston, A. C., Klebanov, P., ... Japel, C. (2007). School readiness and later achievement. *Developmental Psychology, 43*(6), 1428–1446. doi:10.1037/0012-1649.43.6.1428 PMID:18020822

Duncan, G. J., Ziol-Guest, K. M., & Kalil, A. (2010). Early-childhood poverty and adult attainment, behavior, and health. *Child Development, 81*(1), 306–325. doi:10.1111/j.1467-8624.2009.01396.x PMID:20331669

Eccles, J. S. (1999). The development of children ages 6 to 14. *The Future of Children, 9*(2), 30–44. doi:10.2307/1602703 PMID:10646256

Education Encyclopedia. (n.d.). Stages of growth in child development. Retrieved from http://education.stateuniversity.com/pages/1826/Child-Development-Stages-Growth.html#ixzz0j0jMHgRB

Ferrer-Mico, T., Prats-Fernàndezb, M. A., & Redo-Sanchez, A. (2012). Impact of Scratch programming on students' understanding of their own learning process. *Procedia: Social and Behavioral Sciences*, *46*, 1219–1223. doi:10.1016/j.sbspro.2012.05.278

Greenfield, P. M., & Bruner, J. S. (1966). Culture and Cognitive Growth. *International Journal of Psychology*, *1*(2), 89–107. doi:10.1080/00207596608247117

Hale, L., & Guan, S. (2015). Screen time and sleep among school-aged children and adolescents: A systematic literature review. *Sleep Medicine Reviews*, *21*, 50–58. doi:10.1016/j.smrv.2014.07.007 PMID:25193149

Halfon, H., Forrest, C. B., Lerner, R. M., & Faustman, E. M. (2018). *Handbook of Life Course Health Development*. Springer. doi:10.1007/978-3-319-47143-3

Halfon, N. (2009). *Life course health development: A new approach for addressing upstream determinants of health and spending*. Washington: Expert Voices, National Institute for Health Care Management Foundation. Retrieved from http://www.nihcm.org/pdf/ExpertVoices_Halfon_FINAL.pdf

Harris, S. S. (2000). Readiness to participate in sports. In J. A. Sullivan & S. J. Anderson (Eds.), *Care of the Young Athlete* (pp. 19–34). Rosemont, IL: American Academy of Orthopedic Surgeons and American Academy of Pediatrics.

Hutchison, E. D. (2015). *Dimensions of Human Behavior: The Changing Life Course Los Angeles*. SAGE Publishing.

Jansen, P. W., Verlinden, M., Dommisse-van Berkel, A., Mieloo, C., van der Ende, J., Veenstra, R., ... & Tiemeier, H. (2012). Prevalence of bullying and victimization among children in early elementary school: Do family and school neighbourhood socioeconomic status matter?. BMC public health, 12(1), 494. doi:10.100712552-016-9182-9

Jurimae, T., & Jurimae, J. (2001). *Growth, physical activity, and motor development in prepubertal children*. USA: Taylor & Francis Group. doi:10.1201/b14219

Kagan, J. (2010). Emotions and temperament. In M. H. Bornstein (Ed.), *Handbook of cultural developmental science*. New York: Psychology Press.

Kuebli, J. (1994). Young children's understanding of everyday emotions. *Young Children*, 36–48.

Kuther, T. L. (2016). *Physical and Cognitive Development in Middle Childhood in Lifespan Development: Lives in Context*. Canada: SAGE Publications.

Levine, L. E., & Munsch, J. (2016). *Social and Emotianal Development in Middle Childhood in book Child Development From Infancy to Adolescence*. Canada: SAGE Publications, Inc.

Liew, J. (2012). Effortful control, executive functions, and education: Bringing self-regulatory and social-emotional competencies to the table. *Child Development Perspectives*, *6*(2), 105–111. doi:10.1111/j.1750-8606.2011.00196.x

Lindon, J. (2010). Understanding child development 2nd edition linking theory and practice. London: Hachette UK Company.

Luby, J., Belden, A., Botteron, K., Marrus, N., Harms, M. P., Babb, C., ... Barch, D. (2013). The effects of poverty on childhood brain development: The mediating effect of caregiving and stressful life events. *JAMA Pediatrics*, *167*(12), 1135–1142. doi:10.1001/jamapediatrics.2013.3139 PMID:24165922

McLeod, S. A. (2008). Psychosexual stages. Simply Psychology. Retrieved from www.simplypsychology.org/psychosexual.html

Mundy, L. K., Canterford, L., Kosola, S., Degenhardt, L., Allen, N. B., & Patton, G. C. (2017). Peer Victimization and Academic Performance in Primary School Children. *Academic Pediatrics*, *17*(8), 830–836. doi:10.1016/j.acap.2017.06.012 PMID:28652070

National Research Council. (1984). *Development During Middle Childhood: The Years From Six to Twelve*. Washington, DC: The National Academies Press.

Olaya, B., Moneta, M. V., Pez, O., Bitfoi, A., Carta, M. G., Eke, C., ... Kovess, V. (2015). Country-level and individual correlates of overweight and obesity among primary school children: A cross-sectional study in seven European countries. *BMC Public Health*, *15*(475), 1–12. PMID:25952506

Potter, M., Spence, J. C., Boulé, N. G., Stearns, J. A., & Carson, V. (2017). Associations between physical activity, screen time, and fitness among 6- to 10-year-old children living in Edmonton, Canada. *Applied Physiology, Nutrition, and Metabolism*, *42*(5), 487–494. doi:10.1139/apnm-2016-0419 PMID:28177727

Prado, E. L., & Dewey, K. G. (2014). Nutrition and brain development in early life. *Nutrition Reviews*, *72*(4), 267–284. doi:10.1111/nure.12102 PMID:24684384

Rogoff, B., Sellers, M., Pirrotta, S., Fox, N., & White, S. (1975). Age of assignment of roles and responsibilities in children: A cross-cultural survey. *Human Development*, *18*(5), 353–369. doi:10.1159/000271496

Rowland, T. (1981). Physical fitness in children: Implications for the prevention of coronary artery disease. *Current Problems in Pediatrics*, 9. PMID:7026178

Sáez-López, J. M., Román-González, M., & Vázquez-Cano, E. (2016). Visual programming languages integrated across the curriculum in elementary school: A two year case study using "Scratch" in five schools. *Computers & Education*, *97*, 129–141. doi:10.1016/j.compedu.2016.03.003

Salkind, N. J. (2002). *Child development*. USA: Macmillan Library.

Santrock, J. W. (2011). *Child development*. New York: McGraw Hill Companies, Inc.

Schore, A. N. (1999). *Affect regulation and the origin of the self: The neurobiology of emotional development*. Florence, KY: Psychology Press.

Shonkoff, J. P., Boyce, W. T., & McEwen, B. S. (2009). Neuroscience, molecular biology, and the childhood roots of health disparities: Building a new framework for health promotion and disease prevention. *Journal of the American Medical Association*, *301*(21), 2252–2259. doi:10.1001/jama.2009.754 PMID:19491187

Shonkoff, J. P., & Phillips, D. A. (2002). (Eds.). From neurons to neighborhood: The science of early childhood development. Washington, DC: National Academy Press. Retrieved from http://files.eric.ed.gov/fulltext/ED446866.pdf

Sinclair, D. (1978). *Human Growth After Birth* (3rd ed.). London: Oxford University Press.

Smith, D. (1977). *Growth and Its Disorders*. Philadelphia: W.B. Saunders.

Sørlie, M. A., Idsoe, T., Ogden, T., Olseth, A. R., & Torsheim, T. (2018). Behavioral Trajectories During Middle Childhood: Differential Effects of the School-Wide Positive Behavior Support Model. *Prevention Science*, *19*(8), 1055–1065. doi:10.100711121-018-0938-x PMID:30121877

Tambalis, K. D., Panagiotakos, D. B., Psarra, G., & Sidossis, L. S. (2019). Association of cardiorespiratory fitness levels with dietary habits and lifestyle factors in schoolchildren. *Applied Physiology, Nutrition, and Metabolism*, *44*(5), 539–545. doi:10.1139/apnm-2018-0407 PMID:30321490

Thompson, R. A. (2009). Emotional development. In R. A. Schweder (Ed.), *The Chicago companion to the child*. Chicago: University of Chicago Press.

Tucker-Drob, E. M., Briley, D. A., & Harden, K. P. (2013). Genetic and Environmental Influences on Cognition Across Development and Context. *Current Directions in Psychological Science*, *22*(5), 349–355. doi:10.1177/0963721413485087 PMID:24799770

Van Itallie, T. (1979). Obesity: Adverse effects on health and longevity. *The American Journal of Clinical Nutrition*, *32*(12), 2723–2733. doi:10.1093/ajcn/32.12.2723 PMID:389027

Vitaro, F., Boivin, M., & Bukowski, W. M. (2009). The role of friendship in child and adolescent psychosocial development. In K. H. Rubin, W. M. Bukowski, & B. Laursen (Eds.), *Social, emotional, and personality development in context. Handbook of peer interactions, relationships, and groups* (pp. 568–585). New York: The Guilford Press.

Watanabe, N. (2017). Accelerated Cognitive Development—Piaget's Conservation Concept. *Journal of Educational and Developmental Psychology*, *7*(2), 68–74. doi:10.5539/jedp.v7n2p68

Weinberg, R., Tenenbaum, G., McKenzie, A., Jackson, S., Anshel, M., Grove, R., & Fogarty, G. (2000). Motivation for youth participation in sport and physical activity: Relationships to culture, self-reported activity levels, and gender. *International Journal of Sport Psychology*, *31*, 321–346.

Wijnhoven, T. M. A. (2015). Overweight and obesity in primary-school children: a surveillance system for policy-making in Europe from 2007 onwards [PhD Thesis]. Wageningen University.

Willoughby, C., & Polatajko, H. J. (1995). Motor problems in children with developmental coordination disorder: Review of the literature. *The American Journal of Occupational Therapy*, *49*(8), 787–794. doi:10.5014/ajot.49.8.787 PMID:8526224

Yan, J. H., & McCullagh, P. (2004). Cultural Influence on Youth's Motivation of Participation in Physical Activity. *Journal of Sport Behavior*, *27*(4), 378–390.

Yavuzer, H. (2011). *Okul Çağı Çocuğu*. İstanbul, Turkey: Remzi Kitapevi.

Yücelyiğit, S., & Aral, N. (2018). Technology Addiction in Children. In R. Efe (Ed.), *Educational Sciences Research in the Globalizing World*. Sofia: St. Kliment Ohridski University Press.

ADDITIONAL READING

Carter, A. S., Briggs-Gowan, M. J., & Davis, N. O. (2004). Assessment of young children's social-emotional development and psychopathology: Recent advances and recommendations for practice. *Journal of Child Psychology and Psychiatry, and Allied Disciplines*, *45*(1), 109–134. doi:10.1046/j.0021-9630.2003.00316.x PMID:14959805

Magro, S. W., Utesch, T., Dreiskämper, D., & Wagner, J. (2019). Self-esteem development in middle childhood: Support for sociometer theory. *International Journal of Behavioral Development*, *43*(2), 118–127. doi:10.1177/0165025418802462

McDowell, D. J., Kim, M., O'neil, R., & Parke, R. D. (2002). Children's Emotional Regulation and Social Competence in Middle Childhood. *Marriage & Family Review*, *34*(3-4), 3–4, 345–364. doi:10.1300/J002v34n03_07

Sørlie, M. A., Idsoe, T., Ogden, T., Olseth, A. R., & Torsheim, T. (2018). Behavioral Trajectories During Middle Childhood: Differential Effects of the School-Wide Positive Behavior Support Model. *Prevention Science*, *19*(8), 1055–1065. doi:10.100711121-018-0938-x PMID:30121877

Suveg, C., & Zeman, J. (2004). Emotion Regulation in Children With Anxiety Disorders. *Journal of Clinical Child and Adolescent Psychology*, *33*(4), 750–759. doi:10.120715374424jccp3304_10 PMID:15498742

van Aken, M. A. G., & Riksen-Walraven, J. M. (1992). Parental Support and the Development of Competence in Children. *International Journal of Behavioral Development*, *15*(1), 101–123. doi:10.1177/016502549201500106

Van Heerden, A., & Wild, L. G. (2018). Grandparent support and mental and behavioural health in middle childhood. *Social Development*, *27*(2), 366–380. doi:10.1111ode.12275

Yücelyiğit, S., & Aral, N. (2018). Technology and Children. In R. Efe, I. Koleva, & E. Atasoy (Eds.), *Recent Researches in Education* (pp. 418–425). UK: Cambridge Scholars Publishing.

Zimmermann, P., & Alexandra, I. (2019). Attachment Disorder behavior in early and middle childhood: Associations with children's self-concept and observed signs of negative internal working models. *Attachment & Human Development*, *21*(2), 170–189. doi:10.1080/14616734.2018.1499212 PMID:30014786

KEY TERMS AND DEFINITIONS

Bullying: Undesired, aggressive and imbalanced behavior among school aged children.

Concrete operational Stage: The period that spans middle childhood between ages 6-12 in Piaget's Cognitive Development Theory.

Conservation: The ability to understand that something stays the same in quantity even though its appearance changes.

Identity: The personality of the individual with all of his/her distinguishing character.

Industry versus Inferiority: The period that spans middle childhood between ages 6-12 in Erikson's Psychosocial Development Theory

Self-esteem: Personal value of an individual.

Self-efficacy: An individual's belief in his or her capacity.

Latency Stage: A stage in Freud's Psychosexual Development Theory which spans from the age of six years until puberty.

Sedentary behaviors: Physical inactivity.

Puberty: Becoming sexually mature.

Chapter 12
Late Childhood Period

Fatma Elif Kilinç
https://orcid.org/0000-0001-8372-5047
Ankara Yıldırım Beyazıt University, Turkey

Burçin Aysu
Ankara Yıldırım Beyazıt University, Turkey

ABSTRACT

Development expresses the process that starts from impregnation and continues till the end of life. Man is a living creature that is born with a number of hereditary features and can improve them. The features of development areas should be known in order to recognize man as a whole. In the classification regarding the development periods of people, the late childhood period is between the ages of eleven to eighteen and it is also expressed as puberty. In this period the process of transition to adulthood is observed. Further, the teenager starts to become independent from his/her family and acquires personal values and purposes in this period. In this study, the late childhood period is explained in detail.

DOI: 10.4018/978-1-7998-2952-2.ch012

INTRODUCTION

The features of development areas should be known in order to recognize man as a whole. In the classification regarding the development periods of people, late childhood period is between the ages of eleven until eighteen and it is also expressed as puberty. In this period the process of transition to adulthood is observed. Further, the teenager starts to become independent from his/her family and acquire personal values and purposes in this period. Last childhood period will be presented the knowledge that forms the fundamentals of the period will be supported by the research about each developmental area.

BACKGROUND

One of the important periods in personality development is the puberty years which are also the late childhood period. The puberty can also be expressed as the life cycle in which the autonomy development is the most important (Özdemir and Çok, 2011). The teenager, who is in his late childhood period, requires help about gaining the values which guide to his behaviors and learning his social responsibilities. Family is the social institution that meets such necessity and affects the teenager's life.

During the late childhood period, which is also known as puberty years, an individual reaches the reproduction capacity and the physical development is also at the forefront. For Atabek (2002) the late childhood period is a transition period from being a child to having a child, and the most distinct feature of the period is the physical and sexual development. In this section, the late childhood period (puberty) is to be examined.

Physical Changes in the Late Childhood Period

Physical and hormonal changes occur in the late childhood period (puberty); further, rapid increases in height and weight are seen (Erden and Akman, 1997). During this period, the child organism, which does not have reproduction capacity, transforms into a productive individual. The age and duration of puberty vary form one child to another. Together with the effects of gonad hormone, the tall stature, which is 80% of the mature height at the onset of puberty, reaches to 99% of the mature height in two-four years period. As the puberty starts for the girls earlier than the boys, the rate of growth also appears earlier in the girls. The peak in the rate of growth is the period when the growth is at its fastest rate and for girls it is 9cm/year and for boys it is 10,5 cm/year on average. In the majority of the girls and boys the increases in weight are seen in preadolescence period when they are eight-ten years old due to fat storage in their bodies. The reasons for the increase in weight in the puberty are; for boys the muscle growth and the increase in skeleton mass and for girls the fat storage. The thickness of the subcutaneous layer of fat diminishes in boys, but it keeps increasing in girls (Özer and Özer, 2000). Accordingly, the increase in weight in girls starts when they are 11 and reaches to high level when they are 12-13 years old. Although the increase in the weight is limited in boys when they are 11-13 years old, there is a rapid increase when they are 13-14 years old. In those ages, the growth in breast tissue, menarche (menstruating for the first time), ovulation and hair growth are seen in girls and the growth in testis and penis, hair growth and gynecomastia are seen in boys. The menarche age is between 10-16 ages, on average 12,5 and the growth of the penis is between 10-16 ages and the growth of the testis is between 9-17 ages. Ovulation is seen after the menarche. The size of breast and pubic hair increase and the sexual maturation is seen is girls. In boys, the growth in the genital organs occurs. There is the production of mature sperm. The hair on the face and body increases. The gynecomastia diminishes. There is a change in their voice. Girls and boys can want to establish relations with the opposite sexes. Inappropriate body changes can ruin the teenager's psychological balance. Personal friendships come to the forefront for teenagers, romantic relations start (Santrock, 2014).

Girls need 2200 kcal when they are between 11-14 years, and 2400 kcal when they are between 15-18 years; boys need 2500 kcal when they are between 11-14 years and 3000 kcal when they are 15-18 years. 15% of the total caloric intake should be met from sufficiently consumes protein. Girls can be sensitive in terms of iron deficiency with the start of the menarche. Fibrous foods are needed (Törüner and Büyükgönenç, 2011).

Intra-Familial Relationships in the Late Childhood Period

The teenager, who is in his late childhood period, requires help about gaining the values which guide to his behaviors and learning his social responsibilities. Family is the social institution that meets such necessity and affects the teenager's life.

Man is a social being and is consistently in contact with his environment. Child begins to live with his family since his birth; the family is his first social environment. In his family, the child experiences how to establish connection with people (Aral and Baran, 2011). Family is the smallest unity of society in which the individuals grow and develop. It is also the most important milieu and social institution for the child's mental development. The child learns the socializing, his identity as a member of the society and his healthy mental development in this environment. Mother-father-child relationship and father-mother attitudes are the key factors in the child's social development (Cansız, 2004). Family is the safest harbor for the child, the young and the adults to shelter against the positive and negative factors of life. The family is also an institution where man can shelter against the difficulties of life, acquire his personal development, and gain the strength and ability of coping with the negative events. Recently, how the early years of one's life play a crucial role in the development of his personality has been noticed. The effects of the first few years of life on man are noteworthy, because the ability about how a man perceives a problem when he meets them in his life or how he can make decisions for the solution of such problems starts in the childhood period. The love, care and sincerity an individual meets can determine the peace in his later life. The more an individual faces rejection, inhospitality and indifference the harder it gets to develop an adaptable personality structure (Kabasakal, 2007).

Taking the fact that children model the relation between mother and father at most into the consideration, it can be stated that the communication pattern of the family members establish a framework for the feelings, attitudes and behaviors children portray. The children grow and develop in and as a part of a dynamic system formed by the family members' physiological interaction. Families bringing up children are affected by the culture, socio-economic situation of the society, tradition, religious belief and values and the way family members themselves are raised. The healthier the parents' physical characteristics, mental conditions and their interactions with each other become, the healthier the children's raising gets (Sadıkoğluet al., 2007).

The teenager does not have a clear idea about his own duty and status in the society he lives. Not giving him the adult duties and responsibilities and the status uncertainty. Horrock regards the family as an institution which fulfills the functions of "a tool for the determination of the status" and "a tool for describing the experiment". The family as a tool for the determination of the status determines the social position of the child and the role he should undertake in order to reach that position (Yavuzer, 2011).

The Importance of Parents-Child Relationship in the Late Childhood Period

The process, in which man transforms from a baby acting with its instincts to a socialized adult, requires a long education and it is believed that while giving this education the way mother and father approach is much more important than many social environment factors in terms of the teenager's personality development. A peaceful, warm and lovely atmosphere at home affects the teenager's emotional and social development and his personality development affirmatively, thus it contributes to the teenager's strength of looking at the life confidently and at people friendly. In the unpeaceful and problematic environments, where there is no love and understanding, the teenagers have difficulty in developing confidence both for themselves and for the others around them (Cansız 2004). The most important event, which makes the family environment stressful and even takes this stress further, is the problems between the parents. The teenager, who witnesses the violence and violent events between his mother and father, can show the disorders related to psychological adaptation and behaviors (Kabasakal, 2007).

Children constantly observe their parent's relationships with themselves, with each other and with the other people. The children acquire the behaviors about cooperation and social relations, which play an important role in human affairs, as a result of their observations and interactions with their parents. The teenagers, who have a healthy relationship with their parents, can express their feelings and thoughts easier and without disputing when they face with problems (Özdal, 2003).

The things a teenager sees at home in his family have an enormous and very deep effect on the shaping of the maturing teenager's personality structure. The atmosphere in the family stems from the relationship between the family members. Certain relationships between the teenager and his parents determine the type of authority with the attitude of the one in power and all the effects of those things on the teenager determine how he perceives them. The effects of behaviors on the teenager depend on the strength of pressure applied by the parents. Such power indicator can be a continuing strong patriarchal family pressure or a father's pressure on the low-level. This issue is not generally mentioned in the family. The multifaceted and complex feature of the relationship between the parents and the child should not be ignored although it looks very compatible at first sight (Yavuzer, 2011).

The clearest difference between the childhood period and the puberty in terms of emotion is that; the children express their feelings such as anger and joy immediately (spontaneously) with apparent behaviors, but during the puberty such feelings are hidden or masked. It can be stated that girls generally reach the emotional maturity in puberty earlier than the boys. Therefore, it can be claimed that the teenager girl behaves more maturely than teenager boy in terms of controlling her feelings and anger. The need and concern for emotional development can vary in children and teenagers. The children get bored of unpleasant feelings such as pessimism, uneasiness and anger. There is a need for mother and father who give confidence and tell that such feelings are typical to the period the children are in and also they are permanent. The most distinct problem of the childhood and puberty is not to be understood completely. The teenager demands intimacy from his parents, when his independent behaviors are prevented, he explodes with temper and tension and can conflict with his parents. It is because of the fact that when parents ask for something apart from the teenager's expectations, they humiliate him, frequently criticize him, compare him with his peers can cause to a worrying relationship between the parents and the teenager (Kabasakal, 2007).

In parallel with the reaction against the parents' control in puberty, the need for authority support cause to emotional stress. In other words, besides his rebellious attitudes, the teenager also needs his parents' support. This bipolarization is a reason that increases the teenager's inner conflict. The sense of discipline which depends on the adults' pressure and prohibition on the teenager can turn this stage, which should be positive and constructive, into a negative period which is full of conflicts (Yavuzer, 2011).

In many studies, it is frequently indicated that the psychologically and psychiatrically problematic family environments are a preparative factor for suicide. The quality of family environment can be conceptualized with the perceived family support and its unity. It is possible to say that the risk of suicide is high for those whose family unity is destroyed and who get insufficient family support. The school problems and some environments where the stress tolerance is low can push the child to extreme risk. This situation raises the following negative feelings; lack of self-confidence, twist in the wind, feeling alone and hopelessness. Thus, the child is deprived of emotional supports which are the most important things for the prevention of the suicide. The child, who is affected by the physical or psychological features or parents' negative behaviors, can show his reactions by inclining to the suicide (Kabasakal, 2007).

According to the psychoanalytic theory, when a child reaches puberty, he meets the reawakening of the family romance which is an early childhood problem. The way he follows in solving the problem directs to the romantic relationships from his own generation. Erikson claims that on the positive side of the interpersonal relationships, which appear in this period, there is a sense of identity, but on the negative side of it there is role confusion. Accordingly, the teenager comes across the duty of organizing the different roles (a son, a student, an athlete, a friend, a tracker) about himself significantly in this period. The teenager gains the sense of psycho-social identity to the degree that he succeeds in this effort. If parents make their child reach the puberty with the feelings of confidence, freedom, sociability and skillfulness in this stage, the possibility of the teenager's developing a significant sense of identity increases. To sum up, it can be stated that the relationships between parents and child since the babyhood are very important for gaining an integrated sense of psycho-social identity in the puberty (Özdemir, 2009).

The children, according to their ages, can experience different behaviors from their parents in the intra-family relationships. Mothers and fathers can attribute fewer permanent features to their children who are at small ages. As the children grow, the mother and father describe their behaviors with more permanent personality features. Also, the family, according to the level of culture and the conditions they are in, can be biased against their children's attitudes, behaviors and their personality features by connecting them to the children's physical structure or habits.

Such prejudices are common in the society and they prevent families from evaluating their children objectively. For example, some prejudices such as; boys are more prankish than girls, fat children are lazy or girls are more dependent on their families than boys cause the children to be misevaluated by their parents. The parents' evaluation about their children shapes the children's perception about themselves and their relationships with their parents (Kulaksızoğlu, 2005).

In puberty the values of the teenager and the adult differentiate, but mother and father become the most important role models with whom the teenager can identify himself. The adults' attitudes in the family should balance between the teenager's rights and responsibilities. In the family the different attitudes towards the teenager increase his unbalance and indecision (Yavuzer, 2011).

In the puberty the time the teenager spends with his friends increases and he is affected by his friends more than he does in the childhood period. However, the issues, which the teenagers are affected by his peers, are limited and his parents are still the reference guide for the subjects that he regards important.

The teenager finds his parents' thoughts important about the subjects related to the school, choice of profession and his future.

The demands and new quests of the young in the puberty cause a change in the family system and shake the power balance in the family. As usual, the parents make an effort to protect their balance and authority against the change in the family and they resist against the teenager's desires (Kulaksızoğlu, 2005).

Conflict of Generations in the Late Childhood Period

The conflict between the generations is not peculiar to these days. There have been disagreements and conflicts between the young and their mother and father since the ancient times. A generation means the age difference between the mother-father and their children and it is approximately 25 years period of time.

In our country, we see a very rapid change and development at all levels of the society. Together with this change, the established values in our society lose their collectiveness. The role of internet, television, social media and printed press is very important in the change of general value judgment. This rapid change is apparent mostly on those who come to big cities from the villages. In the village life the interpersonal relationship is close and sincere, the moral values are more effective and collective, but in the big cities, of which structure is larger and more different, the human affairs are not as warm as they are in the villages, also moral and social value judgments are less effective in proportion to the villages, all these differences create negative effects especially on the young. There is a difference between the values adopted by the family and immediate environment and the values of new environment in the city and the young cannot make out which value judgments and to what extent he should or should not obey. Having more different values than the society of adults is a reason for conflict and disharmony in itself (Kulaksızoğlu, 2005).

Puberty is, in some ways, a stage of searching for identity, proving oneself and gaining independence. If the teenager, who is in the struggle for gaining an identity, does not get the confidence he demands from the adult, then he feels uneasy. In fact, parents and other adults are not in a total agreement about what the teenagers should and should not do. The teenagers does not have economic independence and this situation is an obstacle for them to enter one generation earlier adult society, be independent and have an identity specific to himself. Not earning the money in order to continue his living can prevent the teenager from developing an independent personality. The lengthening of the education period retards the teenager's beginning to work and taking responsibilities, therefore, it can also be a reason of unease for the teenager. Generally, in our country the teenagers receive more education than his parents and others in his immediate environment, and this situation causes the conflict of generations as the generations in the family have different culture and understanding and they even use different languages (Yörükoğlu, 2011).

During the childhood period an individual is close to his family and is in the position of receiver, but it is seen that in the puberty this attitude changes and the teenager begins to move away from his family. The teenager moves away from his family in order to understand his own perception and way of thinking about the life. Instead of the influences coming from his parents in the field of thinking, the teenager is interested in new areas that involve the others' ideas. From now on, the teenager tries to adapt the way of behaviors he sees in his family to the behavior measures of the group. The demand for being liked and approved by the others, a severe desire for independence and getting rid of the connections which make him subject to the adults and necessity for proving his personality lie at the root of the teenager's such behaviors. The teenager's struggle for his autonomy is against not only his family but also all authority (Yavuzer, 2011).

Today, the rapid development and enlightenment shorten the time between the generations. The young generations are more active about learning new behaviors against the changing environment and new conditions in proportion to the earlier generations. In such a condition, the young and the adults cannot evenly adapt to the common environment they live in (Kulaksızoğlu, 2005).

Two generations' socialization in different way creates differences between the generations in terms of thinking, belief and action. Therefore, the social and cultural patterns, which the mothers and fathers internalize, differentiate, even if just a bit, with the ones their children learn. As people get older the socialization decreases or a certain decrease is seen in adaptation to the society, and these situations are other reasons which enlarge the gap between the generations. Due to this different perception and thinking between the generations, a sufficient and friendly relation cannot be established between the parents and the teenager (Yavuzer, 2011).

The other important factor causing to the conflict between the generations is the effort of teenager, who gains new opportunities as he grows, to make himself being regarded as an adult. The teenager attempts to make his personality being accepted by the others. The teenager, who is infantilized due to his behaviors, rebels frequently. Another factor causing to the conflict is parents' having adaptation problems about the new status of their children. The change in the parents' role from a guide role in the characteristic of socialization institution to regarding their children partly in same statue with themselves creates this difficulty. The teenager learns quickly as he is on the stage of development, but it is difficult for adult to learn new concepts and rules. The adult does not undergo basic personality changes any more (Yavuzer, 2011).

Intra-family supportive communication is an important factor which provides the development of social competence and the teenagers' gaining of a positive identity. If a peaceful exchange of views occurs between the parents and the young, it becomes easier to continue the relationships in a healthy way. If the parents expect the understanding and respect from their children, first they should show them to their children; because the most effective way to teach something to the teenager is to be a positive model for him. Even the relationship between the teenager and the parents is positive, sometimes there can be conflicts. The conflicts stem from the difference between thinking, emotions, interest and purposes. For example, various problems can appear about the teenager's choice of friend, school, profession and spouse, way of dressing, the activities he does in his free time, spending money and fulfilling his responsibilities. The problems between the teenager and his family can cause him to exhibit negative behaviors by affecting his personality negatively (Çopur et al., 2006)

Educational differences or contradictions between the educations increase the conflicts between two generations. Such differences and contradictions stem from either the children coming from the low-level of socio-economic environment receive higher education and go beyond their fathers' education or the educations, which the two generations receive, are different than each other. This situation causes to have different expectations, values and behaviors (Yavuzer, 2011).

The reason of the conflicts between the parents and the teenagers is the teenagers' being very insistent during their development stages about having an autonomous life and on the contrary the parents' attempts to adopt the importance of harmony in the family and being together to the teenagers (Çopur et al., 2006).

About the main reasons of the conflicts of generation the teenagers mentions the following issues; they are criticized by their parents, their parents are too much interested in their health, the parents are too meticulous about the cleaning at home, a subject is dwelled on too long, they are advised excessively, they are coddled too much, their parents do not understand them, their parents do not let them come home late, they are scolded at home, their parents want to know everything about them, the teenagers are told by their parents that they are very messy, their success at school is criticized, their parents put the pressure on them, their parents discuss near them, the works at home are not done in time and their parents are not knowledgeable enough (Çopur et al., 2006).

Choice of Profession in the Late Childhood Period

Although a profession can be defined as a job someone has in order to earn a livelihood and have a comfortable life, it is not a permanent business which is done only for earning money. The profession is a continuous (lifelong) job which requires certain formation, ability, mastership, love and interest.

Choice of profession takes a very important place in man's life. While choosing a profession, an individual also chooses a certain work environment and lifestyle for himself. He works hard to reach the top in the profession he chooses. It is observed that the individuals work productively and succeed in the fields which they choose in parallel with their personal interest and abilities. However, it is also known that the individuals, who choose professions contradicting with their abilities and interests, are unwilling to work and they always tend to change their jobs. Therefore, the qualifications of a profession an individual has should harmonize with his own characteristics and pleasures. The choice of profession is one of the most important decisions a person made about his life. Every individual determines the job he professes in the future by confirming his choice of profession during the certain part of his life. If an individual gets confused about the choices when he reaches the time of deciding his profession, changes his decision frequently and cannot confirm his choice of profession, it can be thought that he has the career indecision. There can be various reasons for having the career indecision. If the individuals have high potential in every field, they can have approach-approach conflict; accordingly, they can have difficulty in making professional decision. Sometimes, the interests and abilities of a person are not separated and such a condition prevents the person from making his choice of profession clear (Kuzgun, 2000).

Family has effects on the teenager's choice of profession from different aspects. The children of the families, whose economic and cultural levels are high, have the opportunity of choosing their professions according to their abilities and interests. In the families, whose economic and cultural opportunities are insufficient, generally the professions, which bring money in a short period and provide a position, are preferred. In patriarchal families and in the families, of which members are in a struggle for continuing the father's profession, it is wished to support and continue the father's profession instead of the children's success. Therefore, the children are conditioned to choose the father's profession since the early periods of their lives. Also, some families wish the realization of their desires, which they could not come true in their own lives, in their children's lives. The father or mother puts pressure on their children so that they can enter the area of profession where the father or mother could not enter. Further, in some families, the members of the family affect the children about the reputation of their own professions or some other professions. Thus, the families affect the child's profession in the future indirectly. However, the criterion which should be taken into consideration is not only the respectability of the profession but also its degree of suitability for the individual's qualifications. Comparisons stemmed from the parents' wrong attitudes can also affect the young's choice of profession negatively.Comparing the child with someone else means the rejection of his personality. Demanding the young to make similar choices with the neighbor's child or sibling studying in a university causes to give rather wrong decisions. In such an important subject, instead of being in a compeller and oppressive attitude, parents' motivating attitudes would diminish the problems in the father-mother-teenager triangle and ensure the teenager to make the correct choices in parallel with his interest and abilities (Aytekin, 2005).

Besides the determination of the profession he would perform, the decisions about the choice of profession, is very important for an individual's planning his life and determining his life standards. Therefore, the decisions about the profession should not be given instantly and it should be given as a result of the professional development process (Yeşilyaprak, 2013). The factors affecting the teenager's process of giving decisions are as such; the social factors like the teenager's personality characteristics, his values, accumulation of knowledge, his experiences, family, friends and immediate environment also the age, gender, having enough time or not, self-reliance, focus of control and level of stress (Öztabak and Erdoğan, 2017). Therefore, the studies about the career concern that the teenagers have during their professional development process should be carried out (Çetin Gündüz and Nalbantoğlu Yılmaz, 2016; Erdem, and Bilge, 2016), various response programs within the frame of counselling programs in the process starting from the early ages to the puberty (Demir and Yıldırım, 2017; Traş and Gökçe, 2017) can be organized. Moreover, the group guidance related to the choice of profession with the aim of informing the teachers and families can be organized (Traş and Gökçe, 2017). In a study carried out by Şeker and Kaya (2019) it was found that professional deciding program, which was developed as data processing based, contributed to the students' ability of deciding positively. In the study done by Okutan and Akbaş (2019) the anxiety the students who are between 15-24 ages, have during their career development process in Turkey were related with some concepts such as exam anxiety, social anxiety, and unemployment anxiety.

Parents are the most important people who affect the teenagers' professional and educational plans. The parents can affect their children's professional development processes and behaviors for professional deciding with their supportive attitudes and behavior for the professional development, in addition to their democratic, authoritative and accepting behaviors they exhibit in the family relations. For example, the parents can direct their children to recognize the professions in order to help their professional development, discuss the obtained knowledge with their children, talk to their children about the professions they are interested in and do different activities so that their children can realize and develop their own abilities and interests. Also, as the parents' level of knowledge about the factors such as the process of the choice of professions, ability in the choice of profession, interest and values and in the subject of recognizing the professions increase, they can exhibit more positive attitudes. As a result of it, the children may have lesser professional indecisiveness (cited in Hamamco and Hamurlu, 2005).

Working Teenagers

The organizations such as International Labor Organization and UNICEF carried out studies for the improvement of children's working conditions in the short term, completely preventing the working of children in the long term and signed various international conventions as a result of those studies (Anonymous, 2010). In the convention of the rights of the children of the United Nations dated 20th November 1989 those under 18 are defined as "child". Further, the protection of child against the economic exploitation, not employing him in a way that is risky, can damage his physical, mental, spiritual, moral and social development or prevent his education are determined as the fundamental right (Bildik, 1998).

Second item of the regulation about the rules and procedures of employing the child and young laborers; "In accordance with the 71st article of the labor law dated 4857, this regulation involves the rules and procedures about the forbidden works for the child and young laborers who are under 18, the permitted works for the young laborers who turned 15 but under 18, the light works for children who turned 14 and completed the elementary education and the working conditions. According to this regulation; the young labor represents the one who turned 15 but still under 18 and the child labor represents the one who is 14 but does not turn 15 and completed the elementary education.

Reasons Pushing the Individuals in the Late Childhood Period (Puberty) to Work

The main reason for the use of child labor is poverty. If the economic power of the family is not sufficient, the children work with the aim of contributing to the family income. Not receiving education due to various reasons or leaving any levels of education unfinished are among the reasons for entering the working life. The problems such as the poverty and unemployment, which emerge as a result of the immigration, can be reasons for children's start in the working life. Due to the insufficient legal regulations about the small-scaled manufacturing and service sector and the lack of control cause the children's informal working (Karataş, 1993).

Bildik (1998) examined the problematic areas of working teenagers and effective psycho-social factors in his study and found that 43% of the teenagers' families were immigrant due to various reasons, 50,6% of them were primary school graduate, 11,7% of them were secondary school graduate, 36,4% of them were secondary school dropout, 1,3% of them were high school graduate and 1 one them was open high school graduate. 53,2 of them failed the class due to absence because of the indifference, dislike the school environment, failure in the lessons or negative friend environment, they also did not continue their education due to the school-related difficulties such as the disagreement with their teachers and school friends, not passing the exam for higher education, 19,5% of them left the school because of the problems with their families such as;they were not sent to school by their families just because they were girls, families' indifference to the child, unable to focus on the lessons due to the parents' serious marital problems, not starting the school in order not to live with the family or one of the parents, the death of the parents, leaving the home due to the conflicting relations with the stepmother and starting a job, extreme family pressure for the success at school and theprevention of the education by the stepfather. It was also determined that 61,1% of them were from the families with more than one kid(4 or more), many of the parents were primary school graduate, the fathers of 67,5% of them bring money to home regularly, the fathers of 32,5 of them were unable to work every day due to diseases, worked when they find a job and were unable to bring money to home because he was unemployed. When the reasons for entering the working life were examined, the followings were found; 55,8% the employment anxiety, 31,2% the obligation to support the income of the family, 7,8% the needs of puberty such as; moving away from the house, becoming freer, obtaining the economic independence, knowing the life, having new friends, conflicting relations with the teachers and getting rid of the school discipline, 5,2% emulating to the friends or family members.

Bildiket al., (2004), carried out a study with the aim of determining the psychosocial problems and developmental needs of the teenagers working in full-time jobs, and in this study he found the conflicts between the teenager and the parents, parents' marital conflicts, single-parenthood and the problems related to the work as the teenagers' psychosocial problems. Further he also detected the solution offers as such; the prevention of the exploitation at work, wage rise, not working till the late hours, having rest, having holiday and having free time to spend with friends. The average ages of the subjects in the study was 16,35. They consisted of 55 boys and 22 girls. 61% of them were from nuclear families, 10,4% were from extended families and 28,6% of them were from divided families. 31,2% of the subjects complained from ergonomic risks such as; carrying heavy things, lifting heavy things, continuously working on foot, working in a certain position and working in very small room. 22,1% of them complained from physical problems (blurred vision, stomachache, headache, backache, knee pain, leg pain, heel pain, chest pain, drop in the blood pressure, weakness, tiredness, fatigue).

The Effects of Working Life on the Individual who is in Late Childhood Period

Children work:

1. Full-time and periodically
2. Out of school time
3. Seasonal (Karataş, 1993).

The working life affects the teenagers' development from different aspects and the majority of such effects have negative features. The children's workings are regarded negative in the worldwide because of the negative effects of long working hours and heavy works on the physical, mental and emotional development of the children. Working can lead to the decrease in the children's school performance and make them move away from their schools and miss out their education. The negative effects of the working life on the teenager:

- Participation in the working life can cut back the adaptive resources of the teenagers who invested time and energy on his school, family and friends.
- The teenagers' working in some jobs which conflict with his autonomy, self-esteem, responsibility and goal-direction can cause them to have stress.
- In addition to the teenagers' depression, which is a normal part of their development, they can face with important events which require additional adaptations.
- The teenagers can have health and behavioral disorders due to the fact that their jobs are stressful (Anonymous, 2010).

Peşken Süslü (2002) did a study with the aim of examining the sense of identity of the teenagers who are working and students. For the study, teenagers aged 15-16 and 17-18 working in the clothing sector and 240 students aged 15-16 and 17-18 studying in the 1st grade or final year students in Kartal High school in 2000-2001 academic year were reached. In the study it was found that the scores of the sense of identity of the student teenagers in the study group were higher than the scores of the sense of identity of the working teenagers.

The Teenagers Inclined to Crime

Crime, which is defined as the departure from the social rules or the situation of objection, involves the activities which are contrary to the values, rules and common behaviors of the society and go beyond such patterns. In contemporary societies the activities, which are regarded as crime, are determined with the laws (Bakırcıoğlı, 2010; Yavuzer, 2011).

According to the 3rd article of the Child Protection Act numbered 5395 which came into force on 15th July 2005, the definition of child is as such: "The person who is under 18 even if s/he matures earlier ages" (Günşen İçli, 2009).

Connections and components of various factors such as the factors forming the teenager's personality, natural, social, economic and cultural conditions of the environment he grows cause the behaviors directed to crime. The mental retardation, epilepsy and psychological disorders which are inherited can be examples of the biological reasons, obsessions during the development, and immature sexual motives can be given as the psychological reasons, last but not the least, family can be taken as an example of social reasons (Bakırcıoğlu, 2010).

- Psychological disorders and the situation of being disabled can pave the way for committing a crime when they come together with the environmental reasons. For example, the attitude of people towards an individual with physical disable can turn into a reason for crime.
- Early start of the puberty is regarded among the biological factors for the child's in clining to crime.
- It is thought that self-representation is a factor for child's committal.
- Bad socialization process in the family, which is the first social unit an individual meets since his birth, can drag the child to criminal behaviors. Negative relations in the family and family members' negative attitudes and behaviors to each other can push the child to crime (Günşen İçli, 2009).
- School is an organized unit of the society and it has a critical role not only in informing but also in the socialization of the child and the developments of values and norms. In case the school cannot fulfill its functions, an individual's success, development, adaptation to his environment and mental health are influenced negatively. The effects of school on the socialization of the individuals are as such: It plays a role in shaping the social values and norms. It provides the individual's interaction with friend groups. It provides opportunity for the individual to meet people whom he can model or imitate. The school is responsible for taking the necessary precautions for the students who have discipline and absence problem and are unsuccessful in their lessons. Many precautions the schools take provide the prevention of behavioral problems.
- If the peer group, which the child is in, is inclined to the crime and involves risk factors, this situation can cause the child to incline to the crime.
- Negative feelings stemmed from as a result of the poverty and the opposition between the rich and the poor in the society can push the individual to crime (Ereş, 2009; Günşen İçli, 2009).

Gültekin Akduman and Baran (2011) carried out a study to compare the children,who were in the 12-15 age group, got involved in crime and who were in the same age group but did not get involve in crime, in terms of being abused by their peers, and to determine the influential familial factors. They found that the children, who got involved in crime, experienced the peer abuse more and the children, who had three or more siblings, got higher scores in all the aspects of general peer abuse than the other children.

The Disabled Teenager

According to the research carried out by General Directorate of Family and Social Researches in 2010, it was determined that 48,1% of the disabled teenagers were visually handicapped, 26,6% of them were physically handicapped, 29,6% of them were hearing handicapped and 6,2% of them were mentally disabled.

Having children is one of the nicest feelings in a person's life. As a subsidiary factor and a product of their marriage, the couples wish to have a child. They dream about their baby and make preparations for it. Families' dreams and their preparations are for a normal and healthy baby. The families disappoint if any developmental difference occurs, the function of the family structure changes, the plans of the family members are influenced and an immense anxiety can emerge in the family (Şahin et al., 2011; Özay, 2004).

Living with a disabled individual creates many negative feelings in the family members. The families worry whether their children can develop like the other children and can be an independent adult or not (Şahin et al., 2011).

The reactions of the families with disabled individual can be summarized under three headings;

1. Shock, Denial, Pain, Depression: Just after the birth, the parents' learning that their child is disabled causes helplessness. The families, who experience the process of helplessness and excessive crying, try to escape from this reality and can deny it by ignoring. Also, they mourn and go into depression due to the excessive pain stemming from not having a normal child. Some families feel this pain and disappointment throughout their lives. The pain can act as a transition period for families to get used to this reality. As a result of the process of pain, the depression can emerge when the families direct their anger to themselves.
2. Anger, Guilt, Shame: As a result of their anger the parents ask the question "Why me?" or direct their anger to someone else, for example; to the doctor who is responsible for the birth or who diagnose this situation, or to the center which provides the early response service. The guilt is the most difficult feeling that families can cope with and they can believe that they caused this situation or they were punished due to their mistakes in the past. They can regard the condition of their children as their own fault. Further, they may have difficulty in going out in public as they are ashamed of their children.
3. Acceptance and Harmony: Even though the previous negative feelings never disappear, they parents are in struggle for doing something for their children. As time goes by, the parents learn to accept themselves and their children as they are. For the families the acceptance includes recognizing their children, understanding them and solving the new problems. The parents, who become stronger as they overcome the obstacles, accept both themselves and their children as they are. The adaptation process can be completed in a couple of days or it can continue for years. During this process, it is important to understand the parents and respect to their feelings (Özen, 2011; Şahin et al., 2011).

Seligman stated that in connection with the child's developmental periods, the feelings of the families can change and they can relive the feelings they experienced during the adaptation process. He also expressed that such reactions have four important turning points:

1. The first period when there are first doubts about the inability or the diagnosis is learnt.
2. The period when the decisions about the child's education is given.
3. The period when the school life of the child ends.
4. The period when the parents become unable to care their child due to elderliness (Cited from Şahin et al., 2011).

The relations with the peers and group friendship gain importance for the teenagers in this age period. The relations with the opposite sex increase, the development of sexual identity is completed, cognitive capacity rises, abstract thought takes the place of concrete thinking and the value system belonging to one's own is formed. The increase in weight and height is seen in this period, sexual features are gained. The teenager develops new relations with his family and environment in this process. The difficulties experienced in this critical period of life, leaves important influences in the teenager's mental structure (Bolat et al., 2011).

Bolat et al. (2011), investigated depression and anxiety levels and the features of the self-concept of the teenagers who are with sight handicapped by birth. The study was done on total 80 people who are aged 11-14, sight handicapped and sighted. In the study it was found that the anxiety levels of the sight handicapped teenagers were higher than the sighted teenagers (Bolat et al., 2011).

The sexual development in the mentally disabled child and the young follow the same stages with the children who develop normally, but depending on the level of mental development it can be slower and follow from behind. The sexual training, which starts with the birth, continues until the adult age and even the death. The questions asked from the early ages about the sexual subjects constitute an important part of the sexual training. However, it is forgotten that the mentally and physically disabled child and the young have sexual rights, and their needs for knowledge about the sexual subjects, their feelings and thoughts can be ignored. Oppressive and negative attitudes observed since 1960s, but the harmonization movement made the disabled's rights of sexual life and receiving sexual training a current issue. Moreover, the necessity of the preparation of healthy sexual training programs for the social relations, the danger of sexual harassment and AIDS, the development of self-reliance and preventing the problems in the social life were defended (Artan, 2003).

In a study carried out by Mandan Sürücü (2009), it was observed that the knowledge of moderate level mentally disabled girls aged 11-17 were insufficient about the difference between the man and the woman, the naming of the parts of body, difference between becoming lover and connubiality, the process of pregnancy, birth and birth control and perceiving the sexual abuse.

An individual moves away from the society due to the defect in his physical functions and the effect of this defect on his locomotion. The importance of the family, due to its place on the healthy development of the children, remains stable, it is a unit where the first behaviors for the social life, rules and roles, basic habits and happiness are taught within the daily relations (Şahin et al., 2011).

- The individuals with special needs should be ensured to establish correct social relations for social behaviors which are important in the society.
- Positive behaviors, which are approved in the society, should be taught to the disabled individuals.
- Giving less responsibility to the disabled individuals than their peers, sympathizing with them and showing mercy can cause them to develop a weak and dependent personality or an insufferable cruel personality. The disabled individuals should learn, at young ages, not to ignore other's rights.
- The disabled individual needs experiences which help him be aware of his personality. Valuing to themselves should be taught to them
- The disabled individuals should accept their condition and believe that they are valuable. Also, they should learn to live with their inabilities.
- The sense of belonging should be given to the disabled individuals; they should be accepted as they are and respected (Barut, 2011).

Problems the Teenagers in Need of Protection Have

There are many problems in front of the healthy development of the children in need of protection. The children's development can be affected negatively or positively by the followings; their interaction with the environment the microsystem harbors, their families and friends, mass communication which constitutes the mesosystem, the neighbors, their peers, education institutions, health, the institutions about the social security and the connections inside such institutions, senior socio-economic policies involved by the macro system, the beliefs, values and ideologies of the country. The children in need of protection are constantly is relation and interaction with many systems (Yolcuoğlu, 2009).

The negativities an individual has in the puberty, which is an important period of the life, cause him to have physical, emotional and psychological depressions. The teenagers, who do not live with his parents due to various reasons and are obliged to live in orphanages, have many problems. The teenagers living in the orphanages may have difficulty in establishing social relations due to the lack of attention which is the need of this period. Being deprived of positive and confidential behavior models mat create shyness in creating the attachment with other people. Also, being separated from the family environment or coming from the broken families affect the personality of the teenagers in the orphanages negatively. In a study carried out by Gürsoy et al., (2005) it was detected that the levels of self-representation of the teenagers staying in the orphanages were lower than the teenagers who did not live in the orphanages.

Various prejudices and stigmatization against the youth growing up in the social service institutions influence the teenagers negatively and cause to the lack of self-reliance (Algül, 2009).

In his study done on 87 teenagers, who stayed in the orphanages and aged 12-19, Aşık (2006) found that the thoughts and desires about the future were the issues in which teenagers had the most difficulty. It was also determined in the study that the teenagers used some methods such as developing themselves, emotional relief and not taking the problem serious/cracking jokes in order to cope with this situation. Problems about the friendship with the opposite sex and learning sexual problems were common in the boys; also psychological development problems were common in those whose academic success was low. Further, as the psychological development problems increased, externalizing the negative feelings increased, too and as the thoughts and desires about the future increased, not taking the problem seriously/cracking jokes and emotional relief increased more.

This research received no specific grant from any funding agency in the public, commercial, or not-for-profit sectors.

REFERENCES

Algül, S. (2009). *Erzurum'da yetiştirme yurtlarında kalan korunmaya muhtaç ergenlerde sosyodemografik özellikler ile psikiyatrik belirti ve tanı dağılımı*. Atatürk Üniversitesi Tıp Fakültesi. Tıpta Uzmanlık Tezi.

Anonim. (2010). Türkiye'de Ergen Profili 2008. Başbakanlık Aile ve Sosyal Araştırmalar Genel Müdürlüğü.

Aral, N., & Baran, G. (2011). *Çocuk gelişimi*. İstanbul: Ya-Pa.

Artan, G. (2003). Engelli çocuk ve gençlerin cinsel eğitimi [Sexual education of children and youth with disabilities]. *Bilim ve Aklın Aydınlığında Eğitim Dergisi [J. Ment. Sci. Educ. Light]*, (4), 18-20.

Aşık, E. (2006). *Yetiştirme yurdunda kalan gençlerin ergenlik sorunları ve baş etmeleri*. Abant İzzet Baysal Üniversitesi Sağlık Bilimleri Enstitüsü Yüksek Lisans Tezi.

Atabek, E. (2002). *Erken Büyüyen Çocuklar*. İstanbul: Altın Kitaplar Yayınevi.

Aytekin, A. (2005). Meslek seçimini etkileyen sosyo-ekonomik ve kültürel faktörler Isparta örneği, Süleyman Demirel Üniversitesi.

Bakırcıoğlu, R. (2010). *Çocuk ve Ergende Ruh Sağlığı*. Ankara: Anı.

Barut, Y. (2011). Özel gereksinimli çocuk ve ailelerine yönelik psikolojik danışma ve rehberlik hizmetleri. Özel Gereksinimli Çocuklar ve Özel Eğitime Giriş. Edt: Ayşegül Ataman. S:87-101, Ankara: Gündüz Eğitim ve Yayıncılık.

Bildik, T. (1998). *Çalışan ergenlerin sorun alanları ve etkili olan psikososyal faktörler*. Ege Üniversitesi.

Bildik, T., Büküşoğlu, N., & Kesici, H. (2004). Çalışan ergenlerin psikososyal sorunları ve kendiçözüm önerileri. *Ege Tıp Dergisi*, *43*, 105–111.

Bolat, N., Doğangün, B., Yavuz, M., Demir, T., & Kayaalp, L. (2011). Doğuştan tam görme engeli olan ergenlerin depresyon, kaygı düzeyleri ve benlik kavramı özellikleri. *Türk Psikiyatri Dergisi*, *22*(2), 77–62. PMID:21638229

Cansız, R. (2004). *Çocuk yetiştirme modellerinde cinsiyetçi yaklaşımlar*. Ankara Üniversitesi Sosyal Bilimler Enstitüsü, Kadın Çalışmaları Anabilim Dalı.

Çetin Gündüz, H., & Nalbantoğlu Yılmaz, F. (2016). Lise Öğrencilerinin Kariyer Kaygılarını Belirlemeye Yönelik Ölçek Geliştirme Çalışması. *Mersin Üniversitesi Eğitim Fakültesi Dergisi*, *12*(3), 1008–1022.

Çopur, Z., Erkal, S., & Şafak, Ş. (2006). Annelerin 12 yaş üzeri çocuklarıyla ilişkilerinde yaşadıkları çatışmaların incelenmesi, I. In Ulusal ev ekonomisi kongresi. Academic Press.

Demir, V., & Yıldırım, B. (2017). Sanatla Terapi Programının Üniversite Sınavına Hazırlanan Öğrencilerin Depresyon, Anksiyete ve Stres Belirti Düzeylerine Etkililiği. *Ege Eğitim Dergisi*, *18*(1), 311–344. doi:10.12984/egeefd.280267

Erdem, A. Y. & Bilge, F. (2016). Lise öğrencileri için meslek seçimine ilişkin akılcı olmayan inançlar ölçeği'nin geliştirilmesi. *Türk Psikolojik Danışma ve Rehberlik Dergisi*, *3*(29), 95-114

Erden, M., & Akman, Y. (1998). *Gelişim-öğrenme-öğretme: eğitim psikolojisi*. Ankara: Arkadaş.

Ereş, F. (2009). Tolumsal bir sorun: Suçlu çocuklar ve ailenin önemi. *Aile ve Toplum Eğitim Kültür ve Araştırma Dergisi*, *17*, 88–96.

Gültekin Akduman, G., & Baran, G. (2011). Suça karışan ve karışmayan 12-15 yaş grubundaki çocuklarda akran istismarının çeşitli ailesel özelliklerle ilişkisi. *Anadolu Psikiyatri Dergisi*, *12*, 76–83.

Günşen İçli, T. (2009). *Çocuk Suç ve Sokak*. Başbakanlık Aile ve Sosyal Araştırmalar Genel Müdürlüğü.

Gürsoy, F., Aral, N., Yıldız Bıçakçı, M. (2005). Yetiştirme yurdunda kalan ve kalmayan ergenlerin benlik tasarım düzeylerinin incelenmesi. *Eğitim Araştırmaları Dergisi*, 24-34

Hamamcı, Z. & Hamurlu, M. (2005). Anne babaların meslek gelişimine yardımcı olmaya yönelik tutumları ve bilgi düzeylerinin çocuklarının mesleki kararsızlıkları ile ilişkisi. *İnönü Üniversitesi Eğitim Fakültesi Dergisi*, *6*(10), 55-69.

Kabasakal, U. (2007). Ergen ve genç intiharında aile faktörü. *Doğu Anadolu bölgesi araştırma ve uygulama merkezi*, *6*(1), 32-37.

Karataş, K. (1993). Çocuk işgücü sorunu: Nedenleri, sonuçları ve çözüm önerileri. *Hacettepe Üniversitesi Sosyal Hizmetler Yüksekokulu Dergisi*, *11*, 85–102.

Kulaksızoğlu, A. (2005). *Ergenlik psikolojisi*. İstanbul: Remzi.

Kuzgun, Y. (2000). *Meslek Danışmanlığı Kuramlar ve Uygulamalar*. Ankara: Nobel.

Mandan Sürücü, S. (2009). 11-17 yaş orta düzeyde zihin engelli kız ergenlerin temel cinsel bilgi ve cinsel istismarı algılamalarının belirlenmesi [Yüksek Lisans Tezi]. Marmara Üniversitesi Eğitim Bilimleri Enstitüsü.

Okutan, E., Akbaş, M. G. (2019). 15-24 Yaş arası Öğrencilerin Kariyer Kaygılarını İncelemeye Yönelik Literatür Araştırması. *Siyaset, Ekonomi ve Yönetim Araştırmaları Dergisi, 7*(1), 33-41.

Özay, H. (2004). Ailenin çocuğa ilişkin kabul ve beklenti düzeyleri. In *Özel Eğitim Kongresi Bildirileri Özel Eğitimden Yansımalar*. Ankara: Kök.

Özdal, F. (2003). İlköğretim dört ve beşinci sınıfa devam eden anne-babası ile yaşayan ve baba yoksunu olan çocukların kaygı düzeylerinin incelenmesi [Yüksek Lisans tezi]. Ankara Üniversitesi.

Özdemir, Y. (2009). Ergenlik döneminde benlik kurgusu gelişiminin kültür ve aile bağlamında incelenmesi [Doktora tezi]. Ankara Üniversitesi.

Özdemir, Y., & Çok, F. (2011). Autonomy Development in Adolescence. *Türk Psikolojik Danışma ve Rehberlik Dergisi*, *4*(36), 152–164.

Özen, A. (2011). Aile eğitimi. In İ.H. Diken (Eds.), Özel Eğitime Gereksinimi Olan Öğrenciler ve Özel Eğitim (pp. 89-107). Ankara: Pegem.

Özer, D. S., Özer, K. (2000). *Çocuklarda Motor Gelişim*. İstanbul: Kazancı.

Öztabak, M., & Erdoğan, İ. (2017). Lise Öğrencilerinin Düşünme Stilleri ile Karar Verme Stilleri Arasındaki İlişkinin İncelenmesi. *HAYEF Journal of Education*, *14*(2), 33–57.

Pekşen Süslü, D. (2002). *Çalışan ve öğrenci ergenlerde kimlik duygusu kazanımının araştırılması*. Ankata Üniv. Eğitim Bilimleri Enstitüsü. Yüksek Lisans Tezi.

Sadıkoğlu, G., Gökdağ, R., Alper, Z., Özçakır, A., & Özdemir, H. & Uncu, Y. (2007). Aile psikolojisi ve eğitimi. Anadolu Üniversitesi Yayınları.

Şahin, S., Işıtan, S., & Gündüz, S. (2011). *Özel gereksinimli çocuklar ve eğitimleri*. Ankara: Eğiten Kitap.

Santrock, J. W. (2014). *Child Development*. New York, NY: Mc Graw Hill.

Şeker, G., & Kaya, A. (2019). Mesleki karar verme programının onuncu sınıf öğrencilerinin mesleki kararsızlık düzeyleri üzerinde etkililiğinin sınanması. *Gazi Üniversitesi Eğitim Fakültesi Dergisi*, *39*(1), 457–476.

Törüner, E. K., & Büyükgönenç, L. (2011). *Çocuk Sağlığı*. Ankara: Gökçe.

Traş, Z., & Gökce, Z. (2017). Lise Öğrencilerinin Kariyer Karar Verme Güçlükleri ve Ego Kimlik Statülerinin Bazı Değişkenlere Göre İncelenmesi. *Türk Eğitim Bilimleri Dergisi*, *15*(1), 58–69.

Yavuzer, H. (2011). *Çocuk Psikolojisi*. İstanbul: Remzi Kitabevi.

Yeşilyaprak, B. (2013). *Eğitimde Rehberlik Hizmetleri Gelişimsel Yaklaşım*. Ankara: Nobel.

Yolcuoğlu, İ. G. (2009). Türkiye'de çocuk koruma sisteminin genel olarak değerlendirilmesi. *Aile ve Toplum Eğitim Kültür ve Araştırma Dergisi*, *18*, 43–58.

Yörükoğlu, A. (2011). *Çocuk ruh sağlığı*. İstanbul: Özgür.

Section 3

Current Issues in Child Development

Contemporary subjects about child development are included in this section. Accordingly, the following subjects are discussed in detail; Bilingualism in Early Childhood, Early Literacy Skills, Early Intervention in Child Development, Family Support Programs, Yoga for Babies and Children: Introduction, Definition and Importance of Yoga, Benefits of Yoga, Yoga in Education Contribution of Outdoor Play Activities to Infants and Toddlers Developments, Interactive Story Reading and Language Development in Early Years, Perinatal and Early Childhood Development: Education, Innovation and Sustainable Development Goals Problems, Adaptation of Cognitive Abilities Scale into Turkish A Validity and Reliability Study: Cognitive Abilities Scale Preschool Form.

Chapter 13
Bilingualism in Early Childhood

Halil Uzun
Kilis 7 Aralik University, Turkey

Gülen Baran
Ankara University, Turkey

ABSTRACT

Bilingualism that arises from different reasons (such as wars, migrations, geographical proximity, and trade, tourism, marriages, living to better standards, work and education) affects children, so children have to grow up in environments where more than one language is spoken. There are various arguments about whether being a bilingual for a child is advantageous or disadvantageous. The studies carried out in this field have shown that the disadvantage can be converted into advantageous situations on condition that the process about the acquisition or learning of the second language has been managed well. From this point of view, the importance of concepts such as language development in early childhood period, views on language development, bilingualism, language acquisition and teaching in early years will be explained in detail in the bilingualism in the early childhood section.

INTRODUCTION

For various reasons, children grow up in environments where more than one language are spoken. Other languages in addition to the official ones are spoken in many countries in the world. Upon getting in touch with the social environment, the children who acquire the language from their families, their closest ones, see that there are people who use other languages to communicate and they interract with these people. This interraction is often experienced in pre-school educational institutions which can be considered as the first social environment of a child after the family. A bilingual child sometimes experiences some problems in terms of communication and social skills with other children in the class. As a criterion for defining bilingualism, Bloomfield states that both languages should be used very well, whereas Haugen, on the contrary, argues that meaningful expressions in other languages are sufficient for bilingualism (Bloomfield, 1933; Haugen, 1966; Baker, 2011). There are various arguments about whether being a bilingual for a child is advatageous ar disadvantageous. The studies carried out in this field have shown

DOI: 10.4018/978-1-7998-2952-2.ch013

that the disadvantage can be converted into advantageous situations on condition that the process about the acqusition or learning of the second language has been managed well (Garcia, 2011; Kormi-Nouri et al., 2012; Pelham & Abrams, 2014; Paap et al., 2015).

In this chapter, the importance of concepts such as what is language, importance of communication, language development in early childhood period, bilingualism, language acquisition and teaching will be explained in detail.

BACKGROUND

Language and Communication

Language, both oral and written, is an important communication tool for people to learn and teach their feelings, thoughts, experiences and cultural values (Sevinç, 2003; Samovar et al., 2015; Bonvillain, 2019). One of the functions of language is to represent thoughts and the other is to convey these thoughts. Conveying thoughts to another person and receiving the thoughts of another person leads to mutual communication. Communication a process by which information is exchanged between individuals through a common system of symbols, signs, or behavior. İt is a process that has existed since the beginning of life and has been dynamic with the evolution of humanity. All living creatures made message transfer with a unique way of communication. Only humanity used an extraordinary and fascinating way in this transfer. Although voice and voice-based call systems exist in many animal species, the transition from voice to word and verbal expression is evolutionary. This behavior, which makes man superior to other living beings, is called "ability to use language via speech" or simply "speech-language" (Sless, 1981; Sperber & Wilson, 1986; Topbaş, 2010; Enderby et al., 2013; Yule, 2016).

Language Development

Language development is a process that starts with birth and continues for the whole life. From the first days of life, children begin to perceive the sounds around them, make sounds and gain the basic structure of the language spoken in the society in which they live. Language, along with all its rules, is naturally learned during the experiences. Children learn the language by listening to models, imitating these models, perceiving feedback, and sharing their experiences and thoughts. While children acquire language, their first models are parents, other family members, and then other individuals they interact with in social and school environments. Enriched language environments that will be offered to children as well as models are supportive factors in gaining and using the language creatively. Language, which is one of the most important interaction tools of early childhood, is required for all activities. Language is the most important means of communication between peers, children and adults (Yazıcı, 2003; Gleason & Ratner, 2005; Aitchison, 2012). Chomsky (2004) argues that children's ability to learn language is due to a genetically programmed organ that is located in the brain. Once children are born and are involved in linguistic environments, they immediately start to develop a language. However, to do that, children must make use of the only tool they have available to them which is their inborn mental grammar. Chomsky, characterizes this mental grammar as Universal Grammar. Through Universal Grammar, newborn babies have available to them the grammar of any language existent in the World (Kıran ve Kıran, 2002). Language is used as a means of communication. Language consists of basic units such as sounds, symbols and words. Language development is the acquisition, storage and use of words, numbers, symbols in accordance with the rules. Language plays an important role in the child's learning. In language development, it is important that voices can be heard and that they have the ability to use the language. It is claimed that children learn the language by imitating adult conversations. The language development of the child is based on the need to communicate, to attract the attention of others and to communicate their wishes, feelings and thoughts (Aral et al., 2000). The famous linguist Ferdinand de Saussure (1985) defines language as a system of signs and indicators.

Children are born ready for language development and learning. From the moment they are born, they have the ability to distinguish between the human voice and other voices, especially the mother's voice. Children's capacity to understand and develop the language begins at an early age and continues throughout life. In other words, linguistic features are part of human development (DeCasper & Fifer, 1980; Yazıcı & Can Yaşar, 2006). Research has shown that there is a relationship between the child's innate language abilities and the brain. The brain is the administrator of the language system. There are two centers in the brain called the Broca and Wernicke for speaking and understanding the spoken language. The Broca is the focal point of the brain's speech-related ability and controls the vocalization in the speech. Wernicke is concerned with the recognition and distinguishing of speech sounds. In general, the Wernicke field is sensory and the Broca field is motor (Morgan, 1991). Chomsky states that the child's understanding and speaking of language cannot be explained only as analogy and imitation. He claims that language skills are present in the brain at birth and children are born with a special ability that they can understand through the discovery of the rules of the language system over time. Chomsky calls it the Language Acquisition Device (LAD). According to Chomsky, the LAD contains the common universal rules that are inherent in all languages at birth and gives the child the capacity to solve the system that will enable them to use these common rules. With this capacity, children formulate hypotheses about the grammar of the language of the culture they live in, and gradually discover the rules of grammar and use these rules to form their own sentences. Children learn not only a series of words while learning the language, but also grammar rules that allow them to express them as meaningful sentences (Kıran & Kıran, 2002; Chomsky, 2004)

Receptive and Expressive Language

Language is generally divided in to two categories: receptive and expressive.

Receptive language is essentially understanding the expressions and words of others. It is defined as the reception and understanding of verbal stimuli through sensory-neural network and auditory-perceptual processes (Dönmez et al., 1997; Karacan, 2000; Gamez et al., 2019). Children begin to develop this skill first.

Expressive language is the child's ability to express themselves. It is the source of communication of thoughts, ideas and feelings which transfer verbally or nonverbally. Expressive language means to assemble ideas into words or grammatically correct and valid sentences which make some sense (Emmorey, 2001; Ahmad & Shahid, 2019).

Stages of Language Development

Children vary in their development of speech and language skills. However, they follow a natural progression or timetable for mastering the skills of language. These stages are divided into two categories as pre-linguistic and linguistic.

Pre-Linguistic Language Development

Pre-linguistic language development is when a child is learning to control the sounds he can produce and to string these sounds together in vocal play. In this stage, the child is not yet able to manipulate these sounds into proper words. There are four categories of pre-linguistic development that can be distinguished (Stark, 1986; Oller, 2000; Hoff, 2013).

Reflexive Crying and Vegetative Sounds: In the newborn period (0-6 weeks) the baby makes reflexive sounds. After a while, these sounds differ, and in relation to the stimulus, they often change to crying of hunger and discomfort. Crying is probably part of the normal development of the central nervous system. Many parents say they can hear a difference in tone between a cry for feeding and a cry caused by pain. In crying and in making these vegetative sounds, an infant's vocal cords vibrate, and the airflow through the vocal apparatus is stopped and started. Thus, even these unpromising sounds include some features that will later be used to produce speech sounds (Hoff, 2013).

Cooing and Laughter: Cooing and laughter occur at 6-16 weeks of age. These are vocalizations that the baby makes when it's happy or content and can be made up of vowel or consonant sounds. Coos are sounds that babies make when they appear to be happy and contented. Social interaction in particular seems to elicit cooing. The first coos that infants make sound like one long vowel. Infants continue to produce cooing noises for many more months, and the quality of these coos changes with age. One change is the variety of different vowel-like sounds that infants start to produce. Also, instead of producing a single sound, infants produce a series of different vowel-like sounds strung together but separated by intakes of breath. Although these infants are not talking yet, more features of speech sounds are present in these vocalizations than there were in earlier ones. Babies produce their first laughter around the age of 16 weeks (Dönmez et al., 1997; Hoff, 2013).

Vocal Play: This period, which is called vocal play time or expansion phase, corresponds between 16 weeks and 30 weeks for infants. During vocal play, the baby begins to string together longer vowel or consonant sounds. Infants seem to gain increasing control over the production of their growing repertoires of sounds, and they combine their different sounds into increasingly long and complex series (Hoff, 2013).

Babbling: Finally, babbling occurs around the ages of 6-9 months. At this time, the child begins to produce a series of consonant-vowel syllables and may develop utterances, such as ma-ma and da-da. When infants babble while looking at an object, they create a state of receptivity for learning at the same moment caregivers are likely to provide specific information about the target object (Albert et al., 2017; Elmlinger et al., 2019).

Linguistic Language Development

Linguistic language development is the stage of language development signaled by the emergence of words and symbolic communication. Prior to this stage, most of the sounds a child produces are no more than the practice of sound manipulation and sound sequencing in order to gain the motor skills necessary to create words. There are six periods of linguistic language development. Before a child masters the ability to form words, they will first begin to use specific sound combinations consistently with specific meaning. This is the early one word period that begins around 12-19 months of age. Children's first words have a simple syllable structure: They are either single syllables or reduplicated syllables, such as baba and dada. First words also use a small inventory of vowels and consonants (Ingram, 1989). An example of this would be a child saying 'baba' every time he wants a bottle of milk. Even though this is not the exact same as the word bottle, the child is using "baba" in the same manner as you would use the word "bottle." The later one word period begins around 14-24 months of age. In this stage, the words used by the child are readily identifiable, and he begins to name and label people and objects in his environment. Next comes the two word period of language development. As the name implies, this is when he will begin to combine two words together to make simple phrases, such as "mommy go". The two word period typically begins from 20-30 months of age. The three word period begins around the ages of 28-42 months of age. During this period, a child adds at least one more word to their phrases and begins to use pronouns. They may also begin to use articles and simple prepositions. At around 34-48 months of age, the four word period begins. At this time, the child will begin to use combinations of four to six words. They will use more prepositions, and adjectives begin to appear in speech. The last period of linguistic language development is the complex utterance period. It begins around the ages of 48-60 months. At this time, a child regularly produces phrases longer than six words in length, and they begin to express concepts of past and future time. Examples are 'Daddy comes home from the trip tomorrow' and 'I saw a dog at the park yesterday.' They may also begin to use contractions, such as "can't" or "don't." Researchers do not agree on when this period is completed and adult sentence structure is achieved. Opinions range from 5 years of age to 12 years of age (Stark, 1986; Dönmez et al., 1997; Senemoğlu, 2007; Hoff, 2013; Mariscal & Benavides, 2017).

Components of Language

The ability of the child to acquire the language depends on the basic systems that form the structure of the language. In the literature, it is seen that the language system is divided into four main areas and each area has some basic rules. These are phonology, syntax, morphology and semantics (De Saussure, 1985; Charlesworth, 1993; Aksan, 1995; Vardar, 1998; Galda et al., 1997; Booij, 2012; Hoff, 2013).

- Phonology: It reveals how the sounds used in a language combine to form meaningful parts and how this formation occurs.
- Syntax: It reveals the rules about the functions of words in a sentence (name, conjuncture, etc.) and how the words form a meaningful sentence in a sequence. Syntax determines the order of the words that make up sentences.
- Morphology: Morphology is the study of the internal structure of words and forms a core part of linguistic study today.
- Semantics: Semantics is the study of meaning. The study of semantics looks at how meaning works in language, and because of this it often uses native speaker intuitions about the meaning of words and phrases to base research on.

These components work together to create meaningful communication among individuals.

Theories of Language Development

Language is a cognition that truly makes us human. This ability is remarkable in itself. What makes it even more remarkable is that researchers are finding evidence for mastery of this complex skill in increasingly younger children (Rowland & Noble, 2010). There are many different theories on how people's language development occurs. The most widely accepted ones in the literature are presented below.

Behavioral Theory

According to the developmental theory of Russian psychologist Lev Vygotsky (1962), the primary function of private speech is not language exploration but behavioral self-guidance. This account of private speech derives from Vygotsky's theory that an individual's cognitive skills develop first in social interaction and then later are internalized. According to this theory, language development in children begins with the classification and shaping of many sound stimuli from the environment. The relationship between mother, father and other people close to the child contributes significantly to the language development of the children. Babies imitating the sounds they often hear is also important in the acquisition of language (Erden & Akman, 2004). When babies make sounds or similar sounds of language spoken in their environment, and when this behavior is rewarded by adults, they repeat these sounds and learn it. In his classic book Verbal Behavior, Skinner applied the principles of behavioral learning to the acquisition of language, but did not conduct research to prove his particular assumptions (Skinner, 1957; Cömertpay, 2006). Many linguists also believe that operant techniques such as imitation, modeling and reinforcement can help explaining the child's language acquisition, but cannot fully explain it. In addition, the view that the behavioral approach assumes that the parents correct the mistakes in the child's syntax and reinforce the correct syntax has been rejected by the opponents of this approach. They suggested that the parents corrected the semantics, not the child's syntax (Staats, 1971; Gander & Gardiner, 2004; Hoff, 2013).

Social Interaction Theory

The interaction theory of language development is a compromise between the nativist theory and the behaviorist theory of language development. The interaction theory recognizes that both environmental and biological factors are important in language development (Piaget, 1956). They state that observational learning is important in social interaction. Through social interaction they learn to utilizeexisting cultural artifacts and to create newones to regulate their biological and behavioralactivity. Mediation happens primarily throughlanguage use, organization, and structure, occur-ring during participation in cultural, linguistic, and historically formed settings such as family life, peer group interaction, and institutional contextslike schooling, social leisure activities, and work-places. Interactionists believe that language acquisition occurs as a result of the natural interaction between children and their environment, more specifically, their parents or caregivers (Selçuk, 1997; Erden & Akman, 2004; Ellis, 2019).

Psycholinguistic Approach (Native Language Approach)

The native language approach (The view linking language development to biological basis) is another view on language acquisition. This approach argues that language is genetically transmitted and the people have already had language acquisition tools previously. Linguists such as Noam Chomsky and Lenneberg base language development on biological basis and they don't ignore the environmental conditions' effects on language development. Theorists who explain language development based on biological and psychological foundations are called psycholinguistic theorists. In psycholinguistic theory, there are two different processes in learning to speak: comprehending the meaning of words and making meaningful sounds or speaking. These processes are intertwined and develop in parallel with cognitive development (Özbay, 2003; Erden & Akman, 2004; Topbaş, 2010; Atay, 2011).

Cognitive Theory

Jean Piaget, who has made important studies in the field of cognitive development and genetic epistemology, has argued that thought and language development in children is not through continuity but occurs through stages and is effectively structured in individual environment relations. According to Piaget, cognitive development is at the forefront. Language development is a coordination of general cognitive changes and cognitive development is not affected by language. The development in early childhood consists of symbolic thinking. The voluntary declaration of symbols makes it possible for the child to say everything he imagines. The first words of the child are not yet real symbols. Because these symbols are related to the objects and events seen. Children use basic symbols when they start talking about things that aren't available around them. According to Piaget, the child does not speak in a social communicative manner until he is seventh, because he is egocentric and thinks in egocentrism (Piaget, 2007; Hoff, 2013).

THE DEFINITION AND IMPORTANCE OF BILINGUALISM

Many different definitions have been proposed for bilingualism in the literature. This variation in definitions is mainly due to bilingualism being a broad and comprehensive concept and the fact that it is affected by the diversity of demographic, geographic, sociological, political and social-cultural factors (Wei, 2000). The word "bilingualism" is a combination of the Latin words "bi" meaning two and "lingualism" meaning linguistics. Bilingualism can often be expressed and defined in various categories and contexts such as ideal - partial and regular - compound, depending on factors such as proficiency and function (Romaine, 1995; Ortega, 2014).

People develop skills to adapt to the characteristics of the environment they live in. Bilingualism often occurs in societies where different languages are spoken (Harris, 1992; Karahan, 2005). Although being bilingual may seem as a disadvantage at first glance, studies in the field indicate the opposite. When a monolingual child is learning his/her mother tongue, he/she only learns the sound image of one word (tree), while a bilingual child learns two sound images (tree and der Baum) and therefor two words. Thus, the child learns that an object/thought can be named using two, three or even more words, not just a single sound image or word. Therefore it can be said that bilingual children who learn both languages more or less simultaneously at a young age can easily comprehend that a signified image/object or thought can be named by multiple indicative images/sound images. This supports the assumption that bilingual children can develop their abstract thinking skills sooner and faster than monolingual children, which is something that has been proved by researches in the field (Rolffs, 2009). According to scientific research, children who learn two languages at an early age form a certain system in their brains. This system makes it easier for them to later acquire more languages. Linguistic studies have shown that children who grow up as bilinguals have a higher ability to solve and comprehend problems, have better use of languages, and can learn a third language easier than monolingual children. Children are not born bilingual; they acquire two or more languages as they grow up. Therefore, children must be encouraged to learn both languages equally from an early age. In order to achieve this goal, children who grow up bilingual should be consciously supported by both educational institutions and their families. If no effort is made in this regard, these children can grow up to be unsuccessful and incompetent in both languages. It is well known that children who grow up bilingual or multilingual possess different thinking skills compared to monolingual children. Knowing the language of the country they live in as good as their native language instills a distinct sense of trust and belonging in Bilingual individuals have different views of the world than those who monolinguals, which allows them to view the world and events through other people's eyes and understand them better (Riehl, 2011).

Reasons for Bilingualism

Bilingualism can occur for a variety of reasons, in terms of individual and social aspects. The most general classification for the reasons in the literature is as follows:

Geographical Proximity

The geographical proximity of two communities naturally breeds relations such as trade, tourism, etc. In such cases, language is seen as a barrier in effective communication until the individuals of both communities learn each other's language. This inevitably leads to bilingualism. Furthermore, this proximity on some occasions can also lead to marriage which creates bilingual families. A bilingual family generally results in bilingual children.

Historical Factors

Those who settled in different regions, as a result of historical events such as conquests and colonialism, influenced a lot of aspect of life. One of these, naturally, is the language spoken in society. The official correspondence was always made in the language of the stronger society after the language of the stronger society put pressure on the language of the less powerful society (Romaine, 1995). This is evident in most African countries. The colonial masters bequeathed their own language as the official language and designated the Saharan African languages as sub-languages. Some countries, such as Cameroon, subsequently abandoned the teaching of the two existing formal languages in schools. The states that took part in colonialism then drew the boundaries of the states they exploited and did not take into account the historical, cultural, geographical or demographic factors when doing so. This can be said to have caused many countries, most of which are in Africa and the Middle East, to become bilingual.

Migrations

Collective or individual migrants fleeing from wars or some people seeking for personal gain (jobs, better living standards, etc.) migrate to different linguistic societies. Immigrants, who are obliged to learn the languages of their host communities to communicate and find jobs, are thus forced to become bilingual (Romaine, 1995; Ortega, 2014).

Religion

In various religions such as Islam, the language of the written source is considered sacred, and it is emphasized that the religion can only be practiced in this language. Therefore, clergy and religious people have to learn the holy language and thus bilingualism arises (Romaine, 1995).

Local or International Relations

People have to learn and use local languages to communicate with each other in countries where more than one language is spoken. Speaking a foreign language has become indispensable for politicians, traders and diplomats in establighing relations between countries. Therefore, many language-teaching programs have been developed for these needs. For example, there are foreign language training programs to teach English, French, Spanish, German and even Chinese and Italian in Cameroon (Romaine, 1995; Wei, 2001; Ortega, 2014).

Types of Bilingualism

Although bilingualism can be classified according to language pairs, Weinreich put forward three types of bilingualism in terms of how an individual thinks about the concepts of the language encoded in his/her brain. These are compound bilingualism, coordinate bilingualism and sub-order bilingualism (Romaine, 1995).

- *Compound* bilingualism: is the case of having two distinct linguistic codes connected to a single string of meanings. This concept of compound bilingualism generally refers to individuals acquiring two languages simultaneously in a common environment, without creating autonomous contexts for those languages, that is to say, using the same context. In compound bilingualism, the individual creates a universe of common meaning for both languages, without realizing and exploring the conceptual differences between these languages. Thus, for example, the word "bread" in Turkish and the word "bread" in Swedish have the same connotations for those who are compound bilinguals.

Various studies have indicated that the compound bilingualism occurs especially when parents are also bilingual, or when parents speak to the child in two languages in a mixed manner without functionally separating the languages from each other. A child who grows up in such environments does generally not have difficulty speaking either language; however, he/she is often unable to master the discriminated characteristics of these languages.

- *Coordinate bilingualism:* is the case where both languages show an autonomous, but a related structure in the brain in terms of rules and meaning. The concept of coordinate bilingualism refers to individuals who have acquired both languages in their differentiated special contexts. In coordinate bilingualism, the child interacts with both languages on the same level when growing up.

Therefore, for example, when the child hears any word (such as "friend"), she/he comprehends that word by distinguishing its semantic equivalent in the language to which that word belongs. For them, the semantic content of a word will differ in terms of the universe of meaning specific to the languages. Bilinguals can become coordinate bilinguals when their parents have different mother tongues and each of the parents speak to them regularly in their native language. In this way, the child can construct a cognitively autonomous linguistic mechanism for both languages and use these different but related mechanisms to serve the situation and communicative wishes. Coordinate bilingualism can also arise when both parents speak the same mother tongue, but the official language of the country the family lives in is different. In such cases and "normal circumstances", in other words, if there is no obstruction, the child will acquire his/her mother tongue competently and naturally adopt the official language of the country to use at home or in any required communication situation. The differences between the two languages are clear and natural for a child growing up in such environments.

- *Subordinate bilingualism:* is the case where one of the languages is stronger than the other. The weaker language is interpreted according to the stronger one. In other words, it is the structuring of the two languages in a non-autonomous relationship in the brain. One language plays the role of "core" to shape the other in terms of structural rules and the universe of meaning, and is therefore the dominant language (Romaine, 1995).

The New Encyclopedia Britannica mentions two types of bilingualism. These are:

- *Coordinated Bilingualism:* In this type, the individual learns the languages in separate environments, and the words of both languages are separated from each other by preserving the meaning of each word. Cameroonian children learning English at school is an example for this type of bilingualism which can also be considered as reductive bilingualism.
- *Unified Bilingualism:* In this type, the individual learns both languages simultaneously from the same source. Therefore, a narrative formed by the merging of languages emerges in the brain. This is the case for children born with bilingual parents or those from two different language environments. It should be noted that this classification has led to the emergence of various bilingual education programs.

Larsen and Long (1994) divided bilingualism into two main categories:

- The model that allows students to continue learning the second language while growing up in their first language environment.
- The program that allows those who speak their local language to receive their primary education in a second language. The program content is increasingly provided in the intended language after the first grades.

When looking for an answer to the question of who is bilingual, the broadest and comprehensive definitions have been provided by Wei (2000). According to Wei (2000), bilingual individuals are classified as shown in Table 1.

Table 1. Types of Bilingual

Types of Bilingual	Description
The additive bilingual	An individual who combines two languages in a complementary and unifying form
The accelerated bilingual	An individual who develops the second language due to the increased use
The balanced bilingual	An individual who uses both languages in roughly equal proportion
The compound bilingual	An individual who has learned both languages simultaneously and in the same context
The coordinate bilingual	An individual who learns two languages in distinctly separate contexts
The hidden bilingual	An individual who conceals his or her knowledge of the given language due to a certain attitude and tendency
The diagonal/crossing bilingual	An individual who is in a non-standard language or an unrelated dialect of the standard language
The dominant bilingual	An individual who is very proficient in one language and has certain competencies in another language or languages
The dingy bilingual	An individual who immigrated to another country, learned the language of that country, and had little opportunity to preserve his or her first language
The early bilingual	An individual who acquired both languages in early childhood
The functional bilingual	An individual who has mastered languages for a particular task, even if not with full fluency
The horizontal bilingual	An individual who is bilingual in two separate languages with similar or equal status
The early-stage bilingual	An individual whose one language has not developed yet and is in the early stage of bilingualism
The tardy bilingual	An individual who became bilingual after early childhood
The high-level bilingual	An individual who can use two or more languages at a level close to his or her mother tongue
The low-level bilingual	An individual who can use a few words or phrases in a second language
The natural bilingual	An individual who is bilingual without any special training as well as without looking at the relationship between language-related translation and meaning
The practical bilingual	An individual who speaks and occasionally writes in a language other than to understand that language
The perceptual bilingual	An individual who reads and understands something written in another language but cannot write or speak in that language
The recessive bilingual	An individual who has some difficulties in understanding and expressing himself/herself due to various usage deficiencies
The secondary bilingual	An individual who gained the second language through learning in addition to the first language
Semi-bilingual	An individual with inadequate knowledge in two languages
The simultaneous bilingual	An individual who has acquired both languages since he/she started to speak
The subordinate/dependent bilingual	An individual who attempts to use his/her own language by reducing the patterns of the second language while using his/her own language
The subtractive bilingual	An individual who has the skills required to acquire the first language and the ability to acquire the second language
The successive bilingual	An individual who acquires the second language after developing the first language at a certain stage
The vertical bilingual	An individual who is bilingual in a standard language or a dialect related to the standard language

The Difference Between Acquiring and Learning a Language

This section covers the difference between acquiring and learning a language, theories about second language teaching, stages of second language acquisition and second language education in childhood, and the techniques used in these trainings.

There are two main processes that play a role in the development of linguistic competence and skills. The first is the acquisition process, which is carried out unconsciously, involuntarily, instinctively, and equivocally through senses. The second is the learning process, which is carried out consciously, voluntarily, and manifestly using information (Kraschen & Terrel, 1983; Zengin, 2003).

Language is acquired through a natural ability that exists in humans. Children acquire their mother tongues by hearing and imitating what they hear. In other words, this process does not entail conscious and systematic learning or teaching. On the other hand, learning a language requires voluntary effort. Foreign language "teaching" in schools now begins in pre-school. The words taught to children in a foreign language in kindergarten are generally forgotten if they are not used in other environments. However, the language acquisition process is continuous. Parents need to know the difference between these two situations and have a clear understanding of what they can expect from their children. Teaching a second language to children between the ages of three and six is of course not the same as the children growing up in a bilingual environment. Studies have shown that children aged four to six are slower in second language learning than children aged seven to nine. The reason for this is that children between the ages of four and six are not able to master their native language, the social environment to share what they learn is limited, and their literacy skills are not fully developed. Most children forget or mix up the words they learn in kindergarten after a certain period of time. In addition, children of this age group can often be taught one-or two-word phrases, which do not meet the child's need for communication. Children who grow up in two different language-speaking environments actively continue to hear and use both languages. Therefore, the languages they acquire become more permanent (Schelletter, 2002; Karakaya, 2008).

Approaches and Theories of Second Language Teaching

There have been different opinions among scientists in the field of second language teaching, and as a result, different approaches and theories regarding second language teaching have been put forward.

Comparative Theory

The comparative theory was developed by Fries and later Lado and suggests that the process of language learning takes place through the comparison of the first learned language with the second learned language. Therefore, the language learner looks for similarities between the two languages. The similarities between the structures of the second language and the mother tongue make the second language easier to learn, while the differences in the languages cause difficulties in learning. According to this theory, the linguistic, lexical, operatic, syntactic, vocal, and scientific features of the first language are transferred to the second language and the language learner follows a path called transfer-strategy during the learning process (Ergenç, 1983; Gui, 2019; Zengin, 2003).

Grammar-Translation Method

The debate on language analysis and the use of language dates back to the 19th century. When it comes to the analysis of language, the basic rules of grammar come to mind. The grammar-translation method is based on language analysis. It has been widely practiced since the middle ages, and was initially used mostly in the field of teaching Latin in the West and Arabic in the East, and then adapted into other languages (Benati, 2018; Hengirmen, 2001). While the teaching of grammar rules gained importance for this method at the beginning, the aim of the method turned towards language teaching over time. The use of the grammar-translation method was common in the 19th and early 20th centuries, as it was based on the memorization of grammar rules and words. However, the prevalence gradually decreased and the grammar-translation method lost its importance. The method was also effectively implemented in Turkey until the 1950s (Demircan, 2005).

Natural Approach

The natural approach is the product of a new teaching idea put forward by Tracy Terrell. In the formation of this approach, Krashen's ideas were extremely influential. The natural approach took its place in the field of language teaching with the book titled "Natural Approach" published by Terrell and Krashen in 1983. According to this approach, natural communication activities of language are used in language teaching. Therefore, language input and exposure to the language are crucial. In the next stage, the student is expected to be ready to produce the language, but at this point, the student must have been exposed to the language on a sufficient level to feel the need to produce the language. Krashen and Terrell regarded communication as the primary function of language and therefore focused on communicative skills. In this context, it can be said that the natural approach is a flood approach (Richards & Rodgers, 2014). The natural approach is against traditional approaches and is not based on any language theory. According to this approach, meaning and comprehension are the most important elements in language.

Direct Method

The emergence of the direct method was based on psychologist Gestalt's holistic views, Herbart's educational views and Humboldt's language and culture views. This method is an extension of the natural method and is against the grammar-translation method, in which the language is not actively used. Naturalism and the use of real-life language are the outlets in the development of the method. Studies regarding this method were carried out in the early 20th century. The natural use of language in the classroom by staying away from the rules of grammar is important in the implementation of this method. Thus, the focus of attention shifted from the rules of grammar to speech and vocalization. Within the scope of this method, lessons begin with a short story or dialogue and are supported by verbal movements and pictures. The mother tongue is never used during lessons (Demircan, 2005; Yorulmaz, 2009). Although the direct method was successful in private schools such as Berlitz, it failed in secondary schools due to practical difficulties and various conditions (Richards & Rodgers, 2014).

Berlitz's Practice

Berlitz's practice is not a method, but rather the application of a specific method. The basic skills in Berlitz's practice are comprehension - listening, comprehension - reading, speaking, writing, which is the application of the direct method. According to the Berlitz's practice, which mainly includes daily speaking, teachers should speak the target language as their main language. Although success has been achieved with the Berlitz's practice in language schools, it can be said that this success belongs to the direct method (Hutagalung, 2018; Oebel, 2008; Yorulmaz, 2009).

Auditory-linguistic Method

The auditory-linguistic approach was developed by linguist Bloomfield who transferred his behaviorist views into language teaching and is based on behaviorism in terms of learning theory and structuralism linguistically. This method is important for phonological studies as its main idea is that language is primarily a speech. Linguists have drawn attention to the study of structurally related elements of language within this method (Richards & Rodgers, 2014; Yorulmaz, 2009).

Verbal Approach (Situational Language Teaching)

The verbal approach was proposed by Harold E. Palmer and A.S. Hornby to add a scientific base to language teaching. According to this approach, which focuses more on vocabulary teaching, grammar should not be taught in specific stages, but by examining the difficulties and mistakes that students come across. In the verbal approach, lexicography is the basic prerequisite for understanding what one reads. This approach emerged on a scientific basis and, for the first time, introduced the concept of the syllabus in the field of language teaching. The structure of language in the approach is regarded as the center of speech ability since the basis of the verbal approach is based on British structuralism (Daskalovska, 2018; Richards & Rodgers, 2014; Yorulmaz, 2009).

Communicative Approach

Communicative competence is important in the perception of language as a whole, emphasizing the importance of communication in language teaching and other skills in language teaching. The aim of the communicative approach is to demonstrate that language is not only made up of structures but also to teach language by combining grammar and functions. The main aim of the communicative approach, which is based on communication, is to develop communicative competence. Communicative competence is a term that was created based on Chomsky's competence theory but in contradiction to it. Another important figure in this approach is Henry Widdowson. Widdowson focused on the necessity of using language for different purposes and put forward views that emphasized the relationship between linguistic systems and their communicative values (Richards & Rodgers, 2014; Yorulmaz, 2009).

Cognitive Approach

The cognitive code - learning approach was proposed as an alternative to the auditory - linguistic approach and is considered a critique of the auditory - linguistic approach in the light of the rules of linguistics and psycholinguistics. However, the approach which came into existence in the mid-1960s was overshadowed in the 1980s by communicative approaches. One of the key figures in the cognitive approach is Ausubel. Cognitive theory focuses on the student's understanding of foreign language structure and the use of this structure easily. The aim of this approach is to provide students with a proficiency in language skills in a foreign language just as they possess in their native language. Accordingly, instead of learning certain sentences, students learn a mechanism that can produce sentences that are appropriate for situations they have never encountered before (Demircan, 2005; Yorulmaz, 2009).

Task-Based Teaching

Prabhu was the first to discover and implement the method of task-based teaching (Wang, 2006). Prabhu, who conducted his studies on the implementation of this method in elementary schools in Bangalore, South India, stated that, instead of focusing on the language the students use, the focus should be placed on the desired tasks that would bring the desired success in learning. Krashen argued that individuals should be exposed to the target language in the long term and that grammar rules should not be taught throughout this process (Yorulmaz, 2009). Another advocate of this method, Willis, stated that task-based teaching is more advanced and logical than communicative language teaching. Meaningful tasks play a supporting role in communicative learning (Richards & Rodgers, 2014).

Simultaneous Acquisition Stages of Two Languages in Early Childhood

The simultaneous acquisition of two languages in the first few years of a child's life takes place in three stages:

In the first phase, the characteristics of the language development stage of monolingual children can be observed in the language development of bilingual children. According to various researchers, this stage is in fact monolingualism. Initially, the child produces words in both languages with a single sound unit system. The child then uses the words of both languages to combine them as if they were a single language as he/she cannot make an equivalent translation. One or two words from both languages can be seen in the child's speech up to the age of two. During this period, the child uses words from both languages and begins to use the languages in parallel without identifying the language or the person using the language, in other words, without paying attention to who speaks which language and whether they understand it or not. Although children might know the meaning of an object in the other language, they cannot actually say its name in that language. Not being able to name an object in both languages is due to the children's inability to understand languages as a system that has different definitions for situations, events or objects (Watson, 1995; Nicoladis, 1998; De Houwer, 1999).

In the second phase, children, around the age of two, gain a wide vocabulary set in both languages to name actions, objects and situations. At this stage, although the vocabulary of the two languages has started to be dissociated, children use the same syntactic rules in both languages. Children begin to identify both languages with those who speak them, however they use words that have the same meaning in both languages one after the other. (Watson, 1995; De Houwer, 1999).

In the third phase, the lexical and syntactical separation of the languages now begins, although the transfers between the two languages continue to somewhat carry on. During this phase, children are exposed to both languages in accordance with the "one language, one parent" rule, at this stage, they identify the language with the person who speaks that language and using the same method communicate with adults (Watson, 1995; De Houwer, 1999).

Foreign Language Education in the Preschool Period

Children are born prone to learning languages. Children record what they hear as if they were recording it on tape. However, after this period, they lose this recording feature (Halıcı, 2004). It is a well-known fact that language acquisition and learning processes differ in people according to age and that children can learn languages more quickly and easily than adults. However, which philosophies and theories should be used on which age group is still discussed. From this point of view, the level of readiness of children and adults during the teaching-learning period, their reactions in teaching activities and their motivation in learning situations are different, therefore the methods used to teach a foreign language to adults and children should be different. Different facts should be taken into account for each group and targets and target behaviors should be determined in line with these factors. A foreign language is a second language and can only be built on the native language. The most important point in foreign language education is not to make the mother tongue a second language. Success in learning a foreign language depends on reaching a certain maturity in the native language (Kotil, 2002).

The world of children is very different from the world of adults. It is not enough to be a language expert to teach children as guidance on child psychology is also needed. Games, fairy tales, songs, cartoons, and rhymes have an important place in the world of children. When approached with this understanding, children tend to learn foreign language rules quickly. Play education is an activity-based education that keeps children's interests awake and piques their curiosity. In the play education system, children actively participate in the planning, preparation, and evaluation processes. Children have fun playing, focus their attention for a long time, and express themselves easily in multiple ways. Therefore, play education is a tool that can be used effectively in the foreign language education process (Tuğrul, 2002). For example, it is possible to make children carry out great motor skills by saying "run, jump, walk" in English. Information can be given via songs as if it was part of the game (Halıcı, 2004).

Colorful books that are suitable for children, with cartoon characters in them, can be used in language education. Objects should be taught by either showing the object itself or a picture of the object. Written texts should be avoided in such books. The first stage of foreign language learning is listening which begins by repeating words. Children are prone to learning by the deductive method. Instead of breaking a word into pieces, it should be taught as a whole. Short sentences are allowed to be used during the teaching process. Picture books are very useful as they make it easier for the children to understand what is said in the picture by solving words with pictures (Angı, 1999). In brief, foreign language education should be presented to children in a child-centered structure, by games, rhymes, storytelling, drama technique, songs and technological tools (video, tape, and computer) during the pre-school period.

SOLUTIONS AND RECOMMENDATIONS

Bilingualism / multilingualism exists in many countries around the world. According to the researches, when the languages spoken in the countries are examined, it is seen that 20-25% of the languages are spoken besides the official language. This situation requires governments to make some arrangements. They provide bilingual services to them, such as health, safety and education in public services. Therefore, in education, different bilingual education practices are seen in many countries around the world. Many governments are implementing such arrangements for the healthy development of children.

Acquiring two different languages in their natural environments provides many advantages. Two different languages develop two different ways of thinking; therefore, children use the social advantages of getting to know different cultures. If children learn both languages from native speakers, they will generally acquire an accent-free speech. However, this only happens if the children have the opportunity to hear and speak both languages equally.

In order to support the child's development in social emotional, cognitive, psycho-motor, language area and self-care skills in early childhood, a native language learning program, which will be prepared considering the basic principles of language development, acquisition and language teaching, may be beneficial for bilingual children in early years. Teachers who will implement this program can be selected from teachers who are fluent in both languages and can be trained and assigned to work in related regions. Quantitative research methods can be used to determine which two languages children speak in choosing the regions where the program will be implemented. In-service trainings can be provided to teachers working in regions where multiple languages are spoken to effectively implement the language training program.

CONCLUSION

Bilingualism is an indispensable phenomenon brought to us by the changing world. In the past, people whose second language skills (listening, speaking, reading, and understanding) were equivalent to native language skills were called "bilingual (Bloomfield, 1933), and today it is thought that a certain level of competence in both languages is enough to evaluate that person as bilingual. (Bialystok, 2001). Some adaptations are being made for this phenomenon in the world. For example, a proposed application model for bilingualism may be the HIPPY program. In Israel in 1969, the Hebrew University National Council for Jewish Women was designed to help young children become more responsible, enthusiastic and successful school students, and to raise awareness of mothers about their own strength and potential as teachers of their children. This model is a home-based family education program and is implemented in many different countries such as the USA. The main objective of the program is to facilitate intercultural harmony. In order to achieve this, they developed a family-centered program based on teaching the official language as a means of communication, enabling people to understand each other starting from early childhood (Kuday, 2007). Similarly, programs such as Head Start support the language development and social cultural cohesion of children and their families. The Head Start program was designed to help break the cycle of poverty, providing preschool children with a comprehensive program to meet their emotional, social, health, nutritional and psychological needs. Head Start programs have served more than 36 million children since 1965, growing from an eight-week demonstration project to include full day/year services and many program options in USA. Head Start has maintained two culturally and linguistically specialized programs: American Indian/Alaska Native Head Start programs and Migrant and Seasonal Head Start programs. More generally, Head Start legislation, regulations, and program guidelines all contain requirements and recommendations for supporting Dual Language Learners (DLL) children. DLL children represent a large subgroup of the Head Start population: in 2007-2008 DLL children made up 29% of Head Start participants. In conclusion, programs such as HIPPY and Head Start for low socio-economic and disadvantaged families and children contribute to the needs of families and bilingual children's language development. For the healthy development of children and social and cultural cohesion, it may be appropriate to adapt these programs to 21st century standards and to implement them in the regions needed around the world.

REFERENCES

Ahmad, I., & Shahid, H. (2019). A study of the problems in expressive language learning faced by hearing impaired children. *Academic Research International, 10*(1).

Aitchison, J. (2012). *Words in the mind: An introduction to the mental lexicon*. John Wiley & Sons.

Aksan, D. (1995). Her Yönüyle Dil Ana Çizgileriyle Dilbilim. Ankara: TDK.

Albert, R. R., Schwade, J. A., & Goldstein, M. H. (2018). The social functions of babbling: Acoustic and contextual characteristics that facilitate maternal responsiveness. *Developmental Science, 21*(5), e12641. doi:10.1111/desc.12641 PMID:29250872

Angı, R. (1999). Yabacı dil öğrenirken-öğretirken. *Okul Aile Dergisi, 1*(12), 18–20.

Aral, N., Baran, G., Bulut, Ş., & Çimen, S. (2000). Çocuk gelişimi 1. cilt. İstanbul: Ya-Pa Yayınları.

Atay, M. (2011). Erken çocukluk döneminde gelişim I. Ankara: Kök Yayıncılık.

Baker, C. (2011). *Foundations of bilingual education and bilingualism* (Vol. 79). Multilingual matters.

Benati, A. (2018). Grammar-Translation Method. In The TESOL Encyclopedia of English Language Teaching. Academic Press.

Bialystok, E. (2001). *Bilingualism in development: Language, literacy, and cognition*. Cambridge University Press. doi:10.1017/CBO9780511605963

Bloomfield, L. (1933). *Language*. New York: Holt, Rinehart & Winston.

Bonvillain, N. (2019). *Language, culture, and communication: The meaning of messages*. Rowman & Littlefield.

Booij, G. (2012). *The grammar of words: An introduction to linguistic morphology*. Oxford University Press.

Charlesworth, R. (1993). Understanding child development. Albany, NewYork: Delmar publishers Inc.

Chomsky, N. (1986). *Knowledge of language: its nature, origin and use*. New York: Praeger.

Chomsky, N. (2004). Knowledge of Language as a Focus of Inquiry. In B. Lust & C. Foley (Eds.), First language acquisition: The essential readings (pp. 15 24). Malden: MA: Blackwell Pub.

Cömertpay, B. (2006). Dramanın 5-6 yaş grubu çocuklarının dil edinimine etkisi, Sosyal Bilimler Enstitüsü Okul Öncesi Eğitimi Ana Bilim Dalı. Unpublished master thesis, Çukurova Üniversitesi.

Daskalovska, N. (2018). Historical Perspectives on Teaching Speaking. In The TESOL Encyclopedia of English Language Teaching. Academic Press. doi:10.1002/9781118784235.eelt0710

De Houwer Annick, D. (1999). *Two or more languages in early childhood: some general points and practical recommendations*. Retrieved from http://www.cal.org/resources/digest/earlychild.html

De Saussure, F. (1985). Genel dilbilim dersleri. Ankara: Birey ve Toplum Yayınları.

Decasper, A., & Fifer, W. P. (1980). Of human bonding: Newborns prefer their mothers voice, *Science. Sayı, 208*, 1174–1176. PMID:7375928

Demircan, Ö. (2005). Yabancı dil öğretim yöntemleri. İstanbul: Der Yayınları.

Dönmez, B. N., Abidoğlu, Ü., Dinçer, Ç., Erdemir, N., & Gümüşcü, Ş. (1997). *Okul öncesi dönemde dil gelişimi etkinlikleri*. Ankara: Sim Yayıncılık.

Ellis, N. C. (2019). Essentials of a theory of language cognition. *Modern Language Journal, 103*, 39–60. doi:10.1111/modl.12532

Elmlinger, S. L., Schwade, J. A., & Goldstein, M. H. (2019). The ecology of prelinguistic vocal learning: Parents simplify the structure of their speech in response to babbling. *Journal of Child Language, 46*(5), 998–1011. doi:10.1017/S0305000919000291 PMID:31307565

Emmorey, K. (2001). *Language, cognition, and the brain: Insights from sign language research*. USA: Psychology Press. doi:10.4324/9781410603982

Enderby, P., Judge, S., Creer, S., & John, A. (2013). Examining the need for and provision of AAC methods in the UK. *Advances in Clinical Neuroscience & Rehabilitation, 13*, 20-23. Retrieved from https://www.thecommunicationtrust.org.uk/media/3414/other_ways_of_speaking_final.pdf

Erden, M. & Akman, Y. (2004). Gelişim ve öğrenme. Ankara: Arkadaş Yayınları.

Ergenç, İ. (1983). Yabancı dil öğretimi ve olumsuz aktarım. *Türk Dili Aylık Dil ve Yazın Dergisi, Dil Öğretimi Özel Sayısı, 32*, 379–380.

Galda, L., Cullinan, B. E., & Strickland, D. S. (1997). Language, literacy and child. Orlando: Harcourt Brace & Company Puplishing.

Gamez, P. B., Griskell, H. L., Sobrevilla, Y. N., & Vazquez, M. (2019). Dual Language and English-Only Learners' Expressive and Receptive Language Skills and Exposure to Peers' Language. *Child Development, 90*(2), 471–479. doi:10.1111/cdev.13197 PMID:30566240

Gander, M. J., & Gardiner, H. W. (2004). Çocuk ve ergen gelişimi. Ankara: İmge Kitabevi.

Garcia, O. (2011). *Bilingual education in the 21st century: A global perspective*. John Wiley & Sons.

Gleason, J. B., & Ratner, N. B. (2005). *The development of language*. Pearson/Allyn and Bacon.

Gui, L. (2019). Research on Teaching Chinese as a Foreign Language from the Perspective of Ecological Linguistics. In *Proceedings of the 5th International Workshop on Education, Development and Social Sciences (IWEDSS 2019)*. Academic Press. Retrieved from https://webofproceedings.org/proceedings_series/ESSP/IWEDSS%202019/IWEDSS19334.pdf

Halıcı, P. (2004). Çocuklarda yabancı dil eğitimi [Doktora semineri]. Ankara Üniversitesi.

Harris, R. J. (1992). *Cognitive processes in bilingual*. Amsterdam, Netherland: Elsevier Science Publishers.

Haugen, E. I. (1966). *Language conflict and language planning: The case of modern Norwegian*. Harvard University Press. doi:10.4159/harvard.9780674498709

Hengirmen, M. (2001). *Yabancı dil öğretim yöntemleri ve tömer yöntemi*. Ankara: Engin Yayın Evi.

Hoff, E. (2013). *Language development*. USA: Cengage Learning.

Hutagalung, S. I. S. (2018). *Using Berlitz Method to Improve Students' Vocabulary at Happy Holy Kids Cinere Preschool* [Doctoral dissertation]. Universitas Kristen Indonesia.

Ingram, D. (1989). *First language acquisition: Method, description, and explanation*. Cambridge, UK: Cambridge University Press.

Karacan, E. (2000). Bebeklerde ve çocuklarda dil gelişimi. *Klinik Psikiyatri Dergisi*, *3*, 263–268.

Karahan, F. (2005). Bilingualism in Turkey. In *Proceedings of the 4th ınternational symposium on bilingualism*. Cascadilla Press.

Karakaya, Z. (2008). Dil edinimi "Okul öncesi dil ve oyun eğitimi." Samsun: Eflatun Matbaacılık.

Kıran, Z. & Kıran, A. (2002). Dilbilime giriş. Ankara: Seçkin Yayıncılık.

Kormi-Nouri, R., Moradi, A. R., Moradi, S., Akbari-Zardkhaneh, S., & Zahedian, H. (2012). The effect of bilingualism on letter and category fluency tasks in primary school children: Advantage or disadvantage? *Bilingualism: Language and Cognition*, *15*(2), 351–364. doi:10.1017/S1366728910000192

Kotil, Ç. (2002). *Yabancı dil eğitimi*. Nisan: Çocuk ve Aile.

Krashen, S. D., & Terrel, D. T. (1983). *The natural approach: language acquisition in the classroom*. Hayward, CA: Alemany Press.

Kuday, F. S. (2007). *Aile destekli kurum merkezli eğitim alan ve hiç okul öncesi eğitim almayan 3–6 yaş çocukların bilişsel gelişimlerinin karşılaştırılması,* (unpublished master thesis) Marmara Üniversitesi, Eğitim Bilimleri Enstitüsü, Eğitim Bilimleri Ana Bilim Dalı, Okul Öncesi Öğretmenliği Bilim Dalı, İstanbul

Larsen, F. D., & Long, M. H. (1991). An introduction to second language acquisition research. New York: Longman Inc.

Mariscal, S., & Benavides, A. A. (2017). Typical Language Development of Monolingual Spanish-Speaking Children. In *Language Development and Disorders in Spanish-speaking Children* (pp. 3–36). Cham: Springer. doi:10.1007/978-3-319-53646-0_1

Morgan, C. T. (1991). *Psikolojiye giriş*. Ankara: Hacettepe Üniversitesi, Meteksan A.Ş.

Nicoladis, E. (1998). First clues to the existence of two input language: Pragmatic And Lexical Differentiation In Bilingual Child. *Bilingualism: Language and Cognition*, *1*(2), 105–116. doi:10.1017/S1366728998000236

Oebel, G. (2008). *So-called" Alternative FLL-Approaches*. GRIN Verlag.

Oller, D. K. (2000). *The emergence of the speech capacity*. Mahwah, NJ: Erlbaum. doi:10.4324/9781410602565

Ortega, L. (2014). *Understanding second language acquisition*. Routledge. doi:10.4324/9780203777282

Özbay, Y. (2003). Gelişim ve öğrenme psikolojisi. Trabzon: Akademi Kitabevi.

Paap, K. R., Johnson, H. A., & Sawi, O. (2015). Bilingual advantages in executive functioning either do not exist or are restricted to very specific and undetermined circumstances. *Cortex*, *69*, 265–278. doi:10.1016/j.cortex.2015.04.014 PMID:26048659

Pelham, S. D., & Abrams, L. (2014). Cognitive advantages and disadvantages in early and late bilinguals. *Journal of Experimental Psychology. Learning, Memory, and Cognition*, *40*(2), 313–325. doi:10.1037/a0035224 PMID:24294916

Piaget, J. (1956). *The language and thought of a child*. New York, NY: Harcourt Brace and Co.

Piaget, J. (2007). Çocukta dil ve düşünme. Ankara: Palme Yayıncılık.

Richards, J. C., & Rodgers, T. S. (2014). *Approaches and methods in language teaching*. Cambridge university press.

Riehl, C. M. (2011). *Çocuklarda ikinci dilin önemi*. Retrieved from http://parkakademi.com/haber.html

Rolffs, S. (2009). Yine iki dillilik üzerine. *Die Gaste*, *5*(Ocak-Şubat).

Romaine, S. (1995). Bilingualism (2nd ed.). Malden.

Rowland, C. F., & Noble, C. L. (2010). The role of syntactic structure in children's sentence comprehension: Evidence from the dative. *Language Learning and Development*, *7*(1), 55–75. doi:10.1080/15475441003769411

Samovar, L. A., McDaniel, E. R., Porter, R. E., & Roy, C. S. (2015). *Communication between cultures*. Nelson Education.

Schelletter, C. (2002). The effect of form similarity on bilingual children's lexical development. bilingualism: Language and cognition. *Language and Cognition*, *5*, 93–107.

Selçuk, Z. (1997). Eğitim psikolojisi. Ankara: Pegema Yayınları.

Senemoğlu, N. (2007). *Gelişim öğrenme ve öğretim: Kuramdan uygulamaya*. Gönül Yayıncılık

Sevinç, M. (2003). Erken çocuklukta gelişim ve eğitimde yeni yaklaşımlar. İstanbul: Morpa Kültür Yayınları.

Skinner, B. F. (1957). *Verbal behavior*. New York: Appleton-Century-Crofts. doi:10.1037/11256-000

Sless, D. (1981). *Learning and visual communication*. Routledge.

Sperber, D., & Wilson, D. (1986). *Relevance: Communication and cognition* (Vol. 142). Cambridge, MA: Harvard University Press.

Staats, A. W. (1971). Linguistic-mentalistic theory versus an explanatory S-R learning theory of language development. In D. I. Slobin (Ed.), *The ontogenesis of grammar* (pp. 103–152). New York: Academic Press.

Stark, R. E. (1986). Prespeech segmental feature development. In P. Fletcher & M. Garman (Eds.), *Language acquisition* (2nd ed., pp. 149–173). Cambridge, UK: Cambridge University Press. doi:10.1017/CBO9780511620683.010

Topbaş, S. (2010). Dil ve kavram gelişimi. Ankara: Kök Yayıncılık.

Tuğrul, B. (2002). Oyun çocuğun en ciddi işidir, *Çoluk Çocuk Dergisi,* 30-31.

Vardar, B. (1998). Açıklamalı dilbilim terimleri sözlüğü. İstanbul: ABC Kitabevi.

Vygotsky, L. (1962). *Thought and language.* Cambridge, MA: MIT Press. doi:10.1037/11193-000

Wang, C. (2006). *Designing communicative tasks for college English courses: a dessertation submitted as a partial fulfillment for the degree of m.a. in English language and literature, school of foreign languages and literature.* Chongqing University & Yangtze Normal University, China. Retrieved from www.asian-efl-journal.com/thesis_Wang_Cheng_jun.pdf

Watson, C. (1995). *One Language or two? helping families from other one language or two? helping families from othercultures decide on: how to talk to their child with language delays.* Spanish Workshop for Children. Retrieved from http://www.spanishworkshopforchildren.com/resources/resources.htm

Wei, L. (2000). *The bilingualism reader.* UK: Routled.

Yazıcı, Z. (2003). Okul öncesi eğitimin okul olgunluğu üzerine etkisinin incelenmesi. *M. E. B. Dergisi,* 155-156.

Yazıcı, Z. ve Can Yaşar, M. (2006). Erken çocukluk döneminde dil kazanımı. *Mesleki Eğitim Dergisi Sayı, 8*(15), 126–138.

Yorulmaz, M. (2009). *Göreve dayalı öğretim yönteminin Türkçenin yabancı dil olarak öğretiminde dinleme becerisinde kullanılmasının avantajları.* Unpublished master thesis, Trakya Üniversitesi, Edirne.

Yule, G. (2016). *The study of language.* Cambridge university press.

Zengin, D. (2003). Sıfatların başka sözcük türlerine kayması, *Ankara Üniversitesi Dil ve Tarih Coğrafya Fakültesi Dergisi Sayı, 43,* 83–99.

ADDITIONAL READING

Appel, R., & Muysken, P. (2005). *Language contact and bilingualism.* Amsterdam University Press.

Baker, C. (2011). *Foundations of bilingual education and bilingualism* (Vol. 79). Multilingual matters.

Chomsky, N. (1986). *Knowledge of language: its nature, origin and use.* New York: Praeger.

Hakuta, K. (1986). Mirror of Language. In The Debate on Bilingualism. New York, NY: Basic Books, Inc.

Hamers, J. F., Blanc, M., & Blanc, M. H. (2000). *Bilinguality and bilingualism*. Cambridge University Press. doi:10.1017/CBO9780511605796

Hoff, E. (2013). *Language development*. Cengage Learning.

Myers-Scotton, C. (2006). *Multiple voices: An introduction to bilingualism*. Wiley.

KEY TERMS AND DEFINITIONS

Bilingualism: (or multilingualism) is the phenomenon of speaking and understanding two or more languages.

Communication: the exchange of meanings between individuals through a common system of symbols.

Early Childhood: the period from birth to eight years old, is a time of remarkable growth with brain development at its peak.

Language: a system of conventional spoken, manual, or written symbols by means of which human beings, as members of a social group and participants in its culture, express themselves.

Language Acquisition: is the process by which humans acquire the capacity to perceive and comprehend language (in other words, gain the ability to be aware of language and to understand it), as well as to produce and use words and sentences to communicate.

Language Development: is the process by which children come to understand and communicate language during early childhood.

Language Education: teaching and learning a language. It also includes the efforts made by a learner to improve his/her native language. Language education is a branch of applied linguistics.

Types of Bilingualism: Depending on the age in which an individual has acquired a second language or in what way or how is he/she acquired more than one language, they are classified into different types of bilingualism.

Chapter 14
Early Literacy Skills

Adalet Biltekin
Ankara University, Turkey

Müdriye Yildiz Biçakçi
Ankara University, Turkey

ABSTRACT

Early literacy skills are related to many developmental areas of children such as language development and motor development. The development of children's early literacy skills affects their future development of literacy. Therefore, teachers should promote children's early literacy skills. In addition, the activities of conscious parents with their children positively affect their early literacy development. Based on these, it is important to address the issue of early literacy. For this reason, the definition of early literacy and concepts related to early literacy, the importance of early literacy, the factors which affect early literacy skills, the development process of early literacy, educational approaches related to early literacy, promotion of early literacy skills are included in this book chapter.

INTRODUCTION

Being literate, that is, being able to acquire literacy skills, is a process which requires many preliminary skills and these preliminary skills are defined as early literacy skills (Karaman, 2014).

It has been approved by many researchers that early literacy skills were acquired earlier in kindergarten and associated with many developmental areas such as language development and motor development (Teale & Sulzby, 1986).

Studies have shown that early literacy skills affect the development and success of children about literacy (Burke et al., 2009; Catts et al., 2002; Spira et al., 2005, Utchell et al., 2016).

For example, Spira et al. (2005) conducted a study in a preschool with low-income families, and it was concluded that about 70% of children who lacked these pre-skills in the first grade had significant problems about reading in the fourth grade. It was also found that the lack of early literacy skills made it difficult for the child to close this gap with their peers over time. In another study, reading performance of children with language disorders was evaluated in the longitudinal study. As a result, it was stated that

DOI: 10.4018/978-1-7998-2952-2.ch014

children's experience and knowledge about literacy in preschool period affect their reading achievement in primary school (Catts et al., 2002).

In another longitudinal study, children's early literacy skills in kindergarten age and literacy skills in primary school were evaluated. According to this study, it is concluded that performance related to early literacy skills affects reading performance up to the third grade of primary school (Utchell et al., 2016). In another study, early literacy skills of kindergarten children and reading performance of children up to second grade were evaluated. As a result of the studies and evaluations, it was concluded that early literacy skill performance of children was predictive of reading performances in the second grade of primary school (Burke et al., 2009). These studies show that early literacy skills are related to reading skills in primary school. There are many studies concluded that children's achievement level in primary school is related to following academic achievement (Hernandez, 2011; Juel, 1988, Lloyd, 1978). For instance, a longitudinal study showed that children's' learning experience up to the third year of primary school affect their possibility to drop out or successfully attend of high school (Lloyd, 1978). According to another longitudinal study which was conducted in Texas with 54 children, 80% of children with low literacy skills had also low literacy skills in the fourth grade and 87% of the children with moderate literacy skills in the first grade had also moderate literacy skills in the fourth grade (Juel, 1988). According to Hernandez (2011), one out of every six children in the third grade who do not have sufficient reading skills did not graduate without failure of a class and this failure rate was four times higher than children with good reading skills.

In order to develop children's' literacy skills and provide them to succeed in their following education life, teachers should be aware of early literacy and especially contribute to the development of kindergarten children in the most appropriate way (Snow et al., 1998). It is important to address and draw attention to the issue of early literacy. For this reason, the definition of early literacy and concepts related to early literacy, the importance of early literacy, the factors which affect early literacy skills, the development process of early literacy, educational approaches related to early literacy, promotion of early literacy skills are included in this book chapter.

THE DEFINITION OF EARLY LITERACY AND CONCEPTS RELATED TO EARLY LITERACY

Many definitions of the term literacy have been made until recent years. United Nations Educational Scientific and Cultural Organization (1997) defined the concepts of literacy and semi-literacy with a committee of experts gathered to standardize educational statistics in 1951. This committee recommended 'A person is considered literate who can both read with understanding and write a short, simple statement on his everyday life. ' as the term of literacy and 'A person is considered semi-literate, who can read with understanding, but not write, a short simple statement on his everyday life.' as the term of semi-literacy. Until the end of the nineteenth century, literacy was defined as the ability to read and write an article, but in broader terms it was defined as a knowledgeable or educated person in a particular field.

When examining the history of studies about early literacy; in the early 1900s, the idea that literacy started in the first grade of formal education was dominated. The most important reason for this view to prevail is the view of developmental psychologist Gesell, the most important factor for a child to learn to read is to reach the required maturity. In these years, preschool teachers generally did not provide reading-related activities until the children reached this maturity. However, they would laid the groundwork for children by reading books and similar materials to encourage, explore, develop problem-solving skills, sing, or discuss a topic based on the theme of the unit (Morrow, 2005).

According to Morphett and Washburne in 1931; the ideal time for a child to learn to read and write was to reach the age of 6 years and 6 months (as cited in Sampson & Briggs, 1981). This ideal time is called 'reading readiness' (Smith & Chapel, 1970). However, after the 1930s and 1940s, instead of spontaneously expecting this ideal time for children, educators focused on the idea of promoting their preliminary skills to facilitate to be literate. For this reason, materials that promote literacy skills of children have increased in preschools (Morrow, 2005).

In 1966, the concept of 'emergent literacy' was defined by New Zealand researcher Marie Clay. According to Marie Clay, emergent literacy skills are defined as the behavior of children who are not yet literate by using writing and reading materials to imitate reading and writing. Nowadays, the concept of emergent literacy is used more broadly than this definition. At the present time, it is known that the development of literacy began from birth and even the first experiences in oral language, books, music and mass media are effective. Emergent literacy covers all areas of development that can contribute to children's reading and writing skills, including all possible developmental areas, such as cognitive, emotional, social and intellectual development (Canadian Council on Learning, 2006).

Emergent literacy is used interchangeably with early literacy. The term emergent literacy or early literacy, which is more commonly used in literature in recent years, can be defined as all the pre-skills which are children should acquire before they can acquire literacy skills (Ergül et al., 2014; Karaman, 2014; Whitehurst & Lonigan, 1998). Early literacy skills of children are acquired before the kindergarten age and are associated with many developmental areas such as language development and motor development. These pre-skills are defined as early literacy skills up to formal education in primary school (Teale & Sulzby, 1986).

Whitehurst and Lonigan (1998), in their research, revealed two views on early literacy skills. The first view is that there is no clear-cut boundary between literacy and pre-literacy processes, but early literacy skills begin before formal education. The other view is that early literacy skills are different from readiness to read and write and that verbal language development, reading and writing development are interrelated.

According to the American Library Association's (2009) definition, early literacy is not to give babies literacy education prematurely or to teach them how to read. Early literacy is the ability of babies to entertain through books as a spontaneous part of their development and to have positive interactions with their parents. Moreover, early literacy skills of children play a critical role in enriching the literacy experience. According to the definition of Cedar Mill Community Libraries (2017), early literacy is defined as everything that children need to know about writing and reading before they can actually read or write.

THE IMPORTANCE OF EARLY LITERACY

There is a development of children's learning to read and write, just like the development of running, walking and controlling their own body at an early age. In the preschool and primary school period, the literacy process begins with basic topics such as how the words are composed of letters and sounds, and how the page is organized. Basic skills about literacy in the early years of children's education are cornerstone of literacy skills in secondary and high school (Moore et al., 1999). Early literacy skills begin with verbal and non-verbal self-expression skills, including children's speech and signs in early years. Over time, children's language skills develop, as well as children's early literacy skills through experiences such as reading books or talking to adults around them (Shrier, 2013). For these reasons, it is important to support children's early literacy skills in early ages.

In one study, in the kindergarten period the concepts of writing (such as writing from left to right) and early literacy skills such as alphabet knowledge and in the early years of primary school reading performances were examined. Accordingly, it was concluded that children's reading achievement in the early years of primary school was related to their performance in early literacy skills (Plaza & Cohen, 2006). Duncan et al. (2007) analyzed the relationship between children's skills in early years and subsequent academic achievement in a longitudinal study. In the study (Duncan et al., 2007), it is concluded that the preliminary skills related to reading at the beginning of primary school are significant to predict the academic success of the students in the following years. For this reason, it is stated that it is important to develop pre-skills including early literacy skills in preschool education programs and the importance of children in future academic life is emphasized. In another study (Claessens et al., 2009), it was concluded that children who have problems in learning to read and write during and after primary school are related to problems related to acquiring some preliminary skills, including early literacy skills.

When children start primary school, their language and early literacy skills should be sufficient level, so that children are willing to learn reading and writing. In particular, it is important to support early literacy skills of disadvantaged children, such as those at risk of reading difficulties or having a low income, because these disadvantages can be reduced or prevented (Snow et al., 1998).

It is important to address the issue of early literacy in preschool education because preschool period is important to develop children's' early literacy skills. Moreover, verbal language skills are the basis of reading and writing and these skills mostly learned during preschool period. Early literacy skills have an important role on children's' school success and literacy skills in the following years. Therefore, a child who has not a sufficient level of verbal language skills cannot catch up with his peer success easily (Biemiller, 2006).

The evaluation and programming of early literacy skills is also important. Assessing children's early literacy skills is essential for a comprehensive and detailed, good early literacy program. Evaluating children's early literacy development is important for identifying children who may need special education in this regard; moreover, it will guide teachers preparing programs (McAfee et al., 2004)

THE FACTORS WHICH AFFECT EARLY LITERACY SKILLS

Early literacy skills are affected by many factors and there are risk groups depending on these factors. These risk groups are likely to be slower than the peer development of the child's early literacy and consequently have difficulty reading and writing in later life. These risk groups may be due to economic reasons, inadequate environment, developmental deficiencies, parents' reading difficulties as children, speaking in a language or dialect different from that of the academic program, and inadequate home environment (Justice & Kaderavek, 2004). In this section, factors that affect early literacy skills are explained under the headings of reasons related to socio-economic status, reasons related to family, home environment and inadequate environment and reasons related to developmental disabilities.

Reasons Related to Socio-Economic Status

Early literacy skills of children can be affected by the socio-economic status of their families for many reasons. Children of families with low socio-economic status often have less materials and less experience to promote their early literacy skills. In addition, even if there are activities at home that can support the early literacy skills of these children, doing them inadequately and inefficiently may cause the child to fail to develop the potential and increase the risk of reading difficulties. Furthermore, children with low socioeconomic status increase the risk of cognitive impairment due to factors such as preterm birth and low birth weight, and the low rate of attendance to preschool institutions, thus affecting the development of early literacy skills of children. In this case, socio-economic reasons affect factors such as the environment of the child and the opportunities that the family can offer to the child, so it is an important factor (Buckingham et al., 2014). A study concluded that the socio-economic well-being of the family's neighborhood affects the early literacy skills of preschool children because economic disabilities limit the opportunities that children can acquire to develop early literacy skills at home (Froliand et al., 2013).

A study on families with low socio-economic status related to the development of early literacy skills was also conducted in Chile. According to the results of this study, the practices of mothers in families with lower socio-economic status to support children's early literacy skills are insufficient therefore children are at the risk group in the development of early literacy skills. In this case, it is necessary to prepare support programs targeting parents and preschool teachers of children in the risk group and to increase the frequency of children's exposure to written materials, improve letter knowledge and daily speaking skills. In addition, increasing the number of libraries in low socioeconomic neighborhoods and increasing the number of books suitable for children's development are among the measures that can be taken for the risk group (Mendive et al., 2017).

In a study conducted with Arabic-speaking kindergarten children and their families in Israel, it was concluded that the socio-economic status and home environment of the family affected children's early literacy skills. Educational games in the home environment, games for reading and writing, materials for playing mathematical games, books for children, educational CDs, DVDs, computers and tools that can support children's early literacy skills at home, and their relationship with children's early literacy skills analyzed. Information such as educational status, financial status and occupations of the parents were examined as criteria determining the socio-economic status of the family. Accordingly, children whose parents are more educated and have more skilled professions have shown higher literacy skills (Aram et al., 2013).

The socio-economic status of a family, obviously, affects the family's home environment and environment positively or negatively. Therefore, the second reason affecting early literacy skills is explained under the heading of reasons related to family, home environment and inadequate environment.

Reasons Related to Family, Home Environment and Inadequate Environment

Family, home environment and inadequate environment play an important role among the factors that affect early literacy skills.

The socio-economic well-being of the neighborhood where families live, the facilities at home (parents and children reading books together and the number of books at home) and early literacy skills of preschool children were analyzed in a study. According to this study, it is concluded that the socio-economic welfare of the family's neighborhood indirectly affects the early literacy skills because socioeconomic welfare affects the opportunities at home. In addition, opportunities at home affect the early literacy skills of preschool children. However, it was stated that the education level of the parents was related to the number of books in the house and the reading of the books with the parents and children, the parents with higher education had more books in their homes and they more frequently read with their children (Froiland et al., 2013). In a study conducted in the Netherlands, the literacy environment in the homes of 22 children (approximately four and a half years old) whose native language is German was examined. Accordingly, it has shown that the environment that supports literacy at home, such as the frequency of reading to children and the frequency of reading by parents, plays an effective role in predicting children's early literacy skills. In addition, the frequency of children's exposure to the book was estimated by examining whether or not children would recognize the book from the covers of the picture books. It was concluded that children who are more familiar with the book have a better environment to improve their literacy skills. In short, this study showed that families' reading with their children and the frequency of children's exposure to books may be a factor affecting early literacy skills (Davidse et al., 2011).

Parents' educational background, especially the mother's educational background and language skills, may be linked to the child's early literacy skills. If mother's vocabulary is large, her child's early literacy skills are positively affected and can improve vocabulary. According to a study in Turkey conducted with around three-year-old children, maternal vocabulary level is related to children's' vocabulary level (Baydar et al., 2014).

Children's early literacy skills and learning environment are interrelated. In a longitudinal study (Rodriguez & Tamis-Lemonda, 2011), the relationship between children's early literacy skills and learning environment at 15, 25, 37 and 63 months of age was examined. Accordingly, in all age groups, it was concluded that the learning environment and children's early literacy skills were effective.

Even a quiet and organized environment at home is important for the development of children's early literacy skills. According to a study by Johnson et al. (2009) with children approximately six years old and their mothers, regular and quiet environment at home and the mother's ability to read are associated with early literacy skills of children. It was concluded that the ability of mothers to read is related to the areas of expressive language and phonological awareness in early literacy skills of children. Moreover, variables related to the home environment, such as the numbers of books owned by children and the frequency of fun when children spend time with books on their own were significantly correlated with early literacy scores of them.

It was aimed that with the Project EASE (Early Access to Success in Education) in Minnesota 248 kindergarten children and their families who attend, to increase the frequency and quality of language interactions through book-centered activities and to inform parents about their children's literacy skills. For this reason, parents have been informed about a number of topics that will improve the literacy skills of children, such as strengthening the vocabulary, recognizing letters, and expanding the understanding of the story. Pre-test and post-test were applied to the participating children to test the findings of the EASE Project, which is a one-year intervention program. According to the findings, the intervention program has been very effective in improving children's language skills, and in this context, book-centered activities at home have had a great impact. In addition, the intervention program had a strong effect on children with poor performance on the pre-test. This study shows that improving the home environment and informing parents play an important role in improving children's early literacy skills (Jordan et al., 2000).

It is another important factor that families consciously carry out activities to develop children's early literacy skills. For example, one of the most usual activities for children to develop their early literacy skills at home is reading together. How parents read books to children is an important factor and parents need to be made aware of this (Erdoğan et al., 2016; Ergül et al., 2015).

Reasons Related to Developmental Disabilities

One of the factors affecting early literacy skills is developmental disabilities. Especially preschool children with speech and language disabilities may have problems in gaining early literacy skills and may have difficulty in reading and writing when they start school. Difficulties in learning the name of the letters, persistence of baby talk, inability to recognize the letters of the child's own name, indifference to activities such as reading books or rhymes together may indicate that the child may have difficulty in language and speech and also in development of early literacy skills (Roth et al., 2006).

According to a study by professionals working in the field of language and speech difficulties, children with special educational backgrounds and those at risk; it was concluded that determination of early literacy skills of children in risk group is important for learning strengths and weaknesses of children such as phonological awareness and letter knowledge etc. (Justice & Kaderavek, 2002). In another study, early literacy skills of children (at the age of between four and seven) who have speech and language difficulties, who have only speech disabilities and who have no disabilities as a control group were observed for two years. According to the results, it was seen that children with language and speech difficulties had difficulties in the development of phonological awareness, reading and spelling skills. In addition, only phoneme awareness was statistically significant difference in children with language and speech difficulties compared to other groups (Nathan et al., 2004).

THE DEVELOPMENT PROCESS OF EARLY LITERACY

Being literate is a process and there is a process of development of the pre-skills, early literacy that children acquire up to the primary school that supports literacy. In addition, early literacy development is closely related to children's language, physical, emotional and cognitive development (Strickland & Schickedanz, 2004). In this section, the development process of early literacy skills is categorized as development of early literacy from birth to three years and development of early literacy from three to six years.

Development of Early Literacy from Birth to Three Years

Even in the period from birth to three years old babies have early literacy skills. During the baby's six-week-to-one-year-old process, babies begin to distinguish between the faces and sounds of those who care for them. When parents read a book to a baby in this period, the baby can establish positive relationships with the parent through the book (United States Department of Education Office of Communications and Outreach, 2005). Moreover, babies can hold his head up in this period. Babies can sit on lap without support, grasp books, take them to his mouth and throw books. In cognitive and communication skills; they smile, hum and babble. They love and want the parent's voice. Babies like the picture of their faces and make sounds like syllables. They react to their name and stroke the pictures to show their interest (Reach out and Read, 2013).

In this period, babies can imitate the tone of their voices while adults are talking. Babies learn to respond to facial expressions and gestures. They learn to establish the relationship between the word they hear and the meaning of the word. They make humming and babbling sounds and play sound game with rhyming and meaningless words with their parents or caregivers. They can play games like 'peek-a-boo' together. They can grasp thick cover books and alphabet blocks with their hands. By seeing the cover of a book, they can know which book it is. They can imitate to read books. They can know how to hold a book. Reading together with adults can become the routine of their lives. They can say the names of some objects in the book. They can talk about the characters in the book. They can look at the pictures in the books and realize that the symbols mean things in real life. They can listen to stories. They can want to read or write with adults. They can pay attention to certain texts, such as the first letter of their names. They can make doodles with the aim of trying to write or draw pictures. They can write doodles and letters in form similar to letters (Snow et al., 1998).

When babies have an environment that is exposed to all kinds of materials related to literacy, they develop print awareness around the age of two or three. In order to increase the awareness of writing, it is necessary to draw attention to how the texts are read (such as reading from left to right, top to bottom) when reading books with babies. In addition, the awareness of infants to recognize their own names or to draw attention to the structure of letters are activities that support print awareness (Ranweiler, 2004).

Development of Early Literacy from Three to Six Years

Between the ages of three and four, babies enjoy listening to story books and talking about story books. They begin to understand that writing has a meaning. They try to read and write. They begin to recognize familiar signs and labels. They start playing rhyming sound games. They can recognize some letters and match the sounds of the letters. They can use familiar letters to represent written language such as their names or 'I love you' (Reach out and Read, 2013).

At the age of five, they can imitate reading as if they really knew how to read and write. They start to get eager to read. They can tell simple stories over and over. They begin to explain in descriptive language and ask questions. They can recognize the letters and know which sound the letters represent. They can be familiar to rhyme. They can recognize that the texts are written from left to right and from top to bottom. They can begin to match the spoken language and the written language. They can start to write letters and familiar words. They begin writing stories with letters and words in some parts (National Association for The Education of Young Children, 1996).

At the age of six, many first-year students who begin primary school can read and retell familiar stories. They can do various activities to help them read, such as re-reading, predicting what will happen in the story, asking questions, or making predictions based on visual clues. They can read some texts easily. They can guess new words by matching sounds and letters, understanding part of the word or understanding from what is told in the story. There may be an increase in the words which they learn on their own. While trying to spell words, they can try to pronounce the dominant sounds in the words carefully. They can start writing about topics that are important for them at this age. They can try to write some punctuation and spelling rules appropriately (U.S. Department of Education Office of Communications and Outreach, 2005).

EDUCATIONAL APPROACHES RELATED TO EARLY LITERACY

Professionals such as psychologists and educators working on the concept of early literacy have adopted a variety of approaches from the past to the present. These approaches are explained under the titles of reading readiness, emergent literacy, whole language approach, phonics and balanced literacy approach.

Reading Readiness

In a 1931 study by Morphett and Washburne, they investigated the age of maturation sufficient for literacy, and concluded that if a child's mental age is 6 years and 6 months, the child is a better literacy process than children of other ages. In short, the idea that child should not start reading and writing until they are 6 years old and 6 months old has become widespread (Sampson & Briggs, 1981).

According to a psychologist named Arnold Gesell, the maturation of children is an important factor for learning to read and write, and formal education of literacy education should begin in the first grade of primary school. Gesell et al. (1940) listed the characteristics that children who develop at the same level as their peers between the age of six and six and a half to learn to read and write. These characteristics were normal vision and hearing, mental maturity between six and six and a half years, good motor coordination (such as the ability to use the hand to draw), the ability to use and understand the language at a sufficient level, correct pronunciation, mature behavior appropriate to the age, to be able to follow a story of average length, to have enough attention to carry out a task assigned to him, to have the skills to adapt to class. He was recommended that children who fail to fulfill these skills should not be enrolled in primary school except in exceptional circumstances. Gesell called this state of being mature as 'reading readiness'.

Emergent Literacy

Marie Clay was the first person to use emerging literacy in 1966 (Morrow, 2005). Children learn a lot about literacy before they learn to read and write with formal education. In fact, this is very usual, because every day the child is exposed to story books, calendars, lists, signs, and finally tries to figure out what the writing means. Children's experience of actively developing literacy skills is called emergent literacy (Berk, 2008).

According to the definition in another study, emergent literacy is all kinds of preliminary skills, knowledge and attitudes about literacy in the process before the child learns to read and write (Whitehurst & Lonigan, 1998).

Emergent literacy is the view that verbal language, listening, reading and writing skills are interactive and that each skill affects each other in development. In this child-centered approach, more attention is given to the child's problem-solving skills than direct instructions given to the child. Children are also exposed to the book as early as possible (Morrow, 2005).

Whole Language Approach

Children who have the same development as their peers can easily learn their mother tongue without formal education or going to school, but may have difficulty learning the written language at school. the whole language approach based on this dilemma is not only an approach, it is a philosophy of education that is based on the principle that natural language is the basis for learning how to read and write (Smith, 1994).

The researcher Bergeron (1990), who defined the whole language, compiled 64 separate studies. According to Bergeron, reading and writing are only possible literally, through functional and collaborative experiences in the learning process in order to increase the interest and motivation of the child. In this study, Bergeron emphasized the support of early literacy and literacy through activities and materials such as independent reading, predictable reading, reading corner and library in a classroom where education is planned with this approach. In addition, it is important that teachers providing education based on this approach, instead of supporting children's individual skills one by one, should let them to brainstorm children in their natural environment, let children work with their own interests and collaborate with their peers.

Phonics

Instruction of phonics has six approaches, but there is no clear distinction between them: synthetic phonics, analytic phonics, analogy-based phonics, and phonics through spelling, onset-rime phonics instruction and embedded phonics (Armbruster et al., 2006).

Synthetic phonics approach is taught by first specifying which sound the particular letters represent and combining these sounds to form different words. Gradually, much more vowel, consonant and letter combinations are taught. In the analytic phonics approach, it is essential to look at the whole word and break it into phoneme (Armbruster et al., 2006; Ewing & Maher, 2014). Furthermore, analogy-based phonics is to learn how to read unknown words by using similar parts of known words. According to phonics through spelling, children learn the word by separating it into phonemes and form words by writing letters for phonemes. In onset-rime instruction, children learn sounds of the letter or letters in a one syllable word by learning onset (consonants preceding the vowel) and rime (sounds of the remaining part). Finally, embedded reading is the learning of phonics during reading and writing based on the relationship between reading and phonic (Armbruster et al., 2006).

Balanced Literacy Approach

According to the International Reading Association's (1999) study, adhering to one approach is insufficient to teach children to read. Therefore, he argues that teachers should use appropriate approach which children need to learn. Therefore, the balanced approach is the balance between whole language approach and phonics. According to this approach, the approaches are not chosen randomly, the approaches that every child's needs are different. For example, learning with a phonics approach is not appropriate for a visually better learning child. Instead, it would be more appropriate to teach a child who is better in auditory learning, with a phonics approach.

In one study, well-organized and systematic direct instruction was compared to the whole language approach that increased motivation and creativity. Accordingly, it was concluded that these two approaches have strengths, but it is not effective to adopt only one approach. Therefore, it is concluded that the eclectic approach, which is a balanced literacy approach, is a more effective instruction (as cited in Freeman, 2007, p. 14).

PROMOTION OF EARLY LITERACY SKILLS

In this section, suggestions for activities to support children's early literacy skills are given under the headings of recommendations to families to promote children's early literacy skills, recommendations to teachers in the educational environment, recommendations for integration of technology into the school curriculum and recommendations with book reading activities.

Recommendations to Families to Promote Children's Early Literacy Skills

First of all, parents should choose books and materials suitable for the child's development and age (Reach out and Read, 2013). Because this process starts from birth (Barclay, 1995) and studies show that the conscious activities of parents with their children affect the early literacy development of their children (Dodici et al., 2003; Phillips et al., 2008).

Second of all, parents' self-reading and reading with children play an important role in promoting children's early literacy skills. Parents in particular should allow their children to see their parents reading. In addition, when reading books to children, it is important to draw attention to sounds and familiar words, to explain unknown words, and to ask appropriate questions for children's development, about the story (Reach out and Read, 2013). Moreover, parents can also play games together to learn new words and draw attention to the sounds and syllables of words. For example, activities such as nursery rhymes, asking the sounds of familiar words, or asking the meaning of new words support children's early literacy skills. Labeling the names of some objects and writing simple few-word sentences and giving picture clues to what they are writing are recommendations to promote children's' early literacy skills (Whitehurst, 2003).

Recommendations to Teachers in the Educational Environment

The development of early literacy skills is usually dependent on pre-school education. Perspectives and views of teachers working in preschool institutions and their activities also affect children's early literacy skills. It is important for teachers to raise awareness by pointing out words in a material read aloud to support children's literacy development. Various printed materials (e.g. newspapers, magazines, books, labels, signs) and their purpose should be explained in a way that the child can understand. Children can learn the letters kinesthetically such as drawing letters with play dough, singing songs, playing games. Every child in class should be exposed to their own name and familiar words so they can see and notice these words. This increases the child's awareness of these words. Illustrations and pictures give clues as to what the child is writing in the written material, so a lot of illustrated written material should be used in the classroom (Hullinger Sirken & Staley, 2016).

Unlike children who develop at the same level as their peers, some children may need special education. For example, it is important to diagnose children with language and speech difficulties early, to intervene early and to support early literacy skills. In this case, the language and speech therapist should support the early literacy skills of the child's known or suspected language and speech related problems. Teachers in pre-school should cooperate with language and speech therapists, families, and caregivers and ensure that they support early literacy skills not only during therapy, but at home, at school, or wherever possible (Roth et al., 2006).

Recommendations for Integration of Technology into the School Curriculum

With the development of technology, there has been a need to integrate the school curriculum with technological developments, and therefore studies are being carried out on the effective use of children's early literacy skills. In a study conducted in the Netherlands, teachers from five different educational institutions were asked for opinions about the relationship between early literacy and the use of technology. According to the professionals involved in the study, they agreed that technology is associated with early literacy skills and should be given importance. In addition, educators agreed that electronic books have benefits for children's early literacy skills (Voogt & Mckenney, 2016).

At the present time, there are applications that can be used as a tool for the development of early literacy skills in tablet computers and smart phones. It has been researched how such applications can be integrated with the curriculum while teaching a subject related to early literacy. According to Northrop and Killeen (2013), in order to integrate with the curriculum by using such practices, the subject should be explained according to the normal curriculum. Secondly, teachers should tell about the application and how to use it. Thirdly, children should be guided on how to use the application on the tablet and in the selection of the target subject taught. Finally, children should be exposed to tablets and applications while learning centers and leisure activities are carried out.

Recommendations with Book Reading Activities

This section deals with using audio books, dialogic reading, and public libraries.

The use of singing audio books supports children's early literacy skills and Barclay (2010) has provided some strategies to promote early literacy in audio books. Accordingly, before singing the song, the words and the subject mentioned in the song should be explained to the children. Then, teachers should sing along with the children and ensure that children make a connection between writing and song. Moreover, it should be ensured that children develop phonological awareness and writing skills.

Reading books with children improves children's early literacy skills. However, it is important how the child will be more productive when reading a book. In this context, dialogic reading, which made the child an active listener, has positive effects on children's language, early literacy skills and attitudes towards reading (Whitehurst et al., 1988). According to dialogic reading; first, the book should be selected appropriate for the child's development, the most productive and appropriate place and time should be selected. The front cover of the book should start reading with the children facing and the pages of the book should be facing the children during the reading of the book. The title of the book should be read aloud and followed by finger pointing. This may allow the child to comprehend the direction of the text and the space between words. In addition, asking open-ended questions about the picture of the book makes it easier for the child to express himself, use the language and pay attention to the story. Moreover, reading the book author, illustrator and publisher before reading the story also contributes to the child's awareness of the books. (Ergül et al., 2015; Ergül et al., 2016; Zevenbergen & Whitehurst, 2003).

Specially designed public libraries are also an important factor in supporting children's early literacy skills. According to the International Federation of Library Associations and Institutions (IFLA), public libraries should support children in the process of learning to read and encourage children to use all kinds of written material (Gill, 2004). For this reason, the public libraries that support children through a wide range of materials and activities contribute to both encouraging children to read, as well as accessing information, learning to explore and developing their imagination. It is possible to see examples of children's sections of public libraries in many countries of the world. There are also activities where these libraries are organized to improve early literacy skills. For example, many Canadian libraries websites have activities to develop early literacy skills both for parents and children. (Campbell-Hicks, 2016; Prendergast, 2013).

REFERENCES

American Library Association. (2009). *Early Literacy*. Retrieved from http://www.ala.org/united/products_services/booksforbabies/earlyliteracy

Aram, D., Korat, O., Saiegh-Haddad, E., Arafat, S. H., Khoury, R., & Elhija, J. A. (2013). Early literacy among Arabic-speaking kindergartners: The role of socioeconomic status, home literacy environment and maternal mediation of writing. *Cognitive Development*, *28*(3), 193–208. doi:10.1016/j.cogdev.2012.10.003

Armbruster, B. B., Lehr, F., Osborn,J. (2006). *Put reading first: The research building blocks for teaching children to read: Kindergarten through grade 3*. Retrieved from https://www.readingrockets.org/guides/put-reading-first-research-building-blocks-teaching-children-read

Barclay, K. H. (1995). Literacy Begins at Birth: What Caregivers Can Learn from Parents of Children Who Read Early. *Young Children*, *50*(4), 24–28. doi:10.12691/education-6-6-14

Barclay, K. H. (2010). Using Song Picture Books to Support Early Literacy Development. *Childhood Education*, *86*(3), 138–145. doi:10.1080/00094056.2010.10523134

Baydar, N., Küntay, A. C., Yagmurlu, B., Aydemir, N., Cankaya, D., Göksen, F., & Cemalcilar, Z. (2014). "It takes a village" to support the vocabulary development of children with multiple risk factors. *Developmental Psychology*, *50*(4), 1014–1025. doi:10.1037/a0034785 PMID:24188041

Bergeron, B. S. (1990). What does the term whole language mean? Constructing a definition from the literature. *Journal of Reading Behavior*, *22*(4), 301–329. Retrieved from doi:10.1080/10862969009547716

Berk, L. E. (2008). *Child development* (8th ed.). Illinois: Pearson.

Biemiller, A. (2006). *Vocabulary Development and Instruction: A Prerequisite for School Learning. In The Handbook of Early Literacy Research*. New York, NY: Guilford Press.

Buckingham, J., Wheldall, K., & Beaman-Wheldall, R. (2013). Why poor children are more likely to become poor readers: The school years. *Australian Journal of Education*, *57*(3), 190–213. doi:10.1177/0004944113495500

Burke, M. D., Hagan-Burke, S., Kwok, O., & Parker, R. (2009). Predictive validity of early literacy indicators from the middle of kindergarten to second grade. *The Journal of Special Education*, *42*(4), 209–226. doi:10.1177/0022466907313347

Campbell-Hicks, R. (2016). Early literacy programmes in public libraries: Best practice. *The Australian Library Journal*, *65*(2), 121–129. doi:10.1080/00049670.2016.1121770

Canadian Council on Learning. (2006). How parents foster early literacy. Lessons in Learning. Retrieved from http://www.child-encyclopedia.com/sites/default/files/docs/suggestions/how-parents-foster-early-literacy.pdf

Catts, H. W., Fey, M. E., Tomblin, J. B., & Zhang, X. (2002). A longitudinal investigation of reading outcomes in children with language impairments. *Journal of Speech, Language, and Hearing Research: JSLHR*, *45*(6), 1142–115. doi:10.1044/1092-4388(2002/093) PMID:12546484

Cedar Mill Community Libraries. (2017). *Early Literacy*. Retrieved from https://library.cedarmill.org/kids/early-literacy/

Claessens, A., Duncan, G., & Engel, M. (2009). Kindergarten skills and fifth-grade achievement: Evidence from the ECLS-K. *Economics of Education Review*, *28*(4), 415–427. doi:10.1016/j.econedurev.2008.09.003

Davidse, N. J., de Jong, M. T., Bus, A. G., Huijbregts, S. C., & Swaab, H. (2011). Cognitive and environmental predictors of early literacy skills. *Reading and Writing*, *24*(4), 395–412. doi:10.100711145-010-9233-3 PMID:21475735

Dodici, B. J., Draper, D. C., & Peterson, C. A. (2003). Early parent—Child interactions and early literacy development. *Topics in Early Childhood Special Education*, *23*(3), 124–136. doi:10.1177/027112140 30230030301

Duncan, G. J., Dowsett, C. J., Claessens, A., Magnuson, K., Huston, A. C., Klebanov, P., ... Duckworth, K. (2007). School readiness and later achievement. *Developmental Psychology*, *43*(6), 1428–1446. doi:10.1037/0012-1649.43.6.1428 PMID:18020822

Erdoğan, N. I., Atan, A., Asar, H., Mumcular, F., Yüce, A., Kıraç, M., & Kilimlioğlu, Ç. (2016). Ebeveyn ve öğretmenlerin birlikte hikâye okuma etkinliklerinin incelenmesi. *İlköğretim Online*, *15*(1). doi:10.17051/io.2016.05211

Ergül, C., Akoğlu, G., Sarica, A., Tufan, M., & Karaman, G. (2015). Ana sınıflarında gerçekleştirilen birlikte kitap okuma etkinliklerinin" etkileşimli kitap okuma" bağlamında incelenmesi. *Mersin Üniversitesi Eğitim Fakültesi Dergisi*, *11*(3). doi:10.17860/efd.88429

Ergül, C., Karaman, G., Akoğlu, G., Tufan, M., Sarıca, A. D., & Kudret, Z. B. (2014). Okul öncesi öğretmenlerinin "Erken Okuryazarlık" kavramına ilişkin bilgi düzeyleri ve sınıf uygulamaları. *İlköğretim Online, 13*(4). doi:10.17051/io.2014.71858

Ergül, C., Sarıca, A. D., & Akoğlu, G. (2016). Etkileşimli kitap okuma: Dil ve erken okuryazarlık becerilerinin geliştirilmesinde etkili bir yöntem. *Ankara Üniversitesi Eğitim Bilimleri Fakültesi Özel Eğitim Dergisi*, *17*(2), 193–206. doi:10.21565/ozelegitimdergisi.246307

Ewing, R., & Maher, M. (2014). *Phonics: its place in the literacy story*. ALEA. Retrieved from https://www.alea.edu.au/documents/item/943

Freeman, J. L. (2007). Using a balanced literacy approach in elementary schools. (Master Dissertation). Available from UNI Scholar Works. (UMI No. 690)

Froiland, J. M., Powell, D. R., Diamond, K. E., & Son, S. H. C. (2013). Neighborhood socioeconomic well-being, home literacy, and early literacy skills of at-risk preschoolers. *Psychology in the Schools, 50*(8), 755–769. doi:10.1002/pits.21711

Gesell, A., Halverson, H. M., Thomson, H., Ilg, F. L., Castner, B. M., Ames, L. B., & Amatruda, C. S. (1940). *The first five years of life, the preschool years*. Oxford, England: Harpers.

Gill, P. (2004). *Halk Kütüphanesi Hizmeti: Gelişim için IFLA/UNESCO İlkeleri* (B. Yılmaz, Trans.). İstanbul, Turkey: Pandora. (Original work published 2001)

Hernandez, D. J. (2011). *Double Jeopardy: How Third-Grade Reading Skills and Poverty Influence High School Graduation*. Retrieved from https://eric.ed.gov/?id=ED518818

Hullinger-Sirken, H., & Staley, L. (2016). Preschool Through Grade 3: Understanding Writing Development: Catie's Continuum. *YC Young Children, 71*(5), 74–78. Retrieved from http://www.jstor.org/stable/ycyoungchildren.71.5.74

International Reading Association. (1999). *Using Multiple Methods of Begining Reading Instruction*. Retrieved from https://www.literacyworldwide.org/docs/default-source/where-we-stand/multiple-methods-position-statement.pdf?sfvrsn=d04ea18e_6

Johnson, A. D., Martin, A., Brooks-Gunn, J., & Petrill, S. A. (2008). Order in the house! Associations among household chaos, the home literacy environment, maternal reading ability, and children's early reading. *Merrill-Palmer Quarterly Journal of Developmental Psychology, 54*(4), 445–472. doi:10.1353/mpq.0.0009 PMID:19526070

Jordan, G. E., Snow, C. E., & Porche, M. V. (2000). Project EASE: The effect of a family literacy project on kindergarten students' early literacy skills. *Reading Research Quarterly, 35*(4), 524–546. doi:10.1598/RRQ.35.4.5

Juel, C. (1988). Learning to read and write: A longitudinal study of 54 children from first through fourth grades. *Journal of Educational Psychology, 80*(4), 437–447. doi:10.1037/0022-0663.80.4.437

Justice, L. M., & Kaderavek, J. N. (2002). Using shared storybook reading to promote emergent literacy. *Teaching Exceptional Children, 34*(4), 8–13. Retrieved from doi:10.1177/004005990203400401

Justice, L. M., & Kaderavek, J. N. (2004). Embedded-explicit emergent literacy intervention I. *Language, Speech, and Hearing Services in Schools, 35*(3), 201–211. doi:10.1044/0161-1461(2004/020) PMID:15248791

Karaman, G. (2014). Erken Okuryazarlık Becerilerinin Tarihsel Gelişimi, Kuramsal Temelleri ve Kapsami. In F. Temel (Ed.), *Her Yönüyle Okul Öncesi Eğitim 4 –Dil ve Erken Okuryazarlık* (pp. 12–24). Ankara, Turkey: Hedef Press.

Lloyd, D. N. (1978). Prediction of School Failure From Thirdgrade Data. Educational and Psychological Measurement. *Mental Health Study Center National Institute of Mental Health, 38*(4), 1193–1200. doi:10.1177/001316447803800442

McAfee, O., Leong, D., & Bodrova, E. (2004). *Basics of Assessment: A Primer for Early Childhood Professionals*. Washington, D.C.: National Association for the Education of Young Children.

Mendive, S., Lissi, M. R., Bakeman, R., & Reyes, A. (2017). Beyond mother education: Maternal practices as predictors of early literacy development in Chilean children from low-SES households. *Early Education and Development*, *28*(2), 167–181. doi:10.1080/10409289.2016.1197014

Moore, D. W., Bean, T. W., Birdyshaw, D., & Rycik, J. A. (1999). Adolescent literacy: A position statement. *Journal of Adolescent & Adult Literacy*, *43*(1), 97–112. Retrieved from https://eric.ed.gov/?id=ED437640

Morrow, L. M. (2005). *Literacy Development in the Early Years Helping Children Read and Write* (5th ed.). Boston: Allyn & Bacon: Pearson.

Nathan, L., Stackhouse, J., Goulandris, N., & Snowling, M. J. (2004). The development of early literacy skills among children with speech difficulties. *Journal of Speech, Language, and Hearing Research: JSLHR*, *47*(2), 377–391. doi:10.1044/1092-4388(2004/031) PMID:15157138

National Association for the Education of Young Children. (1996). *Technology and Young Children Ages 3 Through 8*. Washington, DC: NAEYC.

Northrop, L., & Killeen, E. (2013). A Framework for using iPads to build early literacy skills. *The Reading Teacher*, *66*(7), 531–537. doi:10.1002/TRTR.1155

Phillips, L. M., Norris, S. P., & Anderson, J. (2008). Unlocking the door: Is parents' reading to children the key to early literacy development? *Canadian Psychology*, *49*(2), 82–88. doi:10.1037/0708-5591.49.2.82

Plaza, M., & Cohen, H. (2006). The Contribution of Phonological Awareness and Visual Attention in Early Reading and Spelling. *Dyslexia*, *13*(1), 67–76. doi:10.1002/dys.330 PMID:17330736

Prendergast, T. (2013). Growing readers: A critical analysis of early literacy content for parents on Canadian public library websites. *Journal of Library Administration*, *53*(4), 234–254. doi:10.1080/01930826.2013.865389

Ranweiler, L. (2004). *Preschool Readers and Writers: Early Literacy Strategies for Teachers* (N. Brickman & H. Barton, Eds.). Ypsilanti, Michigan: High/Scope Press.

Reach Out and Read. (2013). *Milestones of Early Literacy*. Retrieved from http://www.reachoutandread.org/FileRepository/RORmilestones_English.pdf

Rodriguez, E. T., & Tamis-LeMonda, C. S. (2011). Trajectories of the home learning environment across the first 5 years: Associations with children's vocabulary and literacy skills at prekindergarten. *Child Development*, *82*(4), 1058–1075. doi:10.1111/j.1467-8624.2011.01614.x PMID:21679179

Roth, F. P., Paul, D. R., & Pierotti, A. M. (2006). *Let's Talk For People With Special Communication Needs. Emergent Literacy: Early Reading and Writing Development*. ASHA. Retrieved from https://www.asha.org/public/speech/emergent-literacy.htm

Sampson, M. R., & Briggs, L. (1981). What Does Research Say About Beginning Reading? *Reading Horizons: A Journal of Literacy and Language Arts*, *21*(2). Retrieved from https://scholarworks.wmich.edu/reading_horizons/vol21/iss2/6

Shrier, C. (2013). *ABC's of Early Literacy: The importance of developing early literacy skills*. Retrieved from https://www.canr.msu.edu/news/abcs_of_early_literacy_the_importance_of_developing_early_literacy_skills

Smith, C. B. (1994). *Whole language: The debate* (W. Lewis, Ed.). Bloomington, Indiana: EDINFO Press and ERIC Clearinghouse on Reading, English, and Communication.

Smith, D. E., & Chapel, J. R. (1970). Reading readiness. *Reading Horizons: A Journal of Literacy and Language Arts, 10*(2), 3. Retrieved from https://scholarworks.wmich.edu/reading_horizons/vol10/iss2/3

Snow, C. E., Burns, M. S., & Griffin, P. (Eds.). (1998). *Preventing Reading Difficulties in Young Children*. Washington, DC: National Academies Press.

Spira, E. G., Bracken, S. S., & Fischel, J. E. (2005). Predicting improvement after first-grade reading difficulties: The effects of early language, emergent literacy, and behavior skills. *Developmental Psychology, 41*(1), 225–234. doi:10.1037/0012-1649.41.1.225 PMID:15656751

Strickland, D. S., & Schickedanz, J. A. (2004). *Learning about Print in Preschool: Working with Letters, Words, and Beginning Links with Phonemic Awareness*. Newark, DE: International Reading Association.

Teale, W. H., & Sulzby, E. (1986). *Emergent Literacy: Writing and Reading. Writing Research: Multidisciplinary Inquiries into the Nature of Writing Series*. Norwood, NJ: Ablex Publishing Corporation.

United Nations Educational Scientific and Cultural Organization. (1997). *50 Years for Education*. Retrieved from http://unesdoc.unesco.org/images/0011/001102/110264eb.pdf

United States Department of Education Office of Communications and Outreach. (2005). *Helping Your Child Become a Reader*. Washington, DC: U.S. Department of Education.

Utchell, L. A., Schmitt, A. J., Mccallum, E., Mcgoey, K. E., & Piselli, K. (2016). Ability of Early Literacy Measures to Predict Future State Assessment Performance. *Journal of Psychoeducational Assessment, 34*(6), 511–523. doi:10.1177/0734282915621221

Voogt, J., & Mckenney, S. (2016). TPACK in teacher education: Are we preparing teachers to use technology for early literacy? *Technology, Pedagogy and Education, 26*(1), 69–83. doi:10.1080/1475939X.2016.1174730

Whitehurst, G. J. (2003). *Read Together, Talk Together: A Dialogic Reading Program for Young Children*. New York, NY: Pearson Early Learning Group.

Whitehurst, G. J., Falco, F. L., Lonigan, C. J., Fischel, J. E., Debaryshe, B. D., Valdez-Menchaca, M. C., & Caulfield, M. (1988). Accelerating language development through picture-book reading. *Developmental Psychology, 24*(4), 552–558. doi:10.1037/0012-1649.24.4.552

Whitehurst, G. J., & Lonigan, C. J. (1998). Child development and emergent literacy. *Child Development, 69*(3), 848–872. doi:10.1111/j.1467-8624.1998.tb06247.x PMID:9680688

Zevenbergen, A. A., & Whitehurst, G. J. (2003). Dialogic reading: A shared picture book reading intervention for preschoolers. In A. Van Kleeck, S.A. Stahl, & E.B. Bauer (Eds.), *On reading books to children: Parents and teachers* (pp. 177-200). Mahwah, NJ: Lawrence Erlbaum Associates. Retrieved from http://citeseerx.ist.psu.edu/viewdoc/download?doi=10.1.1.31.5935&rep=rep1&type=pdf

ADDITIONAL READING

Crosby, S. A., Rasinski, T., Padak, N., & Yildirim, K. (2015). A 3-year study of a school-based parental involvement program in early literacy. *The Journal of Educational Research*, *108*(2), 165–172. doi:10.1080/00220671.2013.867472

Evans, M. A., Shaw, D., & Bell, M. (2000). Home literacy activities and their influence on early literacy skills. *Canadian Journal of Experimental Psychology/Revue Canadienne de Psychologie Expérimentale*, *54*(2), 65.

Gischlar, K. L., & Vesay, J. P. (2018). Literacy curricula and assessment: A survey of early childhood educators in two states. *Reading Improvement*, *55*(3), 106–117.

Kirk, S. M., & Kirk, E. P. (2016). Sixty minutes of physical activity per day included within preschool academic lessons improves early literacy. *The Journal of School Health*, *86*(3), 155–163. doi:10.1111/josh.12363 PMID:26830501

Ready, D. D., & Chu, E. M. (2015). Sociodemographic inequality in early literacy development: The role of teacher perceptual accuracy. *Early Education and Development*, *26*(7), 970–987. doi:10.1080/10409289.2015.1004516

Sonnenschein, S., & Munsterman, K. (2002). The influence of home-based reading interactions on 5-year-olds' reading motivations and early literacy development. *Early Childhood Research Quarterly*, *17*(3), 318–337. doi:10.1016/S0885-2006(02)00167-9

Steiner, L. M. (2014). A family literacy intervention to support parents in children's early literacy learning. *Reading Psychology*, *35*(8), 703–735. doi:10.1080/02702711.2013.801215

Tabors, P. O., & Snow, C. E. (2001). Young bilingual children and early literacy development. In Handbook of early literacy research (Vol. 1, pp. 159-178). Academic Press.

KEY TERMS AND DEFINITIONS

Arnold Gesell: An American psychologist who argues that children must reach a certain maturity in order to learn to read and write.

Dialogic Reading: It is the strategy of parents to read more effectively when reading with children.

Emergent Literacy: It has the same meaning as early literacy. It is all the experience that contributes to the literacy skills of children before formal education.

Marie Clay: New Zealand psychologist person who defined the term of emergent literacy.

Phonics: Educational approach focusing on the ability to read the sounds of letters.

Phonological Awareness: The ability to combine, separate and manipulate the sounds of words.

Print Awareness: The ability to recognize the functions of the text and the ability to distinguish how the text is organized, such as writing from left to right, top to bottom.

Reading Readiness: The ideal age at which a child is mature enough to learn to read and write.

Chapter 15
Early Intervention in Child Development

Nazan Kaytez
Sağlık Bilimleri Fakültesi, Çankırı Karatekin Üniversitesi, Turkey

ABSTRACT

Early intervention refers to programs developed for infants and young children with developmental delays or inadequacies or who are at risk group, and their families. The aim of the early intervention is providing services for children with disabilities, minimizing their developmental delay and disability levels and guiding parents for positive development of children. Early intervention includes extremely important studies for children in need of developmental support and their families. The effective early intervention treatment requires a gradual progress, teamwork with a multidisciplinary approach, and family should be considered as the most important element of the treatment and participate in the decision-making at every phase, alongside effective use of the Individualized Family Service Plan and updating the plan when it is necessary. It should be noted that early intervention is an ongoing process. In this process, the family and the child should be considered as a whole and supported in a holistic way.

INTRODUCTION

Early childhood period is the most important period in a child's life. In such an important period, when the development process is adversely affected, it causes children to experience developmental delay or developmental delay risk. Early childhood is very important for children with developmental delay. The earlier a child with developmental delay is diagnosed, the greater the likelihood of benefiting from early intervention programs to address the child's needs (Bruder, 2000). Therefore, determining the risk of delay in children as early as possible and initiating the necessary early intervention studies is of great importance in minimizing the problems that may occur in later periods (Akoğlu & Şipal, 2012).

Early intervention consists of programs developed for families of infants and young children who have developmental delays or inadequacies or are at risk group. In the most general definition, they are programs aiming to minimize or prevent the effects of cognitive physical and social-emotional limitations by supporting the development and learning of children who have developmental disabilities

DOI: 10.4018/978-1-7998-2952-2.ch015

or whose development is at risk from birth (Bayhan, 2012; Blackman, 2002). Early intervention and prevention are more focused on the long-term social and economic benefit of the child, the family, and therefore the community, rather than solving a problem. In this direction, children whose development is appropriately supported in the early stages of their lives are more likely to adapt easily to social life in the future and are less affected by negative environmental conditions. Therefore, the most important objective of prevention and early intervention studies is to change the balance between risk and protective factors (Aytekin& Bayhan, 2015).

Early intervention programs with family participation aim to reduce the effects of adverse situations for children and their families, as well as to teach healthy and effective parenting practices or gain skills necessary for those practices (Kılınç & Aral, 2015). Early intervention programs serve purposes such as providing parents with social support, providing information on topics like the importance of family support in the early years in the development of the baby, child development and discipline, providing education and job opportunities. For these purposes, early intervention programs must be developed systematically and conducted in a planned manner. In this process, it is very important that professionals from different disciplines and families act as a team and cooperate. (Karoly, Kilburn & Cannon, 2005). In this section, information regarding the importance of early intervention programs, their historical development, early intervention stages, early intervention models and the importance of teamwork in early intervention are presented.

Importance of Early Intervention

Early intervention is a practice aimed at supporting children's development in the best possible way with a comprehensive program including children and their families, including the family and the environment, for children between the ages of zero and eight who have developmental disabilities or are at developmental risk. According to another definition, early intervention means practices that are focused on long-term benefit to the child, family and society (Knitzer and Lefkowitz, 2006).

The aim of early intervention was to prevent the developmental disability, reduce the negative effects of developmental risks for the family and the child, and support the child and his/her family to maximize what the child can do (Bayhan, 2012). The objectives of early intervention are generally as follows.

- Providing social support to the family
- Informing the family about child development
- Developing parenting skills of families
- Providing parents with educational opportunities
- Supporting the development of infants and children with special needs and minimizing developmental delays
- Providing children with learning experiences according to their needs
- Trying to meet the needs and support the development of children of families especially with low socioeconomic status, living in rural areas or in institutional care and (Akoğlu & Şipal, 2012; Blackman, 2002).

For these purposes, early intervention services should be started as soon as possible. Early intervention services are very important especially in the first three years in terms of preventing and reducing the risks that may occur in the development of children. Early intervention services have very important benefits for both children and families (Knitzer & Lefkowitz, 2006). The benefits of early intervention for children and families are listed as follows.

- Early intervention programs positively support the development of children in a short period of time.
- They help to reduce developmental differences between children and their peers.
- They help children adapt more easily to society.
- They enable children to benefit from the same educational environments as their peers.
- They support children's educational process.
- They offer many learning opportunities for children and their families.
- They enable families to learn about their children's performances.
- Families learn how to support their children's development.
- Families understand what an important role they have in the development of their children.
- The quality of life of families with children whose development is different from their peers increases with early intervention programs (Aytekin & Bayhan, 2015; Blackman, 2002; Yıldız-Bıçakçı, Dursun & Bolatbaş, 2018).

Furthermore, it was determined that the implementation of early intervention programs resulted in lower costs for the development and education of the child. In particular, countries addressing the gains of early intervention programs emphasize costs and emphasize that it is important to transform the financial investment earned into different sources for children. Besides, it is stated that the fee spent on special education is reduced thanks to early intervention (Brookes, Berlin & Fuligni, 2000).

Early intervention programs are particularly important for prevention. Three levels of prevention are mentioned in early intervention programs. The purpose of primary level prevention is to raise the awareness of the families on issues such as healthy nutrition, pregnancy follow-up and thus to reduce the developmental disability that may occur in the child (Meraki & Yıldız - Bıçakçı, 2015). The purpose of secondary level prevention is to reduce the degree of disability diagnosed in early childhood and shorten its duration. The purpose of tertiary level prevention is to prevent and reduce the complications (physical and behavioral) related to the disability that will lead to the need for institutional support. This prevention approach in early intervention services clearly demonstrates the importance of early intervention (Yıldız-Bıçakçı, Dursun & Bolatbaş, 2018).

Early intervention services are generally planned and carried out to include health, development, education and social services. Some intervention programs include academic disciplines and bureaucratic areas. Early intervention services offered by combining different areas are multidimensional and provide more effective benefits. That impact again reveals the importance of early intervention (Brookes, Berlin & Fuligni, 2000).

History of Early Intervention

Below is information on the development of early intervention in the world and in Turkey.

Historical Development of Early Intervention in the World: The idea of early intervention emerged in the United States in 1954, when the supreme court ruled against racial discrimination on the grounds that it is a universal right of all children to receive appropriate education. Since the 1960s in many countries, especially in the United States developments and studies in the field of early intervention have begun to gain speed and importance (Bayhan, 2012).

After the 1960s, creative and effective programs were developed in the field of early intervention. In 1965, diagnostic, developmental-medical evaluation and treatment services came to the fore especially in the United States. In this direction *Social Security Act* was revised and with that law all children under the age of twenty were provided with these services (Meisels & Shonkoff, 2000). In the 1960s, early intervention programs were developed for low-income families and their children. The first early intervention program developed is *Head Start Program*. This program was developed to meet the social, emotional, health and nutritional needs of families and children with low socio-economic status (Diken et al. 2012).

In the 1970s, interdisciplinary training programs were developed for children with developmental disabilities and various funds were created to train staff to work with these children. With *Education for All Handicapped Children Act* that entered into force in the United States in 1975, the first official and legal studies on the field of early intervention have started. With this law, the right to education of all children with developmental disabilities who have reached the primary school age was guaranteed by the state and early intervention services were provided to those children. However, it was seen as a major deficiency in the law that this service was not made compulsory for children between the ages of three and five (Odom & Wolary, 2003).

In 1986, significant changes on the *Education for All Handicapped Children Act* have been brought to the agenda. The name of the law was changed to *Individuals with Disabilities Education Act* in 1990. Together with these regulations, early intervention services have been secured by the state in many parts of the United States, and significant budgets have been allocated. This law has led to increased awareness studies on children with developmental disabilities or at risk (Pletcher & Younggren, 2011). In 1997 *Individuals with Disabilities Education Act* was updated again. In the new regulation, the principles on ensuring the full participation of the families in the decisions and education of their children and the role of the families have been strengthened. In 2004 *Individuals with Disabilities Education Act* was revised and the issues such as the evaluation of the child and family, determination of the services to be provided, people who would prepare the individualized education plan, people who would take part in the process, the intervals of the regular evaluation and planning of the transition process were addressed. All these legal developments in the United States have affected other countries in the world. As a matter of fact, after all these developments in the United States, legal arrangements have been made regarding early intervention in many countries (Ceylan & Yalman, 2018).

Historical Development of Early Intervention in Turkey: Studies in the field of early intervention in Turkey began with the work of a group of academics in 1982. In this context, the first early education program in Turkey *Early Support Project* has been implemented. With this project, both institution and home-centered programs have been developed for children from families with low-income and low education level (Kağıtçıbaşı et.al., 2005).

Issued in 1983, with number 2916, *Law on Children in Need of Special Education* was an important step in the field of early intervention. It is stated in this law that it is essential to start early to education of children with developmental disabilities. In 1989, in order to evaluate the developmental skills of children in early childhood and to prepare educational programs according to their needs, *Portage Early Education Program,* and in 1996 *Small Steps Early Education Program* began to be implemented in Turkey. Until the 1990s, studies in the field of early intervention in our country were mostly limited to individual studies of universities and non-governmental organizations (Ceylan & Yalman, 2018).

The most important development in the field of early intervention in our country came into force with *573 numbered Decree Law on Special Education* in 1997. With the two decisions in the first part of the decree; *"Special education services in early childhood are provided in homes and institutions on the basis of informing and supporting the family."* (Article 6) and *"Preschool education is compulsory for children diagnosed with need of special education."* (Article 7), important steps have been taken in the field of special education in early childhood (MoNE, 1997). With the *Law on Disabled People* which entered into force in 2005, important decisions have been made with regard to early diagnosis and preventive services. However, the decisions were found to be limited to the procedures and principles for the field of application. With the legal decisions in the *Special Education Services Regulation* that entered into force in 2006 and was updated in 2012, the procedures and principles related to the field of application have been tried to be clarified. Especially in this regulation, decisions have been made regarding the services to be provided to children with developmental disabilities between the ages of zero and six (MoNE, 2006). In 2000s the concentration was on studies that focused on development and adaptation of early intervention programs for children with developmental disability or at risk (Ceylan & Yalman, 2018) Developed by UNICEF and the Ministry of National Education in 2003, *My Family Project* has been implemented as a project aimed at providing parents with children between the ages of zero and six years with knowledge and skills related to child development. Launched in 2007 *Basic Education Program for Family* was developed and implemented in order to support the parenting skills of parents with children between the ages of zero and six. In 2012 *An Equal Start Project for Rural Children* was developed by AÇEV. The aim of this project is to support the development of early childhood children living in villages, to provide institutional education opportunities and to provide families with skills in child development and education (Diken et. al. 2012).

When we look at the development of early intervention programs in Turkey, we see that both public institutions and non-governmental organizations considering the welfare of children, developed many projects, that are still practiced today. However, the lack of state security and problems of implementation throughout the country of the programs are some of the important issues that need to be solved (Guralnick, 2011; Odom et.al., 2010).

Early Intervention Steps

Early intervention studies should be carried out systematically in the framework of early intervention steps. In order to provide the necessary support to infants and young children within the scope of early intervention, it is important to carry out the studies in a planned manner (Bayhan, 2018). Early intervention steps are listed as follows.

- Application and Detection
- Formation of Early Intervention Programs
- Implementation of Early Intervention Programs
- Evaluation of the Program
- Transition Time (Aytekin & Bayhan, 2015).

Application and Detection: In order to ensure the effectiveness of early intervention at the highest level, it is of great importance to provide the service as early as possible. Hospital staff, doctors, other health professionals, and parents are the ones who regularly communicate with and observe the most with newborns, infants and young children (Zhang, Schwartz & Lee, 2006). Any of these individuals may find that infants or young children have a condition to be diagnosed for developmental delays to be assisted by early intervention services. There are two main ways in which children and families can apply for early intervention. These are;

- ***The first application sources;*** these resources usually include doctors, hospitals, and other programs. For example, if there is a suspicion by the medical staff about the child's development during routine health check-up or at any hospital visit, the child and family are directed to apply for early intervention.
- ***Suspicion of parents;*** parents who have any doubts about their child's development may apply for early intervention and request an assessment of their child. In this case, the child is referred for evaluation (Bayhan, 2018).

Formation of Early Intervention Programs: This step consists of four parts: assessment, diagnosis, creating an individualized family service plan and service coordination.

- ***Evaluation;*** family approval is required for evaluation. After the consent of the family is obtained, it is evaluated whether the child should be referred for the diagnosis of incompetence and whether it is appropriate for the child to be included in the scope of early intervention. The individual needs of the child and appropriate early intervention services to support these needs are identified. The child's family is evaluated in detail. The evaluation process should be completed within 45 days at the latest for the child to be included in the early intervention (Bayhan & Şipal, 2011).
- ***Diagnosis;*** in this step the decision is made for whether the infant or young child is eligible for early intervention services. Completing the diagnostic step as soon as possible is crucial for timely start of services. The diagnostic process in our country is carried out in the form of medical and educational diagnostics. As soon as developmental delays or risk conditions are identified, a comprehensive medical diagnosis should be performed first. In medical diagnosis, biological, environmental and existing risk factors that may delay the development of the child are reviewed. The scope of medical evaluation may vary according to risk factors. If necessary, brain imaging may include assessments such as electroencephalogram (EEG), genetic testing and/or metabolic testing. Medical diagnostic evaluation is performed by a trained and specialized pediatrician or by medical professionals working with neuro-developmental pediatrics, pediatric neurologists, pediatricians in the sub-branches of pediatrics (Aytekin & Bayhan, 2015).
- Educational diagnosis is made by the special education evaluation board in Guidance and Research Centers. In educational diagnosis, the child is directed to the appropriate educational institution by taking into account the competencies in the areas of development, educational performance, needs, process of benefiting from educational services and individual development report. (Kağıtçıbaşı et.al., 2005).
- ***Creating an individualized family service plan;*** The individualized family service plan (BAHP) is a written document outlining the early intervention services that the child and his/her family will receive. The family of the child should be included in the process when BAHP is formed. Because families are especially important in providing early intervention services. A child's BAHP should include:

- Current performance level of the child, physical, cognitive, language, social-emotional development skills and adaptive needs
- Family information, such as resources, priorities and interests of both parents and other family members in close relationship with the child (must be approved by the family)
- Main outcomes or outcomes that parents hope to achieve for both the child and the family
- Basic early intervention services that the child will receive
- Where early intervention services are to be provided in a natural environment (home, a community, etc.) (if the services cannot be provided in a natural environment, there should be a sentence explaining the reasons for it)
- When and where the child will receive early intervention services
- How many days or sessions worth of services the child will receive and how long each session will last;
- Whether early intervention service will be provided individually or as a group;
- Name of the service coordinator who will oversee the implementation of the BAHP
- Steps to be taken to support the child's transition from early intervention to another program when the time comes (Aytekin & Bayhan, 2015; Bayhan & Şipal, 2011; Zhang, Schwartz & Lee, 2006).

Once it is confirmed that the infant or young child will be covered by early intervention, a meeting is held with parents to identify appropriate services for the child and develop BAHP. At the meeting organized to develop BAHP, communication with the family should be established and the family's culture and other characteristics should be taken into consideration. The group that will develop BAHP is called the BAHP team. This team consists of the parents of the child, the supporters or other family members that the parents want, the service coordinator assigned to the family, the individuals who participate in the evaluation process of the child and the individuals who will provide early intervention services to the child if appropriate (Bayhan, 2018).

- ***Service coordination;*** After the first BAHP meeting, the referral unit receives parents' consent to provide the early intervention services identified in the BAHP. The service coordinator serves as the sole point of contact for the family and helps deliver early intervention services to the child and the family (Bruder, 2010).

Implementation of Early Intervention Programs: At this stage, the goals in the individualized family service plan are implemented by the early intervention team in line with the family's environment and resources. How and where these practices will be carried out varies according to the family resources, needs, age and development of the child. Early intervention programs are implemented in the form of hospital-based, home-based, institution-based and combined home-institution-based programs (Kağıtçıbaşı et.al., 2005).

Program Evaluation: At this stage, the applied program is evaluated. The impact of the program on the child and the family and whether the set goals have been achieved are examined. Teamwork is very important when evaluating. Decisions are made with the consensus of the team members. During the evaluation, the BAHP six-month review and annual BAHP update steps are followed.

- ***BAHP 6 Month Review;*** the BAHP is reviewed at least every six months to see if there is any change in BAHP content, whether there is any change in the child's developmental performance and family concerns or requirements. It can be done sooner if the family wants it or if it's required by certain conditions (Bayhan, 2018).
- ***Annual BAHP Update;*** the BAHP needs to be updated annually. The annual update meeting is attended by the service coordinator, the parents, other family members if requested by the parents, and the people who provide the services. At the meeting, the BAHP team members review the child's progress towards measurable results or outcomes listed earlier in the BAHP. All necessary changes are made by the BAHP team with the consent of the parents (Aytekin & Bayhan, 2015; Bayhan & Şipal, 2011; Zhang, Schwartz & Lee, 2006).

Transition process; transition process is preparing infants or small children in the early intervention process for the transition to special education programs or other appropriate services or programs when they are at the age of three. The child's transition plan must be made at least 90 days before the child turns three. 90 days prior, with the consent of the family, the referral unit organizes a meeting with parents to make the transition plan. At the transition plan meeting, the steps and services required to ensure that the child and family move from early intervention to the next program environment *are* focused on. Transition for children, is carried out in three ways; *daily transitions between different tasks, transition between applications* and *transition from one program to another* (Bayhan, 2018).

There are some stages to make the transition easier. These stages are *planning* and *communication.* At the planning stage, preparing the child, using a systematic approach and adaptation are very important. During the communication phase, families should be informed about what can happen during and after the transition. (Kağıtçıbaşı et.al., 2005).

Early Intervention Models

Early intervention models include the way the system will be presented to the family (development system model, transactional model, hybrid model, etc.), where it is presented (home-based, institution-based, home and institution-based), or the type of collaboration (interdisciplinary, multidisciplinary, transdisciplinary) provided by the service providers. (Tuna, Sümer & Diken, 2018). Information about the models according to the way early intervention services are provided is given below.

Development Systems Model: In the development systems model, the components that can be considered as critical and the process to be followed are modeled according to a developmental perspective for the success of early intervention services. According to the development systems model, there are two important factors for the success of early intervention services. These are the patterns of interaction the child has with the family, integration and participation (Guralnick, 2011).

- ***Interaction patterns of the child with his/her family;*** patterns of interaction with the child's family play an important role in the development of the child. The interaction patterns of the family with the child are listed as parent-child interaction, parents organizing the child's experiences, providing care and safety. Providing these interaction patterns supports the development of the child. There are also developmental risks that may arise from the inadequate interaction patterns of the child with the family. These are biological and environmental risks (Tuna, Sümer & Diken, 2018). *Biological risks;* if the child is biologically at risk or is inadequate, it is necessary to intervene with the sources of stress arising from the child's characteristics. These sources of stress can be eliminated by providing the family with information about the inadequacy of their children and by focusing on the family's priorities, taking into account the difficulties the family has experienced. *Environmental risks;* personal characteristics of the parents, economic difficulties of the family and insufficient social support. Because of these reasons, sometimes there may be difficulties in interaction patterns (Zhang, Schwartz & Lee, 2006).

- ***Integration and participation;*** the services provided in the developmental systems model should enable the child to integrate. In this respect, it is necessary to cooperate with the institutions and organizations where the child can interact with his/her peers and receive services in line with his/her needs. In the comprehensive intervention program, how to plan for integration and participation should be considered (Tuna, Sümer & Diken, 2018).

Transactional Intervention Model: The transactional intervention model focuses on both the individual and his/her experiences for the development of the individual. Three basic techniques are used in the transactional intervention model. These include improvement, redefinition and retraining (Sameroff & Fiese, 2000).

- ***Improvement technique;*** aims to change the parents with the intervention of the child. This technique focuses on the characteristics and behaviors of the child and does not focus on family and cultural characteristics. In the healing technique, professionals only work with children. The conditions that adversely affect the child's interaction are tried to be improved so that the response of the family and the environment to the interaction is expected to change. However, in some cases improvement technique alone may not be sufficient (Diken et al. 2012).
- ***Redefinition technique;*** in cases where children do not meet the expectations of parents and parents cannot cope with this situation, the redefinition technique is used. In the redefinition technique, it is aimed to turn the families' perspectives towards their children into positive.
- ***Retraining technique;*** this technique is especially used to support the development of children at risk due to environmental conditions. For this purpose, home visits are made to families, and education is provided on issues such as playing with their children and providing care for children. It is thought that risk factors will disappear when environmental conditions are improved with retraining technique (Akoğlu & Şipal, 2012; Sameroff & Fiese, 2000).

Integrated Intervention Models: Integrated intervention models are one of the intervention models developed by using multiple theories and applications together. Early intervention is an area where many theories can be integrated in line with the needs of the clients. *Natural intervention model* and *relationship-based intervention model* are examples of integrated intervention model (Tuna, Sümer & Diken, 2018).

- **Natural intervention model;** it is a model developed by using learning and development theories together. In the natural intervention model, the teaching of certain skills to children is based on transforming the child-initiated interactions into teaching opportunities in natural environments. Many natural teaching techniques and strategies are used in line with the instructional goals of the child. Based on development theories, various practices are included and thus, it is aimed to strengthen adult-child interaction (Ocak & Arda, 2014).

- **Relationship-based intervention model;** these are interventions developed by using ecological and developmental theories together. From the perspective of ecological theory, in the relationship-based intervention model, it is important to determine the resources and risk factors provided by the child's family and environment. According to developmental perspectives, it is aimed to support the development of the child by regulating the environment with rich stimuli, strengthening adult-child interactions and reducing environmental risk factors (Tuna, Sümer & Diken, 2018).

Programs in Early Intervention Services

Early intervention services are carried out in different ways to support the multi-dimensional development of the child. These services are classified as hospital-based, home-based, institution-based and combined home-institution-based programs. These programs in early intervention differ according to the scope, target group and quality of the program (Yıldız-Bıçakçı, Dursun & Bolatbaş, 2018).

- ***Hospital-based programs:*** Hospital-based early intervention programs are of great importance in terms of monitoring and supporting the development of children between the ages of zero and three, especially in the health system, and early diagnosis of situations that may pose developmental problems and obstacles in all areas. In recent years, there has been an increase in hospital-based developmental early intervention programs with the increase in child development staff within the Ministry of Health (Ou & Reynolds, 2006). The fact that child development specialists and other occupational groups serving the child and the family were appointed within the scope of health personnel provided important opportunities for the preparation, implementation and monitoring of multidimensional early intervention programs for the child and the family (Hadadian & Koch, 2013). Determining the medical and developmental diagnoses of children with developmental delays in the hospital environment is very important for the rapid planning of interventions. In this respect, the importance of hospital-based early intervention programs becomes more prominent. Hospital-based early intervention programs are handled in three dimensions. The first of these is the medical and developmental identification of the child by an appropriate team, the second is the educational and developmental support programs offered to children in the hospital, and the third is the education-oriented hospital schools. The hospitalization process affects both the child and the family (Yıldız-Bıçakçı, Dursun & Bolatbaş, 2018). As a result of the observation of the effects of the hospital process on children and families, early intervention programs are established in order to support children especially socially and emotionally and to facilitate child adaptation to the hospital. The main purpose of the early intervention programs for hospitalized children is to prepare children for this process and to ensure the right of development and protection, especially children's right to play. For this purpose, various activities are planned for children in hospital play rooms. For example, animation shows are organized in the service environment, and various activities (drawing, story reading, box games, etc.) are planned for children who cannot leave the room (Ocak & Arda, 2014).
- ***Home-based programs:*** Home-centered early intervention programs are programs that target the family in supporting the child. The basic element of early intervention services is that early intervention services should be provided within the framework of a family-centered approach (Bolatbaş & Yıldız-Bıçakçı, 2018). Home-based early intervention programs are carried out with the aim of providing social support to the family, informing the family about the importance of early years

and family support in the development of children and improving their parenting skills (Bayhan, 2018). The general objectives of home-based early intervention programs can be listed as follows.

- Supporting the development of the child,
- Supporting the development of the child in the home environment,
- Ensuring that the child's home environment is regulated,
- Supporting the family in multiple ways,
- Helping the family to get to know their child,
- Explaining to the family how important they are in the development and education of the child
- Strengthening the communication between the child and the family (Bayhan, 2018; Bolatbaş & Yıldız-Bıçakçı, 2018; Crawford & Weber, 2017).

Home-based early intervention programs are conducted especially with disadvantaged parents. In home-based early intervention programs, professionals create an early intervention program for parents to take greater responsibility in the home environment and see the family as the most important part of the program. The effectiveness of the program depends on the parents' approach. It is very important for the success of the program that the parents accept the early intervention program and make efforts in the implementation process (Howard et. al., 2011). In home-based programs, professionals make regular home visits in order to become a model for families in the teaching process or other interventions, to advise, to ensure the success of the intervention, to monitor the child's development and to provide developmental support. These programs began in Turkey with parents' education. Mother-Child Education Program (AÇEP), Father Support Program (BADEP), Family Child Education Program, Ministry of National Education General Directorate of Lifelong Learning Parent-Child Education Program can be given as examples of home-based early intervention programs (Bolatbaş & Yıldız-Bıçakçı, 2018).

- ***Institution-based programs:*** Institution-based early intervention programs started as early childhood education practices. One of the first examples is the Head Start program. In institution-based practices, the child may be given one-to-one education in line with his/her developmental level and needs or may be directed to kindergarten within the scope of child integration. These programs aim to support the child in many different ways by offering an environment outside the home (Meraki & Yıldız-Bıçakçı, 2015). Institution-based early intervention programs play an important role especially in the socialization of the child. The time spent by the child in the institution is taken as the basis for the educational applications. Educational practices may vary depending on the quality of the program and the targeted group of children. In Turkey, preschool education and special education institutions play an important role in terms of institution-based early intervention. Early intervention services given to families are usually carried out in an institution-based way in Turkey (Yıldız-Bıçakçı, Dursun & Emre-Bolatbaş, 2018).
- ***Unified home-institution-based programs:*** Both institution and home-based services can be carried out together in order to make positive contributions to the life of the child according to the disadvantage. Most early intervention programs combine institution-based activities with home visits. In these types of programs, the support given at the institution is combined with the support given by the parents at home, as more than a few hours of support is required for children in need of early intervention. Small Steps Early Intervention Program, Mobile Kindergarten Project, Toy Libraries and Women's Leadership in Women and Children's Centers are examples of the com-

bined home-institution-based early intervention programs implemented in Turkey (Bayhan, 2018; Crawford & Weber, 2017).

Some Early Intervention Programs Implemented in Turkey

The following is information about some of the early intervention programs implemented in Turkey.

Portage Program: Developed in 1969 in Portage, Winsconsin, United States, an early intervention program for parents living in rural areas to teach skills to children with special needs, but it is also used to support the development of normally developing, at-risk or vulnerable children. Portage, which is a home-based education model, aims to make an active contribution to the development of the child based on the education of parents (Unal, 2014). The Portage Program, developed for children aged 0 to 6 years and their families, consists of three parts: the Development Scale, the Activity Suggestions Guide, and the User Guide Booklet that provides information about the Portage Program and its materials. The program is implemented in the form of home visits. Under the program, the families are taught new skills and strategies for raising children with home visits. The program to be applied is shown to the family by the home visitor. Until the next visit, the family applies the education program shown to them to their child and records the results of the behavior evaluation on the registration form left by the home visitor. In the next interview, the home visitor reassesses the child about the target behavior (the behavior that the child wants to gain). If the child does not gain the target behavior, the application example for the old behavior is shown to the family again (Şahin, 2017).

Small Steps Early Education Program: It is a home-based early education program designed for the needs of children between the ages of zero and four with developmental delays and their families. The program, which has been implemented in Turkey since 1996, was first prepared for children with Down syndrome in the USA. The prerequisite for the Small Steps Early Education Program (KAEEP) is to have never participated in any training program and not have received training from another institution during the program (Birkan, 2002).

The content of the program consists of teaching principles, assessment inventory and teaching skills in the areas of development. The program, which aims to enable parents to learn and apply the basic teaching principles in the most accurate way, is designed to enable children with developmental disabilities or those at risk to participate in social life, to gain the teaching skills of their families and to make adaptations in individual education programs when necessary. The program adopts the view that the most important teachers of children are their parents. As a result of the program implemented, it was observed that the knowledge and skill levels of parents increased (Kırcaali-İftar, 2000).

Family-Child Education Program: The program, which was developed by Gazi University Faculty of Vocational Education in 1988 with the name *"Zero-Four Age Mother-Child Education Program,"* has been applied since 2003 with the name *"Zero-Six Age Family Children Education Program"*. The aim of the program is to improve the knowledge and skills of family members on child development and education, to support the development of the child and to enrich the home environment. The target group of the program is children and their families between the ages of zero and six who live in socio-economically disadvantaged regions.

The program consists of a total of 12 sessions. For mothers with children between the ages of zero and three, the program is carried out as weekly 40-minute home visits, group meetings once a month, and group meetings for mothers with three-six-year-old children and home visits every 15 days. The content of the program consists of two parts. These are;

- *Family Education Program (AEP):* The program covers topics such as child development, communication with the child, positive behavior development in children, games and toys, health and nutrition.
- *Developmental Education Program (GEP):* It is a weekly program prepared in accordance with the developmental characteristics of the child. The educator teaches the mother the activities of the program through a home visit and follows the mother's practices (Temel, 2001).

Mother and Child Education Program: Mother and Child Training Program (AÇEP), was developed by faculty members of Boğaziçi University and experts of AÇEV as a solution to the early childhood education problems in Turkey. The main purpose of the AÇEP is to support the multi-faceted development of children living in adverse conditions through their mothers, to develop the educational potential of the mother and to make the home environment the most suitable environment for children. AÇEP's target group is six-year-old children and their mothers who are socially and economically unfavorable and cannot attend nursery school. AÇEP consists of three parts (Cömert-Erdem, 2016). These are;

- *Mother Support Program:* Each week, mothers are informed about topics such as child care, toilet habits, disciplinary methods, parental attitudes through group discussion in the form of a chat.
- *Reproductive Health and Family Planning:* Mothers are informed about the reproductive system, protection of reproductive health and family planning methods.
- Mental Education Program (ZEP): The aim of this section is to support the child's mental development and prepare him/her for school. There are 25 forms and 8 story books that the child should practice during the program. The mother learns how to apply the forms from the educator and applies them with her child at home. AÇEP educator organizes home visits and observes the work of mother and child in the home environment (Ünal, 2014).

Interaction Based Early Childhood Intervention Program: Interaction Based Early Childhood Intervention Program (ETECOM) is a relationship-based approach. Relationship-based practices improve the quality of the primary caregiver's (parent, expert/educator) interaction with the child. ETECOM's effectiveness between children with different disabilities or who are at risk of developing zero-six years old and parents has been and continues to be tested in Korea, USA and Turkey. Based on these studies, it has been shown that ETECOM studies have been effective on children with different conditions such as down syndrome, autism spectrum disorder, cerebral palsy, mental disability, delayed language and speech, developmental retardation of unknown cause and their parents. ETECOM is a very systematic program that is easy to follow and implement. The educator who has the application certificate can apply the program to the child himself/herself and apply it with the child's primary caregiver. The program is seen as a home or institution-based early intervention program since it can be implemented both at home and in the institution. The program aims to develop or develop the interactive behaviors of the specialist/educator or parent, and to help the child gain or develop 16 basic behaviors in the areas of cognitive, social and emotional development. ETECOM is held once or twice a week in one and a half hour sessions. Two strategies are studied in each session. Strategies in the program are completed in about six months and significant improvements are observed in the child's development (ETECOM, 2018).

Neuroplay Early Intervention Program: Neuroplay is a home-based family education centered autism early intervention program developed for children aged 12 to 42 months. It's an intensive game therapy process based on parent-child interaction. In the early intervention method of Neuroplay, the child's home is taken into consideration. It is aimed to improve the social disorders in children with social play shaped in the home environment. The Neuroplay early intervention program provides the family with detailed information about which material and how to play a strategy to support children with autism suspicion. With this program, parents learn how to play with their children. In the first stage, the child is followed up for five consecutive days, followed by weekly and monthly follow-up. In this process, the attitudes and behaviors of the parents are adapted to the home environment by changing sessions in the clinical setting. The interaction process of the child is supported in his/her natural environment (home) until he/she starts primary school (Yıldız-Bıçakçı, Dursun & Emre-Bolatbaş, 2018).

Teamwork in Early Intervention

Interdisciplinary cooperation is very important in early intervention studies. Collaboration refers to the collaboration of multiple professionals. The cooperation process involves the professionals working in different fields to identify the needs of the child and the family and develop solution strategies for them. In other words, people with different knowledge, experience, talent and professional identity work for a common purpose and work in coordination within a planned period of time. Teams are communities in which employees with different skills, training and opinions combine their creative power, effort and knowledge to achieve their goals effectively. However, coordination of different units (such as health, education, social services) that will provide early intervention services is also very important (Sandall, Smith, McLean & Alison, 2002). As far as cooperation is concerned, it is of paramount importance that the family is always considered a part of the team in early intervention efforts. It is not possible to achieve successful and lasting results in cases where the family is not included. Many professionals from different disciplines should adopt a careful working approach when collaborating with each other and with the family at the same time. A careful and effective working approach can help children and their families. There are some elements that need to be considered in order to establish an effective team and perform teamwork (Haward, et.al, 2011). These are listed as follows:

- Open communication
- Building mutual trust
- Developing common values and attitudes
- Empowering/improving team members
- Conducting effectively managed team meetings
- Providing feedback on the team's operation
- Appropriate assessment at the right time
- Comprehensive and effective intervention

In teamwork, team members are expected to share some common responsibilities. These shared responsibilities include assessing the child's developmental status and developing and implementing a program that includes the family. It is very important to have a good team in all these processes (Bayhan, 2016).

In the early intervention team, there should be professionals and services from different disciplines in line with the needs of the child and the family. These disciplines and services include child development, special education, early childhood education, audiology, language and speech therapy, physiotherapy, occupational therapy, nutrition and dietetics, social services, family education/therapy and counseling, health services, medical services, assistive technologies, nursing services and psychological support services. The presence of people from different disciplines and the exchange of ideas in the early intervention team is extremely important for the benefit of the child. Because a well-planned early intervention program with a good team will positively affect the child's development. The presence of teamwork in early intervention is an indication that children in need of early intervention receive quality support (Reynolds, Temple, Robertson & Mann, 2001).

CONCLUSION

Early intervention consists of programs developed for families of infants and young children who have developmental delays or inadequacies or are at risk group. The idea of early intervention was first introduced in 1954 in the United States by the supreme court ruling against racial discrimination, which states that it is a universal right of all children to receive appropriate education. Early intervention and prevention are more focused on the long-term social and economic benefit of the child, the family, and therefore the community, rather than solving a problem. Early intervention programs with family participation aim to reduce the effects of adverse situations for children and their families, and to teach or gain skills for healthy and effective parenting practices. For these purposes many early intervention programs have been developed in the world and Turkey. These programs have many benefits for children and families. In order for early intervention programs to be useful for children and family they need to be developed and implemented systematically. In this process, it is very important that professionals and families from different disciplines act as a team and cooperate. Otherwise, it is not possible to obtain successful and lasting results from early intervention studies.

REFERENCES

Akoğlu, G., & Şipal, F. (2012). Düşük Sosyo Ekonomik Düzeydeki Çocuklara Yönelik Erken Müdahale Yaklaşımları: Dünya'dan ve Türkiye'den Uygulamalar. *Toplum ve Sosyal Hizmet Dergisi*, *23*(1), 173–184.

Aytekin, Ç., & Bayhan, P. (2015). Erken Müdahalede Uygulama Basamakları. *Hacettepe University Faculty of Health Sciences Journal*, *2*(2), 1–14.

Bayhan, P. (2012). Erken Müdahale Programlarında Aile Merkezli Uygulamalar ve Geçiş: Önemi, Özellikleri, Uygulaması. In N. Metin (Ed.), Özel Gereksinimli Çocuklar İçinde (pp. 29-52). Ankara: Maya Akademi.

Bayhan, P. (2018). Erken Müdahaleye Genel Bakış. In P. Bayhan (Ed.). Erken Müdahale İçinde. Ankara: Hedef Yayıncılık.

Bayhan, P., & Sipal, R. F. (2011). Early Intervention Services in Turkey: Perspectives Of South-Eastern Families. *International Social Work*, *54*(6), 781–799. doi:10.1177/0020872811404262

Beard, K. Y., & Sugai, G. (2004). First Step to Success: An Early Intervention For Elementary Children At Risk for Antisocial Behaviour. *Behavioral Disorders*, *29*(4), 396–409. doi:10.1177/019874290402900407

Bekman, S., Aksu-Koç, A., & Erguvanlı-Taylan, E. (2004). *Güneydoğu Anadolu bölgesinde bir erken müdahale modeli: Yaz anaokulu pilot uygulaması*. İstanbul: Anne Çocuk Eğitimi Vakfı Yayınları.

Birkan, B. (2002). *Küçük Adımlar Kursunun Gelişimsel Geriliği Olan Çocuğa Sahip Annelerin Küçük Adımları Uygulama Becerilerini Kazanmalarına Etkisi*. Eskişehir, Turkey: Anadolu Üniversitesi Yayinlari.

Blackman, J. A. (2002). Early Intervention: A Global Perspective. *Infants and Young Children, 15*(2), 11–19. doi:10.1097/00001163-200210000-00004

Brookes, G., Berlin, J., & Fuligni, S. (2000). Early Childhood Intervention Programs: What About The Family? In J. P. Shonkoff & S. J. Meisels (Eds.), *Handbook of Early Childhood Interventin* (2nd ed., pp. 549–588). United States of America: Cambridge University Press. doi:10.1017/CBO9780511529320.026

Bruder, M. B. (2010). Family Centered Early Intervention: Clarifitin Our Values for New Milenium. Topic in Early Childhood. *Special Education, 20*(2), 105–115.

Ceylan, Ş., & Yalman, F. (2018). Erken Müdahale Tarihçesi. In P. Bayhan (Ed.). Erken Müdahale İçinde (pp. 38-61). Ankara: Hedef Yayıncılık.

Cömert, D., & Erdem, E. (2016). *Erken çocukluk eğitiminde aile eğitimi*. Ankara: Eğiten Yayıncılık.

Diken, İ., Bayhan, P., Turan, F., Sipal, F., Sucuoglu, B., Ceber-Bakkaloglu, H., ... Kara, O. K. (2012). Early Childhood Intervention and Early Childhood Special Education in Turkey within the Scope of the Developmental System Approach. *Infants and Young Children, 25*(4), 346–353. doi:10.1097/IYC.0b013e318268541d

Guralnick, M. J. (2011). Why Early İntervention Works: A Systems Perspective. *Infants and Young Children, 24*(1), 6–28. doi:10.1097/IYC.0b013e3182002cfe PMID:21532932

Hadadian, A., & Koch, K. R. (2013). Issues in Labeling Young Children with Developmental Delay: Whose Responsibility is it? *International Journal of Early Childhood Special Education, 5*(2), 187–199.

Jackson, S., Pretti-Frontczak, K., Harjusola-Webb, S., Grisham-Brown, J., & Romani, J. M. (2009). Response to İntervention: Implications For Early Childhood Professionals. *Language, Speech, and Hearing Services in Schools, 40*(4), 424–434. doi:10.1044/0161-1461(2009/08-0027) PMID:19801404

Kağıtçıbaşı, Ç., Sunar, D., Bekman, S., & Cemalcılar, Z. (2005). *Erken Müdahalenin Erişkinlikte Süren Etkileri: Erken Destek Projesinin İkinci Takip Araştırmasının Ön Bulguları*. İstanbul: Anne Çocuk Eğitimi Vakfı Yayınlar.

Karoly, A. L., Kilburn, R. M., & Cannon, J. (2005). *Early Childhood Interventions, Proven Results, Future Promise*. Santa Monica: Rand Corporation. doi:10.7249/MG341

Kılınç, E. F., & Aral, N. (2015). Erken Müdahale İçin Bir Model: Anne Çocuk Etkileşim Programı. *Hacettepe Üniversitesi Sağlık Bilimleri Fakültesi Dergisi, 1*(2), 63–76.

Kircaali-Iftar, G. (2000). *Küçük Adımlar Gelişimsel Geriligi Olan Cocuklara Yonelik Erken Egitim Programinin Aile Goruslerine Dayali Olarak Değerlendirilmesi*. Eskişehir, Turkey: Anadolu Universitesi Yayinlari.

Knitzer, J., & Lefkowitz, J. (2006). *Helping The Most Vulnerable Infants, Toddlers, and Their Families*. New York: National Center for Children in Poverty, Columbia University.

Meraki, Z., & Yıldız Bıçakçı, M. (2015). Bebeklik ve İlk Çocukluk Döneminde Erken Müdahale, Bebeklik ve İlk Çocukluk Döneminde (0-36 Ay) Gelişim. In M. Yıldız Bıçakçı (Ed.), Duyuların Gelişimi ve Desteklenmesi (pp. 179-210). Ankara: Eğiten Kitap.

Milli Eğitim Bakanlığı. (1997). Özel Eğitim Hakkında Kanun Hükmünde Kararname. 06.06.1997 tarih ve 23011 sayılı Resmî Gazete, Ankara.

Milli Eğitim Bakanlığı. (2006). Özel Eğitim Hizmetleri Yönetmeliği. 31.05.2006 tarih ve 26184 sayılı Resmî Gazete, Ankara.

Ocak, Ş., & Arda, T. B. (2014). Okul Öncesi Dönemde Önleyici Müdahale Edici Programların Karşılaştırılması. *Hacettepe Üniversitesi Eğitim Fakültesi Dergisi*, *29*(4), 171–188.

Odom, S. L., Boyd, B. A., Hall, L. J., & Hume, K. (2010). Evaluation of comprehensive treatment models for individuals with autism spectrum disorders. *Journal of Autism and Developmental Disorders*, *40*(4), 425–443. doi:10.100710803-009-0825-1 PMID:19633939

Odom, S. L., & Wolery, M. (2003). A Unified Theory of Practice in Early İntervention/Early Childhood Special Education: Evidence-Based Practices. *The Journal of Special Education*, *37*(3), 164–173. doi: 10.1177/00224669030370030601

Ou, S. R., & Reynolds, A. J. (2006). Early Childhood Intervention And Educational Attainment: Age 22 findings from the Chicago Longitudinal Study. *Journal of Education for Students Placed at Risk*, *11*(2), 175–198. doi:10.120715327671espr1102_4

Pletcher, C. L., & Younggren, N. (2011). Early intervention-IDEA Part C: Service Delivery Approaches And Practices. In C. Groark & S. P. Maude (Eds.), Early Childhood İntervention: Shaping The Future For Children With Special Needs And Their Families (pp. 1-39). USA: ABC-CLIO, LLC.

Reynolds, A. J., Temple, J. A., Robertson, D. L., & Mann, E. A. (2001). Longterm Effects Of An Early Childhood İntervention on Educational Achievement And Juvenile Arrest: A 15-Year Follow Up Of Low-İncome Children in Public Schools. *Journal of the American Medical Association*, *285*(18), 2339–2347. doi:10.1001/jama.285.18.2339 PMID:11343481

Şahin, S. (2017). Portage Erken Çocukluk Dönemi Eğitim Programı. In Z.F. Temel (Ed.), Erken Çocukluk Eğitiminde Yaklaşımlar ve Programlar İçinde (pp. 327-353). Ankara: Vize Yayıncılık.

Sameroff, A. J., & Fiese, B. H. (2000). Models of Development And Developmental Risk. In C. H. Zeanah Jr., (Ed.), *Handbook Of İnfant Mental Health* (pp. 3–19). New York: The Guilford Press.

Sandall, S. R., Smith, B. J., McLean, M. E., & Alison, B. R. (2002). Qualitative research in early intervention/early childhood special education. *Journal of Early Intervention*, *25*(2), 129–136. doi:10.1177/105381510202500210

Temel, Z. F. (2001). *Okul Öncesi Eğitime Aile Katılımı. Gazi Üniversitesi Anasınıfı-Anaokulu Öğretmen El Kitabı*. İstanbul: YA-PA Yayınları.

Tuna, M., Sümer, M., & Diken, İ. H. (2018). Erken Eğitsel Müdahale Modelleri ve Temel Kuramlar. In P. Bayhan (Ed.) Erken Müdahale içinde (pp. 64-81). Ankara: Hedef.

Ünal, F. (2014). Aile eğitimi. A.B. Aksoy (Ed.). Her Yönüyle okul öncesi eğitim içinde (pp. 273-303). Ankara: Hedef Yayıncılık.

Yıldız-Bıçakcı, M., Dursun, Ş. S., & Bolatbaş, D. E. (2018). Erken Müdahale Hizmetleri. In P. Bayhan (Ed.), Erken Müdahale içinde (pp. 84-102). Ankara: Hedef.

Zhang, C., Schwartz, B., & Lee, H. (2006). Collaborative Services for Infants and Toddlers with Disabilities: Perspectives From Professionals in an Urban Setting. *Early Child Development and Care*, *176*(3-4), 299–311. doi:10.1080/03004430500092282

KEY TERMS AND DEFINITIONS

Early Intervention: It consists of programs developed for families of infants and young children who have developmental delays or inadequacies or are at risk.

Individualized Family Service Plan (BAHP): It is a written document outlining the early intervention services that the child and his/her family will receive.

Transition Process: Transition process is preparing infants or small children in the early intervention process for the transition to special education programs or other appropriate services or programs when they are at the age of three.

Evaluation: It is an important step in planning the early intervention process and determining the path to follow.

Chapter 16
Family Support Programs

Devlet Alakoç Pirpir
Selcuk University, Turkey

Canan Yildiz Çiçekler
https://orcid.org/0000-0001-6820-661X
Necmettin Erbakan University, Turkey

ABSTRACT

Family support programs are defined as the programs which are organized to provide parents to take part in all the stages of their child's development and education periods. Through the family support programs, it is aimed for parents to gain their self-confidence about their child's development and education and increase their knowledge and skills they need. Therefore, it is very important to consider social and cultural views of the parents, prioritize the voluntariness and privacy, and establish a clear communication within the frame of respect while organizing programs aimed at this field. Many family support programs are being applied both in our country and abroad.

INTRODUCTION

Definition of Family Training and Its Importance

Family training is the training efforts that are shown to develop and strengthen the family institution- the smallest unit of the society- and contribute to the family members and family life. In other words, the family training is to provide educational and practical opportunities, with the aim of increasing the life standard of the family members and their unity, so that the family members can gain necessary knowledge and skills. Generally, the parents interiorize the way they were brought up and make an effort to raise their children in accordance with those behaviors and attitudes. However, in parallel with the socio-cultural and economic developments in the societies, the ways of raising the children differentiate. In developed societies training programs which enhance the parents' knowledge and skills about the child-care and growing are applied. There are many longtime studies about the effect of the pre-school training practices, in which parents take part, on the parents and children. It was concluded

DOI: 10.4018/978-1-7998-2952-2.ch016

that the level of learning, academic success, school adaptation behaviors and future employment of the children, whose parents from low socioeconomic levels participated in such programs, increased and their levels of early pregnancy and crime decreased. Considering the parents, it was found that the needs of the parents, who participated in such training programs, were met and their levels of self-reliance and skills of controlling their lives increased (Yılmaz Bolat and Özgün, 2013; Güler, 2010; Gür and Kurt, 2011; Temel, 2003; Tezel Şahin and Özyürek, 2010a; Ünlüer, 2010). In a study carried out by Ömeroğlu-Turan, Tezel Şahin, Turla and Can (1999) to examine the effects of "Mother Child Training Program" on the mothers' skills about child development and training, and on the organization of home environment, it was found that as the mothers' levels of knowledge increased they reflected it on their home environment. Also, in another study done by Evirgen (2002) with the aim of determining the effects of School-Supported Mother Training Programs on the mothers, the mothers, who participated in the study, reported that they developed affirmative attitudes in their communication with their children, they talked with their children more, they read stories to their children, they played together with their children, they interiorized more positive discipline methods and they perceived themselves in a more positive way. Further, in a study carried out by Elibol (2007) with the aim of determining the contribution of the training given to mothers with 12-36 months children to the mothers' self-efficacy and to their children's development; it was detected that the training given to the mothers influenced the mothers' self-efficacy and their children's development scores affirmatively. As a result of the studies conducted by Durualp, Kaytez and Kadan (2016) and Demircioğlu (2012), it was observed that the knowledge of child development and education of mothers who received parent education increased and family functions and child nurturing attitudes changed positively.

Today, the parent training is regarded as one of the important ways of gaining communicative skills required to advance the perception of child raising and intra-familial dependence. Except for types of behaviors that parents gain with the daily life experiences, the need for training programs which contribute them to carry out their "motherhood and fatherhood" tasks properly has been rising. Because of the fact that the parents meet all the needs of their children, especially during the pre-school period, they are the closest people to their children and help them to gain many basic skills. Therefore, the concept of value the children experience in their families and the personality they display in accordance with the behaviors they regard as role models have an important position in their academic life and in keeping up with the social life. The element, which provides the continuity in the school-home relation, is the parents' participation in the training. As a result of that participation, the knowledge and skills that children gain at school are supported and strengthened, thus progression is gained in training. In this way it gets easier for the child to adapt to social life and his academic success increases (Ersay, 2013; Gür and Kurt, 2011; OBADER, 2013; Özel and Zelyurt, 2016; Ünal, 2017).

Aims of Family Training

Family training is the programs which are organized with the aim of ensuring the parents to take part in all the stages of their children's development and training periods. The family training program is all of the activities developed and arranged so that the parents can bring up their children in a healthy and equipped way. Family support programs aim at informing the parents, making the parents acquire new skills, fastening child's development period, giving consultation to family problems and activating the family ties (Cömert and Erdem, 2010; Dere Çiftçi and Aksoy, 2017; Tezel Şahin and Özyürek, 2010a). The real aim of the training which is carried within the scope of family training is to make parents gain self-reliance, increase the knowledge and skill the parents need in their children's development and activate the fathers in training programs in which the mothers are usually active. Also, informing the parents about the various areas of the concept of parenthood, increasing their level of awareness, making them acquire different skills and raising their awareness about their legal rights and responsibilities are among the aims of family training programs. Further, one of the primary aims in family training is to prevent any possible problems before they emerge. Training for knowledge and skills, which help parents overcome the problems when they face a crisis situation, should be given them by giving training to the parents about the basic subjects that can emerge in the family. With this method; various problems, which possibly emerge in the family, can be prevented and parents can gain a level of awareness to solve the problems before they deepen (Arabacı, 2017; Gür and Kurt, 2011).

FAMILY TRAINING MODELS

Home-Based Family Training

In home-based family training, firstly, the trainers carrying out the home-visiting take the family members' needs into consideration, secondly, the subjects which are thought to be included in the training are mentioned to the families in the predetermined period of time, and thirdly, the trainers make parents perform the practices. The level of success of the study is expected to be directly proportional with the success of the home visitor who come home. In the home-based family training; it is aimed to raise the parents' awareness and increase the stimulant level of the home environment in the home-visitings with the aim of supporting the development of disadvantaged children who are at risk, providing the necessary food that their body need, make the necessary arrangements to improve the conditions in children's environment so that they can show a healthy development. The followings are also among the targets of the program; preparing children for the primary education with various activities to support their mental development and with exercises for literacy preparation which improve their language skills, contributing the people, who are responsible for the care of the children, to establish healthy connections with the children, and raising the woman's position in the family (Tezel Şahin and Özyürek, 2010a).

Institution-Based Family Training

Institution-based family training is a program in which parents attend to parents training by coming to the institution in certain days of the week. Further, it is a program that supports the parents' communication with their child via the materials which can be used by parents with their children outside the institution. Sometimes, parents attend such programs with their children and while one group receives parenthood training, the other group, consisting of children, receives training appropriate for their levels of development. In institution-based family training programs, the trainers can give training to many parents at the same time. In this community environment; it is an advantage not only for the trainer, but also for the parents because it provides opportunities for parents to socialize and share their ideas with the other parents. Parents' coming to the institution in a certain day and time provides opportunity to reach many people in a short time. Parents, who have peer children, can communicate with each other and have the opportunity to talk about the possible problems which can emerge in their children. However, due to the rising demand to the training, it gets difficult to provide the continuity. The idea of institution-based training is interiorized more in Turkey (Tezel Şahin and Özyürek, 2010a; Tezel Şahin and Özyürek, 2010b).

Family Training Integrated with Institutional Pre-School Education

Starting from the early years of the life, the support on child's development and its reaching to the target level closely depend on the relation of different system areas where the child lives, and the benefits of those areas for the child's development. When viewed from this aspect, it is understood that the collaboration between the parents and school is necessary in order to obtain the efficacious of the education, given to the child in the pre-school period and provide the continuity in education. It is required to include the family, which has an important place in children's life, in the content of pre-objective and the parents should be regarded as the stakeholder of the educational institutions in order to accomplish the training program given with the aim of ensuring the children's multi-directional development. Training and raising of the children, who are the most important part of the foundation of the society, should not be left to chance, their development and training should be supported in a deliberatively created environment. In the family training; not only the mother, the father or the child should be dealt with separately; instead, the family-as a whole system-should be approached and the institutional training should be integrated with the family training. It should be kept in mind that providing training to human, family and society is an important part of the social and economic development of the country (Tezel Şahin and Özyürek, 2010a; Tezel Şahin and Özyürek, 2010b).

Family Training through Distance Learning

The primary aim of the distance learning is to provide training service to individuals who have regional limitations and are deprived of training opportunities by overcoming the limitations in the training service partly or completely. The training environment appropriate for a great number of individuals' needs can be met in this way (Tezel Şahin and Özyürek, 2010b). The distance learning method provides the individuals to learn the knowledge by themselves. Also, compared to face-to-face education, the distance learning method is more flexible and can be adjusted according to the individuals' conditions. In this training method it is desired to provide training service to the regions with limitations and ensure the continuity of the individuals' training (Tezel Şahin and Özyürek, 2010a).

Mobile Services

Within the scope of mobile services, the training service is provided through mobile play buses in the regions where the schooling rate is low. While putting the project into action, special attention is paid to start from the areas with limited training services and, within the bounds of possibility of the buses, from the zones where the number of children, who are deprived of preschool education, is high. Through the mobile services; it is ensured to actualize the training programs, which progress in the same direction with existing preschool training programs and support all the developmental areas of the children, and to plan and perform the training programs for the parents. One of the first practices of the mobile preschool institution in the worldwide is the school bus project which got started in England in 1969. In this project it was aimed to integrate the groups, which consisted of less number of individuals compared to other societies with common culture and language, with the society and within the content of this project it was planned to take the training service to the areas where people lived. Firstly, the needs of various areas especially those of women and children were defined, then some small projects were presented. A charity was established by combining the projects put forward in 1970s in a common understanding and a bus network spreading to large areas were established (Tezel Şahin and Özyürek, 2010b). To Tezel Şahin and Özyürek (2010a) the objectives of the mobile services are as such; providing the equality in opportunity to the children who cannot benefit from preschool institutions in the early childhood period and preparing them for primary education, supporting the development and education of the children, who carry on their living in various areas, in parallel with their needs, raising the level of awareness of the parents and other family members by informing them about the children's education, planning necessary training materials and training programs, carrying out necessary medical screening by making children, who benefit from mobile services, go through doctor's examination, and supporting the establishment of a permanent preschool education institution in the areas where mobile kindergarten service is provided as a result of the efforts for providing the mobile services to the large areas.

From Child to Child

From child to child training program was started in 1977 with the efforts of English government and World Health Organization (WHO). This training program was developed as a connection of the studies put forward for International Year of the Child by Maternal-Child Health Institution which is located London. From child to child training program is regarded as a common message network in sharing the knowledge about the health among the children and family members. Benefitting from this network brings good results for the family members and the children. The children are informed about the subjects related to health such as personal hygiene, healthy and balanced nutrition, vaccinations, development monitoring, diarrhea, anemia and first aid, and environmental subjects, and the transfer of those subjects to the children's families and friends are provided through this approach. The informing process is carried out with various interesting and vivid activities such as art activities, songs, acting and games. The main purpose in from child to child training approach is not only learning but individuals' sharing the knowledge they learn with the others. Having strong school-society ties is important in order to encourage the children to participate in the training process actively, give tangible information and examples, help the children to interiorize the information they learn and help the children apply the information they learn quickly through this training approach. Turning the interaction between the children into positive behaviors and attitudes is a desired result. In this training approach, which can be used mostly in information transfer about health and nutrition, it is aimed to make the families of the children, to whom information is transferred, and the society, in which the children live, learn how to improve the health and environmental conditions (Tezel Şahin and Özyürek, 2010b; Mishra, 2006).

FAMILY SUPPORT PROGRAMS APPLIED IN TURKEY

Family Support Programs Organized by Mother-Child Training Foundation

Mother-Child Training Foundation aims to actualize equality of opportunity in training, to include parents and immediate environment in the training besides supporting the child, and, develop and apply the programs in order to make training lifelong as a non-governmental organization. Below there is a compilation carried out by reaching the family support programs organized by Mother-Child Training Foundation and related application examples (Kartal, 2013).

Mother-Child Training Program

Mother-Child Training Program is an early childhood and preschool training program prepared by academics from Boğaziçi University. The program is carried out in Public Training Centers by the experts who have had the application training. The trainings of the experts are given by Mother-Child Training Foundation (MCTF). The program lasts 25 weeks; it is carried out one day a week and in 3-hour sessions. The program targets mothers of the children who are disadvantaged in terms of socio-economic aspects, do not receive preschool education, and has not started school yet. It is among the general objectives of the program to increase the mothers' sensibility for their child's development and help the mothers organize a home environment supporting the child's development. The special targets of the programs are as follows; supporting the development of the children who live under disadvantaged conditions in terms of socio-economic aspects and cannot carry on their education, helping parents establish healthy relationships with their children, increase the time that the children continue their education by raising their success at school and improving the women's trainer role as a mother. It consists of three subsections; reproductive healthcare program, mother support program and mental support program. In addition to that, it is examined under four basic subjects: child discipline, discipline methods, communication and health. In the program which, is applied in the group training, home-visiting is also carried out by the group leaders during the application (Apaydın and Canel, 2011; Mutlu, 2018).

As the program targets not only the development of child but also mother and child together, as a multi-purpose program, it includes both adult training and preschool training. Mother-Child Training Foundation (MCTF) is for both home-based and institution-based training. There are three parts in the content of each training:

Mother Support Program: In this part the conversations are organized in total 25 subjects such as; physical, cognitive and social development of the children, mother and child relation, nutrition, diseases, different discipline methods and so on. The meetings are held every week and the subjects are dealt with in a discussion. The mother Support Program aims to make mothers more conscious, more competent and happier individuals (Alpaydın and Canel, 2011).

Reproductive Health and Family Planning: In this part of the program it is aimed to inform the women about reproduction system and train them about reproduction health and family planning (Alpaydın and Canel, 2011).

Mental Development Program: In this part, which aims to support the cognitive development of the child from all aspects and prepare him/her for the school, the children are given 25 forms and 8 story books which they should apply during their training period. The forms are applied as one form in one week by the mothers and their children in the home environment. The mothers learn how to apply the forms in the meetings. Mother-Child Training Foundation (MCTF) teacher observes the mother and child's application in the natural environment during the home-visiting (Alpaydın and Canel, 2011).

The subjects found in three parts of all training contents are determined as such: mother support program, the importance of preschool period and the role of the mother, characteristics of growing and development, cognitive development, personality development in the preschool period, healthy nutrition, health and care, avoiding from the accidents, toilet habit, sexual training, the importance of play and its contribution to the development, mother-child relations, negative methods of discipline, positive methods of discipline, discipline (revision), accepting the child's behaviors, listening to the child, body language, woman, generalizing the learnt methods with others and techniques of studying (Alpaydın and Canel, 2011). Mother-Child Training Program (MCTP) adopts a holistic approach by reaching the child together with his/her parents and immediate environment. Mother has been chosen as the important person in the home environment in order to influence and contribute to all the developmental areas of the child. This application is affected by the ecological approach which states the effect of child's environment to his/her development. The program has many objectives as it targets lots of groups. It aims to raise the preschool preparation level, building an environment, which supports the healthy life, top level social development and better nutrition habits, by providing the mental development in the child. In parallel with those objectives, the program contributes to all developmental areas and it requires enhancing the role of parents (Bekman, 2000).

Preschool Child Parent Training Program

In Preschool Child Parent Training Program (PCPTP) it is aimed to strengthen the parents' relationship with the school, make the education given in the school continuous supplementary, focus preschool preparation by contributing to children's cognitive development, language and mathematics skills. PCPTP, which was started to be applied in İstanbul in 1999-2000 academic year, was carried out with the collaboration of General Directorate of Preschool Education and Mother Child Training Foundation (MCTF) with the target of supporting 61-72-month-old preschool children's preparation to primary education and strengthen the relation between the school and parents. Preschool Child Parent Training Program was developed in 1999 by Mother Child Training Foundation experts Deniz Şenocak and Dilek Erdoğan under the supervision of Prof. Dr. Sevda Bekman from Boğaziçi University Faculty of Education Department of Primary Education. As it is continued to apply the program, the needs are determined and they are reorganized by being renewed. Preschool Child Parent Training Program, which continues 24 months, is carried out under two different types of application; Child Training Program and Mother Support Program. This program is continued with the trainings that teachers carry out every day in the classroom and meetings with parents once a month (AÇEV, 2013; Arı, Deniz, Erişen and Çeliköz, 2006; Kartal, 2013).

Family Letters

Family letters are information sets which are created to support the parents, who wish to have a baby, during the years starting from pre-pregnancy period till the child becomes 3-year-old. The application of family letters which is carried out as a training project is supported with seminars. It is aimed to answer parents' questions about the pregnancy and 0-3 age period through information sets. The family letters program, which is run with the institutions supporting the Mother Child Training Foundation, is a social responsibility project. During the program; seminars about the different subjects such as the child development and training, pregnancy period, birth, nutrition and health are organized by the expert trainers from Mother Child Training Foundation for the mothers-to-be and mothers. These training seminars are free and they are given in three-hour sessions to the groups consisting of 50-100 people.

In the training seminars "Family Letters" information sets, which tell the subjects to be dealt with in detail, are distributed to the mothers-to-be and the mothers. The participants can ask questions at the end of the seminar. "Family Letters" consist of thirty letters which give information about the first, second and third age of the baby. The first 6 letters belong to the pregnancy, 12 of them belong to the first age and other 12 of them belong to the second and third age. In those letters information is given in detail about the subjects parents need (nutrition, development of the baby, diseases, vaccinations, etc) in the pre-pregnancy period, pregnancy period and postnatal period (AÇEV, 2007; Kartal, 2013).

Mother Support Program

Mother support program aims to inform mothers about child development and training, and strengthen child-mother relationship. The program depends on contributing to all developments areas of the children and improving the parenthood skills of mothers who have 3-11 aged children. Due to the differences in every age group of the children, the program is divided into two modules; mothers with 3-6 aged children and mothers with 7-11 aged children. Family support program, which continues for 12 weeks, is carried out with group meetings which are held every week. The program consists of 8 sessions together with 6 primary sessions (week). This duration can be enlarged till 12 weeks by the group leaders in parallel with the participants' needs, desires and the conditions. Subjects such as nutrition, health, developmental features (mental, social and motor developmental areas), play activities which develop the child's creativity and the importance of play are told in the Mother Support Program. Further, the other subjects such as the relationship between the mother and child, interpersonal communication, authority-discipline, turning negative behaviors into positive ones, expressing your feelings, listening to child and regarding him/her as an individual are also emphasized in this program. After all those subjects, the mothers' relations with their husbands and other people in their environment are dealt with. Sometimes, especially the subjects of woman and mother are emphasized in the meetings so that the mothers can express their feelings. Developing the mothers' self-reliance feelings and make them gain the self-concept by supporting the mothers to feel themselves efficient and useful during the program are among the important targets of the program (AÇEV, 2007; Bekman, 2000; Kartal, 2013).

Father Support Program

Fathers' role in the child development and training is as important as the mothers'. Based on this view, Father Support Program was started as a pilot project by MCTF in 1996. The primary aim of the Father Support Program is to make fathers, who come from low socio-economic environment, conscious about their children's development and training. In this program it is aimed to make the fathers feel themselves happy and efficient by removing the lack of self-reliance, which stems from being unemployed and work for low wages, their thought that they are bad fathers, and their concerns. In Turkish culture it is a known case that men do not share their problems related to home, family, income and similar subjects. Therefore, the fathers' stress loads increase. Father Support Program (FDP) gives the fathers an opportunity for sharing their problems and lives with each other's. FDP which targets the fathers, who moved to İstanbul and have low socio-economic level, is in the characteristics of strengthening the relation between the family members and preventing the psychological problems. Fathers' participation in this process provides an opportunity for them to realize that they are important about the child development and training, and share the responsibilities with their wives. Thus, the position of women and intrafamilial relations are expected to be in positive direction. The Father Support Program aims to reach fathers who have 2-10 aged children, are literate and from all education levels. The application place of this program is the primary schools and public training centers. The program, which is carried out one day in a week and lasts for 2-2,5 hours, continues during 13 weeks. Every week the participants share their views about different subjects. The groups consist of 15-20 fathers. The meetings progress with face-to-face training techniques and provide interactive learning by taking the fathers' own lives as the subjects (Atmaca Koçak, 2004; Deretarla Gül, 2013; Gözün Kahraman and Yuvacı, 2017; Kartal, 2013; Kıldan, 2010).

Seminar on Becoming Parents

Children need their families' support in all the periods of their development. Parents' relation and communication with their children starting from the early ages, their support and attitudes towards their children are of great importance. It is required to raise parents' awareness about those subjects, inform them and build a strong family. "Seminar on Becoming Parent" developed by Mother Child Training Foundation is a short term and scientific based training program that aims to expand the social awareness and make parents conscious. Target population of the seminars is parents with 0-14 aged children. In the seminars there is information about parents' attitude, child development, ways of positive discipline and communication with the children. The seminars last 3 hours on average and they are carried out in the form of meetings with groups consisting of 70-100 people. The trainers use different teaching methods and techniques (telling, case study, question-answer, story-telling and animation) during the seminars. At the end of the seminars, the participants are given a booklet in which the content of the seminars are told in detail and a questionnaire form. Then, the questionnaire forms are evaluated and put into report form (AÇEV, 2007; Kartal, 2013).

FAMILY SUPPORT PROGRAMS CARRIED OUT WITH THE SUPPORT OF UNIVERSITIES, MINISTRY OF NATIONAL EDUCATION AND NON-GOVERNMENTAL ORGANIZATIONS

Gazi University Parent-Child Training Program

The program has been applied by Social Service Children Protection Institution, Ministry of Agriculture and Eğitim-Sen with the support of UNICEF and under the coordination of Ministry of National Education. The target population of the program is mothers with 0-2, 2-4 and 4-6 aged children. Home-visiting, which lasts 40 minutes in a week, and the group meetings, which are held once a month, are organized for the mothers with 0-2 aged children. Also, the group meetings and home-visiting made every 15 days for the mothers with 2-4 and 4-6 aged children are organized. The application frequency of the program, which consists of 12 sessions, is arranged as once a week depending on works in the institution or once o month according to the difficulties in reaching the rural areas. The content of the program consists of two sections; Family Training Program (FTP) and Developmental Training Program. Between the years of 1993-2003, "0-4 Age Mother Child Training Program" was applied with 284 teachers in 59 cities by giving training to 28.854 mothers. "0-6 Age Mother Child Training Program" has been put into pilot practice since 2003 in six cities; İstanbul, Ankara, İzmir, Samsun, Mersin and Diyarbakır and 120 practitioners were educated for this (Kartal, 2013). A study was carried out by Alkan Ersoy, Kurtulmuş and Çürük Tekin (2014) with the aim of determining the effect of Mother Child Training Program given to mothers with two children aged 3 on the mothers' attitude of raising children and organizing a home environment. It was found in this study that the program had an effect on the mothers' behaviors of preparing appropriate home environment for supporting their children's development and training, and on their attitude of raising children.

Ministry of National Education General Directorate of Technical Education for Girls Parent-Child Training Project

The program, which aims to educate mothers-to-be, girls, and family members involving the other women and man, was started by Ministry of National Education General Directorate of Technical Education for Girls in 1997-1998 academic years. The target population of the program is mothers-to-be, girls and women from all ages. The project application of the project is organized by Schools of Technical Education for Girls. Training in the program is given in the home-visiting to adults, at schools or in mobile courses. The teachers working in vocational schools for girls give training about their field in this project. If it is needed, personnel from healthcare and police department can be assigned as the trainers. In parallel with the needs of the area where the program is implemented, the programs are carried out as hourly, daily and weekly. Besides the subjects determined by Ministry of National Education General Directorate of Technical Education for Girls, the subjects chosen in accordance with the conditions of the area can be selected while preparing the program content (Kartal, 2013).

Parent School

Within the scope of this program, the trainings are given to parents with the aim of making them interiorize the behaviors they should exhibit towards their children and know their children of all ages in real terms. It is aimed to help parents develop positive attitude and behaviors towards their children through this program. Further, it is aimed to guide the parents about different subjects related to child and youth such as; how they can establish healthy communication with their children, mentioning the features about the children development, effective parent training and sexual training. To this end, developing the parents' basic skills about effective communication, conscious parenthood, different developmental periods and needs belonging to those periods is targeted. Individuals participating in the program are informed with booklets and posters. The training environment has been decorated with auditory and visual items. The content of the program, which lasts 32 hours, consists of children's developmental characteristics and training for them, sexual training, communication with children, behavioral and adaptation disorders in children (Alpaydın and Canel, 2011; Kartal, 2013).

My Family Parent Child Training Program

This program is grounded on the family training model prepared with cooperation of MCTP. 0-6 aged children, who continue their living under economically inadequate conditions, and their parents are included in the program. The training is given to mothers with 0-3 aged children with 40-minute home-visiting in a week and group meetings held once a month. At least 4 home-visitings for the mothers with 0-3 aged children and at least 2 home-visitings for the mothers with 3-6 aged children are carried out. Those home-visitings start after the fourth week. Program duration is 18 weeks. Every week, the trainings, which last two hours in a week, are given. The frequency of the sessions varies as once or twice in a week depending on the difficulties in reaching the related areas and working of the institutions in which the program is applied. The content of the program consists of two topics; Parents Training Program (PTP) and Developmental Training Program (DTP). Parents Training Program is prepared for the mothers with 0-2, 2-4 and 4-6 aged children. The content may change in accordance with the age groups, but generally it consists of the following subjects; child development, communication with child, healthcare and nutrition, developing positive behaviors in children. Also, the Developmental Training Program is prepared according to children's developmental characteristics. It is applied weekly. The trainers give the necessary subjects to the mothers during the home-visitings and group meetings. The trainers also watch the mothers while they apply the training they have received. What is more, creating products from the surplus materials and making toys are taught to mothers. Program is applied both in Public Training Centers and other centers connected to Social Services Children Protection Institution (Kılıç, 2010). Parent education programs implemented in Turkey are arranged as a zero- eighteen years-old parent training program by Coordinatorship of Lifelong Learning General Directorate in the Ministry of Education (Dere Çiftçi and Aksoy, 2018).

Ministry of National Education General Directorate of Lifelong Learning Parent Training Course Program (Parents with 0-18 Aged Children)

0-18 Age Parent Training Course Program aims to make the teenagers use their potential by through strengthening the relations in the family by developing the parenthood skills of the parents. To this end, the followings are targeted through this program; strengthening the family ties, introducing the activities which should be done with the aim of supporting the developmental characteristics and areas of all age ranges, supporting the children in terms of social, emotional and academic aspects by developing domestic relations qualitatively, developing parents' personal life quality, the consciousness of domestic democracy, parents' skill of coping with risky conditions, providing the parents with the skill of managing conflicting relations, ensuring reliable motherhood and increasing the fathers' support in family training. 0-18 Age Parent Training Course Program consists four separate and complementary sections based on the children and teenagers' developmental periods (0-3 age Parent Training Course Program, 3-6 age Parent Training Course Program, 7-11 age Parent Training Course Program, 12-18 age Parent Training Course Program). Apart from this, there 3-6 age and 7-11 age father support training course programs, 3-6 age parent development course programs and 3-6 age illiterate mother support training course programs. 0-18 age parent training program has successive sub-groups for four different age ranges; each of the sub-groups consists of 14 sessions. 0-3 and 3-6 age sub-groups also have 14 sessions which are applied in two ways; face-to-face trainings and home-visitings (Millî Eğitim Bakanlığı Hayat Boyu Öğrenme Genel Müdürlüğü, 2014).

INSTITUTION AND SOCIETY-BASED TRAINING MODELS

Women and Child Centers

Together with the support taken from Foundation for Evaluation of Women's Work, women and child centers aims to make women take part in the working life, gain their economic freedom and raise their life qualities. The main objective in the project is to give training to children, who come from poor families and cannot receive preschool education in the earthquake regions, and provide opportunities for women to work and earn money. FEWW, local people, other non-governmental organizations and municipality work in cooperation so that the objectives in the project can be reached. FEWW is a non-governmental and non-profit organization which was established in order to support the improvements in the lives and settlement of women living in the low-income areas. The women and mothers' feature of being innate trainers is transformed into leadership qualification with the trainings given to them. Therefore, this model provides services with the participation and cooperation of all the people who wish the peace of their neighborhood and children under the mothers' leadership. According to the training understanding the child is centered and enriched setting is prepared in accordance with the child's family, his environment and needs. The factors in this setting are in interaction with each other. The trainer, who presents the discovering and learning to the child thanks to this interaction, takes on a task of guide by presenting the necessary opportunities. Thus, the children regard themselves as valuable individuals by realizing their own power. In this process, trainers' becoming guides, who help, lead and provide opportunities, by leaving behind the traditionalism and authoritativeness is very important. The target population of the program is 3-6 aged children and parents with children in that age range. The environments, in which the trainings are given, are shaped in accordance with the children's needs. Both the environment, prepared by the children, as a whole can be used and group rooms or drama, art, playing house, story, puppet, concept and nature centers which support children's development and learning can also be used. What is more, there are workshops that children can participate in with their parents (Kartal, 2013).

Small Steps Early Training Program

Small Steps Early Training Program is run as home-based and aims to meet the needs of 0-4 aged children, who have developmental delay, and their parents. Small Steps Early Training Program (SSETP) was first brought to our country from Australia. It has been being applied in our country since 1996. The copyright of the program was bought by Mentally Handicapped Support Foundation. Protocols were signed with Anadolu, Ankara, Gazi and İstanbul Universities. The target population of the program is the children, who have the risk of developmental delay and have this diagnosis, and their parents. The children who have been included in the program so far are from 0-48 month range. Not receiving any training earlier and not participating in any other training during the program are stipulated. 78 children and their parents; 23 children in 1996-1997, 39 in 1997-1998 and 16 in 1998-1999 were included in the program. Evaluation of the parents were carried out with pre and post evaluation method, Back Depression Inventory, Scale for Evaluating Parents' Stress and Scale for Evaluating Parents' Needs. In the evaluation of the children Small Steps Developmental Skills Inventory was employed. This scale consists of control lists which are used to evaluate 0-4 aged children's development and prepare personal training plan. The scale was created by considering the skills in four developmental areas; large muscle skills, small muscle skills, receptive language and social skills, and skills in the age development

(Kartal, 2013). Meral, Özbey and Sola (2011) carried out a study for the effectiveness of Small Steps Early Training Program and they concluded that the mothers stated that they became more conscious in their children's training and realized their deficiencies better, and it was also observed that the children learned various skills easier.

FAMILY SUPPORT PROGRAMS APPLIED IN THE WORLD

Head Start Program

Head start program can be applied by parents, trainers and volunteers. In this program it is aimed to raise parents' awareness about the subject that changes should be made in the children's lives in order to increase their life quality. Here, it is emphasized how much the parents are important in their children's lives. The Head Start Program involves different contents such as parents' participation, health and social services and vocational development. Within the scope of this program, it is targeted to provide awareness to parents about acting consciously while raising their children and becoming more effective. In the institutions, where the Head Start Program is applied, the designs related water, construction, temperature, color and sound and the practices of painting and drama are cared specifically. In the institutions vaccination and medical screening are carried out and medical programs related to mind, body and nutrition are created (Kartal, 2013). Head Start Program is prepared for socio-economically disadvantaged children and it is an institutional preschool program. Further, it is designed to meet the children' emotional, social, psychological, health and nutritional needs and close the gap between these children and their peers who have more advantages (Garces, Thomas and Currie, 2002).

Further, in Head Start Program it is aimed to support the children's training, health and social services and provide service to disabled children by including the parents of the children from preschool period in this process. Head Start application centers continue their activities in local societies and cities using the techniques of observation and being a model (Tezel Şahin & Özyürek, 2010a). In the program it is targeted to support the children's growth and development, provide support to the parents in their children's development and training, providing educational, medical and nutritonal services to the children, and associate the children and their parents with desired public services (Henry, Gordon and Rickman, 2006).

Early Head Start

Early Head Start program is a program in which supports for the developments of mothers and babies are provided to the mothers, who have low income and are expecting baby, from pregnancy period till postnatal period. Early Head Start program was first applied in 1995, and mothers-to-be and their babies form the target population. Similar to the Head Start Program, the application can be carried out as home-based or institution-based; also, they can be applied as home and institution-based. What is more, full-day care is also provided for disabled children within the scope of the program. The program can be applied by professionals, semi-professionals or volunteers and it is applied during two years. Early Head Start program was prepared by Head Start Program experts with the aim of removing the problems about the poor parents and the nutrition, development and raising their children or decreasing the problem to the minimum levels. As a result of the Early Head Start programs positive effects were seen in children's cognitive, language, social and emotional developments and these results also influenced the children's success at school. The reasons of creating the Early Head Start programs are to ensure the children and parents' development, provide staff and social development (Fenichel and Mann, 2001; Kalkan, 2010; Kartal, 2013; Love, Kisker, Raikes and Tarullo, 1997). As a result of the trainings applied in Early Head Start project, positive effects were seen on the parents about becoming self-sufficient (http://www.acf.hhs.gov).

Even Start Program

Even Start Program was first applied in 1989, and now it has been being applied in USA and Australia. It involves forming the environments, which are necessary for the children's training, by government. In Even Start Program the trainings are given under four different categories; training for adults, childhood training, parents support training and training in which parents and children are educated together. Training can be applied by professional and semi-professional personnel. In Even Start Program, the training are given in the center, also home-visitings are carried out once a month and families are supported for applying the training, given in the centers, in the daily life. Event Start Program is created for families with low income in order to increase parents and young kids' academic success specifically in reading. Similarly, the researchers, teachers and managers accept that high qualified early childhood training is of great importance for all the children (http://www2.ed.gov/; Kalkan, 2010; Kartal, 2013).

Home Activities for Toddlers and Their Families (HATAF)

The program, which was developed by Avima Lombard in 1973, lasts two years. After the success of HIPPY program, which was applied to the children in 3-5 age range, this program was developed in order to generalize the program and carry out the training of the mothers with 1-3 aged children at home. In the program, the applications are performed with mothers who have less than twelve years education and have 1-3 aged children. The applications involve home-visiting and group meetings. The mothers are informed about workshops, making toys and being effective while playing with the children. Attention is paid to use affordable materials in the training. The program is applied by the coordinators who have university degree, have information about early childhood training and are educated about the program. Apart from the coordinators, semi-professional home-visitor, who is recommended by local health and social service authorities, also includes in the program. Semi-professionals, who are chosen for the program, are selected among the mothers who have already participated in the study and they receive one-week training about HATAF (Kartal, 2013). Within the scope of program appropriate techniques are taught to the parent so that they can play with their children aged 1-3 for the following reasons; developing both mother and toddler's language skills, raising the parents' awareness about the toddler's cognitive, emotional and social needs, developing the parents' skills of using natural learning situations at home and in the society for toddler's mental and personal development, increasing the parents' understanding of the importance of positive support in toddler's learning process (http://www.jewishvirtuallibrary.org).

AVANCE Parent Child Program

In AVANCE Parent-Child Program, the training is given to the children of the parents with low income level or poor parents in Texas. The thing which is emphasized in this program is that children's parents are the most important teachers in their lives. Within the scope of this program the parents are given support in the various subjects such as parent training, early childhood training, training at home environment, training for making play and toy, healthcare, nutrition, etc. What can the parents do in the first three years of their children is focused through the program. Also, it is targeted to help families prepare their children for the school, and remove the negative effects of the factors which influence the learning of the child such as; cultural differences, the parents' level of income, home security and the parents' level of education, and provide better futures for the children. To this end, the following skills are determined to be developed in the mothers; being able to talk with their children more, being able to communicate more, being able to spend more time, being able to exhibit positive approach towards their children and being able to create a more effective home environment (Alpaydın and Canel, 2011; Kalkan, 2010; Kartal, 2013). It is emphasized in AVANCE Parent-Child training program that home environment is a setting that provides the fulfillment of the learning; also, the presence of the rich stimuli in the environment are also underlined. In the program the training are given to the participants about a review of the parenthood, physical needs of a small child, children diseases, understanding the children's behaviors, cognitive development, language development, emotional needs of the children, social needs of the children, self-awareness and creating target, nutrition, small child, pregnancy and needs of a baby (http://www.avance.org).

PAT Parents as Teachers

PAT Parents as Teachers program is a program which targets mothers-to-be and their children starting from pregnancy period till age five. Together with the babies' born; their medical monitoring are carried out, and the children with development delay or with learning difficulty are detected and necessary precautions are taken (Kartal, 2013). Through the program; it is aimed to make the parents prepare a strong substructure for their children's success at school, prevent the child abuse, increase the parents' self-reliance and ensure the cooperation between home and school (Family Strengthening Policy Center, 2007). Parents as Teachers is a training program which is planned in the birth or before the birth and involves home visits. During the home visits the visitors help the parents in order to inform them strengthen their skills about the child development and preparation of small children for the school (Wagner and Clayton, 1999). What is more, the parent trainer shares the positive parenthood attitude and behaviors with the parents, reinforce them and allocate resources. Generally, each visit lasts 45-60 minutes and the visits are carried out once a month. However, personal family needs are evaluated during the advisor meetings which are planned regularly and if there is a determined problem, the families are visited more than one in a month (Tolani, Brooks-Gunn, and Kagan, 2006).

Home Instruction Program for Parents of Preschool Youngsters HIPPY

Home Instruction Program for Parents of Preschool Youngsters was first started to be applied in 1969 and was developed by Avima D. Lombard. The duration of the program is two years and it can be applied by trainers, semi-professionals and volunteers. Home Instruction Program for Parents of Preschool

Youngsters is a home-based program which is carried out for the parents with 3-5 aged children to support their children's training. In this program, the children are given training at home and it is aimed to support the parents so that they support their children's cognitive developments. The program is applied with home visits. With the aim of preparing the children for preschool training; the program is applied with reading story book for fifteen minutes, puzzle or teaching any other determined play by using the HIPPY education materials (Kartal, 2013). In the application of the program, different books were prepared for the children in different age ranges. Further, the materials are prepared by taking the principle of from simple to complex into consideration in order to support the different developmental areas of the children (Family Strengthening Policy Center, 2007; Powell and Cassidy, 2001). HIPPY curriculum involves 30-week activity packet for 3-5 aged children, nine story books and twenty developer shapes for each year. In addition to those primary materials, some other needs such as scissors, candle and paint are provided for each of member families. Daily activities in the program are incrementally appropriate for children. The packets are designed for parents and children within an open draft format in order to provide a successful learning experience at home and guide the parents. The parents and children help the program in the learning methods. Families gain confidence with their participation in the HIPPY program. At the end of the program, the parents easily teach and develop the learning and activities for their children (http://www.hippyusa.org). HIPPY is created with the aim of helping the parents, who have limited training, to prepare their children for the school. In HIPPY the home visitors do not directly work with the children and the home visits are carried out when the children are not at home. Instead, the home visitors teach the parents how to use the activity paper and story books prepared in the program center. During the home visits, which last 30-60 minutes, the parents take the role of child and the professionals take the role of parents. Lombard believes that the parents can learn what kind of teacher they can be during the role playing in a comfortable atmosphere and if they continue to apply what they learn, they can help their children in academic success after two-year program. Also, in the group meetings, which are held in HIPPY, activity packets of the next week are introduced to the parents, enriched group activities such as questions about participation of HIPPY families, artistic drafts and school management presentations are shared, those rich activities are designed by local coordinators in order to meet the parents' need (Baker, Piotrkowski, and Gunn-Brooks, 1999).

Healthy Start

Healthy Start program is a training program aiming to provide healthy behaviors which should be fulfilled in order to prevent child obesity and other risk factors for long-termed diseases in the latter life (http:/ www.healthy-start.com). It was started to be applied in 2001 and it is applied to 3-5 aged children for a year. The aim of the program is to decrease the risks related to health like obesity by making children gain healthy nutrition behaviors throughout children's whole life via medical experts and trainers. Various stories are read, songs are sung to children, and games are played with them in order to fulfill those targets. Also, the children apply such activities at home with their parents (Kartal, 2013). The program aims to support the development of the children, develop the positive parenthood, increase the parent-children interaction and parenthood skills, provide a regular doctor and medical home service, prevent child abuse and neglect (Gomby, Culross, and Behrman, 1999).

This research received no specific grant from any funding agency in the public, commercial, or not-for-profit sectors.

REFERENCES

Alkan Ersoy, Ö., Kurtulmuş, Z., & Çürük Tekin, N. (2014). Aile çocuk eğitim programının annelerin çocuk yetiştirme tutumlarına ve ev ortamını düzenlenmelerine etkisinin incelenmesi. *Kastamonu Eğitim Dergisi*, *22*(3), 1077–1090.

Alpaydın, Y., & Canel, A. N. (2011). *Aile eğitimi programları üzerine notlar*. Ankara: Aile ve Sosyal Politikalar Bakanlığı.

Anne Çocuk Eğitim Vakfı (AÇEV). (2007). *Ekonomik ve toplumsal kalkınma için erken çocukluk eğitimi: Önemi, yararları ve yaygınlaştırma önerileri*. İstanbul: Anne Çocuk Eğitim Vakfı Yayınları.

Anne Çocuk Eğitim Vakfı (AÇEV). (2013). *Erken çocukluk eğitimi politikaları*.Ankara:Yaygınlaşma, yönetişim ve yapılar toplantısı" raporu.

Arabacı, N. (2017). Aile katılımı. Ayşe Belgin Aksoy (Ed.), Aile eğitimi ve katılımı (s.125-148). Ankara: Hedef CS Basın Yayın.

Arı, R., Deniz, M. E., Erişen, Y., & Çeliköz, N. (2006). *Milli Eğitim Bakanlığı projeler koordinasyon merkezi başkanlığı. Çocukların okula hazırlanması (Anne-Çocuk Eğitimi Temel Araştırması)*. Ankara: T.C. Milli Eğitim Bakanlığı Projeler Koordinasyon Merkezi Başkanlığı.

Atmaca Koçak, A. (2004). *Baba destek programını değerlendirme raporu*. İstanbul: Anne-Çocuk Eğitim Vakfı Yayınları.

Baker, A. J., Piotrkowski, C. S., & Brooks-Gunn, J. (1999). The home instruction program for preschool youngsters (HIPPY). *The Future of Children*, *9*(1), 116–133. doi:10.2307/1602724 PMID:10414013

Bekman, S. (2000). *Eşit fırsat: Anne-çocuk eğitim programının değerlendirilmesi*. İstanbul: Anne-Çocuk Eğitim Vakfı Yayınları.

Cömert, D., & Erdem, E. (2010). *Erken çocukluk döneminde aile katılımı etkinlikleri*. Ankara: Eğiten Kitabevi.

Demircioğlu, H. (2012). *Altı yaşında çocuğu olan annelere uygulanan aile eğitimi programının aile işlevleri ve anne-baba tutumları ile çocuğun gelişimi üzerindeki etkisinin incelenmesi. Yayımlanmamış doktora tezi*. Ankara: Gazi Üniversitesi.

Dere Çiftçi, H., & Aksoy, B. A. (2017) Aile eğitim programları. In A.B. Aksoy (Ed.), Aile eğitimi ve katılımı (pp. 43-70). Ankara: Hedef CS Basın Yayın.

Deretarla Gül, E. (2013). Okul öncesi eğitimde babaların katılımını sağlama. In Y.A. Arnas (Ed.), Aile eğitimi ve okul öncesinde aile katılımı (pp. 223-237). Ankara: Vize Yayıncılık.

Durualp, E., Kaytez, N., & Kadan, G. (2016). Okul öncesi dönemdeki çocukların annelerine verilen aile eğitiminin annelerin bilgi düzeyleri ile çocukların gelişimlerine olan etkisinin incelenmesi. *Abant İzzet Baysal Üniversitesi Eğitim Fakültesi Dergisi*, *15*(4), 1706–1722.

Elibol, F. (2007). *12-36 aylar arasında çocukları olan annelere verilen grup eğitiminin annelerin anne-babalık görevlerinde öz yeterliliklerine yönelik katkısının incelenmesi, Yayınlanmamış yüksek lisans tezi*. Ankara: Hacettepe Üniversitesi.

Ersay, E. (2013). Aile eğitimi programlarının planlanması ve değerlendirilmesi. In F. Temel (Ed.), Aile eğitimi ve erken çocukluk eğitiminde aile katılım çalışmaları (pp. 274-324). Ankara: Anı yayıncılık.

Evirgen, Ş. (2002). *Okul öncesi eğitim çerçevesinde okul destekli anne eğitim programının anneler üzerindeki etkilerinin incelenmesi, Yayınlanmamış yüksek lisans tezi*. İstanbul: Marmara Üniversitesi.

Family Strengthening Policy Center. (2007). *Home visiting: Strengthening families by promoting parenting success. National human services assembly*, Policy Brief, No. 23.

Fenichel, E., & Mann, T. L. (2001). Early Head Start for low-income families with infants and toddlers. *The Future of Children, 11*(1), 135–141. doi:10.2307/1602817 PMID:11712450

Garces, E., Thomas, D., & Currie, J. (2002). Longer-term effects of Head Start. *The American Economic Review, 92*(4), 999–1012. doi:10.1257/00028280260344560

Gomby, D. S., Culross, P. L., & Behrman, R. E. (1999). Home visiting: Recent program evaluations: Analysis and recommendations. *The Future of Children, 9*(1), 4–26. doi:10.2307/1602719 PMID:10414008

Gözün Kahraman, Ö., & Yuvacı, M. (2017). Baba katılımı. In A.B. Aksoy (Ed.), Aile eğitimi ve katılımı (pp. 197-234). Ankara: Hedef CS Basın Yayın.

Güler, T. (2010). Ailenin tanımı, tarihsel gelişimi ve aile çeşitleri. In T. Güler (Ed.), Anne baba eğitimi (pp. 1-14). Ankara: Pegem Akademi.

Gür, S. B., & Kurt, T. (2011). Türkiye'de ailelerin eğitim ihtiyaçları. *Aile ve Toplum, 7*(27), 33–62.

Henry, G. T., Gordon, C. S., & Rickman, D. K. (2006). Early education policy alternatives: Comparing quality and outcomes of Head Start and state prekindergarten. *Educational Evaluation and Policy Analysis, 28*(1), 77–99. doi:10.3102/01623737028001077

Kalkan, E. (2010). Dünyada anne baba eğitimi ve aile katılımı: Amerika Birleşik Devletleri örneği. In T. Güler (Ed.), Anne baba eğitimi (pp. 166-181). Ankara: Pegem Akademi.

Kartal, H. (2013). Erken eğitim ve destek programları. In F. Temel (Ed.), Aile eğitimi ve erken çocukluk eğitiminde aile katılım çalışmaları (pp. 159-221). Ankara: Anı yayıncılık.

Kıldan, O. (2010). Aile içi ilişkiler ve özel durumlarda aile. In T. Güler (Ed.), Anne baba eğitimi (pp. 56-71). Ankara: Pegem Akademi.

Kılıç, Ç. (2010). Aile Eğitim Programları ve Türkiye'deki örnekleri. *Ab ant İzzet Baysal Üniversitesi Dergisi, 10*(1), 100-111.

Love, J. M., Kisker, E. E., Raikes, H. H., & Tarullo, L. (1997). Overview of the Early Head Start research and evaluation project: February 1998. *NHSA Dialog, 1*(4), 181–192. doi:10.120719309325nhsa0104_8

Meral, B. F., Özbey, F., & Sola, C. (2011). Küçük adımlar erken eğitim programının aile görüşlerine göre değerlendirilmesi. *Sakarya Üniversitesi Eğitim Fakültesi Dergisi, 22*, 25–40.

Millî Eğitim Bakanlığı Hayat Boyu Öğrenme Genel Müdürlüğü. (2014). Aile eğitimi kurs programı (0-18 yaş çocuğu olan ailelere).

Mishra, G. (2006). When child becomes a teacher - the child to child programme. *Indian Journal of Community Medicine, 31*(4), 277–278.

Mutlu, H. (2018). *Baba Destek Eğitim Programının aile işlevleri üzerindeki etkisinin incelenmesi. Yayınlanmamış yüksek lisans tezi.* İzmir: Dokuz Eylül Üniversitesi.

OBADER. (2013). *Okul öncesi eğitim programı ile bütünleştirilmiş aile destek eğitim rehberi.* Ankara: Millî Eğitim Bakanlığı Temel Eğitim Genel Müdürlüğü.

Ömeroğlu-Turan, E., Tezel Şahin, F., Turla, A., & Can, M. (1999). *Anne çocuk eğitimi programının annelerin çocuk gelişimi ile ilgili bilgilerine ve ev ortamlarının düzenlenmesine etkisi. In Ulusal Eğitim Bilimleri Kongresi Bildirgeleri* (pp. 306–325). Eskişehir: Anadolu Üniversitesi Yayınları.

Özel, E., & Zelyurt, H. (2016). Anne baba eğitiminin aile çocuk ilişkilerine etkisi. *Sosyal Politika Çalışmaları Dergisi, 36*, 9–34.

Powell, L. H., & Cassidy, D. (2001). *Family life education. An introduction.* London: Mayfield Publishing.

Temel, F. (2003). *Aile eğitim modeli: Dünya'da ve Türkiye'deki uygulamalar. Erken çocukluk eğitimi politikaları: yaygınlaşma, yönetişim ve yapılar toplantısı raporu.* Ankara: Anne-Çocuk Eğitim Vakfı.

Tezel Şahin, F., & Özyürek, A. (2010a). *Anne baba eğitimi ve okul öncesinde aile katılımı.* İstanbul: Morpa Kültür Yayınları.

Tezel Şahin, F., & Özyürek, A. (2010b). Anne-baba eğitimi ve Türkiye'de uygulanan aile eğitim programları. In T. Güler (Ed.), Anne baba eğitimi (pp. 142-158). Ankara: Pegem Akademi.

Tolani, N., Brooks-Gunn, J., & Kagan, L. S. (2006). Parenting education programs for poor young children: A cross-national exploration. In *Proceedings of the UNICEF/New School International Conference*. Academic Press.

Ünal, F. (2017). Aile eğitimi. Ayşe Belgin Aksoy (Ed.), Aile eğitimi ve katılımı (s.11-42). Ankara: Hedef CS Basın Yayın.

Ünlüer, E. (2010). Okul öncesi eğitimde ailenin yeri ve önemi. In T. Güler (Ed.), Anne baba eğitimi (pp. 93-101). Ankara: Pegem Akademi.

Wagner, M. M., & Clayton, S. L. (1999). The parents as teachers program:Results from two demonstrations. *The Future of Children, 9*(1), 91–115. doi:10.2307/1602723 PMID:10414012

Yılmaz Bolat, E., & Özgün, Ö. (2013). Erken çocuklukta aile eğitimi ve aile eğitimine yönelik uygulanan programlar. In Y.A. Arnas (Ed.), Aile eğitimi ve okul öncesinde aile katılımı (pp. 21-51). Ankara: Vize Yayıncılık.

Chapter 17
Contribution of Outdoor Play Activities to Infant and Toddler Development

Sibel Atli
Van Yuzuncu Yil University, Turkey

Gülen Baran
Ankara University, Turkey

ABSTRACT

The natural environment has generally a structure to support all areas of development of children. It can be said that the interior spaces are deprived of naturalness and have less stimulation compared to outdoor spaces. Playgrounds outdoors have forms, textures, heights, and mobility in a natural or artificial landscape have the energy to encourage children to play and move. The first years of the early childhood period are of a vital importance in terms of all areas of development. Pedagogues or parents should provide the activities and environments that will enable the infants and toddlers to get pleasure out of life and that can create a learning skill and give new skills to the children at same time. Therefore, in this study, outdoor and outdoor play activities that will support the development areas of 0-36-month-old children will be discussed.

INTRODUCTION

The natural environment has generally a structure to support all areas of development of children. Studies conducted have shown that the external environment has stimulants such as social-emotional, cognitive, physical and linguistics to support all development of children. It can be said that the interior spaces are deprived of naturalness and have less stimulant compared to the outdoor spaces. Playgrounds in the outdoor, forms, textures, heights and mobility in a natural or artificial landscape have the energy to encourage children to play and move. Outdoor spaces also have a structure that supports children in terms of creativity. Change in seasons, climate change and differences in the weather conditions also

DOI: 10.4018/978-1-7998-2952-2.ch017

differentiate children's games and leave significant developmental impressions on behaviors such as perception, attention and adaptation.

The outer environment is mainly covered by soil that attracts children to compelling and entertaining activities. Materials such as soil and sand give an unlimited freedom to the child. It offers opportunities for the child to experience socially, emotionally and cognitively in line with his/her developmental characteristics, to cooperate with others and to construct his / her own game starting from very early ages. Therefore, it can be said that outdoor play activities also provide positive contributions to the formation of characteristics such as self-management of children, positive psychological self, self-confidence and self-esteem. Today's societies ignore the importance of risk on child development. The most important places for the child to experience risk are outdoor playgrounds. The child experiences in achieving and losing by taking risks in the plays he/she plays in the natural environment. He/she learns to be happy or to overcome unhappiness through play. For example, the child has the opportunity to experience more by climbing a tree in the outdoor, playing on a playground equipment, taking care of a material and setting the preliminary limits by himself/herself. The adult must support these risky plays of the child because these plays are very important in terms of problem solving and self-learning skills for the child.

Learning skills acquired in the infant and toddler period form a basis for the periods in the future. When the first years of life are taken into consideration in development, the importance of supporting the development areas of children emerges in this period. Infants and toddlers can learn much from any kind of experience and observation. It is necessary to observe them and to play games with them to respond to their interests and needs. Pedagogues or parent should provide the activities and environments that will enable the infants and toddlers to get pleasure out of life and that can create a learning skill and give new skills to the children at the same time. For infants and toddlers, the places to provide these conditions are outdoor play activities. Apart from extremely hot or very rainy days, it is seen that it is important to prefer the outdoors to support the development of children as from infancy period. Therefore, in this study, outdoor and outdoor play activities that will support the development areas of 0-36-month-old children will be discussed.

BACKGROUND

The first years of early childhood are critically important for all development areas (Morrison, 2007). It is noted that during the childhood children physically develop very fast, almost complete their linguistic development and generally speak their mother tongue, and realize their self, gender identity and emotions (Oktay &Unutkan, 2005; San Bayhan & Artan, 2011; MEB, 2013). The cognitive development that allows their way to understand the world to be more complex and effective (Kürkçüoğlu, 2010) is also very important for the childhood.

The research has shown that environment has significant influences on the period of 0-36 months, which is important for development. A well-designed playground and natural environment provide opportunities that enable children to develop thinking skills and improve creativity (Greenman, 2005). The natural environment is considered to be a teacher of people. When we raise the children not in the nature, i.e., separate from the nature, the nature will completely become a strange concept (White, 2004). The children who do not grow interaction with nature and outdoor begin to remain insensitive to environmental problems; therefore, the children should be introduced with outdoor game activities from the early ages (Wilson, 2008). The natural world will make children curious, enable them to solve problems, motivate them, force them physically, and allow them to fight against difficulties as an individual. Greenfield (2007) and Stephenson (1998) suggested in their research that games liked by the children most are the outdoor games because the motivating factors of outdoor and environments that allow children to act freely and strive make children happier (Greenfield, 2007). The Real Play Coalition warns that time and place that children have for playing games face a serious threat, which compromises development of skills required by children to survive and be successful in the future (Real Play Coalition [RPC], 2019).There are many studies suggesting that the games that are played indoor in a stationary manner or under supervision of adults require less motion and movement might lead to increase in obesity and cardiovascular disorders in children (Brown et al., 2006; Cardon & De Bourdeaudhuij, 2008; Staempfli 2009).

The playgrounds are widely used in education because they meet the needs of children to play actively. From a psychological perspective, a playground is outside of the child, but it does not belong to the outer world. It is a third area that is beyond the inner and outer fact (Pehlivan, 2005). A playground is one of the important places where children demand since they were born and enjoy spending time. In the outdoor, the children also have the opportunity to discover and create, do and undo, and learn how objects are and for what they exist. The children acquire concrete experience through their sense in the natural environment and have the chance to get to know the life. Therefore, children's field of play is generally an outdoor playground (Demiriz, Karadağ, & Ulutaş, 2003). A well-planned outdoor playground allows children to plan, wait for their turn, share, speak to the teacher and other children, establish a relationship and socialize, i.e., it responds their emotional needs (Oğuzkan & Oral, 1997). The children start to know the environment through their sense organs from the moment when they were born. Therefore, well-known researchers including Maria Montessori, Jean Piaget, Rousseou, and Frobel addressed motion with behavior and underlined the contribution of outdoor activities in the early childhood to developments (Alat, Akgümüş, & Cavalı, 2012). The children who participates into outdoor activities display rapport, reduced antisocial behaviors, cooperation, solidarity, and increased skills for conflict-solving (Burdette & Whitaker, 2005). From the perspective of childcare, the outdoor play activities are one of the most important areas where children interact with nature and experience life. It is because a child is completely deprived of outdoor activities at home after the mother begins to work. Therefore, the program of a day care nursery for children must include outdoor activities (Cosco, 2007; Nedovic & Morrissey, 2013).

However, the research suggests that for many children in different places of the world, time to play is rather a series of planned activities not an experience chosen and managed by them. Moreover, games have gradually become an indoor activity because of the concern of parents about threats such as foreign people, pollution, traffic, use of drugs, and antisocial behaviors (Ünal, 2009; RPC, 2019). Games can be defined as a "nutrition" for development of skills that help use to decide what to do or what should not be done in life. Deep learning and development of high skills can be enriched by games. An English research indicated that children reduced to allocate time to play outdoor games by fifty percent (RPC, 2019). Today's children appear to spend less time for outdoor play activities as compared to the previous generations as a result, they miss important opportunities for development. Therefore, this chapter of the book focuses on the general developmental characteristics of 0-36 month-old şeklinde değişmeli children, influences of outdoor games-having an important role in development- on the development, outdoor activities for infants and toddlers, and how to design outdoor playgrounds because this is important for the child development.

MAIN FOCUS OF THE CHAPTER

General Development Characteristics of Infants and Toddlers

The period when growth and development is the fastest in the human life is the infancy. Development is a dynamic process in which human changes both qualitatively and quantitatively from the fertilization to death. Completion of skills and occurrence of more complex skills occur in a non-accidental order (Howard et al., 2011; Trawick Swith, 2013). Development progress periodic. Each of the periods is qualitatively different from the other and takes place in a specific age range (Özyurt, 2012; Trawick Swith, 2013). In this developmentally organized process, the child's physical, motor, cognitive, social, emotional and language development allow to predict successive processes (Howard et al., 2011).

The infancy period is a quick growth and change process where physical, cognitive, social, emotional and linguistic development can be easily noticed (Marrotz & Allen, 2016). A newborn that who is dependent on his/her family in all aspects turns into an individual at the end of this process, who is able to take care self and build an effective communication with others (Marrotz & Allen, 2016; Shelov & Altmann, 2015). The rapid development of muscles and bones is very effective on the growingly increased coordination in the infancy. The brain that physically coordinates the infant develops simultaneously with bones and muscles. A child that shows a normal physical development from birth to three years old doubles his/her height and his/her weight increases four times. In addition to this quick physical development, they acquire many motor skills towards the end of three years, including use of fingers in coordination, holding a spoon correctly, sitting-standing up, walking, running, urinating, and catching-throwing a ball (Marrotz & Allen, 2016; Shelow & Altmann, 2015; Trawick Swith, 2013).

In the infancy, the infant's language skills also develop significantly and quickly. The receptive communication skills that allow infants to understand conversation of people around develop before they are able to speak. The words that infants initially understand are either their name or "No" because they actively and mostly hear these words (San Bayhan & Artan, 2011). The infants communicate with people around through sounds, words, crying, non-verbal gestures and facial expressions before they turn one year of age. They start to gather words together between 18 and 24 months (Berk, 2013; Shelov & Altman, 2015; Trawick Swith, 2013). They slowly add new words to their vocabulary. An infant starts to tell approximately fifty to four hundred words within several weeks (San Bayhan & Artan, 2011). The reading-writing skills can be now supported from the twelfth month in the infancy. The infants can acquire basic reading skills through their parents, such as holding a book correctly, turning the pages properly, etc. (Trawick Swith, 2013). So, the children will be able to analyze and use the basic structure and function of a language and actively use the language during the infancy and at the end of first year following infancy (Berk, 2013; Gander & Gardiner, 2015). The reading-writing skills can now be supported 12 to 36 months. When the families read a book to infants, they would allow children to acquire basic reading skills such as how to hold a book and how to turn the pages (Trawickh Swith, 2009)

The first years of life are critically important for social-emotional development. This process progresses in steps and continues for a lifetime. Social development occurs as a result of interaction with other people and other areas of development play an active role in shaping of social development (Schunk, 2009). An infant is born with having fundamental basis required for social development. It is instinctive why they choose to look at a human face. This is why they like people to talk to them. They soon start to distinguish the smell, face and voice of persons who primarily take care of them and meet their needs. They learn to indicate their various requirements by crying differently, making a sound and using their body language (Berk, 2013). The children who lack of social skills are rejected by their friends and might not achieve an academic success. Such children may have more social problems than their friends have in their future life (Baran, 2011). This is a period where the foundation of many characteristics is set such as desire to discover, curiosity, searching, independency, and basic confidence required for healthy development of an infant. Therefore, an environment that is full of correct and rich stimulants, and a family and pedagogue that will use it effectively are very important. The infants experience emotions such as fear, sorrow, surprise, joy, and anger during the first six months. The sense of self starts to emerge after the first six months in the infants. The sense of self manifests with emotions including guilty, jealousy, pride, empathy, and embarrassment. The infants are able to laugh loudly after the third and fourth month (Ministry of National Education [MEB], 2013).

The infants acquire basic feeling of trust during the first year. They can display the ability to understand the mood of other people, which is described as "social referencing" at the end of first year. The infants show interest in their peer after they complete six months. The infants' face-to-face game is reduced and they start to be interested in surrounding and peers. From eighteen months to two years, "playing with peers" that is non-verbal such as running and jumping and imitates one and other's actions can be observed in the behavior of infants. An infant that completes one and a half year of age prefers playing with children over playing with adults. They start to play a pretending game towards the end of second year (Berk, 2013; MEB, 2013; Trawick Swith, 2013). They prefer to play with other children rather than playing with adults after one and a half year of age. Symbolic games are observed towards the end of second year. They play in parallel between two and four years of age. When they are two-and-a-half-year-old, they start to imitate the persons they like and play role-based dramatic games. Their symbolic game is more persuasive at three years old (MEB, 2013).

The brain starts to develop before the birth and continues to develop during the adulthood. The most critical part of brain development is seen in the first years after birth. A baby is born with millions of neurons in its brain. The functionality of brain is based on interconnection of neurons through axons. As the infant experiences the world, the neuron connections are also affected, which are required for formation of paths between the different parts of the brain managing physical, cognitive and emotional reactions. Therefore, the environment and stimulants provided to infant are very important for development of brain during the first years (Mustard, 2002; Shonkof & Phillips, 2001). Early years of life are also very important for cognitive development. The behaviors acquired by the infant during this period form the foundation of cognitive development. Although the infant is not initially able to distinguish itself from other objects, they try to discover their body through initial patterns such as sucking, holding and catching, develop a meaningful relationship with other objects and start to distinguish themselves from the outer world. The infants discover the patterns that they have by accident, and try to get to know the environment and form new cognitive structures during the same period (Senemoğlu, 2009). The other cognitive skills such as memory, attention and play are acquired during infancy (Trawickh Swith, 2009).

Contribution of Outdoor Games to Development of Infants and Toddlers

Games can be considered as a universal concept because there is a game wherever a child is present. As the children and games are usually considered as a whole, there are many definitions combining children and games. A general description of games would be: "a natural instrumental or non-instrumental learning activity that is performed with or without a specific purpose, with or without rules, affects all development areas of a child, and is joined by children willingly to enjoy" (Koçyiğit, Tuğluk, & Kök,

2007). Games are a major component of a healthy development. It is a process that is learned by children in activity and where children accommodate with social and physical environment (Uluğ, 2007). It is a test room where children test and try what they hear and see, and reinforce what they learn. Playing games is the freedom of a child (Yörükoğlu, 2002).

In the games, the children have the chance to exhibit many conflicts and problems without the pressure of adults and outer world, and experience the corresponding feelings without suppressing. The children reflect their characteristics of developmental period on the games. For example, if the child is egocentric, this would be reflected on the games, or the child can learn to share through playing games (Koçyiğit, Tuğluk, & Kök, 2007). The playgrounds that are usually constructed on a concrete site for children are intended to support children's gross motor development and have plastic play equipment. The playgrounds lack an arrangement that would allow to interact with nature, to improve creativity of children, and to promote or satisfy their curiosity and interest. The playgrounds do not have any features that draw children who want to play games individually, who want to search outdoor, and who do not want to get involved in intense physical activity (Dyment & Bell 2008; Staempfli, 2009; Alat, Gümüş, & Cavalı, 2012).

The activities performed by children outdoor, particularly running, jumping, climbing, and crawling requiring physical strength, enable a child's body systems (e.g., respiratory, circulatory, digestive, urinary systems) to function properly. Such activities require movement of body, ensure burning of extra fats in the body, strengthen muscles, and enable endocrine glands to function properly, which all especially contributes to development-related functions (Baykoç Dönmez, 2000).

Playing games is the optimum education program for children in terms of social, physical and conceptual development. They use the materials and communicate with other children to learn how to control people around and the feeling of competition and joy. Playing games is also a way of acquiring basic information by children about the world. Such information paves the way for language, arts, social sciences, mathematics and science that they would learn later (Korkmaz, 2009). The children also symbolically reflect and express their emotions they cannot share with other people and the conditions that disturb them in the real life through games (Arnas, 2007).

Infants and toddlers want outdoor activities as well because the first ages are very important to discover the world. The children want to have outdoor activities under the guidance of adults that value outdoor and take care of them because they have the opportunity to interact with nature, are able to be physically active, can take risks, spend their time safely under the supervision of adults, can have rich and emotional experience, and have the chance to create, reorganize and move to some other place. In brief, they find meaningful learning environments through spontaneous games in the outdoor playgrounds. The nature makes feel children valuable as it provides the opportunity to experience the games (Greenfield, 2007). The children who do not have the chance to play outdoor games are more likely to exhibit aggressive behaviors. It is because outdoor activities make children feel to be free. The children watch the clouds, observe the limitless motions of birds, feel that they embrace the world with the rainbow, and have the chance to interact with rain, snow and wind, so the children do not feel restricted (Perry, 2001). Infants and toddlers need outdoor activities as much as other children do because there are many stimulants inducing them to go out. The outdoor activities provide opportunities for them to experience their physical and emotional skills (Greenman, 2005).

As the infants and toddlers interact with natural materials such as sand, soil, water, leaf, trees, etc. during early ages, their hand-eye coordination develops and the development of small muscles is indirectly supported. The games allow children to search vicinity, know the objects and solve problems. Thus, the children learn many concepts such as size, shape, color, dimension, weight, volume, measurement, count, time, place, distance, and space and many intellectual processes such as matching, classification, sequencing, analysis, synthesis and problem solving. The children usually tend to creativity when they find the required environment and materials. They have the chance to experience freely and develop confidence because they constantly reflect their desires and skills on the games (Koçyiğit, Tuğluk, & Kök, 2007).

Outdoor Playgrounds and Arrangement for Infants and Toddlers

For children to play outdoor games, the streets-safe and comfortable areas- and the areas near their home are preferred. A playground must be free of traffic elements, have the texture of nature and sufficient playing equipment (Çakırer Özservet, 2015). The place of games should allow flexible use and creative ideas because excessive control disturbs the nature of games. The children also discover their ability while they play outdoor games. Therefore, they need activity areas where they can move freely (Hamilton, 2002). The children need fewer adults for outdoor activities. In addition, the research suggested that children enjoyed spending time with adults in outdoor places (Brown et al., 2009; Dowdell et al., 2011). According to Kernan (2010), the children enjoy most of the outdoor places where adults feel comfortable and find pleasure to be with children. The potential of outdoor for child development is not completely understood by many teachers and outdoor activities are not prioritized most of the time (Maynard & Waters, 2007; Brown et al., 2009).

The outdoor spaces designed for children must be arranged to allow children to configure their game and set creative games. The playgrounds should encourage playing games and provide opportunity to learn about the nature (Dighe, 1993). There must be gardens and natural areas that have wetlands, greenhouses, grass, plants, vegetable, animals and butterflies where children can observe, study and take care (Wilson, Kilmer & Knauerhase, 1996). 0-36-month children tend to nature of game and discover rather than creative playgrounds as compared to older children. Therefore, a playground must have the equipment and landscape features that encourage to get around, walk, toddle, climb and run. The infants that begin to crawl would enjoy a small hill or a tunnel to discover. A playground that contains a ramp, steps, tunnel, bridge and slide would influence the toddlers and young children. Crawling, creeping and walking on different textures would arouse feelings for explore. The toys they can drag and push, a firm ground where they can use them, or the areas where they can leave a mark when dragging them on the ground, and the swings they can safely use with the help of adults are very important for outdoor activities (Wortham, 1989; Greenman, 2005).

The outdoor playgrounds should contain the following play tools and equipment: seesaws, swings, slides, climbing equipment, spring instruments, big wheels, springboards, tunnels, water sprays, water tables, water channels, water pumps, sand pools, swinging pot, walking board, hammocks, sacks, car tires, old cars and camper vans, bicycles with three wheels, empty cartoon boxes, buckets, shovels, harrows, plastic tunnels, empty barrels, and hoops. This equipment can be positioned to allow children to use them easily (Sevinç, 2004). There must be also spaces containing animals, gardens and bushes where children can observe the nature. The materials in the outdoor playgrounds must be safe, economic, suitable for development of children, practical and interesting (Deretarla Gül, 2012).

The children may feel uncomfortable when the outdoor playgrounds are too large, too small or too busy. The children want a playground to have enough size to provide their privacy, to be outside off the adult spaces and to have three basic features: a playground must allow them to move freely, manage and feel comfortable. The routes of playgrounds should be linked to each other. The playground should have a firm ground for children to play games such as hitting, kicking and bouncing a ball; a grass ground to tumble down over, roll over, and lie down; a space to play with water and sand, and a sloped area to use wheel vehicles (Olds, 2001; Gill, 2005).

Except when it is very hot or extremely rainy weather, outdoor places can be chosen to support development of infants and toddlers (NCAC, 2009). It is very important to developmentally arrange outdoor places. The properties of bushes, trees and plants located in the playgrounds or parks where infants/children play must be considered. Non-toxic plants must be used. The ground must be coated with different materials such as grass, wood or sand; however, the safety of children must also be considered. In the design of slight slopes covered with grass for infants/children to move down or up, different textures children can feel when they lie down or crawl on it, and safe walkways, such places should be constructed considering the possibility of fall or injury of toddlers in particular. The playgrounds should use swings made of soft plastic, tunnels, protruding trees, bushes, safe climbing spaces, logs, ladders, child's house, slide and similar toys, and cushions and water toys to create an environment for supporting development of children (Greenman &Stonehouse, 1996).

Outdoor Activities for Infants and Toddlers

The major outdoor activity for infants and toddlers is to allow them to interact with elements present in nature. It is because the nature has many stimulants to draw attention of children and encourage playing. Numerous stimulants such as rocks, stones, soil, leaf, flowers, etc. invite children to play and learn. The nature comprising of open-ended materials meet children's need for creativity in a natural way because all the natural elements is limitless, has no cost and is not a toy that would cause a commercial concern (White, 2008).

When they go to outdoors, the children should be encouraged to carry out actions to support motor development such as crawling, walking, running, climbing, throwing, jumping, rolling over, sliding, bouncing, and balance. So, the children start to interact with outer environment and learn how to control their body. They can play games suitable for their age group to support skills including locomotor motions, non-locomotor motions, manipulative skills, agility, balance and coordination (Ataov, 2004). When outdoor, the infants and toddlers acquire ability to touch and grab objects of different sizes. They have the chance to creep on the flat ground and climb up the stones and logs. Their skills for position can be developed more. Their language development can be supported more, such as behind-in front of-next to the tree because the nature has many animals and plants. Their names can be told for children to repeat the names (Hapwood Stephens, 2015).

The emotional development and physical development are very important especially for infants and toddlers. Therefore, 0-36-month old children should be allowed to interact to support all of their development areas (Hutchins & Simms, 1999). In this respect, crawling on grass, soil and concrete, touching the leaves, smelling a fruit on the tree, vegetable or flower on the soil, touching the bod of trees, watching natural phenomenon such as rain, snow and wind, playing with water which is the safest material in the nature, playing with mud, and interaction with animals outdoor can be considered important activities to do during the first years of age.

After the child is able to stand, he/she should be allowed to exercise hitting a ball on a suitable ground outdoor. After he/she is 24-30 months old, they should exercise sitting on a seesaw and jumping over a sand hill. When he/she is 30-36 months old, they should exercise different ways of walking by putting obstacles on the challenging grounds (gravelly, sandy and earth grounds) outdoor. Adults should walk with them in the park to attract their attention on the surrounding and speak about the things in nature. The children should be allowed to interact with materials and trees and talk about their properties (Atli & Baran, 2017).

- Example Game 1: Bird calls
- Age: 24-36 months
- Material: Recorder
- Activity area: A natural environment where bird calls can be heard
- Area of development: Social-emotional, language and cognition
- Learning process: An adult should go and sit with the child where they can hear the bird calls early in the morning or when the sun goes down. The adult listens to the bird calls and draws child's attention on the birdsong. The adult catches and imitates different bird calls. Then, the adult ask child to imitate the same. The recorder, if they have one, records the birdsongs and the imitation of child.
- Example Game 2: Balloon and Water
- Age: 18-36 months
- Material: Color balloons, hose and bucket
- Activity area: Firm ground, and a grass area where hose and water can be used
- Area of development: Social-emotional area, cognitive area, and fine-gross motor area

Learning process: The adult go to the grass area with the child. The adult fills the bucket with water using the hose. The child is also allowed to fill the bucket with water using the hose and play with water. The adult blows up the color balloons and puts them in the water bucket. The adult moves the balloons with hose. They play for a while, then the adult deflates the balloons and they fill the balloons with water using the hose. They go to the firm ground and drop the balloons filled with water on the firm ground. The adult draws child's attention to the sound made by the balloons when they hit the ground and to the shape on the ground. They continue to play this game if the child likes to play.

SOLUTIONS AND RECOMMENDATIONS

The outdoor playgrounds allow children canalize their aggressive feeling to the right way, discover their skills, experience the tools, acquire sense of place-direction, and support development of positive ego (Sevinç, 2004). What is important is that children interact with nature in safe areas that would provide the correct support in line with the developmental characteristics of child.

During the preschool period, although many developmental characteristics of child are supported in indoor spaces, they need to explore their potential to move and gain information about their body. Therefore, adults have major roles here. They should keep children away from environments making them stationary and encourage them to move (Özyürek, Özkan, Begde & Yavuz, 2015).

A project implemented in the Portuguese early childhood setting studied with 14 children who were 14-36 months old and reported that outdoor play activities had positive effects on the children's learning and healthy development (Bento &Dias, 2017). It is because children have the opportunity to move freely and have fun in such environment. While children play, they also breathe the clean air, take advantage of sunlight, and improve their physical motion skills, as a result, their bones and immune system gain strength. The physical activities prevent children to gain overweight and reduce the risk for obesity (Maynard &Waters, 2007; Dyment & Bell, 2008). The parents have concerns about the outdoor play activities, for example, children may want to play risky games, put foreign materials into their mouth, and eat poisonous plants. If a child wants to climb up a tree or experience an object, this is in fact very important for development of problem-solving skills of the child (Stephenson, 2002; Tovey, 2011).

The outdoor play activities also provide significant opportunities for socialization of child because as the child feels more comfortable in the outdoor space, he/she is not worried about interacting with more people. There are important interactions between the adult and the child during the outdoor activities; therefore, the child takes the adult as a role model. They also have the chance to communicate and play with other children (Stephenson, 2002). Through their children, the families have the chance to collect materials children can play with, acquire the skills to set a game and create an environment to play from their children, so they can more effectively interact with their children (Bento & Dias, 2017).

This chapter generally outlines the critical importance of outdoor play activities for children and advises that children should be introduced with such games from early childhood. However, the families express their concerns for very early ages such as concern about health, concern about security, lack of sufficient and suitable outdoor spaces, therefore, they do not allow children to sufficiently play in the outdoor places. As indicated above, outdoor activities are very important to infants and toddlers who just meet the life. Thus, a suitable environment should be provided for them to spend time in outdoor spaces.

FUTURE RESEARCH DIRECTIONS

The institutions that deal with family and early childhood education should allocate a period of time for the infants and toddlers to spend time outdoor every day. It is because children spend most of the time indoor during the day and need activities to interact with the world and nature. Therefore, the investigators may conduct research to determine how the families regard outdoor play activities and whether they are aware of the importance of outdoor activities. Cooperation should be established between the early childhood education institutions and the families to discuss collection of natural materials, what kind of games can be played with them, development of outdoor activity programs, and assessment of their influence on the child development.

The professionals should tell families about the benefits of getting dirty, wet, playing with mud and playing risky games for the children, so the families can develop a positive attitude towards outdoor play activities. The municipalities and child development professionals should design together playgrounds considering the development of children. Especially, there is a need to design spaces where the infants can crawl, roll over and lie down as well as outdoor garden toys suitable for physical development of infants and toddlers. The investigators should conduct research to present the current conditions and determine whether the urban playgrounds and natural environments are healthy for the infants and toddlers. They can also develop projects with municipalities for child-friendly outdoor playgrounds.

The 0-36 education programs that effectively include outdoor play activities in their education program and the programs including rather indoor activities should be compared in order to assess the impacts on the development of children. The attitude and conduct of parents, teachers and caretakers for outdoor play activities can be assessed. The safety and security of outdoor playgrounds should be ensured and the security problems families are concerned about should be identified.

The literature may be reviewed and this chapter of the book that presents the benefits of outdoor play activities for children should may be used. The contribution of education programs that intensively use the playing activities during the first three years of life to the development of children may be assessed by the investigators. Furthermore, the child development experts, educators and architectures may develop projects together on how existing outdoor playgrounds could be made more beneficial to the infants and toddlers and how they could be used in the most proper way.

CONCLUSION

As we particularly become urbanized, the outdoor activities are gradually removed from the children's life. In a way, the children are forced to spend more time indoor even at early ages as the number of apartments and concrete areas is growingly increased and families cannot find a suitable place to their children. This can also be considered children who are forced to spend time and play games in a restricted area and with limited materials. The adults and children slowly start to lose their habits and spend most of their time indoor because of the new advances in the technology. If this continues as it is, the children will start to lose their interaction with natural environments in the near future. The natural life that provides children with limitless freedom, opportunity to act, and materials is in fact a communication language that child builds with the world. During the first years of early childhood development, considering that the children need interaction, images, explore, understand and problem solving very much, we can say that outdoor play activities have very important effects for their healthy development. It is because establishing an effective communication language with the child and nature in the early ages would help child to build more powerful attachments with the nature in the later periods.

REFERENCES

Alat, Z., Akgümüş, Ö., & Cavalı, D. (2012). Okul öncesi eğitimde açık hava etkinliklerine yönelik öğretmen görüş ve uygulamaları. *Mersin Üniversitesi Eğitim Fakültesi Dergisi*, *8*(3), 47–62.

Ataov, A. (2004). Water as part of children's environment & children's preferences for waterfront settings. *Paper presented at the First International Conference on Inclusive Environments: Open Space-People Space*. Academic Press.

Atli, S., & Baran, G. (2017). Gelişimin desteklenmesi, In A. Köksal Akyol (Ed.), *Erken Çocukluk Döneminde Gelişim*, *1*, 449–495.

Baykoç Dönmez, N. (2000). *Üniversite Çocuk Gelişimi ve Eğitimi Bölümü ve Kız Meslek Lisesi Öğrencileri İçin Oyun Kitabı*. İstanbul: Esin Yayınevi.

Bento, D., & Dias, G. (2017). The importance of outdoor play for young children's healthy development. *Porto Biomedical Journal*, *2*(5), 157–160. doi:10.1016/j.pbj.2017.03.003

Berk, L. E. (2013). *Bebekler ve Çocuklar Doğum Öncesinden Orta Çocukluğa*. Ankara: Nobel Akademik Yayıncılık.

Brown, W. H., Pfeiffer, K. A., McIver, K. L., Dowda, M., Addy, L., & Pate, R. R. (2009). Social and environmental factors associated with preschoolers' nonsedentary physical activity. *Child Development, 80*(1), 45–58. doi:10.1111/j.1467-8624.2008.01245.x PMID:19236392

Burdette, H. L., & Whitaker, R. C. (2005). A national study of neighborhood safety, outdoor play, television viewing, and obesity in preschool children. *Pediatrics, 116*(3), 657–662. doi:10.1542/peds.2004-2443 PMID:16140705

Çakırer Özservet, Y. (2015). Geleneksel çocuk oyunları için sokak temelli geleneksel oyun alanlarının yeniden keşfi, I. In *Uluslararası Türk Dünyası Çocuk Oyun ve Oyuncakları Kurultayı*. Academic Press.

Cardon, G. M., & De Bourdeaudhuij, M. M. (2008). Are preschool children active enough? Objectively measured physical activity levels. *Research Quarterly for Exercise and Sport, 79*(3), 326–332. doi:10.1080/02701367.2008.10599496 PMID:18816944

Cosco, N. G. (2007). Environmental interventions for healthy development of young children in the outdoors. In C. Ward-Thompson & P. Travlou (Eds.), *Open space people space* (pp. 125–135). London, England: Taylor and Francis.

Çukur, D. (2009). Çocuk dinlenimi açısından oyunun önemi ve konut yakın çevresinde oyun değerini artırıcı mekansal düzenlemeler. *Ege Mimarlık*, 2(69).

Demiriz, S., Karadağ, A., & Ulutaş, I. (2003). *Okulöncesi eğitim kurumlarında eğitim ortamı ve donanımı*. Ankara: Anı Yayıncılık.

Deretarla-Gül, E. (2012). Ailelerin çocuk bahçelerine ve çocuk bahçelerindeki materyallere bakış açılarının incelenmesi. *Ç.Ü. Sosyal Bilimler Enstitüsü Dergisi, 21*(3), 261–274.

Dighe, J. (1993). Children and the earth. *Young Children, 48*(3), 58–63.

Dowdell, K., Gray, T., & Malone, K. (2011). Nature and its influence on children's outdoor play. *Australian Journal of Outdoor Education, 15*(2), 24–35. doi:10.1007/BF03400925

Dyment, J. E., & Bell, A. C. (2008). Grounds for movement: Green school grounds as sites for promoting physical activity. *Health Education Research, 23*(6), 952–962. doi:10.1093/her/cym059 PMID:17956885

Greenfield, C. (2007). A case study of children's and adults' perceptions of 'being outside' in one early childhood centre. Unpublished Qualitative Masters of Education Thesis, Massey University, Palmerston North, New Zealand.

Greenman, J. (2005). Caring spaces, learning places, Environments that work. Redmond, WA: Exchange Press.

Greenman, J., & Stonehouse, A. (1996). *Prime times a handbook forexcellence in infant and toddler care*. Redleaf Press.

Hopwood Stephens. (2015). Outdoor Play for 1–3 Year Olds. New York: Routledge.

Howard, V. F., Williams, B., & Lapper, E. C. (2011). *Özel Gereksimi Olan Küçük Çocuklar*. Ankara: Nobel Yayınevi.

Hutchins, T., & Sims, M. (1999). *Programme planning for infants and toddlers: An ecological approach.* Sydney: Prentice Hall.

Kernan, M. (2010). Space and place as a source of belonging and participation in urban environments: Considering the role of early childhood education and care settings. *European Early Childhood Education Research Journal, 18*(2), 199–213. doi:10.1080/13502931003784420

Koçyiğit, S., Tuğluk, M.N., & Kök, M. (2007). Çocuğun gelişiminde eğitsel bir etkinlik olarak oyun. *KKEFDI- OKKEF, 16*, 325-342.

Marotz, L. R., & Allen, K. E. (2016). *Developmental Profiles: Pre-Birth Through Adolescence.* United States: Cengage Learning.

Maynard, T., & Waters, J. (2007). Learning in the outdoor environment: A missed opportunity? *Early Years, 27*(3), 255–265. doi:10.1080/09575140701594400

MEB. (2013). 0-36 Ay Çocuklar İçin Eğitim Programı, Ankara, Turkey: MEB-UNİCEF.

Morrison, G. (2007). *Fundemantals of Early Childhood Education. United States: Coral Graphic Service.* United States: Pearson Merill Prentince Hall.

Mustard, J. F. (2002). Early child development and the brain – the base for health, learning and behavior throughout life. *Early Child Development to Human Development, 10*, 23–62.

NCAC. (2009). Putting Children First. *The Magazine Of the National Child care Accreditation Council (NCAC), 30*(6), 3-5.

Nedovic, S., & Morrissey, A. (2013). Calm active and focused: Children's responses to an organic outdoor learning environment. *Learning Environments Research, 16*(2), 281–295. doi:10.100710984-013-9127-9

Oğuzkan, Ş., & Oral, G. (1997). *Okulöncesi eğitimi.* İstanbul: Milli Eğitim Basımevi.

Oktay, A., & Unutkan, Ö. P. (2005). Erken Çocuklukta Gelişim ve Eğitimde Yeni Yaklaşımlar. In M. Sevinç (Ed.), İlköğretime hazır oluş ve okul öncesi eğitimle ilköğretimin karşılaştırılması (pp, 145-155). Istanbul, Turkey: Morpa Kültür Yayınları.

Olds, A. (2001). *Child care design guide.* New York: McGraw Hill.

Ozyürek, A., Ozkan, I., Begde, Z., & Yavuz, N. F. (2015). Okul öncesi dönem çocuklar için hareket ve sosyal becerilerin gelişiminin desteklenmesine yönelik eğitim programı. *International Journal of Science Culture and Sport, 3*(11), 2148–1148. doi:10.14486/IJSCS261

Ozyurt, B. E. (2012). Gelişim konularına genel bakış. In A. Kaya (Ed.), *Eğitim psikolojisi* (pp. 1–27). Ankara: Pegem A Yayıncılık.

Pehlivan, H. (2005). *Oyun ve Öğrenme.* Ankara: Anı Yayıncılık.

Perry, J. P. (2001). *Outdoor play: Teaching strategies with young children.* New York: Teachers College Press.

Real Play Coalition (RPC). (2019). Retrieved from https://www.realplaycoalition.com/

San Bayhan, P., & Artan, İ. (2011). *Çocuk Gelişimi ve Eğitimi*. Istanbul, Turkey: Morpa Yayınları, İstanbul.

Schunk, D. H. (2009). *Öğrenme Teorileri: Eğitimsel Bir Bakış*. Ankara: Nobel Akademik Yayıncılık, Ankara.

Senemoğlu, N. (2012). *Gelişim Öğrenme ve Öğretim*. Ankara: Pegem Akademi Yayıncılık.

Sevinç, M. (2004). *Erken çocukluk gelişimi ve eğitiminde oyun*. Ankara: Morpa Kültür Yayınları.

Shelov, S. P., & Altmann, T. R. (2015). Bebek ve Küçük Çocuk Bakımı (0-5 Yaş). Ankara: İmge Kitapevi.

Shonkoff, J. P., & Phillips, D. A. (2001). From neurons to neighborhoods: the science of early childhood development: an introduction. *Zero to Three*, *21*, 4–7.

Staempfli, M. B. (2009). Reintroducing adventure into children's outdoor play environments. *Environment and Behavior*, *41*(2), 268–280. doi:10.1177/0013916508315000

Stephenson, A. (1998). Opening up the outdoors: A reappraisal of young children's outdoor experiences. Unpublished Master Thesis, Wellington.

Stephenson, A. (2002). Opening up the outdoors:exploring there lation ship between the in door and outdoor environments of acentre. *European Early Childhood Education Research Journal*, *10*(1), 29–38. doi:10.1080/13502930285208821

Tovey, H. (201 1). Achieving the balance. In J. White (Ed.), *Outdoor provision in the early years* (pp. 12–22). London: Sage Publication.

Trawich Swith, J. (2013). *Erken Çocukluk Döneminde Gelişim*. Ankara: Nobel Akademik Yayıncılık.

Ulke Kürkçüoğlu, B. (2010). 0-6 Yaş Arası Çocukların Temel Gelişimsel Özellikleri: Bilişsel gelişim ve dil gelişimi, In İ. H. Diken (Ed.), Erken çocukluk eğitimi. Ankara, Turkey: Pegem Akademi.

Uluğ, H. (2007). Kuzey Adana'daki çocuk oyun alanlarının bitki seçimi yönünden irdelenmesi. Retrieved from http://library.cu.edu.tr/tezler/6587.pdf

Unal, M. (2009). The place and importance of playgrounds in child development. *Inonu University Journal of Faculty of Education*, *10*(2), 95–109.

White, J. (2008). *Playing and learning out doors. Making provision forhigh-quality experiences in the outdoor environment*. Oxon: Routledge.

Wilson, R. (2008). *Nature and young children: Encouraging creative play and learning in natural environments*. New York, NY: Routledge.

Wilson, R. A., Kilmer, S. J., & Knauerhase, V. (1996). Developing an environmental outdoor play space. *Young Children*, *51*(6), 56–61.

Wortham, S. C. (1989). Outdoor play environments for infants and toddlers. *Day Care and Early Education*, *16*(4), 28–30. doi:10.1007/BF01619376

Chapter 18
Perinatal and Early Childhood Development:
Education, Innovation, and Sustainable Development Goals

Lucy Kivuti-Bitok
University of Nairobi, Kenya

Joyce Jebet Cheptum
University of Nairobi, Kenya

ABSTRACT

This chapter looks at the systems approach in addressing the dilemma of investing in neonatal health in resource constraints environment. Is investing in neonatal care the key to attaining sustainable development goals? Is it just or even just to invest in Neonatal care in the midst of competing interests? It applies causal loop diagrams as part of systems dynamics approach in seeking to identify the intended and intended consequences of investing in Neonatal Health. A look into countries have done well in Neonatal health and their investment early childhood education is done in order to give insight into the relationship between the investments.

INTRODUCTION

Perinatal deaths continue being a challenge in the world and especially in developing countries where many interventions have been proposed(Michalow et al., 2015)(Rudan et al., 2010)(Jimenez Soto et al., 2012)with varying outcomes. The WHO and UNICEF in a bid to provide leadership in ending preventable perinatal deaths, improve child development as well as human capital, proposed the Newborn impact framework (Figure 1), (WHO & UNICEF, 2014) a reflection of similar consulted efforts to address the challenges.

DOI: 10.4018/978-1-7998-2952-2.ch018

Figure 1.

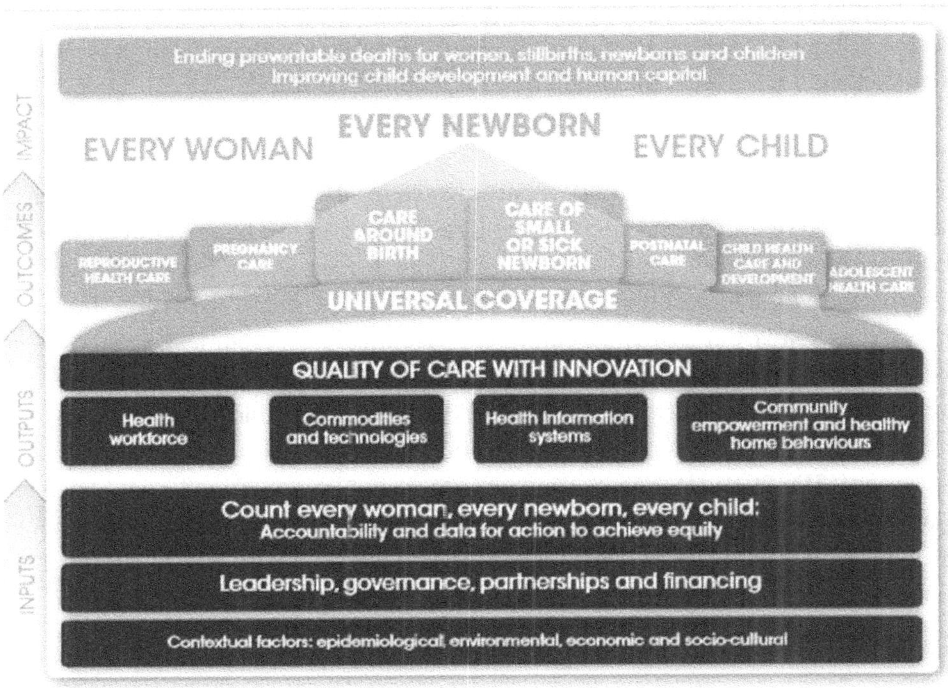

The framework, although good, does not depict the multiple interconnectedness and feedback mechanisms between the different components and hence may not reflect the reality of both intended and unintended consequences of the proposed interventions.

The interdependence and interconnectedness of neonatal health with other systems requires investigation in the context of a wider whole system. The dynamic complexities of contemporary health care setting, the demographic shifts and well as the demand for sustainable development demands increasingly innovative solutions in a resource constrained environment.

Decision making on neonatal health needs to be based on a deep understanding of the whole system behavior and underlying structure. The need to see both the forest and the trees in the grand scheme of things is paramount. Incorporating multiple perspectives not only creates awareness of the intended and unintended consequences of neonatal investment but also demonstrates the need for a multi-sectoral approach through feedback loops and non-linear relationships. Neonatal health's contribution to Sustainable Development Goals(SDG) can only happen in an environment characterized by good governance, transparency, accountability and the rule of law(IISD, 2019)(Hjorth & Bagheri, 2006) (Schmidt, Gostin, & Emanuel, 2015).

System Dynamics(SD), the origin of the current trend of 'whole systems thinking' may provide thinking skills that help to understand the multiple interconnections of complex systems(Wolstenholme, 2005).

This chapter applies SD approach to describe the connectedness between neonatal health and selected significant subsectors: education and training, innovation, income and Universal Health Coverage(UHC) in the context of SDG. The chapter seeks to answer the following questions:

1. Is investing in neonatal care the key to attaining sustainable development goals?
2. Is it just or even fair to invest in neonatal care in the midst of competing interests?
3. What is the effect of investing in education, innovation and training on neonatal health?
4. Is there a connection between income levels and neonatal health?
5. What important aspects of neonatal health should UHC focus on?

A simplified causal loop diagram(CLD) as part of systems dynamics approach is applied in seeking to identify the intended and unintended consequences of investing in neonatal health as well as communicating crucial insights in the grand scheme of things. Investment in neonatal, child and maternal health is likely to have a high impact with Africa as a target (Friberg et al., 2010). The impact of investing in perinatal period as a smart investment in attaining sustainable development goals (SDGs) is viewed as a nexus of UHC, education and innovation and poverty eradication. SDG require systems approach in understanding the complexity of our environment, setting up sensible goals in order to improve the universe and leave a better world for generations to come. In addition, it helps to create a wealthy and prosperous society(IISD, 2019)(Hjorth & Bagheri, 2006) Public health and UHC have an enormous role to play and support SDG (Schmidt, Gostin, & Emanuel, 2015).To support the proposition of the CLD, a look into selected countries that have done well in neonatal health and the relationship with their investment in early childhood development, education and training as well as innovation is analyzed.

SYSTEMS DYNAMICS MODELLING APPROACH

System Dynamics Modelling(SDM) is a sub-sect of systems thinking that has been re-defined as 'a set of synergistic analytic skills used to improve the capability of identifying and understanding systems, predicting their behaviors, and devising modifications to them in order to produce desired effects. These skills work together as a system'(Arnold & Wade, 2015)pg 675.

According to(Dudley, 2011) Systems thinking is applicable in complex problems characterized by:

1. Dynamism- where variables of the system change over time with time delays between cause and effect, actions and reactions. The cause and effect is separated in time and space.
2. Feedback loops with unpredictable behavior.
3. Non linearity- where the causal effect relations between the variables are non-proportional and may be subtle in nature. An increase in a few variables would result in an increase in complexity of the problem nonlinearly.
4. Variability of how different people react to the situations of the system leads to non-predictability of the system and further complicates the non-intuitive inadequacy of policy makers.

Neonatal health challenge fits the bill of complex problem.

STEPS IN SYSTEM DYNAMICS MODELLING

System Dynamics Modelling(SDM) is an approach to studying systems and systems interactions. "System dynamics is a good tool for examining, understanding, and finding solutions to complex problems, especially if these problems involve complex interrelationships, feedbacks and delayed effect(Barlas, 1996).

SD involves a number of steps. First mental models of how systems are structured and interact within and with each other are conceived and analyzed. The mental models are then drawn using lines showing how one variable affects or is linked to the other and the feedback resulting therein. These diagrams are called Causal Loop diagrams (CLD).CLD which are qualitative in nature are cognizant of the fact that most relationships in the world are often nonlinear. The causal loop diagrams are then translated into stocks (which represent all variables capable of accumulating in quantities or values) and flows (which mostly represent actions that cause accumulation or depletion of the stocks). The stocks and flows which are representative of these models are then translated into or drawn using a computer software resulting in the generation of a set of differential equations whose solutions are approximated. These approximated equations are then used in different computer software to virtually simulate different possible scenarios of the model represented hence demonstrate dynamic system behavior over time. Therefore, SD is strategic modelling of non-linear relationships in complex systems. The scope of this chapter focuses on the first two steps with the aim of better understanding neonatal health in selected predefined boundaries in the context of Sustainable development goals.

Models are a great way of combining various types of knowledge that makes it meaningful in a simpler way. Modelling provides a broader perspective and view of planning thus will ensure a holistic approach of concepts. For sustainable development, it is appropriate to make modelling effective, so it is vital that governments consider modelling in all spheres of public health (IISD, 2019). This ought to include social, economic and environmental aspects. Modelling can be applied at all levels of care. It should be diversified so that it is not done independently but across the various sectors. The intended and unintended consequences of various interventions to the wider ecosystem are synthesized in a simplified and holistic way.

Powerful 'small' models can generate, capture and communicate crucial insights to the public and policy makers. The simplified language of feedback loops can capture behavior of the system as a whole in an understandable manner(Ghaffarzadegan, Lyneis, & Richardson, 2011). The identification of endogenous and exogenous factors and the mere realization of the effect of a subsystems behavior to the whole system can challenge the overconfidence of policy makers, eliminate the tendency of silos thinking, persuade different stakeholders to contribute in policy making and reduce policy resistance (Ghaffarzadegan et al., 2011).

Since SD is suited for understanding the problem and enhancing systematic debate between different problem owners (Kotiadis, 2007), full citizenship status is accorded to all stakeholders, all physical and non-physical variable. This makes it easier in making the model and analysing the problem with the aim of better understanding it. Possible predictions are in a qualitative rather than specific numeric output(Richmond & Peterson, 2001).

Some aspects of perinatal health and early childhood development have previously benefited from SD approach. SD approach has been utilized in seeking to understand the dynamics of neonatal care and guidance of policy options in low resource setting (Rwashana, Nakubulwa, Nakakeeto-Kijjambu, & Adam, 2014) (Whitworth, Sewankambo, & Snewin, 2010)(Semwanga, Nakubulwa, & Adam, 2016).

Table 1. Key to the simplified causal loop diagram

Sign	Meaning	Example
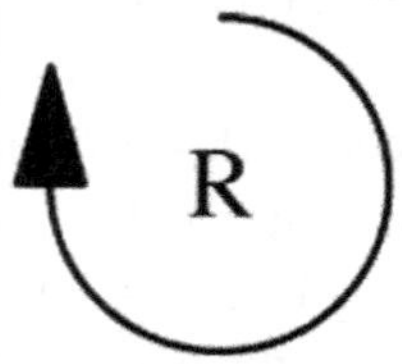	Reinforcing loops represent a growing action where each action adds to another and may be referred to as virtuous cycles when they produce desirable effects or vicious cycles when they produce negative effects	R1 where increase in Neonatal health will lead to an increase in early childhood education enrolment, which will lead to increase in tertiary education, leading to increase in innovation, increase in Income and hence increase in UHC . This then leads to a further increase in neonatal health
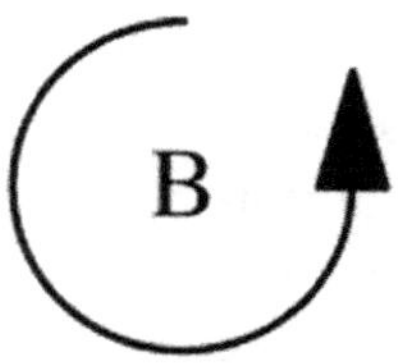	Balancing loops apply where there is an attempt to solve a problem or achieve a goal.	B2- Where an increase in neonatal health aims at reducing utilization of family resources. A reduction in utilization of family resources will lead to a reduction in poverty levels which would lead to an increase in Neonatal Health. B3 –Where an increase in income would lead to a reduction in Poverty levels, which would in turn lead to increase in neonatal health. Healthy neonates would mean an increase in performance and numbers at early childhood education, gradually leading to an increase in tertiary level education and training. This will have an impact on the innovation, leading to an increase in income.
A ——▶ B	A change in A produces a change in B the same direction	Increase in Tertiary education and training leads to an increase in Innovation.
A ——▶ B	A Change in A produces a change in B in the opposite direction	Increase in Population growth would lead to a decrease in Natural Resources.
A —‖—▶ B	A change in A produces a change in B only after a delay	Increase in income would lead to a decrease in poverty levels.

Despite the enourmous potential benefits of SD approach, it is worth noting that there is no single formula that can deal with emerging issues in neonatal health due to its complexity. Even as a subsystem of health, it comprises of subsystems that are interconnected, complicated and have unpredictable non-linear relationships. Despite of having various limitations SDM has documented benefits in addressing public health problems(Homer & Hirsch, 2006) through the integration of various perspectives.

SIMPLIFIED CAUSAL LOOP DIAGRAM

Figure 2. is a simplified Causal Loop Diagram (CLD) that describes how the status of neonatal health impacts and is impacted by childhood health, investment in education, income, innovation and technological advancement as well as UHC, climate change and poverty. It breaks down the complexity in describing how the various variables relate to each other. It looks at investment in social, economic and environmental aspects.

The rise in world population is likely to lead to an unprecedented strain on natural resources. With an approximately 7.2 Billion people in the world, good neonatal health will translate to exponential compilations with the rise in world population leading to both positive and negative, intended and unintended consequences.

Figure 2. Simplified Causal Loop Diagram

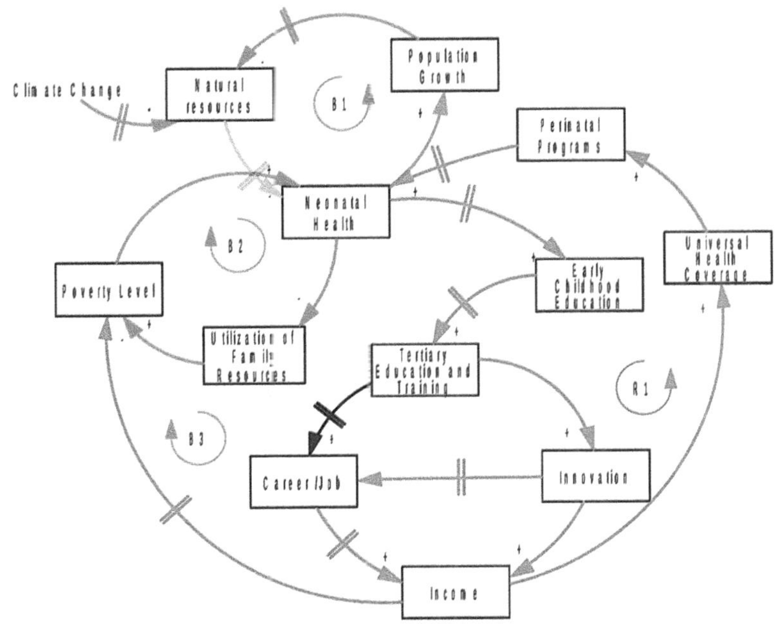

Climate change contributes to neonatal health negatively since it affects the environment which in turn will affect availability of food through poor production. Poor food production or unavailability of food would lead to poor nutrition, a major contributing factor to poor neonatal health. A sick neonate will lead to utilization and depletion of family resources. In situations where there is non UHC, this can lead to family improvisation. According to this B2, lack of resources will lead to poverty thus continuously affect neonatal health. When there is a healthy neonate, there will be positive child development. This leads to early childhood education which contributes to advancement in education and training. This will eventually lead to career development and jobs through employment. Through education, there will be innovations that will be arrived at. Innovations and jobs will lead to source of income. Where there is income, poverty will be eradicated and there will be good neonatal health, but where there is low or no income, the situation will contribute to poor health leading to social gaps. Where there is good income, universal health coverage will be achieved and this will lead to quality prenatal care and thus lead to good neonatal health and also early childhood education.

With the poorest nations recording the highest birth rates and unfortunately 95% of neonatal deaths occurring in developing countries, having a smart investment on the younger population will help in advancing the general health. The highest return on investment occurs during prenatal and early childhood periods. It has been argued that Health status and Human capital are transmitted across generations.

Figure 3 which shows the relationship between health status and human capital, further demonstrates this concept. Health status is 'transmitted' across generations. A healthy fetus will ultimately translate to good health status across the life span. Human skills on the other hand are gained through education systems, job training, work experience and building skills. This therefore implies that human capital increases with age. An increase in human capital is correlated with an increase in earning. It therefore means that concurrent investment in both health and education and training is a smart choice.

Figure 3. Cumulative Investment in Human Capital across Life Span
https://busy.org/@rifat/human-life-cycle-way-we-brown-and-way-we-die

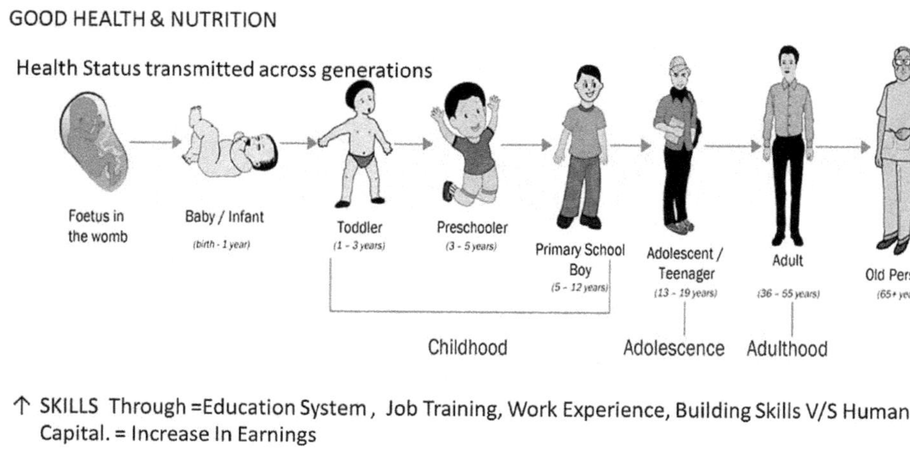

According to Heckman (Heckman, 2012), early childhood development derives success in school and life. The returns to a Unit dollar investment in human capital was highest when invested in Prenatal and early childhood. Figure 4 demonstrates that in regard to age, the earlier the investment, the greater the returns.

Figure 4. Early Childhood Development, a Smart Investment

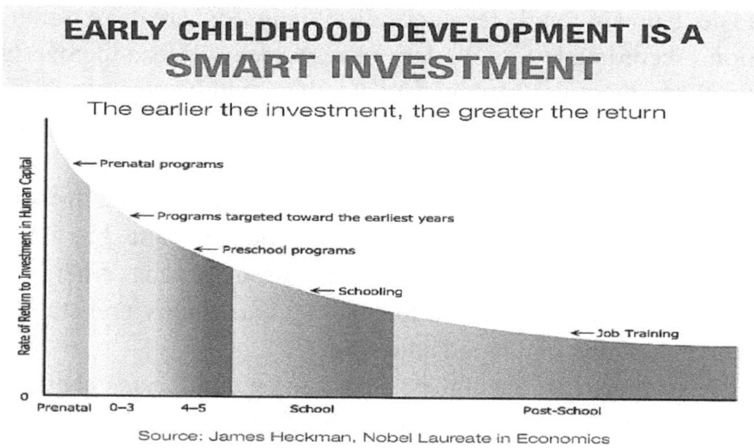

PERINATAL AND EARLY CHILDHOOD DEVELOPMENT

Perinatal period is the period before, during and immediately after childbirth. Perinatal events have a high impact on the later life of an individual both physically and emotionally. The positive outcome of childbirth is influenced by care during pregnancy and during labor and delivery. Every pregnancy carries risks and complications which may arise. The time of childbirth is the period which carries the highest risk for both the mother and the baby. More than 40% of maternal deaths and stillbirths or neonatal deaths occur every year during childbirth (Lawn et al., 2014). Africa bears the highest burden in neonatal and child morbidity and mortality. As a result, it is paramount that quality care is provided to ensure both mother and baby have a good outcome. Poorly managed complications of pregnancy and childbirth affects childhood and eventually the adulthood period. This therefore underpins the critical care required during this period.

It should be noted that the leading causes of neonatal mortality are prematurity, birth asphyxia and prematurity, yet they are preventable. To avert these causes, there should be investment in reproductive, maternal and newborn health. Reproductive health encompasses health for all through the reproductive life cycle. The reproductive life cycle incorporates the infants, children, adolescents, adults and old aged. During each phase of the life cycle, crucial interventions will lead to a healthy population thus it calls for the need to involve all the players in contributing to this.

Perinatal care leads to healthy mothers and babies and prevents stillbirths and disability. Interventions done during the perinatal period have been shown to have the greatest potential in averting complications and death. It is estimated that over 3 million lives of women and babies would be saved by 2025 at an additional running cost of US$1·15 per person (Lawn et al., 2014). This indicates the need for considering perinatal health during government planning so that additional funding could be budgeted for as it will lead to a triple return on investment. Perinatal and childhood development is greatly affected by maternal health. When a mother has poor health, the child may not receive the best care and may not achieve optimum health. For example, when a mother develops severe complications during childbirth, the baby may not be able to breastfeed. This affects the baby's health since breast milk is the best food for the baby that carries adequate nutrients and enhances childhood development through providing immunity and later in life the child's intelligence quotient (IQ). In order to achieve success in perinatal and childhood development, it is important to think about strategies that can improve neonatal health.

The crucial periods for life development are the neonatal period, childhood and adolescence. These periods are key because they entail the formation and development of the brain. Neonatal and childhood periods are marked with higher risks of morbidity and mortality especially in the developing countries. One of the strategies of ensuring a reduction in the risks is through UHC. UHC ensures that all individuals can access affordable health care without financial constraint and it focuses on simple measures with great returns. These include focusing on early developmental ages, nutrition and preventive rather than curative measures of disease. Another way of ensuring healthy individuals is through community empowerment. An empowered community is one that is knowledgeable and is able to make timely decisions in regard to their health maintenance.

NEONATAL HEALTH INVESTMENT IS SMART INVESTMENT

While a number of studies have reported gains as a result of neonatal investment in terms of reduction in neonatal mortality and increase in Gross Domestic Product (GDP) (Chopra et al., 2013; Ten Hoope-Bender et al., 2014; Whitworth et al., 2010) fewer if any studies have given a macroscopic view especially in attainment of SDG. This begs answers to the following questions: Is there extended benefit of investing in neonatal health beyond reduction in mortality and morbidity of neonates and mothers? In an environment of competing interests including high levels of unemployment and non-communicable diseases, is it just and fair to drum for focus on investment in neonatal health? Are there other investment strategies in other sectors that could impact positively or negatively on neonatal health?

While many countries have made notable progress in improving maternal and child health outcomes, it is clear that without more effort in advancing health, sustainable development goals may not be achieved. Investing in neonatal health reduces poverty, is cost-effective and stimulates the economy (WHO, 2010). Despite the existence of high-impact interventions, the implementation has been wanting in regard to scaling up a reduction of morbidity and mortality from preventable causes.

Very few countries have allocated specific funding for neonatal and child health including stillbirths. Projects which exclusively mention newborn have allocated less than $1000 million (Lancet, 2014). To be successful in achieving neonatal health, investment needs to be done. This will include shifting of social norms where death of a neonate or a child is not regarded. Countries need to accelerate progress of attaining neonatal health through political action and leadership. Political willpower will go a long way in realizing neonatal health since there will be support through various government mechanisms such as allocation of resources.

CHILDHOOD DEVELOPMENT INVESTMENT AS A SMART INVESTMENT

Childhood is the period after birth towards adolescence. Childhood development physically, mentally, emotionally and socially is of importance as this will affect the individual's later life. One of the ways of achieving the sustainable development goal (SDG) 3 whose aim is to ensure healthy lives and promote well-being for all by 2030 is to enhance child health. It is estimated that the outcome of a failure to improve birth outcomes by 2035 will be an estimated 116 million deaths, 99 million survivors with disability or those who the potential for have lost development in addition to millions of adults who are at an increased risk of non-communicable diseases after low birthweight (Lawn et al., 2014). It is estimated that 5.3 million children died under the age of five years in 2018 and the risk of a child dying in Africa is nine times higher the risk than for a child born in the European region (WHO, 2018).

Early childhood development is a smart investment. The earlier the investment, the greater the return. This implies that there is a need to input greatly in the early childhood development. The investment should begin during preconception, during the fetal period, infancy and during childhood. In the preconception period, the couple planning for a pregnancy should ensure optimum health. Preconception care ensues a continuum of care, and also tackles any pre-pregnancy health risks and problems that could negatively affect maternal and fetal outcomes (Dean, Lassi, Imam, & Bhutta, 2014). The couple will need to be evaluated for any possible risks that may affect the pregnancy and its outcome. As such, there is a need to identify any modifiable risks. To enhance the journey of pregnancy, there is need for quality antenatal care (ANC). Any complications that develop should be well managed since these complications will affect the state of the mother and fetus. Quality provision of care requires well trained health care providers, availability of resources and effective monitoring and evaluation tools. During pregnancy, a client should attend at least four targeted ANC visits as recommended by the world health organization (WHO). During each visit, care should be individualized and specific services are provided. The client receives health education and promotion messages on various aspects that will guarantee a safe and positive outcome of the pregnancy.

A healthy start is very central to the human life cycle despite the potential complications that may arise during childbirth and can lead to societal problems (Lancet, 2014). Figure 4 illustrates the link between investment and return in human capital. The earlier the investment, the greater the return. This indicates that if investment is done during the prenatal programs, there will be a healthy population that will have greater probability for advancement.

EDUCATION AS A SMART INVESTMENT

The quality of pre-primary education has a greater impact on a child's health, academic advancement and livelihood. There is a close relationship between neonatal health and a child's cognitive development (Figlio, Guryan, Karbownik, & Roth, 2014) and academic achievements (Neilson, Bharadwaj, & Løken, 2013). A failure to invest in neonatal health and early child education and development can trigger economic, social and political costs (World Bank, 2010). Education is a very effective smart investment and is the fundamental driver for personal, national, regional and global development. Education leads to an empowered, healthy and peaceful society. An educated individual has a better understanding and is able to appreciate concepts better. Many studies have established a positive relationship between education and health outcomes (Kuruvilla et al., 2014; Bishai et al., 2016; Akseer et al., 2016). It has been established that maternal education, reduction in poverty and improved perinatal health can lead to improved neonatal health (Rahman, Salameh, Bener, & Ansari, 2010). Increase in school attendance may cost the country more, however, it has been found to lead to an increased gross domestic product (GDP) through empowered and productive individuals. It is established that each additional year of educating an individual boosts the person's income by 10% and increases the country's GDP by 18% (UNICEF, 2015).

Investment in education should be more equitable so that children who are in dire need achieve quality learning education. It is sad to note that most children in conflict areas are unable to access education. Education brings peace and stability. This is because every person will be empowered financially and have their own resources and may also be employed thus busy at work. There is need to invest in education especially in the marginalized areas and the poorest countries as this will help to improve and rebuild their economies. Based on evidence, countries with higher per capita income and low fertility levels are more likely to invest in their children's education thus lead to productive economies with higher returns (Jong-Wha, 2015).

When thinking about education as an investment, it is paramount to think about gender equality as indicated in sustainable development goal 5. In many developing countries, there is gender disparity in enrolment of boys and girls in school where many boys enroll especially in secondary schools and colleges as compared to girls (UNICEF, 2015). Regardless of gender or wealth, there should be equal opportunities and attainment for both boys and girls (Jong-Wha, 2015). Countries should therefore strive towards investing in childhood education and rebuilding their education systems to enable learning.

INNOVATION AS A SMART INVESTMENT

There is increased recognition of the early hours of life for a child and its effect on their milestones (Edmond et.al, 2019). To invest in this period, there is a prodigious need for innovation. Innovation is a new way of doing things. It brings out an advance in an area, new ideas, imaginations and thoughts in a device or a method. For an innovation to be developed, research is involved. In health care practice, evidence-based practice is key to advancement. Despite the findings of Fist and Atun (1998) indicating that in developed countries the investment on research and development is less than 4% in maternal and perinatal health, Frisk et.al (2011) indicate that research and development leads to greater achievement in maternal and perinatal health. Research and development are likely to avert up to 1% of global disease burden. Achievement of UHC goes hand in hand with quality of care and innovation. In neonatal health and child development, cost-effective innovations have played a critical role in SDG achievement. Innovations in maternal, neonatal and child health have contributed to a reduction in morbidity and mortality. Innovations lead to smart investments which have high returns.

Various innovations have been employed in managing neonatal and child health. The simple low-cost innovations have been very successful especially in the low and middle income countries. Some of these low-cost effective interventions include Kangaroo Mother Care (KMC), breastfeeding, cord care, use of insecticide treated nets (ITNs) and administration of Vitamin A. At the same time community based investments have demonstrated a higher impact in reduction of neonatal mortality in comparison with sophisticated technological advancement (Rahman et al., 2010). Technological innovations such as on-point screening have also contributed to efficiency and consequently improved health care.

Innovative approaches to care are beneficial not only to the family, community, health workers and the country at large. For instance, health workers can use technological innovations such as telemedicine in providing care. As such, those who may not physically access the doctors can still access care.

Research, development and innovations mainly occur at tertiary institutions. It therefore means that the countries with high ratio of the population attaining tertiary level education have a high concentration of both innovation and patents. Innovations often translate to a higher income. It is then not surprising that the countries that have these characteristics also have a high per capita income. Figure 7 further demonstrates this hypothesis.

UNIVERSAL HEALTH COVERAGE AND QUAILITY OF CARE

Quality care is care that is deemed accessible, available, acceptable and appropriate to the intended population. Quality care ensures that the recipients of care achieve the best quality of services provided in the health institutions and it has a greater impact in the lives of individuals. As illustrated in figure 1, the lives of women, babies and children matter. This will be achieved if the outputs are addressed. This includes ensuring a strong health care work force. The capacity of health care providers should be built in a manner that they should be able to effectively provide quality care. Provision of commodities is another strategy that will enhance quality care. Lack of resources impedes the provision of services and may lead to a delay in provision of care. Advancement in technology is critical as it helps in handling issues and provides real time evidence-based approaches. Another strategy of ensuring quality care is provided through health information systems. Information is power. It provides an avenue for planning, monitoring and evaluation. Health management information system is an innovative way that ensures there is data reporting from the lowest to the highest level of care.

According to the WHO framework for quality maternal and neonatal care, there should be quality provision of care through evidence-based practices, actionable information systems and functional referral systems (WHO, 2016). To add to that, effective communication, respect, treating patients/ clients with dignity and providing emotional support is crucial. The health providers must be competent and motivated and the physical resources must be available.

SUSTAINABLE DEVELOPMENT

Sustainable development is the development that ensures the present needs are met without compromising on the capacity of future generations in terms of meeting their own needs. It is the development that enables the building of buoyant cities, ensures a reduction of poverty and protects the natural environment. Talking about sustainable development, it is important to understand that the world is complex, complicated, interconnected and a linear relationship does not exist within its networks. This therefore brings the need of managing emerging issues using various approaches. The United Nations (UN), in its 2030 agenda adopted the sustainable development goals (SDGs). The SGDs rest on three pillars: environmental, economic and social pillars. There are 17 sustainable development goals with 169 targets and 232 indicators. These SDGs include no poverty, ending hunger, good health and well-being for all and quality education. In this era, it is critical to think about modelling in order to achieve sustainable development. Modelling is a way of understanding and managing risks and also helps in identifying opportunities (IISD, 2019). System dynamics modelling is an enabling vehicle to understand the world's complexity (Systems approach), identify leverage point(smart investments) and set sensible goals to improve it.

This would result in leaving a better world for generations to come, create a wealthy and prosperous society (inclusive, sustainable) and create an enabling environment characterized by good governance, transparency, accountability and rule of law.

CAUSALITY OF INVESTMENT IN EDUCATION, INNOVATION AND TRAINING AND NEONATAL HEALTH

A few selected countries demonstrate the causality of investment in education, innovation and training and neonatal health as shown in the Figure 5 and Figure 6. There is a correlation between investment in education and neonatal health. Countries which invest heavily in early childhood development have a higher Gross Domestic Product as well as higher records of innovations. The correlation between innovation, income and social mobility cannot be overestimated (Aghion, Akcigit, Bergeaud, Blundell, & Hemous, 2019; Chetty, Hendren, Kline, & Saez, 2014).

Figure 5. Gross Enrollment Ratio in Pre-Primary Education of Selected Countries

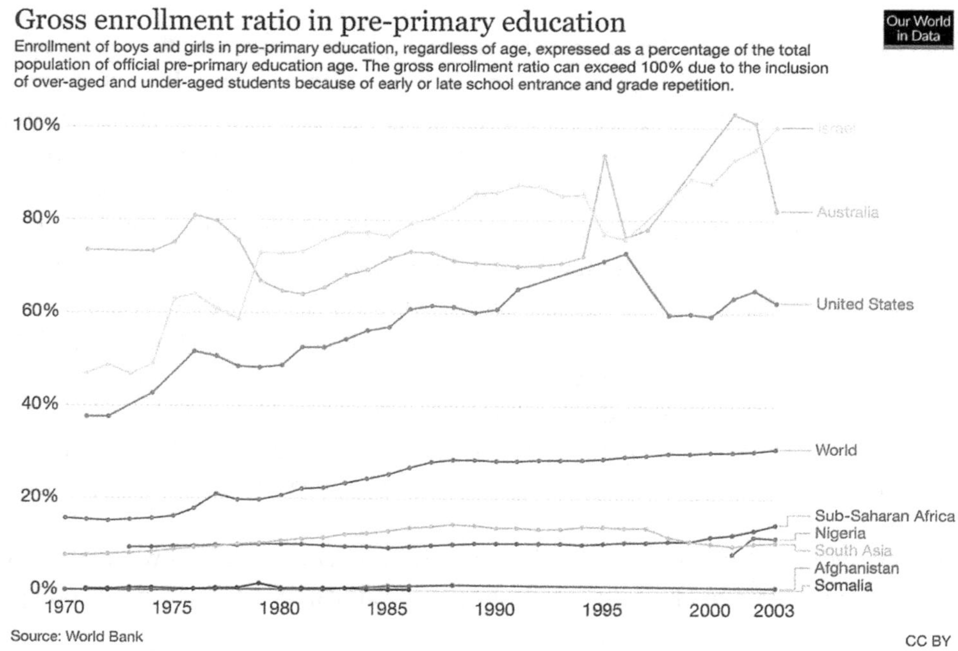

Figure 6. Child Mortality Rate from 1990 to 2017 from Selected Countries

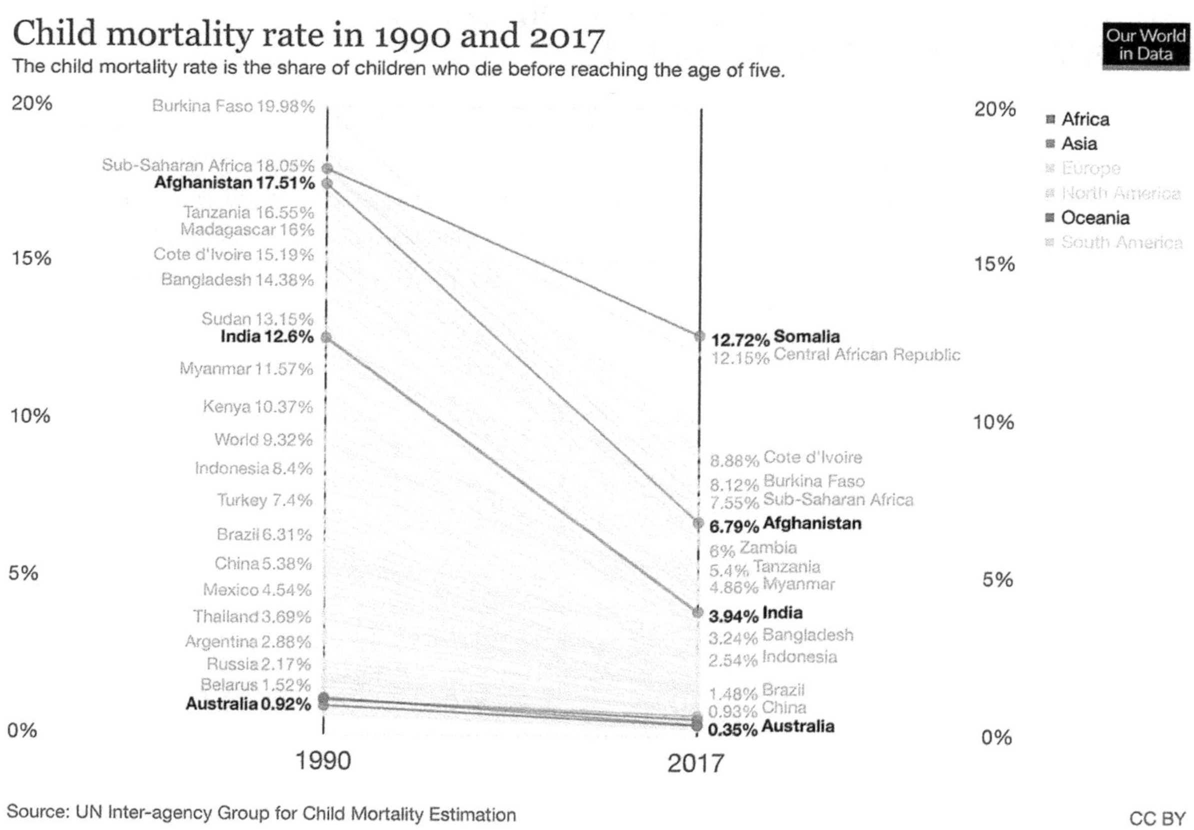

As shown in Figure 7: countries with higher ratios of population with tertiary level education are more likely to have higher number of innovations as well as higher Gross Domestic Product (Figure 8).

Figure 7. Share of the Population With Completed Tertiary Education

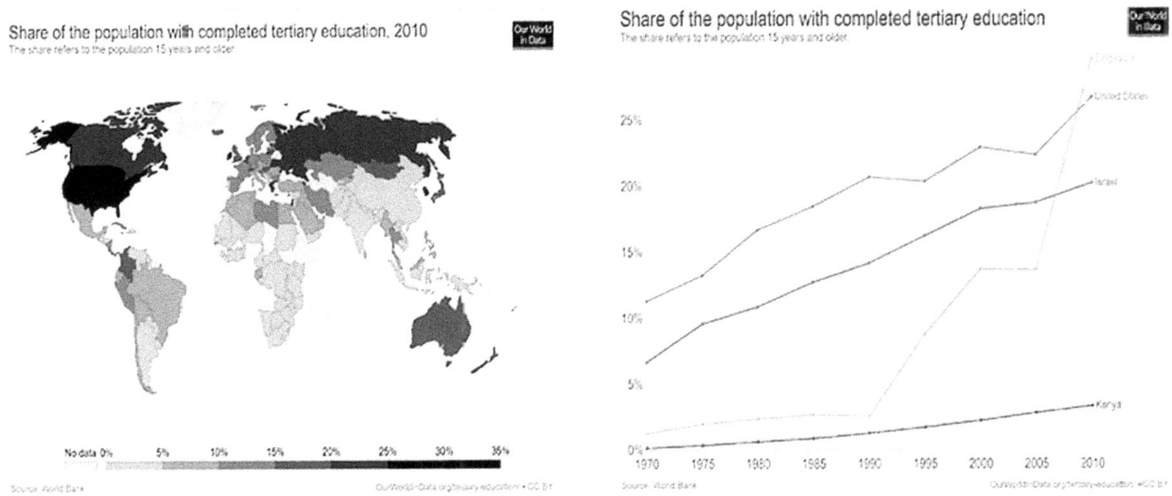

Figure 8. World Most Innovative Countries

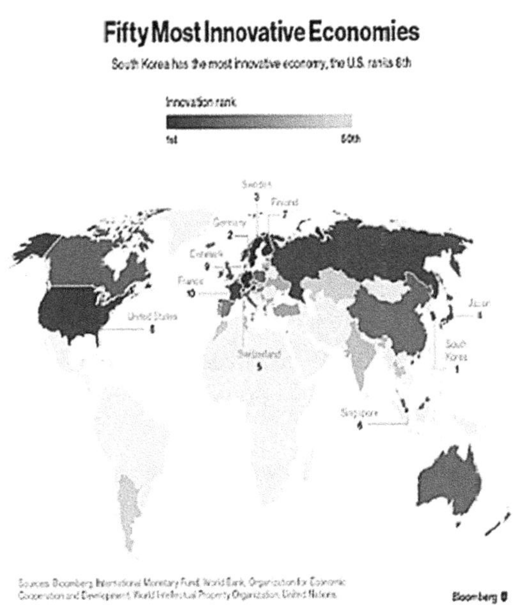

These are the world's most innovative countries

Based on Bloomberg's Innovation Index, 2018

Country	Place change from 2017	Total score
1 South Korea	0	89.28
2 Sweden	0	84.70
3 Singapore	+3	83.05
4 Germany	-1	82.53
5 Switzerland	-1	82.34
6 Japan	+1	81.91
7 Finland	-2	81.46
8 Denmark	0	81.28
9 France	+2	80.75
10 Israel	0	80.64

Source: Bloomberg

Apart from investing in Education and Training in order to achieve good health for its citizens, countries especially in the developing world should consider a greater target on the specific needs of early childhood development and training as these would realize high returns.

At the early childhood and preprimary levels, the main educational concerns depend on Neonatal Health and include;

1. Formation and brain development.
2. Physical health and mental health.
3. Stimulus for brain development
4. Pre-school readiness
5. Social skills and human skills
6. Lifetime effect on cognitive and Physical development

As a priority universal health coverage measures should target;

1. Stress-free and safe environment
2. Diarrhea diseases
3. Communicable diseases
4. Malnutrition
5. Immune depression (Daelmans et al., 2017; da Cunha, Leite, & de Almeida, 2015; Gove & Black, 2016)

IMPLICATION FOR FURTHER RESEARCH

The potential for System dynamics modelling and other system thinking tools need to be explored in management of Neonatal Health. This approach is likely to promote cohesion among multi-sectoral players as each is viewed as an equal citizen in pursuit of health. The complexity of healthcare delivery systems requires innovative yet cost effective in silico simulations to guide policy.

CONCLUSION

The goals of Neonatal health and early childhood development can be achieved through a multisectoral and systems approach. Investment in perinatal health has a positive impact on education, innovation and technology as well as income levels.

Generally, countries that have invested in education and training have a higher per capita income. Developing countries should invest concurrently in universal health coverage, perinatal health and early childhood education in order to maximize return on investment. Early childhood education is a smart investment as it is a great contributor of the macro-economy globally. UHC should have essential services for perinatal and early childhood periods. Governments should make investment in neonatal health and child development a priority, thus should provide sufficient resources and appropriate policies and programs to have the greatest impact and contribute towards achieving sustainable development goals.

System Dynamics approach is a tool that adds value in illuminating the feedback behavior of neonatal health interventions in the context of Universal Health coverage, education and Innovation all aimed at sustainable development goals.

REFERENCES

Aghion, P., Akcigit, U., Bergeaud, A., Blundell, R., & Hemous, D. (2019). Innovation and Top Income Inequality. *The Review of Economic Studies*, *86*(1), 1–45. doi:10.1093/restud/rdy027

Akseer, N., Salehi, A. S., Hossain, S. M. M., Mashal, M. T., Rasooly, M. H., & Bhatti, Z. … Bhutta, Z. A. (2016). Achieving maternal and child health gains in Afghanistan: A Countdown to 2015 country case study. *The Lancet. Global Health*. doi:10.1016/S2214-109X(16)30002-X

Arnold, R. D., & Wade, J. P. (2015). A definition of systems thinking: A systems approach. *Procedia Computer Science*, *44*(C), 669–678. doi:10.1016/j.procs.2015.03.050

Barlas, Y. (1996). System Dynamics: Systemic Feedback Modeling for Policy Analysis. In *System dynamics review* (Vol. 12, pp. 183–210). Academic Press. Retrieved from http://pisis.unalmed.edu.co/cursos/material/3004639/1/Barlas1996Formal-aspects-of-model-validity.pdf

Bishai, D. M., Cohen, R., Alfonso, Y. N., Adam, T., Kuruvilla, S., & Schweitzer, J. (2016). Factors Contributing to Maternal and Child Mortality Reductions in 146 Low- and Middle-Income Countries between 1990 and 2010. *PLoS One*, *11*(1), e0144908. doi:10.1371/journal.pone.0144908 PMID:26783759

Chetty, R., Hendren, N., Kline, P., & Saez, E. (2014). Where is the land of opportunity? The geography of intergenerational mobility in the United States. *The Quarterly Journal of Economics*, *129*(4), 1553–1623. doi:10.1093/qje/qju022

Chopra, M., Bhutta, Z. A., Mason, E., Presern, C., Stenberg, K., Sheehan, P., ... Ostergren, M. (2013). Advancing social and economic development by investing in women's and children's health: A new Global Investment Framework. *Lancet*, *383*(9925), 1333–1354. doi:10.10160140-6736(13)62231-x PMID:24263249

da Cunha, A. J. L. A., Leite, Á. J. M., & de Almeida, I. S. (2015). The pediatrician's role in the first thousand days of the child: The pursuit of healthy nutrition and development. [Versão Em Português]. *Jornal de Pediatria*, *91*(6), S44–S51. doi:10.1016/j.jped.2015.07.002 PMID:26351769

Daelmans, B., Darmstadt, G. L., Lombardi, J., Black, M. M., Britto, P. R., & Lye, S. … Richter, L. M. (2017). Early childhood development: The foundation of sustainable development. *Lancet*. doi:10.1016/S0140-6736(16)31659-2

Dean, S. V., Lassi, Z. S., Imam, A. M., & Bhutta, Z. A. (2014). Preconception care: Closing the gap in the continuum of care to accelerate improvements in maternal, newborn and child health. *Reproductive Health*, *11*(Suppl. 3), S1. doi:10.1186/1742-4755-11-S3-S1 PMID:25414942

Dudley, R. (2011). Overview of System Dynamics Modeling. *Earth01.Net*, 1–8. Retrieved from http://earth01.net/RGDudley/PDF/Dudley - Overview of System Dynamics Modeling.pdf

Edmond, K. M., Strobel, N. A., Adams, C., & McAullay, D. (2019). Effect of early childhood development interventions implemented by primary care providers commencing in the neonatal period to improve cognitive outcomes in children aged 0–23 months: Protocol for a systematic review and meta-analysis. *Systematic Reviews, 8*(1), 224. doi:10.118613643-019-1142-1 PMID:31470911

Figlio, D., Guryan, J., Karbownik, K., & Roth, J. (2014). The effects of poor neonatal health on children's cognitive development? *The American Economic Review, 104*(12), 3921–3955. doi:10.1257/aer.104.12.3921 PMID:29533575

Friberg, I. K., Kinney, M. V., Lawn, J. E., Kerber, K. J., Odubanjo, M. O., Bergh, A. M., ... Black, R. E. (2010). Sub-Saharan Africa's mothers, newborns, and children: How many lives could be saved with targeted health interventions? *PLoS Medicine, 7*(6), e1000295. doi:10.1371/journal.pmed.1000295 PMID:20574515

Ghaffarzadegan, N., Lyneis, J., & Richardson, G. P. (2011). How small system dynamics models can help the public policy process. *System Dynamics Review, 27*, 22–44. doi:10.1002dr.442

Gove, A., & Black, M. M. (2016). Measurement of Early Childhood Development and Learning under the Sustainable Development Goals. *Journal of Human Development and Capabilities, 17*(4), 599–605. doi:10.1080/19452829.2016.1243520

Heckman, J. J. (2012). The Heckman Equation Invest in early childhood development : Reduce deficits, strengthen the economy . The Heckman Equation Make greater investments in young children to see greater returns in education, health and productivity. *The Heckman Equation*, 1–2.

Hjorth, P., & Bagheri, A. (2006). Navigating towards sustainable development: A system dynamics approach. *Futures, 38*(1), 74–92. doi:10.1016/j.futures.2005.04.005

Homer, J. B., & Hirsch, G. B. (2006). System dynamics modeling for public health: Background and opportunities. *American Journal of Public Health, 96*(3), 452–458. doi:10.2105/AJPH.2005.062059 PMID:16449591

IISD. (2019). *Modelling for Sustainable Development New Decisions for a New Age*. 85.

Jimenez Soto, E., La Vincente, S., Clark, A., Firth, S., Morgan, A., Dettrick, Z., ... Widiati, Y. (2012). Developing and Costing Local Strategies to Improve Maternal and Child Health: The Investment Case Framework. *PLoS Medicine, 9*(8), e1001282. doi:10.1371/journal.pmed.1001282 PMID:22879817

Jong-Wha, L. (2015). *Why we must invest more in education*. World Economic Forum.

Kotiadis, K. (2007). Using soft systems methodology to determine the simulation study objectives. *Journal of Simulation, 1*(3), 215–222. doi:10.1057/palgrave.jos.4250025

Kuruvilla, S., Schweitzer, J., Bishai, D., Chowdhury, S., Caramani, D., & Frost, L. … Bustreo, F. (2014). Success factors for reducing maternal and child mortality. *Bulletin of the World Health Organization*. doi:10.2471/BLT.14.138131

Lawn, J. E., Blencowe, H., Oza, S., You, D., Lee, A. C. C., Waiswa, P., ... Cousens, S. N. (2014). Every newborn: Progress, priorities, and potential beyond survival. *Lancet, 384*(9938), 189–205. doi:10.1016/S0140-6736(14)60496-7 PMID:24853593

Michalow, J., Chola, L., McGee, S., Tugendhaft, A., Pattinson, R., Kerber, K., & Hofman, K. (2015). Triple return on investment: The cost and impact of 13 interventions that could prevent stillbirths and save the lives of mothers and babies in South Africa. *BMC Pregnancy and Childbirth*, *15*(1), 1–13. doi:10.118612884-015-0456-9 PMID:25879579

Neilson, C., Bharadwaj, P., & Løken, K. V. (2013). Early life health interventions and academic achievement. *The American Economic Review*, *103*(5), 1862–1891. doi:10.1257/aer.103.5.1862 PMID:29533048

Rahman, S., Salameh, K., Bener, A., & El Ansari, W. (2010). Socioeconomic associations of improved maternal, neonatal, and perinatal survival in Qatar. *International Journal of Women's Health*, *2*(1), 311–318. doi:10.2147/IJWH.S12426 PMID:21151678

Richmond, B., & Peterson, S. (2001). *An introduction to systems thinking*. High Performance Systems., Incorporated.

Rudan, I., Kapiriri, L., Tomlinson, M., Balliet, M., Cohen, B., & Chopra, M. (2010). Evidence-based priority setting for health care and research: Tools to support policy in maternal, neonatal, and child health in Africa. *PLoS Medicine*, *7*(7), e1000308. doi:10.1371/journal.pmed.1000308 PMID:20644640

Rwashana, A. S., Nakubulwa, S., Nakakeeto-Kijjambu, M., & Adam, T. (2014). Advancing the application of systems thinking in health: Understanding the dynamics of neonatal mortality in Uganda. *Health Research Policy and Systems*, *12*(1), 1–14. doi:10.1186/1478-4505-12-36 PMID:25104047

Schmidt, H., Gostin, L. O., & Emanuel, E. J. (2015). Public health, universal health coverage, and Sustainable Development Goals: Can they coexist? *Lancet*, *386*(9996), 928–930. doi:10.1016/S0140-6736(15)60244-6 PMID:26138141

Semwanga, A. R., Nakubulwa, S., & Adam, T. (2016). Applying a system dynamics modelling approach to explore policy options for improving neonatal health in Uganda. *Health Research Policy and Systems*, *14*(1), 35. doi:10.118612961-016-0101-8 PMID:27146327

Ten Hoope-Bender, P., De Bernis, L., Campbell, J., Downe, S., Fauveau, V., Fogstad, H., ... Van Lerberghe, W. (2014). Improvement of maternal and newborn health through midwifery. *Lancet*, *384*(9949), 1226–1235. doi:10.1016/S0140-6736(14)60930-2 PMID:24965818

Transforming the Workforce for Children Birth Through Age 8. (2015). In *Transforming the Workforce for Children Birth Through Age 8*. doi:10.17226/19401

UNICEF. (2015). *Education the most powerful investment in our future - UNICEF Connect*.

Whitworth, J., Sewankambo, N. K., & Snewin, V. A. (2010). Improving implementation: Building research capacity in maternal, neonatal, and child health in Africa. *PLoS Medicine*, *7*(7), 2–4. doi:10.1371/journal.pmed.1000299 PMID:20625547

WHO. (2010). *WHO Investing in our common future Joint Action Plan for Women's and Children's Health*.

WHO. (2016) Standards for Improving Quality of Maternal and Newborn Care in Health Facilities. Retrieved from https://apps.who.int/iris/bitstream/handle/10665/249155/9789241511216-eng.pdf;jsessionid=B32399A6F963AA3917AE1E001D053E71?sequence=1

WHO, & UNICEF. (2014). Every Newborn: an action plan to end preventable deaths: Executive summary. *Geneva: World Health Organization.* Retrieved from http://www.everynewborn.org/Documents/Executive-summary _EN.pdf%0Awww.who.int/about/licensing/copyright_form/en/index.html

Wolstenholme, E. (2005). The potential of system dynamics. *Leading Edge*, (October). Retrieved from http://www.iseesystems.com/resources/Articles/leading_edge_10.pdf

World Bank. (2010). *Child Youth Development.*

Chapter 19
Interactive Story Reading and Language Development in Early Years

Sühendan Er
TED University, Turkey

Müdriye Yildiz Biçakçi
Ankara University, Turkey

ABSTRACT

Early childhood is a precious period when children get their most rapid performance in terms of emotional, physical, cognitive, and language development. During this period, one of the major indications is how children use their mother tongue. In respect of environmental effects, stories have a crucial role. Interactive story reading is one of the most studied formats for increasing language learning in children. Many research studies have shown that children make significant gains in various areas of development through interactive story reading experiences. Parents can support their children's language development and thereby their children's literacy development by interactive story reading. Interactive story reading is developed by Whitehurst and others and it is a process that leads the children improve their analytical thinking, expressing themselves, problem solving skills and inquiring. Moving from this, in this chapter, the effects of interactive story reading on language development of children is discussed through research done in the area.

DOI: 10.4018/978-1-7998-2952-2.ch019

INTERACTIVE STORY READING AND LANGUAGE DEVELOPMENT IN EARLY YEARS

Concepts of Language Development

Language is the most effective tool of communication. Language is used to provide one's needs and wants, to establish relationships with others, to redirect the behavior of the individual and the environment, to provide the existence and properties of the individual, to get to know the world outside, to get information and learn, to inform and explain the ideas of others, to imagine, to create assumptions and to create situations for the future (Yildiz Bicakci and Aral, 2009; Cameron, 2001; Doherty and Hughes, 2009; Mcdevitt and Ormrod, 2010).

Language development is an area that is closely related to both verbal and non-verbal communication (Senemoglu, 2013). Verbal communication involves talking, writing and reading activities and non-verbal communication involves movements such as, gestures, body movements, eye contact or gliding (Knapp and Hall, 2007; Hargie and Dickenson, 2004).

Language is used in three ways as gestures, receptive and productive language. Gestures begin when a child is about 8-9 months old, convey language through facial expression or body movements. For example, a baby who wants its favorite toy extends its arms, opens and closes its hands and somehow wails. Receptive language is the way the child comprehends the language in its environment. Before children start speaking, they indicate that they understand the words that are uttered to them by receptive language in the ninth or tenth month (Bredekamp, 2015). Productive language transforms language into a verbal form during the ninth to thirteenth month (Bee and Boyd, 2009; Pignot Shahov, 2012). Productive language is related to communication, speaking, phoneme, morpheme and mapping.

Communication is to express feelings, thoughts and lives verbally or non-verbally. Speaking is a way of to form acoustic signals bridging sounds and speech organs such as, lips, tongue or palate. Speaking is a motor movement which is a way to convey feelings and thoughts. This transfer enables the child to socialize, gain communication skills, recognize the world, and expand cognitive skills such as problem solving, thinking or binding (Gass and Selinker, 2008).

Language is used to supply needs and wants, to have communication with others, to shape one's own or the immediate environment behaviors, to help put forward one's own existence and personality, to recognize the outside world, to get or transmit information, to imagine, to make assumptions, and to create future plans (Bee and Boyd, 2009).

The system of meaning sounds in a language is called phonology and phonemes are the smallest units of sound which can deliver a change in meaning (Doherty and Hughes, 2009). In the word 'tin', /t/ which is the first sound, is a phoneme. When phonemes are connected in a meaningful way. 'morpheme' is formed. A morpheme is the smallest grammatical unit in a language. For example; when 'T' 'I' and 'N' phonemes are connected, the morpheme 'tin' is formed. Babies can produce morphemes by putting sounds together (Demirci, 2011).

In language development a morpheme is considered as the one-word stage and during this period children try to say something by using only one word (Bee and Boyd, 2009). For example, if a - one and a half year old- child points to a teddy bear and says 'bear', this may mean 'Mummy, look at that teddy bear' or 'Mummy, bring my teddy bear to me, I want to play with it' or so many other things.

Egocentric speech is another important concept as it is a way of speaking of a child that reflects his own point of view. Language development is closely related to brain development and an apparent acquisition is expected when children are at about 3 years old. This is the period when sentences are perceived as a whole, not piece by piece. By 4 years the child uses the language in himself and starts speaking in an egocentric style (Vygotsky, 1978; Demir, 2010).

The theorists Piaget and Vygotsky who deal with cognitive and language development have different views on egocentric speech. Piaget divides the speeches of children into two in the light of his observations and data gathered through experimental studies. The first one is egocentric speech in which the child soliloquise. It is just like a monologue, without any aim of communication. There is no worry about expecting an answer or to be listened to. According to Vygotsky's findings 45% of the children tend to do it by nature. The second one is socialized speaking. In this type, the child needs to have a conversation, in other words he gets angry, demands something or gets information (Erdener, 2009, Vygotsky, 1978).

Telegram speech is known as the 'two words' period in language development. It is usually known as a speech style which is a combination of conjunctions and suffixes (Aral, 2011; Bee and Boyd, 2009). When a child says 'Water, mum' it means that he wants a glass of water.

While children acquire the mother tongue it is inevitable for them to make errors in early word use. In the first years children demonstrate general knowledge and understanding of basic patterns in their native language. Although children have a remarkable ability of acquiring and comprehending language in early years, they make errors and mistakes while enhancing their knowledge and understanding of language. Overgeneralization, overextension, and under extension are three main errors that children make in early word use (Fernandez and Smith Cairns, 2011).

COMPONENTS OF LANGUAGE

Grammar

Morphemes are called the roots and affixes in the smallest words which cover the features of properties in the orderly arrangement of derivation formats. An independent morpheme could be used alone, whereas a dependent morpheme could only be used with an independent morpheme (Owens, 2008). For instance; "bird" is an independent morpheme. However, in the word "birds", the "s" affix is a dependent morpheme and could not be used alone.

Syntax

Syntax defines the order of words in a language which is acceptable or not and what is right or wrong linguistically. It involves the rule that combines the components of forming the structure of the sentence. These rules include; word order, arranging words in sentences and sentence relations between the words, sentence types which are sorted according to the type of sentence phrases. In accordance with these rules, children need to pay attention to the children's audio group. If they do not pay attention to the order, the meaning of sentences may change. Children start to generate their first expressions using rules concerning the syntax, during the two words period (Owens, 2008).

Meaning Information (Semantic) Semantics

Semantics refers to the meaning of words and phrases. In other words, meaning determines the relationship between objects and events through symbols of sounds. When the child begins to use language meaningfully, it establishes meaningful relationships between certain situations and objects with their thoughts and uses the language to create specific meanings of words and phrases. Gaining cognitive concepts, the semantic aspect of language improves (Dilci, 2014).

The biggest indicator of the language of the content is vocabulary. Vocabulary is a subject of study of semantics which is a sub-field of linguistics. Each language has a unique vocabulary which is governed by specific rules. Vocabulary is treated under two headings as word meaning and grammatical function words (Berk, 2013).

Meaning functioning words can be defined as objects, people, experiences and events, and corresponding to the meaning of words in a dictionary to express directly name as cars, trees; verb as go, look at; adjective as a fat, little; adverb as quickly; the pronoun I, you are examples of functional words (Yildiz Bicakci and Aral, 2009).

When the semantics development is considered, we can see that the greater the increase in children's cognitive development, the stronger the increase in language development. As children grow up, their learning domain expands and they learn different words. That process allows them to develop their vocabulary (Berk, 2013).

Pragmatics

Pragmatics covers the use of language in accordance with specific rules (Kandir, 2007b). In other words, pragmatics refers to the use of language in everyday life. It is a component that involves usage, a problem experienced in learning, limited use of vocabulary, phrases, or the incorrect use of proverbs (Demirel, 2002). It is therefore important that the usage component should be formed in language use. The language acquisition process starts when the individual starts to learn to communicate with others (Topbas and Mavis, 2006) because pragmatics is not only associated with writing and speaking, but also social interaction. For example, when a five-year-old child, tells something to a two-year-old child, as a pragmatic information is given to him, he uses simple sentences, he speaks slowly and repeats the words for the two-year-old child's understanding (Zwisler, 2006).

This component involves turn-taking, starting to talk, continuing on the same subject and ending, time, status, speech and self-expression. Depending on the person they're talking to, children start learning using different words and give different responses to the questions from an early age (Owens, 2008).

FACTORS AFFECTING LANGUAGE DEVELOPMENT

Sharing is said to be the first phase in gaining speaking ability. A baby grows up in interacting with the people who are very close to him, so he learns the rules of communication. He lives in the communicative medium of language. In that first phase while experiencing the first sharing the baby realizes the way the objects around him interact with each other. In addition, he can understand the messages and produce his first words. However, these words are non-representative. The second phase of language development identifies cognitive process. The child is able to improve the semantic dimension of language. In this cognitive process the child learns to build abstract facts and objects. This is the way how a child gradually passes into the symbolic stage from the concrete stage (King and Saxton, 2010). This period involves certain aspects which affect children's language development. These aspects can be categorized as genetics, environment and gender.

Genetics

In language development genetics influence the innate capacity to learn languages, developmental problems, neurological development and intelligence. Language is a skill which is innate. Linguist Noam Chomsky (1957) emphasized that people biologically have the gift of learning a language in certain times and in certain ways. He claimed that children are born with a language acquisition device that enables them to perceive the rules and properties of language in phonology, morphology, semantics and syntax (Santrock, 2011) but the innate language learning capacity may vary according to genetic factors. Some children experience fast language development whereas others are slower. A few days after birth babies can distinguish human voice especially their mother's voice among other voices around them. They are sensitive to sound differences. It is accepted that babies are genetically ready before they understand and produce language (Scribner, 2013). Besides inborn language skills, health conditions, neurological development and intelligence of the child are considered under genetics. Acute and chronic health problems will result in speech retardation. The retardation comes out from the child's lack of communication with his environment (Berk, 2013).

Mental disability, cerebral palsy, developmental disorders, autism, vision and hearing disorders affect a child's language development negatively. Babies of four to eight months with hearing disorders clearly show differences in using some words compared with the babies who develop normally. Visual perception is a determiner in language development and children with serious visual disorders have slow development in language skills. Any health or developmental problems in the organs enabling speech influence language development negatively. Problems may occur when larynx and vocal cords maturation are insufficient (Yildiz Bicakci and Aral, 2009).

Neurological development is an indication of language development. When the brain has completed its growth the child starts using language properly. By age three the weight of the brain of a child reaches four fifths of the brain weight of an adult. Although not much satisfactory evidence exists the intelligence level and the sound production of the child until the age two, a strong relationship between sound production and intelligence is found after two years. Children who start speaking early are accepted as normal or above normal intelligence level as language development is closely related to intelligence (Hammer, Farkaz and Macguza, 2010).

Environment

Environment means parents, play, school, peers, culture. A healthy relationship between parents and children affects language development positively (Bredecamp, 2015). Verbal interaction between parents and the baby is the basis of a child's mother tongue acquisition. The environment and verbal stimuli especially the ones produced by the mother influence language development directly. In infancy, the first emotional interaction between mother and the baby begins when they look at each other's face, while touching and talking. A satisfactory interaction in infancy affects the child's future language development (Hammer, Farkaz, and Macguza, 2010).

Playing with the baby is another positive component in language development. Through play a child can learn asking questions about the name of the objects, the actions, or the relation between facts and objects (Kucukkaragoz, 2011). For example, if a mother smiles or touches rather than responding in an ordinary manner soon after her 8-month-old baby babbles, the baby starts to produce more complicated sounds similar to spoken language (Goldstein et.al, 2003). Television and communication instruments either contribute to children's language development or causes delay in language development and articulation disorders (Louw, 2002; Yildiz Bicakci and Aral, 2009). Most linguists emphasize the effect of life experience on language development (Santrock, 2011).

Gender

It is highlighted that girls are better than boys in language development. On the other hand studies in recent years put forward that there is not major difference between girls and boys with respect to language development. Researches shows that mothers talk more with their daughters whereas they touch to their sons to communicate (Berk, 2013; Yildiz Bicakci and Aral, 2009). This situation may be an effect on speech.

THEORIES OF LANGUAGE DEVELOPMENT

Individual language features are associated with brain. The brain is the language system administrator. Speech is basically controlled by two areas in the cerebral cortex. Broca's and Wernicke's areas related to speaking and comprehension of spoken language and are two centers for brain localization. Broca's area is the focal point of the voice skill and the brain controls speech and sounds in speech. Wernicke's area is related to the field of speech recognition and voice. Wernicke's area, sensory, and motor areas can be defined as Broca's area. The child, by cognitive processes, or imitation acquires language as a consequence of communication. Different views about language acquisition are laid out. Psycho-linguistic theory, cognitive-developmental theory, behavioral theories and theories of social interaction describes the acquisition of language in different ways (Selcuk, 2007).

Psycholinguistic Theories

Theorists who describe language development, based on biology and psychology are called psycholinguistics. The most important among them is Noam Chomsky. According to Chomsky's theory, people have inborn capacity to learn language. This mechanism allows children to internalize the spoken language, to learn and understand the rules and then to talk with the appropriate rules. Through this mechanism all children at the same stage, when they come to a certain level of maturity, learn to speak like they learn to walk (Bee, 1992; Dagabakan and Dagabakan, 2007; Keenan, 2002; Selcuk, 2007; Vander Zanden, 1985). Within the framework of this theory, Chomsky stressed that the structure of language can be defined as superficial and deep structures. The surface structure of the language contains rules that regulate the words and sentences. The deep structure of language covers the rules form the basis of the child's own, inborn language system (Senemoglu, 2013).

Interaction Theory

The most important representatives of the theory are Piaget and Vygotsky. Vygotsky claimed that, the child's cognitive development is not a process that is carried out on his own, it is also linked with others, as Piaget suggests. Others may guide the child's mental development, can teach them by giving some information, the child solves its own problems, benefits from the aid of others (Bacanli, 2011; Demir, 2010). Vygotsky has put forward a theory about the mind's social development. He claims that expressive language is significant in child's interaction with his social circle. Vygotsky and Piaget's approaches are different in children talking themselves and their egocentric speech. When children talk to other children, Piaget stresses that egocentric speech decreases when they realize that others have their own different points of view, while Vygotsky claims the children use that to guide themselves (Aral, 2011; Bacanli, 2011; Bee and Boyd, 2009; Demir, 2010). Psycholinguistic theorists analyzed the grammatical structure of speech and cognitive theorists analyze speech in terms of meaning. They stress that language is learned by linking cognitive concepts

Behaviorist Theory

Skinner, who defended the behaviorist theory, has put forward the view that speech, as it is in other behaviors is acquired by functional conditioning. According to this theory, children learn spoken language as they learn anything. Infants begin to learn the spoken language by repeating the sounds that lead to desirable outcomes. When infants repeat sounds and articulate sounds resembling everyday language, they are awarded by the people around them. So, the infant uses sounds that are reinforced more frequently and the use of non-reinforced sounds decreases. At the end of it, talk is shaped (Erden and Akman, 2004). The most important thing in this theory is reward and punishment. When children make a variety of sounds, without intent for a particular purpose and when this behavior is rewarded by the environment. It results in achieving a lasting behavior and settling in his mind. These reactions are acquired over time and create a variety of classifications and rankings. This development is continued by means of reinforcement, such as reward and punishment (Woolfolk, Winnie and Perry, 2011). All in all, according to this theory, children gain the language with conditional status conditioning. For instance, when a year-old baby sees his father, he articulates the sound "baa". The father smiles happily and says, "you said "father" very lovely". The baby articulates "baa" again. The father says, "well done" and kisses the baby. As shown in this example, according to the behaviorist approach, as a result of the baby-daddy interaction, baby has learned a word (Selcuk, 2007).

Social Learning Theory

A Further dimension of the behaviorist approach to language acquisition as a social interaction theory is associated with direct imitation and role modeling. According to this theory, language is gained through parents being models, for the child to imitate, and for parents to reinforce and to give corrective feedback. For example, the mother says to the child "Give me some bread". Children respond by saying "Bre." Mother again says, "Daughter give me the bread." Holding out the bread, child says "Mother, bread". Mother gets the bread and says "Well done, my girl gave her mother the bread". In addition to reinforcement, to imitate the sounds they hear frequently is important for the baby of to acquire the language. The main point of this theory of language acquisition is imitation. Children perform speech by imitating the sounds heard around them_(Demir, 2010; Goldstein et al., 2003; Selcuk, 2007).

STAGES OF LANGUAGE DEVELOPMENT

The development of language monitoring skills improves regularly as well as the development of motor skills. In addition, studies conducted on children language development, found that children all over the world were using the same basic grammar rules on their first period of learning to talk. German linguist Hans Glinzner, in his book German Syntax (Deutsche Syntax) discusses the children's "universal explanation of". He talks about the universal descriptions of children first. In short, although it can vary from child to child language development, the language development process is universal (Kandir, 2003). Language acquisition takes place in two phases, including pre-speech and speech periods. These periods can be explained as follows:

Pre-Speech Period

This is the period where the first communication begins. This period is described as newborn, muttering, cooing and the period also covers mumbling repetition.

Newborn period (Weeping) (0-6 weeks): from birth till a month old, burping, crying and satisfaction sounds can be seen in infants (Bee and Boyd, 2009). Crying expresses discomfort (Santrock, 2011). These cries are differentiated and undifferentiated forms of crying. Undifferentiated crying is reflexive. Differentiated crying is an indication of the needs of the baby. This cry will vary over time. Individual differences in the amount of crying at the end of the second week will begin to be seen. At the beginning of the third week there would normally be less crying (Aral, 2011b, Meggitt, 2006; San Bayh and Artan, 2004). After the fourth week, the baby's crying changes in intensity and pitch, and that means different crying can be seen, as crying indicates that this situation is becoming a communication tool. Children cry differently when they are hungry, when they want to be held by different arms, when they have wind and they cry differently when they wake up and cannot see anybody around them. In this way, the mother can determine the cause of the crying by the differences in crying. Also after the second week of this period, the baby begins to hear voices and when the baby is four weeks. In addition, between one and two month infants, when they articulate "uuuuuuu" vowel sound, it can be seen as a sign of satisfaction. Besides this, the volume or pitch of the sound or the differences observed in the curtain (Bee and Boyd, 2009).

Cooing period (6 weeks - 3 months): there are changes in the sound mechanism, the child starts to articulate simple sounds. This sound is universal. There is no awareness of producing these sounds. During this period, babies' first sounds are usually vowel sounds, but they also remove the consonant sounds very rarely. From the second month, infants start to make syllables made of meaningless vowels (aah, ooh, etc.) with s, k, g which are soft palate and laryngeal sounds. These sounds are articulated when the baby's stomach is full, when baby's bottom is cleaned, and when they are comfortable and happy (Yildiz Bicakci and Aral, 2009). In addition, infants in this period are sensitive to the human voice, they can listen to other people's conversations and record sounds in speech. (Aral, 2011b; Bee, 1992; Cicek, 2002; Gander and Gardiner, 2007).

Humming period (Babbling) (3-6 months): the rounding and continuation of language is seen with the baby's increasing control over the sound mechanism (Aral, 2011b; Yildiz Bicakci and Aral, 2009). In this period, two syllables words are observed. Sounds such as "ba-ba-ba," "de-de-de", "da-da-da", "ma-ma-ma", which contains the repetition of a simple change of vowels and consonants are observed (Santrock, 2011; Yildiz Bicakci and Aral, 2009). These sounds that consist of utterances, and infants are learning them from the people in the vicinity such as their parents and babysitters. Making these nonsensical sounds, infants use the tone of their mother tongue (Gasser and Selinker, 2008). It can be seen that infants make these sounds, in other words they babble, when they are content and alone. (Yildiz Bicakci and Aral, 2009). During this period, the baby's own voice is the motivation for the baby, so the baby's sound is stimulus itself. Baby can make on average seventeen types of sounds during this period (Donmez et.al, 2000; Berk, 2013).

The period of humming again (Lalling) (6-9 months): In this period, the baby can control his muscles and begin to be able to combine vowel letters and consonant letters (Bee and Boyd, 2009). Therefore, diversity is observed in the baby's mouth movements in this period (Aral, 2011b; Yildiz Bicakci and Aral, 2009). The baby repeats the sounds heard with sound production comprising combining diversity and emulate as stimulation for repetition (Bee and Boyd, 2009; Donmez et.al, 2000). This imitation is an indicator for baby's language development and is an important indicator for the acquisition of social skills (Karacan, 2000b). Infants leave behind a set of meaningless syllables which does not belong to their native language over time. Without knowing it, by using repeated syllables that are united, such as (de, ba ba etc.) infants reveal the structure of their native language. During this period, especially the tone of word learning takes place (Bee and Boyd, 2009). Initially, sounds such as p, b, d as labial and presence of gingival "ba - ba - ba", "de - de - de" syllable repetitions are seen (Aral, 2011b; Bee, 1992; Cicek, 2002; Meggitt, 2006; Yildiz Bicakci ve Aral, 2009). The feature that distinguishes this period from the previous period is that between the sounds emitted by the baby, and the lack of long space and extension of a sound for a very long time (Yapici, 2004; Yildiz Bicakci and Aral, 2009).

Generally, language pre-preparation in the preverbal process, or the removal of sound, is in the form of repetition and imitation. Infants learn the meaning of the words before they learn emphasis and their meaning. During this period, infants use a number of nonsense sounds, unique accents, intonations and gestures. The stress infants use during their first months starts to change (Yapici, 2004; Yildiz Bicakci and Aral, 2009).

Speaking Period

This speech period consists of sound words, one word, two-word phrases, and three-word phrases and grammatical rules more appropriate to talk periods.

Audio word period (9 -12 months): During this period, infants imitate the sounds that people generate even if they do not understand. The infant in this period likes exercising various sounds. These voices, distinguished between sounds that are reinforced and acclaimed, will remain part of the sound treasury (Yildiz Bicakci and Aral, 2009). With the sounds of this period, infants start to sound the native language gradually. These sounds do not have meaning at all, but they have a characteristic of fluency. These sounds are defined as jargon. Infants generate more than enough of these sounds before the first is expressed (Yildiz Bicakci and Aral, 2009).

One-word period (12-18 months): before speaking their first words children understand their meanings. Children begin to use their first words in this period. These first words are composed of words they heard from the environment. These first words are usually about the baby's own close people (mother, father), objects moving or movable (balls, cars) and the state or environment (hot, dirty) which are words that are related (Bloom, 1973; San Bayhan and Artan, 2004; Yildiz Bicakci and Aral, 2009). No matter what the infant's society is, the first words of a baby show similarities. It takes a long time for infants to generate their first words. However, their vocabulary develops quickly after they generate ten words or more (Aral, 2011b; Bloom, 1973; Yildiz Bicakci and Aral, 2009). The number of words learned to say in this period does not exceed about thirty words. While most words are initially objects or nouns in time they learn verbs and other words. The reason why the noun learning process comes before the verb learning process is that infants lack the ability of combining until they are eighteen months (Bee and Boyd, 2009).

During this period, infants cannot use words like adults do. Sometimes they use too wide or too narrow a sense of the word. This is called reduction or insufficient extension. For example, it can represent "apple" as the only fruit. However, some infants use words that represent the object more widely. This is called generalization. For example, if an infant has a grandfather, he calls "Grandpa" to all the old people. Although the individual differences among children towards the end of this period, the vocabulary of the infant usually vary between three and one hundred (Avanoglu Kiziltepe, 2004; Bee and Boyd, 2009; Bjorklund, 1995).

Two-word phrases period (18-24 months): in this period infants realize that everything has a name. In the light of this, they improve their vocabulary by adding new words. While the number of children talking about fifty words used by the number of words used when they are saying about three hundred and twenty out of twenty-four months (Bee and Boyd, 2009). This increase takes the form of learning about 10-20 words per week (San Bayhan and Artan, 2004). During this period, children demonstrate a feature called telegraphic speech. It contains mostly nouns and verbs, without linkers or connectors. Over time, the child begins to learn the grammar of their native language and combines words quickly (Cicek, 2002; Meggitt, 2006; Vander Zanden, 1985). The language development of children at this age is seen as a reflection of the concept of cognitive level. Although there are individual differences in this period, the vocabulary of the baby usually varies between one hundred and fifty (Cicek, 2002; Kucukkaragoz, 2011). By the end of this period 70% of the words are monosyllabic (Karacan, 2000a).

More than three-word phrases period (2-3 years old): In this period, child's language skills and vocabulary are learned the fastest. During this period, the onset of auxiliary verbs, pronouns, prepositions, and the two-word combinations without inflections are seen. Two-word combinations in this period can vary in meaning. Children make interrogative sentences by changing the tone of their voice. For example, "Daddy car" like two-word combination, refers to the car and father, as well as to express that his father get in the car or the opposite. In addition, he said changing the tone of the child "Tony shoes" two words composition "Are these Tony's shoes?"

Besides that, children start to use adjectives in combinations of words, with pronouns, adverbs (respectively place, time, and situation), negative structures, and begin to add structures and inflections of the simplest level (Donmez, 2000; Meggitt, 2006, Vander Zanden, 1985).

Grammar rules period (3-6 years): Children generally at three or four years can express their thoughts properly. They learn to whisper. They articulate 90% of vowels and 60% of consonants correctly. The rise and fall in pitch is regular and there is an increase in speed. Loudness and tone of voice is normal and breathing is regular. Language is used in imaginative play so he makes statements to himself, which is egocentric speech. Vocabulary is up to 900 words (Donmez, Abidioglu, Dinçer, Erdenir, and Gumuscu, 2000; Meggitt, 2006). They use elements of sentences and types of words without using nouns and verbs. The rate of understanding speech increases gradually from three years to four years old. The average three-year-old children constantly repeats the words he has learned. Three to four-year-old children can name body parts, animals and people around him. Using nouns decreases in children with increasing age. The ratio of using the nouns of a child three years old, on average, 23.4%, and the rate is 19.7% using pronouns. The rate of asking questions after three years old increases because the child establishes social relations, wants to take attention, gain confidence, and wants consolation or help. Three to four-year-old child internalize grammar rules which are not the same as the adult grammar. Children in this age group, according to its own sentence structure, when they hear ungrammatical sentences; they transform the sentence to a grammatical one and use it properly. That shows that children not only perceive consecutive sentences as sorted words, they also perceive the structure as a whole (Meggitt, 2006). Besides that, researchers have discovered that children around the age of three approaches words differently. As a result of this change, children begin to pay attention to the words of an entire group. For example, they begin to learn kinds of fruit. In this way, the mind develops the ability to connect with new words and their referent in the real world. This capability is also referred to as "rapid mapping" (Bee and Boyd, 2009).

The child has mastered the use of language by four to five years of age. Girls are more skillful than boys in the language. The success rate of sound production is high. Girls articulate 90% of the consonants accurately. They imitate the parent's pitch of the voice. And an increase in the number of words and egocentric speech continues (Bjorklund, 1995). More complex sentence structures are used. They can produce plurals accurately. They are not aware of the compound words that are separate units (Donmez et al., 2000; Torr, 2005). Children can easily understand and use the terms "the top, bottom, inside, outside, behind" as prepositions. An appropriate speech using basic grammar rules can be seen approximately at four and a half years old. The use of the pronoun is seen to increase 20.3% and the use of using nouns is seen to increase 19.3% at about four and a half years while the rate of five-year-old child's speech consists of 23% of the furnishing, and the 41% thing itself and about 36% of the other people (Torr, 2005)

Five to six-year-old children's language use is similar to the use of adult language and conversation takes place at a more understandable level. They use 99% of the vowels 98% of consonants correctly (Donmez et al., 2000). A Five to six-year-old boy can establish six to eight words in appropriate grammatical sentences. The vocabulary of the child at this age is around 2000-2500 words.

School period (7-11 years): Due to the development in preschool, there are no major changes in language development. Children develop their reading and writing skills because the children's vocabulary and grammatical structure skills are improving during the school period. In this period metalinguistic features are seen. Metalinguistic is the child's ability to use and understand his mother tongue. Children can tell the meaning of words by guessing (Omeroglu, 2007). He can understand that a word has different meanings and he can define meanings in multiple ways depending on the nature and characteristics (Bee and Boyd, 2009). He can produce new words and detect passive and verbal phrases. He can also use passive and conditional sentences and at the end of the period he can understood metaphorical sentences. In short, children can speak their thoughts and feelings in a comfortable way to explain a topic in the framework of (Eger, 2006; Omeroglu, 2007).

BILINGUALISM

Bilingualism is stated as acquisition of more than one language for several reasons, under different conditions. The learner becomes competent enough as a native. In other words, bilingualism is defined as knowing two different languages or having reading and writing abilities of two different languages (Alkaya, 2007; Baruca and Bialystok, 2012).

There are different opinions about learning the second language or the acquisition of it. According to one of them, children learn two languages when they are about three to four years old. In the book Speech and Brain Mechanism written by Penfield and Roberts in 1959, they state that the age to learn a foreign language in the easiest way is in the first ten years of life. According to them, flexibility of the brain becomes concrete at the end of puberty, the right and the left hemispheres of the brain become more independent of each other. It indicates that linguistic abilities are more related to the left hemisphere of the brain, which is why there are difficulties in learning a foreign language in puberty. According to this opinion, children, when they are about three to four years old, can speak and understand two languages that are spoken around them. Sometimes they may spell the words slowly and from time to time they may not know the meaning of a word in the other language. The acquisition of two languages happens after some certain steps. In the first step, children learn the words of a language and then the words of the other language start improving. In the second step, the child starts applying the rules of syntax of both two languages in a proper way. In the last step, the child separately perceives both two languages in a certain way when he/she hears (Baruca and Bialystok, 2012).

The opinion about the acquisition of the second language is that the second language is learned in parallel with mother tongue from the birth. The basis of learning the second language in parallel with mother tongue is inevitable as mother and father of the child use two different languages at home. ın other words, when mother and father only speak to their child in their own mother tongues the child learns both two languages as mother tongue, and becomes bilingual (Alkaya, 2007; Scribner, 2013).

The children who live in a community where two different languages are spoken or the ones who have to learn two languages, show slower development when compared to a child that learns a single language from the beginning. Small children have a very keen ear for the language of sound and hearing abilities, so small children in learning a second language is suggested that they are in a superior case (Baruca and Bialystok, 2012).

INTERACTIVE STORY READING

Early childhood period is crucial that children gain cognitive, social, emotional, and language skills in those years. Hence it is the best time for children to gain the reading habit. Children have a natural tendency for reading and they are curious about written materials in their environments such as books, newspapers, magazines at home or signboards in street. They try to make interpretations about the things written. If children's innate inclination to early literacy is supported by adults appropriately, they will gain reading habit easily (Cakmak Gulec and Gecgel, 2006; Dol, 1999). The knowledge and skills of children towards reading and writing or in other words emergent literacy Ergul, Akoglu, Sarica, Tufan and Karaman, 2015; Tercanli Metin and Gokcay, 2014; Whitehurst and Lonigan, 1998) skills should be promoted by parents and teachers (Akyol, 2012; Gonen and Ari, 1989; Gonen, Durmussoglu and Severcan, 2009; Sever, 2017). It is a fact that emergent literacy skills are the predictor of academic success in the future life of children (Bracken and Fischel, 2008; Kotaman, 2007; Kotaman, 2009; Saracho, 2017; Shahaeian et al, 2018; Whitehurst and Lonigan, 1998; Wu and Honig, 2010).

Children's literature is a product that is especially written for children, appealing to their emotions and thoughts, which gives pleasure and also teaches morals to children (Dursunoglu, 2013; Sınar, 2006; Simsek and Yakar, 2014; Sirin, 2007; Tur and Turla, 2005; Yalcin and Aytas, 2014). Reading qualified children's literature to children in early years supports children's receptive and productive language skills (Baker, Scher ve Mackler, 1997; Gonen and Ari, 1989; Karakus, 2006; Kurt, 2008; Er, 2016; Bıcakcı, Er and Aral, 2017; Yavuzer, 1995) and emergent literacy skills which leads to diversification of views of life, and contributes to their social and emotional sensitivity (Akyol, 2012; Er, 2016; Gibbs and Early, 1994; Kolac, Demir and Karadag, 2012; Kraayenoord and Paris, 1996; Sirin, 2000; Veziroglu, 2009; Yalcin and Aytas, 2014; Yurtseven, 2011). Enabling children to meet good quality children's literature(Aytas, 2003; Sever, 2017; Temizyurek, 2003), and reading books with children (Cakmak Gulec and Gecgel, 2006; Isikoglu Erdogan, 2016; Morrow and Gambrell, 2005; Tipton, 2014; Veziroglu, 2009; Whitehurst and Lonigan, 1998; Yurtseven, 2011) not only conveys reading habits but also helps children improve emergent literacy skills.

Interactive reading has several names such as, joint reading, dialogical reading or shared reading and is an important component that improves early literacy and language skills. Interactive story reading is a mutual exchange between adult and the child (Gonzalez, Taylor, Davis and Kim, 2013; Hindman, Skibbe and Foster, 2014; Işikoglu Erdogan, 2016). It is a way that is very different from the traditional way of adult's reading to the listener-the child. It integrates the child into the process of story reading actively (Dixon-Krauss, Januszka and Chae, 2010; Hindman, Connor, Jewkes and Morrison, 2008). Interactive story reading creates a strong bound between the adult and the child through enjoying the process. Moreover, it enables the child to see that the letters and the pictures in the book turn out to be entertaining stories and awaken an interest in reading (Cer, 2016; Deretarla Gul, 2015; Kibris, 2000; Oktay, 2000; Sever, 2017).

Children like to read some stories again and again. In such cases, parents should not avoid reading as this process helps the children understand the story better and then have a stronger interview about the story. On the other hand, a balance between the stories retold and the ones newly read should be constructed (White, 2005).

There are interactive reading sessions in parent education programs and there has been positive effects on home literacy environments (Bicakci, Er and Aral, 2017, 2018; Duursma, 2014, 2016; Levin and Aram, 2012; Tercanli Metin and Gokcay, 2014; Whitehurst et.al., 1988; Whitehurst and Lonigan, 1998). The parents who participated in those programs stated that they have been doing interactive reading with their children more and they realized that their children's language and early literacy skills improved. Moreover, they observed the positive change in their academic progress especially children who are coming from low economic status (Raikes, Pan and Luze, 2006; Shahaeian et.al., 2018).

Process of Interactive Story Reading

Justice and Pence (2005) emphasized that reading stories interactively is much more effective than traditional reading methods. The research done in the area identifies that quality is much more important than quantity (Duursma, 2016; Marjanovič-Umek, Hacin ve Fekonja, 2017; Raikes, Pan and Luze, 2006; Saracho, 2017; Vandermaas-Peeler et al, 2011; Wu and Honig, 2010).

If the part being read is shown to the child, if the child is allowed to ask questions and make interpretations during reading, and the unknown words are explained, then a qualified reading process occurs (Baroody and Diamond, 2010; Duursma, 2016). The process will both help the improvement of early literacy skills and listening skills. By asking questions the child starts thinking analytically and finds reason and result relationship (Baroody and Diamond, 2010; Demir-Lira et al, 2018; Duursma, 2014; Gilkerson, Richards and Topping, 2015; Lyytinen et al, 1998; Marjanovič-Umek et al, 2017; Saracho, 2017; Shahaeian et al, 2018).

Before Reading

Before reading a story the book must be examined by the adult. The story book should not contain any missing pages, foreign words, and abstract terms that the child cannot understand (Gonen, 2000). Adults should take time off for reading books and the children should perceive story reading time an enjoyable event done with adults (Brock and Rankin, 2008).

First the cover of the book is introduced to the child and the name of the book is read. The authors of the book are mentioned in order to arise mindfulness towards the writing and the pictures. There may be questions such as; 'Who is an author/a artist?' and 'What does an author or artist do?' Moreover. the name of the printing house is given. These are the components that mainly constitute a story book as they enable the child to gain awareness towards the occupations, people and institutions (Vukelich, Christie and Enz, 2014).

The cover of the book is used to enable the child to think about the events that will take place in the story. Sample questions are; 'What can the story be about?' 'Who can the boy in the picture be?' The child may make some predictions about the story. Besides the experiences of the child and the events in the story can be connected which makes the story more enjoyable for the child. The definition of the unknown words must be explained to the child if there are any. These are the first steps of interactive story reading.

While Reading

While reading, parents should use facial expressions or change their tone of voice according to the content of the story. Parents can involve children in the story and make story reading process more enjoyable (Vukelich, Christie and Enz, 2014). The pace of the reading must be slow and there has to be talking about the pictures in the book. Before turning the page, asking the child about the upcoming events will let the child to envision and predict. This is a way of enforcing imagination and improvement of problem-solving skills (Sawyer, 2009). The adult will allow the child ask questions about the story and the pictures in the storybook. According to Rosenbalt's (1978) reader response theory the listener links up with his previous experiences which establishes his reason for listening to the story. In this way children improve their emotional skills as they connect their own feelings, thoughts and experiences.

The opportunity of speaking given before reading, will go on. While reading the child must be allowed to see each page and the adult must indicate where he is reading by saying 'Now, I am reading here'. The sentences being read must be underlined by the index finger. This is beneficial as the child realizes where to begin reading, and that the read parts are the words not the pictures, the read pieces are the letters that all symbolize sounds, the words forms sentences, there are gaps between words, and the words are written from left to right. This process also improves the child's early literacy skills which supports print awareness and phonological awareness (Tompkins, 2006).

After Reading

There should be an interactive talk with the child about the story after reading. The adult and the child analyze the story together which enables the child to improve his thinking skills. They will interpret the story again and reveal cause and effect relations. Sample questions can be; 'Which part of the story do you like best?', 'How did you feel when....happened?', 'What do you think about....?', 'If you were...,what would you do?'. Those questions lead the child think analytically and improve their problem solving skills. According to Yopp and Yopp (2006), when children internalize the story, they interpret the story better.

While the child reinforces his higher order thinking skills he may also be asked to role play the story read. This is another way of making the child retell the story. There may be some questions that make the child remember the story, such as; 'What happened first?' 'Where is....going?' etc. If the adult participates in the role play they will share a joyful time. The drama supports the child's creativity, language development and social development (Israel, 2008; Vukelich, Christie and Enz, 2014).

Through the discussion about the story after reading the children can state their feelings and thoughts. Asking questions about the story helps the child improve his language skills and also problem-solving skills. In order to sustain story retelling, the questions must not be compelling and sharp. The child puts the story in the correct order, summarizes the story, and makes his interpretations by using his own words. The atmosphere should strengthen the communication between the adult and the child that contributes to emotional development (Laboo, 2005).

CONCLUSION

The most important feature that distinguishes human beings from other creatures is the presence of intelligence and thinking ability. Intelligence, emotions and thought systems are the elements establishing the human inner self. The individual, who imagines, thinks and touches expresses himself verbally or by behavior. The most important tool that plays a major role in establishing the connection between the environment and the outside world with the inner world of the individual is language (Owens, 2008). Language, defined as a series of symbols, and thoughts that are expressed as a product of the mental activities. Language and thought thrive in simultaneous interaction (Owens, 2008). Language, creating the raw material for thinking, on the one hand, works in parallel with the ability to. So, the more children's conceptual thinking develops, the more his language develops (Bredecamp, 2015).

Language development starts with birth and it is a very rapid process in the early years. While adults support the language development of a child through various activities, they also support their cognitive, social and emotional development. In this process a valuable activity is interactive story reading. It involves not only reading a book to a child but also awakening cognitive skills of the child by asking developmentally appropriate questions to the child, giving way to his questions or letting him make interpretations about the story. It is a known fact that there is a connection between language and cognition; the child thinks and talks. Interactive story reading enables the development and has been gaining importance. It is also a precious idea recommended by the researchers working in the field.

REFERENCES

Akyo, I. T. (2012). Investigation of values in children's picture books. Unpublished masters thesis, Hacettepe University Social Sciences Institute, Ankara, Turkey. (In Turkish) Retrieved from https://tez.yok.gov.tr/UlusalTezMerkezi .

Alkaya, E. (2007). Bilingualism and translation of concepts from Russion to Tatar Turkish. *Türkoloji Araştırmaları*, *2*(2), 41–53. (In Turkish)

Aral, N. (2011). Language development. In N. Aral and G. Baran (Eds.), In Çocuk Gelişimi (pp. 163-192). İstanbul: Ya-Pa Publications. (In Turkish).

Avanoğlu-Kızıltepe, Z. (2004). *Teaching and Learning; A contemporary approach to education psychology*. İstanbul: Ofset Publications. (In Turkish)

Aydın, A. (2008). *Education Psychology*. Ankara: Pegem Publications. (In Turkish)

Aytaş, G. (2003). The role of children's literature in reading development. *Türklük Bilimi Araştırmaları*, *13*, 155–160. (In Turkish)

Bacanlı, H. (2011). *Education Psychology*. Ankara: Pegem Publications. (In Turkish)

Baker, L., Scher, D., & Mackler, K. (1997). Home and Family Influences on Motivations for Reading. *Educational Psychologist*, *32*(2), 69–82. doi:10.120715326985ep3202_2

Baroody, A. E., & Diamond, K. E. (2010). Links Among Home Literacy Environment, Literacy Interest, and Emergent Literacy Skills in Preschoolers At Risk for Reading Difficulties. *Topics in Early Childhood Special Education*, *32*(2), 78–87. doi:10.1177/0271121410392803

Baruca, R. & Bialystok, E. (2012). Bilingual effects on cognitive and linguistic development: Role of language, cultural background, and education. *Child Development*, *83*(2), 413-422. doi:1 doi:0.1111/j.1467-8624.2011.01707.x

Bee, H., & Boyd, D. (2009). *Child development psychology* (O. Gündüz, trans.). İstanbul: Kaknüs Publications.

Berk, L. E. (2013). *Cognition and language development* (A. Dönmez, trans.). (pp. 272–422). Ankara: İmge Publications.

Bicakci, M. Y., Er, S., & Aral, N. (2017). Views of mothers towards interactive story reading to their children. *Education in Science*, *42*(191), 53–68. doi:10.15390/EB.2017.7164

Bicakci, M. Y., Er, S., & Aral, N. (2018). The effects of interactive story reading on children's language development. *Kastamonu Eğitim Dergisi*, *26*(1), 201–208. (In Turkish)

Bjorklund, D. F. (1995). *Children's Thinking. Developmental Function and Individual Differences*. USA: Brooks/Cole Publishing Company.

Bloom, L. (1973). *Language Development: Form and Function in Emerging Grammars*. Cambridge: M. J. T. Press.

Bracken, S. S., & Fischel, J. E. (2008). Family reading behavior and early literacy skills in preschool children from low-income backgrounds. *Early Education and Development*, *19*(1), 45–67. doi:10.1080/10409280701838835

Bredekamp, S. (2015). *Effective applications in early childhood* (H. Z. İnan & T. İnan, trans.). Ankara: Nobel Publications. (In Turkish)

Brock, A., & Rankin, C. (2008). Communication, Language and Literacy from Birth to Five. *Sage (Atlanta, Ga.)*.

Bus, A.G., Belsky, J., Van Ijzendoorn, M.H., & Crnic, K. (1995). Attachment and Bookreading Patterns: A Study of Mothers, Fathers, and Their Toddlers. *Early Childhood Research Quarterly*, 81-98.

Cakmak Gulec, H. & Gecgel, H. (2006). Children's literature. Ankara: Kökpublications (in Turkish)

Cameron, L. (2001). Teaching languages to young learners. United Kingdom: Cambridge. doi:10.1017/CBO9780511733109

Cer, E. (2016). Reality and relativity of child in 0-6 year children's books. Ankara: Eğiten Kitap Publications. (in Turkish)

Chomsky, A.N. (2002). *Syntactic Structures*. Walter de Gruyter.

Cicek, A. (2002). Language development of 0-6 year children. *Erzincan Eğitim Fakültesi Dergisi*, *4*(1), 115–124.

Dagabakan, O., & Dagabakan, F. (2007). Language and theories of language development for children. *Milli Eğitim Dergisi*, *174*, 155–161. (in Turkish)

Demir, T. (2010). An essay on egocentric speaking in speech therapy. *Kastamonu Eğitim Dergisi*, *188*(2), 415–430. (in Turkish)

Demir-Lira, Ö. E., Applebaum, L. R., Goldin-Meadow, S., & Levine, S. C. (2019). Parents' early book reading to children: Relation to children's later language and literacy outcomes controlling for other parent language input. *Developmental Science*, *22*, e12764. doi:10.1111/desc.12764 PMID:30325107

Demirci, K. (2011). On the importance of phoneme theory. *Türkoloji Araştırmaları*, *6*(2), 359–366. (in Turkish)

Demirel, Ö. (2002). *Teaching Turkish*. Ankara: Pegem A Publications. (in Turkish)

Deretarla Gul, E. (2015). Adopt reading habit to preschoolers. In M. Gönen (Ed.), Child Development (pp. 305-314). Ankara: Eğiten Kitap Publications (in Turkish)

Dixon-Krauss, L., Januszka, C.M. & Chae, C. (2010). Development of the dialogic reading inventory of parent- child book reading. *Journal of Research in Childhood Education*, *24*(3), 266–277.

Doherty, J., & Hughes, M. (2009). *Child development*. Pearson.

Dol, A. (1999). Illustrations of 0-6 year children's books. Unpublished masters thesis, Onsekiz Mart University, Social Sciences Institute, Çanakkale. (In Turkish)

Donmez, N., Abidioglu, U., Dinçer, C., & Erdenir, N. ve Gumuscu, S. (2000). Language development activities in early years. İstanbul: Yapa Publications (in Turkish).

Dursunoğlu, H. (2013). Child and literature. In Ö. Yılar and L. Turan (Eds.), children's literature for faculty of education (pp. 27-30). Ankara: Pegem Academy Publications. (in Turkish)

Duursma, E. (2014). Parental book reading practices among families in the Netherlands. *Journal of Early Childhood Literacy*, *14*(4), 435–458. doi:10.1177/1468798414523025

Duursma, E. (2016). Who does the reading, who the talking? Low-income fathers and mothers in the US interacting with their young children around a picture book. *First Language*, *36*(5), 465–484. doi:10.1177/0142723716648849

Ege, P. (2006). Language and concept development. In S. Topbaş (Ed.), Literacy development in childhood (pp. 140-157). Ankara: Kök Publications. (In Turkish).

Er, S. (2016). The importance of parents' interactive story reading. *Başkent University Journal of Education*, *2*(3), 156–160.

Er, S., Aral, N., & Yıldız Bıçakçı, M. (2013). Identifying the relationship between children's language skills and parents' story reading self-efficacy. *Procedia: Social and Behavioral Sciences*, *93*, 1269–1274. doi:10.1016/j.sbspro.2013.10.027

Erden, M., & Akman, Y. (2009). *Educational Psychology*. Ankara: Arkadaş Publications. (in Turkish)

Erdener, E. (2009). Views of Vygotsky on language and thought; A criticism on Vygotsky. *Türk Eğitim Bilimleri Dergisi*, *7*(1), 85–103. (in Turkish)

Ergul, C., Akoglu, G., Sarica, A. D., & Tufan, M., & Karaman, G. (2015). Investigation of shared reading in preschool classes with respect to interactive reading. Mersin University Journal of Education Faculty, 11(3). doi: (in Turkish) doi:10.17860/efd.88429

Fernandez, E. M., & Smith Cairns, H. (2011). *Fundamentals of psycholinguistics*. Malden, MA: Wiley-Blackwell.

Gander, M. J. & Gardiner, H. W. (2004). Child and Adolescent Development. Ankara: İmge Kitabevi Yayınları. (in Turkish)

Gass, S. M., & Selinker, L. (2008). *Second Language Acquisition: An Introductory Course*. United States: Routledge. doi:10.4324/9780203932841

Gibbs, L.J. & Earley, E.J. (1994). *Using children's literature to develop core values*. Bloomington, IN: Phi Delta Kappa Educational Foundation. Retrieval https://files.eric.ed.gov/fulltext/ED366992.pdf

Gilkerson, J., Richards, J. A., & Topping, K. J. (2017). The impact of book reading in the early years on parent–child language interaction. *Journal of Early Childhood Literacy*, *17*(1), 92–110. doi:10.1177/1468798415608907

Goldstein, M. H., King, A. P., & West, M. J. (2003). Social Interaction Shapes Babbling: Testing Parallels between Birdsong and Speech. *Proceedings of the National Academy of Sciences of the United States of America*, *100*(13), 8030–8035. doi:10.1073/pnas.1332441100 PMID:12808137

Gonen, M. & Arı, M. (1989). Investigation of the effects of picture books on language development of 4-5 year old children. *Education in Science*, *13*(72), 21–30. (in Turkish)

Gonen, M. (2000). Investigation of thesis on last 10 years' children's books about the content, graphics, physical properties. Ankara University, Ankara. (In Turkish)

Gönen, M., Durmuşoğlu, M., & Severcan, S. (2009). Examining the views of the preschool education teachers on the content, illustrations and physical characteristics of the picture story books used in education. *Procedia: Social and Behavioral Sciences*, *1*(1), 753–759. doi:10.1016/j.sbspro.2009.01.133

Gonzalez, J. E., Taylor, A. B., Davis, M. J., & Kim, M. (2013). Exploring the underlying factor structure of the Parent Reading Belief Inventory (PRBI): Some caveats. *Early Education and Development*, *24*(2), 123–137. doi:10.1080/10409289.2011.623651

Hammer, C. S., Farkas, G., & Macguza, S. (2010). Language and Literacy Development of Head Start Children: A Study Using the Family and Child Experiences Survey Database. *Language, Speech, and Hearing Services in Schools*, *41*(1), 70–83. doi:10.1044/0161-1461(2009/08-0050) PMID:19948772

Hargie, O., & Dickson, D. (2004). *Skilled Interpersonal Communication Research, Theory and Practice* (4th ed.). London: Routledge Publishing. doi:10.4324/9780203427880

Hindman, A. H., Connor, C. M., Jewkes, A. M., & Morrison, F. J. (2008). Untangling the effects of shared book reading: Multiple factors and their associations with preschool literacy outcomes. *Early Childhood Research Quarterly*, *23*(3), 330–350. doi:10.1016/j.ecresq.2008.01.005

Hindman, A. H., Skibbe, L. E., & Foster, T. D. (2014). Exploring the variety of parental talk during shared book reading and its contributions to preschool language and literacy: Evidence from the Early Childhood Longitudinal Study-Birth Cohort. *Reading and Writing*, *27*(2), 287–313. doi:10.100711145-013-9445-4

Hutton, J. S., Phelan, K., Horowitz-Kraus, T., Dudley, J., Altaye, M., & Holland, S. K. (2017). Shared reading quality and brain activation during story listening in preschool-age children. *The Journal of Pediatrics*, *191*, 204–212. doi:10.1016/j.jpeds.2017.08.037 PMID:29173308

Işikoglu Erdogan, N. (2016). Investigation of the effects of parent-child interactive story reading activities *Kastamonu Eğitim Dergisi*, *24*(3), 1071-1086. (in Turkish)

Israel, S. E. (2008). *Early reading first and beyond*. London: Corwin.

Justice, L. M., & Pence, K. L. (2005). *Scaffolding with storybooks: A guide for enhancing young children's language and literacy achievement*. Newark, DE: International Reading Association.

Kandır, A. (2003). My 3-6 year old child is growing. İstanbul: Morpa Publications. (in Turkish)

Kandır, A. (2007). Gaining language abilities in cognitive development. In E. Omeroglu and A. Kandır (Eds.), Cognitive Development (pp. 131-149). İstanbul: Morpa Kültür Yayınları. (in Turkish)

Karacan, E. (2000). Factors affecting language development. *Sürekli Tıp Eğitim Dergisi*, *7*, 254–258. (in Turkish)

Karakuş, I. Ş. (2006). The effects of children's literature on reading development. Unpublished masters thesis, Gazi University Educational Sciences Institute, Ankara. (in Turkish)

Keenan, T. (2002). *An Introduction to Child Development.* USA: Sage Publication.

King, S., & Saxton, M. (2010). Opportunities for language development: Small group conversations in the nursery class. *Educational and Child Psychology, 27*(4), 31–44.

Knapp, M. L., & Hall, J. A. (2007). Nonverbal Communication in Human Interaction (5th ed.) Wadsworth: Thomas Learning.

Kolac, E., & Demir, T. & Karadag, R. (2012). Views of pre service teachers on using children's literature in language education. *Türkiye Sosyal Araştırmalar Dergisi, 161*, 195–213. (in Turkish)

Kotaman, H. (2007). Impacts of dialogical storybook reading on young children's reading attitudes and vocabulary development. *Reading Improvement, 45*(2), 55–61.

Kotaman, H. (2009). Parents' story reading self efficacy scale. *Türk Eğitim Bilimleri Dergisi, 7*(4), 767–780. (in Turkish)

Kraayenoord, C. E. V., & Paris, S. G. (1996). Story construction from a picture book: An assessment activity for young learners. *Early Childhood Research Quarterly, 11*(1), 41–61. doi:10.1016/S0885-2006(96)90028-9

Küçükkaragöz, H. (2011). Cognitive development and language development. In B. Yeşilyaprak (Ed.), Development and learning psychology (pp. 37-45). Ankara: Pegema Publications. (in Turkish)

Kurt, B. (2008). The effects of children's literature products on basic language skills. Unpublished masters thesis, Gazi University Educational Sciences Institute, Ankara. (in Turkish)

Laboo, L. (2005). Books and computer response activities that support literacy development. *The Reading Teacher, 59*(3), 288–292. doi:10.1598/RT.59.3.10

Levin, I., & Aram, D. (2012). Mother-child joint writing and storybook reading and their effects on kindergartners' literacy: An intervention study. *Reading and Writing, 25*(1), 217–249. doi:10.100711145-010-9254-y

Louw, D. A. (2002). *Human Development. Characteristics of Language of Preschooler.* South Africa: Perskor.

Lyytinen, P., Laakso, M. L., & Poikkeus, A. M. (1998). Parental contribution to child's early language and interest in books. *European Journal of Psychology of Education, 13*(3), 297–308. doi:10.1007/BF03172946

Marjanovič-Umek, L., Hacin, K., & Fekonja, U. (2017). The quality of mother–child shared reading: Its relations to child's storytelling and home literacy environment. *Early Child Development and Care.* doi:10.1080/03004430.2017.1369975

Mcdevitt, T. M., & Ormrod, J. E. (2010). *Child development and education* (7th ed.). Pearson.

Meggitt, C. (2006). *Child Development: An Illustrated Guide.* Oxford: Heinemann.

Morrow, L., & Gambrell, L. (2005). *Using children's literature in preschool: Comprehending and enjoying books.* Newark, DE: International Reading Association.

Oktay, A. (2000). *Magic years of life: Preschool period.* İstanbul: Epsilon Publications. (in Turkish)

Ömeroğlu, E. (2007). Cognitive and language development in school children. In E. Ömeroğlu & İ. Ulutaş (Eds.), Child and adolescent development (pp. 35-61). İstanbul: Morpa Publications (in Turkish)

Pignot Shahov, V. (2012). Measuring L2 Receptive and Productive Vocabulary Knowledge. *University of Reading Language studies working papers, 4*, 37-45

Raikes, H., Pan, B. A., & Luze, G. (2006). Mother-child book reading in low income families: Correlates and outcomes during the first three years of life. *Child Development, 77*(4), 924–953. doi:10.1111/j.1467-8624.2006.00911.x PMID:16942498

Revelle, G., & Bowman, J. (2017). Parent child dialogue with ebooks. In *Proceedings of the 10th International Conference on Interaction Design and Children* (pp. 346-351). New York: ACM. doi:10.1145/3078072.3079753

San Bayhan, P., & Artan, I. (2004). *Child development and education.* İstanbul: Morpa Kültür Publications. (in Turkish)

Santrock, J. W. (2011). *Lifelong development.* Mc Graw Hill.

Saracho, O. N. (2017). Parents' shared storybook reading – learning to read. *Early Child Development and Care, 187*(3-4), 554–567. doi:10.1080/03004430.2016.1261514

Sawyer, W. F. (2009). *Growing up with literature* (5th ed.). USA: Delmar.

Scribner, E. (2013). Exploring Different Factors of Language Development. 2013 Awards for Excellence in Student Research & Creative Activity - Documents. The Keep. Retrieved from http://thekeep.eiu.edu/lib_awards_2013_docs/1

Selçuk, Z. (2007). *Educational Psychology.* Ankara: Nobel Publications. (in Turkish)

Senemoğlu, N. (2013). *Development, learning and teaching. From theory to practice.* Ankara: Gönül Publications. (in Turkish)

Sever, S. (2017). *Teaching Turkish and learning* (6th ed.). Ankara: Anı Publications. (in Turkish)

Shahaeian, A., Wang, C., Tucker-Drob, E., Geiger, V., Bus, A. G., & Harrison, L. J. (2018). Early shared reading, socioeconomic status, and children's cognitive and school competencies: Six years of longitudinal evidence. *Scientific Studies of Reading, 22*(6), 485–502. doi:10.1080/10888438.2018.1482901

Simsek, T., & Yakar, Y. M. (2014). Child and Literature. In T. Şimşek (Ed.), Children's literature from theory to practice (pp. 13- 44). Ankara: Grafiker. (in Turkish)

Sınar, A. (2006). Children's literature studies in Turkey. *Türkiye Araştırmaları Literatür Dergisi, 4*(7), 175–225.(in Turkish)

Şirin, M. R. (2000). *Children's literature in 99 questions.* İstanbul: Çocuk Vakfı Publications. (in Turkish)

Sirin, M. R. (2007). *A critical view towards children's literature.* Ankara: Kök Publications. (in Turkish)

Temizyurek, F. (2003). The importance of children's literature in teaching Turkish. *Türklük Bilimi Araştırmaları, 13*, 161–167. (in Turkish)

Tercanlı Metin, G. and Gokcay, G. (2014). Story reading in infancy and early childhood *Çocuk Dergisi, 14*(3), 89-94. (in Turkish)

Tipton, L. A. (2014). The parent's role: Shared book reading and the child with asd. Unpublished doctorate dissertation, University of California, Riverside.

Tompkins, G. E. (2006). *Literacy for the 21st century* (4th ed.). New Jersey: Pearson Prentice Hall.

Topbaş, S. (2006). The components of language. In S. Topbaş (Ed.), Language and concept development (pp. 21-30). Ankara: Kök Publications. (in Turkish)

Topbaş, S., & Mavis, İ. (2006). Pragmatics development. In S. Topbaş (Ed.), Language and concept development (pp. 123-139). Ankara: Kök Publications. (in Turkish)

Torr, J. (2005). Child Development and Teaching Young Children. In A. Talay-Ongan & E. Ap (Eds.), *Language Development in Early Childhood* (pp. 79–98). Canada: Thomson Learning Nelson.

Tur, G., & Turla, A. (2005). *Child, literature and book in early childhood.* İstanbul: Yapa. (in Turkish)

Vander Zanden, J. W. (1985). *Human Development.* New York: Random House.

Vandermaas-Peeler, M., Sassine, B., Price, C., & Brilhart, C. (2011). Mothers 'and fathers' guidance behaviours during storybook reading. *Journal of Early Childhood Literacy, 12*(4), 415–442. doi:10.1177/1468798411417381

Veziroglu, M. (2009). *Investigation of the harmony of picture books and MoNE early childhood gains.* Ankara: Hacettepe University, Social Sciences Institute. (in Turkish)

Vygotsky, L. S. (1978). *Mind in Society: The Development of Higher Psychological Processes.* Cambridge, MA: Harvard University Press.

White, H. (2005). *Developing early literacy skills in the early years.* London: Paul Chapman.

Whitehurst, G. T., & Lonigan, C. J. (1998). Child development and emergent literacy. *Child Development, 69*(3), 848–872. doi:10.1111/j.1467-8624.1998.tb06247.x PMID:9680688

Woolfolk, A. E., Winnie, P. H., Perry, N. E., & Shapka, C. (2010). *Educational Psychology* (4th ed.). Toronto: Pearson.

Wu, C. C., & Honig, A. S. (2010). Taiwanese mothers' beliefs about reading aloud with preschoolers: Findings from the parent reading belief inventory. *Early Child Development and Care, 180*(5), 647–669. doi:10.1080/03004430802221449

Wu, C. C., & Sterling Honig, A. (2010). Taiwanese mothers' beliefs about reading aloud with preschoolers: Findings from the parent reading belief inventory. *Early Child Development and Care, 180*(5), 647–669. doi:10.1080/03004430802221449

Yalçın, A., & Aytaş, G. (2014). *Children's literature.* Ankara: Akçağ Publications. (in Turkish)

Yavuzer, H. (1995). *Child Psychology*. İstanbul: Remzi Publications. (in Turkish)

Yıldız Bıçakçı, M., & Aral, N. (2009). Language development. In N. Aral & T. Duman (Eds.), Educational psychology (pp. 115- 130). İstanbul: Kriter Publications. (in Turkish)

Yurtseven, L. (2011). Investigation of mothers' story reading effects on children's social skills development. Unpublished masters thesis, İnönü University, Educational Sciences Institute, Malatya. (in Turkish)

Zwisler, R. (2006). Sprachentwicklung. Retrieved from http://www.zwisler.de/scripts/Sprachentwicklung

Chapter 20
Adaptation of Cognitive Abilities Scale Into Turkish:
A Validity and Reliability Study – Cognitive Abilities Scale Preschool Form

Fatma Elif Kilinç
https://orcid.org/0000-0001-8372-5047
Ankara Yıldırım Beyazıt University, Turkey

Neriman Aral
Ankara University, Turkey

ABSTRACT

This study aimed to adapt Cognitive Abilities Scale-2 (CAS-2) Preschool (24-47 months) Form into Turkish by focusing on 24-36-month children and conduct validity and reliability studies. The study utilized the survey model and a study group which consisted of 60 socio-culturally disadvantaged 24-36-month children in Bolu province. CAS-2 Preschool Form includes five subscales: oral language, reading, mathematics, handwriting, and enabling behaviors. Translation and reverse translation procedures were used to adapt the scale developed in English into Turkish. Expert opinion was sought for content validity. Test-retest correlation and item-total score correlation consistency measures were used to examine the psychometric properties of the Turkish form. Results demonstrated that the scale is a valid and reliable instrument to determine cognitive abilities in 24-36-month Turkish children. Infant and pre-school forms are recommended to be used for 3-47-months at every socio-cultural level.

DOI: 10.4018/978-1-7998-2952-2.ch020

INTRODUCTION

The most important challenge that infants have to cope with is learning about and understanding the world they live in (Senemoğlu, 2010). Experiences of infants born in different parts of the world vary based on the conditions of their communities. Although they have completely different environmental experiences and various other differences, infants display basic and very similar characteristics in terms of making sense of their surroundings (Waxman and Leddon, 2011). Cognitive development, which is defined as active mental activities that facilitate understanding and learning about the world (Rutter and Rutter, 1993; Senemoğlu, 2010; Aral, 2011), includes the processes of reasoning, language development, problem solving and obtaining information. Cognition is defined as the manner or process of obtaining information that incorporates perception, recognition, reasoning and judgment (Gander and Gardiner, 1995). Cognitive development is the process in which the manner of understanding the environment and the world becomes more complex and effective from infancy to adulthood and it generates changes in individuals' functions in regard to language, thought and intelligence. In this process, children make sense of various cognitive tasks such as recognizing colors, making comparisons, finding missing parts and distinguishing concepts (Gander and Gardiner, 1995; Ruffin, 2009). According to Piaget, based on their interactions with their environment, children develop some techniques to understand objects and situations. Development of this method of comprehension and knowledge is called "cognitive development (Bertan et al., 2009).

Cognitive development is crucial in preparing children for life so that they can explore, experiment and transform their knowledge into practice. The cognitive system receives input from the environment which is perceived and stored in its memory. During the thought process, these perceptions are recalled from memory to be used and the information is generalized thorough conceptualization for a higher level of thinking. In this direction, outputs are created with new ideas and feedback is received which results in higher cognitive power (Aral, 2011).

Children differ from adults in many areas in early childhood in terms of the characteristics of cognitive development. While there are individual differences among children in this period, children have their own unique perceptions related to the world and a unique frame of mind. Children obtain a number of information from their environment and by perceiving and absorbing this information, they store it in their memory with the help of a cognitive system. When they think about acting in a certain way or make plans about an action, children recall information from memory and use it when appropriate conditions exist. They enhance their thinking through conceptualization, reasoning and generalization (Gander and Gardiner, 1995).

Children need to know about objects and how they work, to learn about the world, to develop language use and vocabulary skills, to develop their use of language and vocabulary and to improve their skills in knowledge creation and problem solving in order to develop their sense making skills for specific purposes by combining new knowledge with previous experiences (Ömeroğlu and Kandır, 2005).

Children possess cognitive abilities such as memory, perception, generalization and assessment of information as well as deduction in order to perform many cognitive processes such as reasoning, thinking, evaluation, problem solving and interpretation. Symbols, schemas, images, rules and concepts are the cognitive components of these abilities. In principle, cognitive abilities generate the stages of problem solving. Problems are solved with the accumulation of information and experience throughput life and the required information and experiences are made possible with the help of the operating system. It is very important that children obtain new knowledge, combine and blend this information with prior knowledge and start producing and developing new skills at an early stage.

CONCEPTS RELATED TO COGNITIVE DEVELOPMENT

Cognitive development consists of perception, attention, concept, memory and recall, thinking and reasoning, problem solving and creativity, each of which develops and works in coordination with each other.

Perception: Perception refers to stimulus awareness. The process of organizing and interpreting the stimuli received from the internal and external world through sensory organs is called perception (San-Bayhan and Artan, 2009; Ülke-Kürkçüoğlu, 2010). Internal and external stimuli received by the peripheral sensory nerves are integrated in the brain and they are the cause of various behavioral reactions. Perception is the sense-making and interpreting the stimuli received by sensory organs. This sense-making is partly based on objective facts and partly on subjective knowledge (Senemoğlu, 2010; Aral, 2011). Perception can be influenced by individuals' frames of mind, past experiences, prior knowledge, motivation levels and many internal factors (Senemoğlu, 2010). Perception emerges as the product of reciprocal interaction and relationship of many cognitive elements. During the process of perception, children develop forms of knowledge such as schemes, images, symbols, concepts and rules. Schemes, images and symbols include perception of information; concepts include rearrangement of information; rules include development and finally, application of information consists of evaluation and use of information (Aral, 2011).

Attention: Defined as the process of focusing on an object or event for a certain period of time, attention is the fundamental element in learning and recall; and learning begins with the process of attention (Banikowski and Mehring, 1999; Senemoğlu, 2011). Among the many stimulants in the environment, only the ones that are the most important and that grab attention are learned. Attention consists of two aspects: attention span and selective attention. Attention span refers to the duration during which an individual is focused on a point and selective attention refers to the process of recognizing the stimulus one is focused on and identifying its specific and essential qualities (San-Bayhan and Artan, 2009).

Factors that stimulate attention are divided into internal and external factors.

- ***External Factors***: External factors are the environmental factors. The strength, magnitude, intensity, mobility and brightness of the stimuli are listed as prominent factors. A loud noise, repeated stimuli, something very different from the environment, moving stimuli generate increased attention (Baymur, 1994; San-Bayhan and Artan, 2009; Aral, 2011).
- ***Internal Factors***: Everyone has a unique internal structure. While a person is interested in certain things, s/he may not be interested in others. Development of attention is based on these personal interests, i.e. basic needs. The first factor that stimulates the state of attention is personality traits. Paying attention to some things while ignoring others is learned through experience. The second factor that stimulates the state of attention is related to interests, wishes and needs at any given moment. The role of internal factors is more prominent on attention (Baymur, 1994).

Imitation: Imitation is defined as the ability to follow and copy an example or model of behavior. If infants observe that the adults around them replicate the reflective movements that they possess when they are born, they will start repeating these behaviors that are in their repertoire (Aral, 2011). At first, imitation is the process of seeing and getting pleasure from other senses, and then it transforms into the manner of maintaining previous interesting reactions (San-Bayhan and Artan, 2009). In the first stage of the development of imitation, when a newborn baby cries, the other baby cries as well. In the second stage (1-4 months), someone who mimics the baby's behavior causes the baby to continue that behavior. In the third stage (4-8 months), coordination begins between looking and behaving and at the same time the baby sees both his or her own behavior and the behavior of others. In the fourth stage (8-12 months), the infants begin to compare the behavior of others with their own, repeating familiar behaviors. In the fifth stage (12-18 months), the infants imitate the behaviors that are new to them. In the sixth stage (18-24 months), the toddlers mimic the behaviors they have learned before. At this stage, the mental representations of children are observed to be developed (Meltzoff and Moore, 1997). Piaget calls this type of imitation as deferred imitation. Infants' ability to imitate movements immediately or after a delay varies depending on the complexity and naturalness of the movement. This kind of imitation is regarded as a sign of progress in cognitive development rather than being direct imitation (San-Bayhan and Artan, 2009). The development of the ability to imitate and the concept of objects in children provide a basis for symbolic thinking.

Concept: Concept is defined as the structure of knowledge, meaningful in the human mind, in which the common and changeable characteristics of different objects and phenomena are represented (Ülgen, 1996). It is also an internal process that symbolizes the common characteristics of concepts, objects or events. This representation is usually done with a word or a name. The concepts are grouped in two classes: concrete or abstract. Development of concepts in children follows a path from concrete to abstract as well as a path from concrete thinking to abstract thinking. At the age of one or two, children begin to acquire concepts as they interact with the objects and people around them. Information in the memory needs to be organized in order to learn concepts. Since children acquire new cognitive abilities and tend to conceptually mobilize them, there is an improvement in the ability to form concepts starting from the age of four (Üstün and Akman, 2003; Bütün-Ayhan and Aral, 2007).

Memory and Recall: Memory is defined as the ability of the individual to reliably, completely and accurately store and re-use the information obtained from previous experiences. Comprehension is fostered when memory, which can also be defined as the ability to repeat the previously learned information, is combined with current experiences (Ömeroğlu and Kandır, 2005; Keleş and Çepni, 2006). Strength of recall plays an important role in the reliability of memory. Recall strength is the ability of the mind to become reliably aware and conscious with a proper stimulus (Ömeroğlu and Kandır, 2005). Memory is composed of different structural components in processing and interpreting information. These include sensory register, short-term memory and long-term memory (Banikowski and Mehring, 1999; Ömeroğlu and Kandır, 2005; Senemoğlu, 2010).

Sensory Register: The individual interacting with the environment constantly senses the stimuli coming through the sensory receptors. The things that an individual sees, hears, tastes, touches or feels are the contents of sensory register. The registering speed of this memory is stated as one million/second (Keleş and Çepni, 2006). The storage period of information in sensory register varies according to the sensory organ with which the information is associated. While visual information is stored for half a second to one second, audio information may be stored between two seconds and four seconds. Sensory register is called instant memory and it passes the incoming information into short-term memory by processing it. It is possible to transfer the information in sensory register to short-term memory only when sufficient attention is given (Banikowski and Mehring, 1999; Aral, 2011).

Short-Term Memory: Short-term memory, where most of the thinking and information processing takes place, converts information from sensory information into behavior or encodes it into long-term memory. Most of the information is stored as audio in short-term memory with limited capacity and time for incoming information display. The amount of information retained in this type memory and the duration of information retention varies according to age. In short-term memory, five to nine items can be recalled without repetition about ten to twenty seconds. Information coming to short-term memory is either neglected, forgotten or kept in short-term memory by repeating or it is transferred to long-term memory by combining it with previous information by means of repetition (Banikowski and Mehring, 1999; Aral, 2011).

Long-Term Memory: Long-term memory is a type of memory where the mastered information is continuously stored. Long-term memory is caused by structural changes in connections in the neurons. The most prominent features of long-term memory, where information can be stored both visually and verbally, include an unlimited capacity, a strong network of connections between the stored information and a lengthy processing period. Although it is difficult to transfer information to long-term memory, the information stored in this memory can be recalled for a lifetime. In long-term memory, words are often stored with their meanings, not only with the sounds they are associated with. It is also possible to store sounds, smells and images in long-term memory. Storing information in long-term memory is only possible with permanent functional, biochemical and structural changes in neural connections of the brain. Long-term memory can be examined in three different parts: episodic, semantic and operational (Banikowski and Mehring, 1999; Senemoğlu, 2010; Aral, 2011).

Although recall is explained as the transfer of information from long-term memory to short-term memory, this is not always possible. A successful recall is realized by retrieving information by using logic, clues and other information (Senemoğlu, 2010).

Thinking and Reasoning: Thinking can be defined as a functional and effective action with a specific goal (Yeşildere and Türnüklü, 2007). Thinking can also be defined as the access to information beyond existing information and going beyond what is available (Aral, 2011). Every problem requires that a new thought is generated in order to solve the problem at hand. The process of thinking can be described as translating objects and events into symbols in the external world. The brain performs a great number of operations on symbols, such as sense-making, hypothesizing, calculating and generating further symbols. Then, the brain can successfully cope with the existing "accurate" situation by converting these symbols back into objects and events in the outside world (Çubukçu, 2004).

Problem Solving: Problem solving can be defined as a thought process that starts with individuals' awareness of a problem and ends with a solution for the problem or it can be described as the effective adaptation to the environment in which individuals live (Heppner and Krouskof, 1987; Ülküer, 1998). The problem is defined as the obstacle that individuals encounter to reach their targets. Problem solving is finding the best way to overcome difficulties and to clear obstacles encountered in achieving goals. In this process, the individual seeks ways to get rid of tension and restore the internal balance of the organism by complying with the conditions and reducing obstacles (Aral, 2011). There are many aspects of problem solving and coping such as problem-focused coping and emotion-focused coping, ability to generate solutions and make decisions, result-oriented thinking and self-confidence in problem solving (Heppner and Baker, 1997). Since problem solving is a complex process, this process is divided into several stages to facilitate teaching and learning. Understanding the problem, making a plan for solution, implementing the plan and evaluating the results are widely accepted problem-solving steps in the problem solving process (Senemoğlu 2010).

Creativity: As a way of thinking, creativity has a very close relationship with imagination and is included in all affective and mental activities, in all sorts of tasks and pursuits. Although there are different definitions of creativity, creativity in general is defined as creating new things from known things by following new, unfamiliar and unique ways and presenting fluid, flexible, original and evocative behaviors and products in various socio-cultural environments (Brockman, 1993; Chien and Hui, 2010; Aral, 2011). Creative individuals are those who explain nonconventional thoughts in an unusual and brief manner, make important discoveries, create new perspectives, judgments and insights and try to explain the world in original ways and writings. Cognitive abilities such as curiosity, patience, ambition, sense of responsibility, leadership skills, adventurous thinking, ability to make discoveries, ability to make experiments and exercises and to make judgments from syntheses are also developed in creative individuals (Dağlıoğlu, 2010).

Intelligence: Intelligence, the complex ability of the human brain, is defined as the ability to learn new things quickly, to understand and use abstractions, symbols, relationships, to invent new and complex combinations, to direct thoughts in a certain channel, to control different data and to criticize (Samurçay, 1983). Intelligence is also known as the capacity of a person to generate a product that finds value in different cultures, the ability to produce effective and efficient solutions to the problems faced in real life and the ability to discover new and complex problems that need to be solved (Saban, 2001). It is emphasized that intelligence develops rapidly in the first years of development and slows down in later years. Factors affecting intelligence are examined in two groups as heredity and environment. According to researchers who argue that intelligence is influenced by heredity, intelligence is an innate ability and is largely determined by the effect of heredity. Studies have discovered a high level of relationship between children's' and their parents' intelligence. However, heredity alone is not sufficient to explain mental performance and environmental factors are also effective in determining intelligence (San-Bayhan and Artan, 2009; Aral, 2011). Richly stimulated environments play an important role in the utilization of mental potential. Failure to adequately stimulate children in the early period causes the child to progress slowly in terms of learning and skills development, therefore, it becomes rather difficult in later years to close the possible gaps.

NEUROLOGICAL BASES OF COGNITIVE DEVELOPMENT

The basis of cognitive development is related to neurological growth and maturation; and early experiences are important for neurological development (Ülke-Kürkçüoğlu, 2010). The brain is the most important center of the nervous system, which enables the control of body movements, regular functioning of organs, learning, thinking and recall and performance of many functions simultaneously. On average, the human brain has hundred billion cells, ten to fifteen billion of which are composed of nerve cells called neurons that facilitate thinking and learning. The rest of the cells are non-neuron cells, called glia, that maintain homeostasis (Keleş and Çepni, 2006). Although glia cells play a role in the maintenance and support of neurons, they also play a critical role in the connection of neurons. Babies are born with almost all their neurons intact. The newborns are equipped with approximately a hundred billion nerve cells responsible for brain functions such as thinking, seeing and feeling and new nerve cells are not generated in later life. Nerve cells are connected to each other by tiny spaces called synapses and form clusters that perform various functions of the brain. The connection between synaptic density and nerve cells in the brain occurs at an extraordinary speed in the first three years of life (Summak and Summak,

2005). In the first years of life, the brain goes through the process of "pruning" the unnecessary neurons and neural connections which are created based on experience. As the brain separates neurons, the child's perception, attention, learning, memory, language, personality and cognition are gradually shaped (Ülke-Kürkçüoğlu, 2010). In the first years of life, brain connections that receive adequate stimulation and reinforcement through experiences become permanent (Summak and Summak, 2005).

From the early years, skills and behaviors in all development areas are rapidly developed through learning as infants experience and interact with their surroundings. Children's cognitive development and adaptation to environment vary depending on their opportunities to receive enriching experiences. Basic knowledge, skills and habits gained through experiences from the early years are strong enough to shape children's social and emotional life and to affect their subsequent learning experiences. It is important to reveal cognitive abilities and improvements and track the changes by observing infants' behavior over time. For this reason, standard screening tools are used to follow children's development and to ensure provision of proper education. However, it is very difficult to examine changes that occur in thinking or concept development. It is also difficult to investigate cognitive development as it involves unobservable processes. However, it is even more difficult to investigate infants' cognitive development because they cannot be questioned about their thoughts. Studies have demonstrated that the assessment tools used in the early years to determine cognitive tasks predict infants' information-processing skills (Kavsek, 2004; Domsch, Lohaus and Thomas, 2009).

Assessment of cognitive abilities is imperative to determine the support needed by infants and to develop appropriate programs in early intervention practices (Feuerstein, Hoffman, Rand, Jensen, Tzuriel, and Hoffmann, 1985). It has been determined that early intervention programs are more effective when assessment tools are utilized to collect information from young children and programs are developed based on these input (Tusing, Maricle and Ford, 2003). Literature review shows that there is only a limited number of measurement tools to effectively examine children' cognitive development in 0-3 age group, the critical period for early intervention.

Since it is difficult to physically measure some variables such as attitude, belief, behavior or intelligence that are explored in the realm of social sciences, measurements are undertaken to determine the quality of these variables in their theoretical structures and relevant characteristics (Karagöz and Ekici, 2004). In these measurements, indirect measurements are conducted based on reactions from individuals by presenting them with items that will stimulate the desired feature (Baykul, 2000). Psychological tests which use indirect measures are often preferred by researchers because of their ease of application, objectivity in scoring and valid and reliable observations (Cronbach, 1951).

This study, based on Cognitive Abilities Scale-2, which is used to measure the cognitive abilities of 3-47 month children in order to determine children's current levels and to identify children who need development, aimed to adapt Cognitive Abilities Scale-2 Preschool Form (developed for 24-47 month children) into Turkish by focusing on 24-36 month children and to conduct the relevant validity and reliability studies for the Preschool Form.

METHOD

This study aimed to conduct the validity and reliability studies for Cognitive Abilities Scale (CAS-2) developed by Bradley-Johnson and Johnson (2001) to measure the cognitive abilities of children between 3-47 months and adapt the scale to Turkish. Survey method was used for this purpose.

The study was carried out in Bolu province with 24-36-month children who had socio-cultural disadvantages. In order to determine the study group, the records of Family Health Centers affiliated with Bolu Provincial Directorate of Health were examined and a list of participants with children that fit the criteria was prepared. Participants in this list were individually called on the phone to explain the purpose of the study and they were asked whether they would like to participate. The study group consisted of the volunteers who agreed to have their children participate in the study. It was decided that "oral language" subscale with the highest number of items (30 items) in the five subscales of the Cognitive Abilities Scale (CAS-2) was taken as the basis and it was determined that twice the number of items in oral language subscale (Tabachnick and Fidell, 2007) would be sufficient for validity and reliability analysis since the universe was small, there were problems in reaching people to implement the scale, the period of application was too lengthy and there might have been relationships between the variable (item scores). Accordingly, 60 children aged 24-36 month were included in the study group. Among these children, 25% were 24-26 months, 26.7% were 27-29 months, 21.7% were 30-33 months and 26.7% were 34-36 months while 55% were female and 45% were male.

The study utilized the Cognitive Abilities Scale (CAS-2), the second version of the first test published by Bradley-Johnson and Johnson in 1987. The scale is not based on a single theoretical approach, but on existing theories (information assimilation, learning and eye perception), Piaget's work and research on early cognitive development. The Cognitive Abilities Scale consists of two forms: the infant form and the preschool form (Bradley-Johnson and Johnson, 2001).

Infant form (3–23 months) consists of exploration of objects, communication with others, initiation and imitation subscales. Exploration of objects subscale (25 items) measures infants' ability to discover their environments, reaching, grasping, understanding cause-effect relationships and problem-solving skills by using their vision, hearing and touch; Communication with others subscale (25 items) measures infants' early receptive and expressive language skills such as independent production of sounds, response to vocal sounds, imitation of sounds, use of gestures and ability to follow directions; initiation and imitation subscale (29 items) measures infants' ability to initiate and model behaviors such as handling books, using paper and pencils, sorting objects by size, and imitating specific behaviors when requested by the practitioner (Bradley-Johnson and Johnson, 2001).

Preschool form (24–47 months) consists of oral language, reading, mathematics, handwriting and enabling behaviors subscales.

Oral language subscale (30 items) measures the ability to use and comprehend language such as understanding position words, knowledge of nouns and pronouns, using plurals, word endings and the correct word order. The first ten items of the oral language subscale include position words and each correct movement is scored as one point. After the first ten items, ten picture cards are used for the following ten items (Items 11-20). Each card is shown to the child to be named. Each item correctly named by the child is scored as two points and one point is given when the points to the card that she/he cannot name. The next four items (Items 21-24) include a set of cups, materials with female and male figures and instructions for children to use pronouns when these items are used by the child. Children are given two points when they answer the questions and one point when they follow the directions such as "pour me tea" or "give the female a spoon". The last six items (Items 25-30) include word order questions and therefore non-verbal scoring is not used. Children are evaluated for using noun-verb combinations, plurals, -ing (present participle), possessives and for their ability to describe simple daily tasks in sentences two to three-word sentences. A score is given for each correct answer.

Reading subscale (16 items) measures early reading abilities such as looking at books, comprehension, naming letters and understanding letter sounds. The first four items in the reading subscale are related to the book titled "Things Mikey Likes". These items assess the child's ability to look at the book properly, turn the pages of the book, show pictures and explain the event in a few sentences. One point is given for each correct answer. The fifth item consists of storytelling and the child is expected to tell a short story which is evaluated with a score of one. In sixth item the child is expected to provide the sound for letter "A" or the first letter of his / her name and a score is given for the correct answer. The last ten items include the sound cards for "M, S, T, L, F and C" and the child is expected to provide the letter sounds in these cards and say the words that start with these sounds. A score of one is given for each correct sound or word the child produces.

Mathematics subscale (22 items) measures early mathematical skills such as arranging objects by shape, matching numbers and quantities, finding equivalent parts, understanding general math concepts (little-more), recognizing numbers and counting. The child is asked to put the rings in the box in the first item of the subscale and one point is given for the correct answer. In the next ten items (Items 2-11), instructions about big-small, empty-full, long-short, little-more and same-different concepts are given to the child for four times in different positions and one point is given for three correct selections. In the next nine items (Items 12-20), the child is expected to show his/her age by using his/her fingers, counting rhythmically, building a tower with nested containers, sorting, recognizing numbers "2" and "5", counting rings and a score of one is given to each correct answer.

Handwriting Subscale (6 items) measures the child's ability in using the correct posture while sitting, grasping and holding a pencil and copying figures involved in letter formation. The first two items in this subscale, a score of one is given for the correct posture and pencil grasp while answering items 3 and 6. In the following items (Items 3-6), the child is asked to copy the circle, the vertical line, the horizontal line and the plus sign and two points are given for each figure that the child copies from the picture.

Enabling Behaviors Subscale (14 items) measures the abilities which support effective learning such as imitating behaviors, recalling sentences and remembering lists of unrelated words. The child is expected to imitate sounds in the first three items in this subscale (Items 1-3) and asked to provide physical imitations in the next three items (items 4-6). A score of one is given for each correct imitation. In the next seven items (Items 7-13), the child is asked to repeat sentences and a score of one is given to each correct repetition. In Item 14, the child hears two lists of words consisting of one word in the first line, two words in the second line, three words in the third line and four words in the fourth line and asked to repeat the words. The child is given points (from one to four points) matching the most number of words they have repeated (Bradley-Johnson and Johnson, 2001).

Both forms (infant and preschool form) measure general cognitive ability and non- vocal cognitive ability. Measuring non-vocal cognitive ability is useful in obtaining valid and reliable results from children who cannot not speak, do not want to speak, or do not understand speech (HaileMariam, 2004; Swanson, Bradley-Johnson, Johnson and O'Dell, 2009). The highest raw score that can be obtained from the questions measuring general cognitive ability by children who are able to speak is 108. The highest raw score that can be obtained from the questions measuring non- vocal cognitive ability by children who cannot speak or who do not want to speak is 70. Total raw scores are evaluated by converting them to standard scores based on age range (Bradley-Johnson and Johnson, 2001).

According to the internal validity of the scale, internal consistency coefficient (Cronbach Alpha) values were .94 for the non- vocal part and .97 for the total scale. Non- vocal part was found to be .93 for the three age groups and .96 for the total scale. Convergences were found in the construct validity of the criteria, since the content validity includes items similar to a normal intelligence test and achievement test. The Cognitive Abilities Scale was compared with the Stanford-Binet Intelligence Test for two and three year olds and values of .79 and .74 were obtained for two and three year olds respectively (Cummings, 2000).

The Cognitive Abilities Scale Preschool Form was adapted to Turkish children within the scope of this study. Preschool Form includes examiner's manual, toy set, picture cards, children's books and instructions (Cummings, 2000) and the period of implementation is 30 minutes. A Preliminary Application Form was developed and implemented on a sample which was selected from the target group who were the audience for the reliability and validity studies of the measurement tool. Content and construct validity were used in the validity studies and test-retest correlation and item-total score correlation consistency criteria were used in the reliability analysis.

FINDINGS

Studies that focus on understanding the way individuals exactly behave when they encounter a task or problem since infancy, the process of information transfer through the system, the way mental strategies work and processes of effective use of information have formed the basis of cognitive development studies (Berk, 2000). Determining mental strategies in early years is important to predict future performance of cognitive development. Therefore, there is a need for standard measurement tools that can be used to measure cognitive abilities since infancy. This study focused on the validity and reliability analyzes of the Cognitive Abilities Scale developed to measure the cognitive abilities of 3-47 month children and on the adaptation of the scale into Turkish culture.

Validity Study: Validity refers to the degree to which the measurement tool measures the qualities it claims to measure without confusing or mixing them with other features (Alpar, 2006; Tavşancıl, 2006; Hovardaoğlu, 2007; Büyüköztürk, 2010). There are several methods for determining the validity of the measuring tool. Although there are different classifications in the literature, content validity, criterion validity and construct validity are known to be the most used validity methods in determining the validity of a measurement tool (Büyüköztürk et al., 2008). The scale should have high validity for all the techniques used to determine the ideal validity for a measuring tool. However, since this is often not possible in practice, it is accepted to determine the validity by any of these techniques (Hovardaoğlu, 2007).

Content validity and construct validity were used in the validity studies of the Cognitive Abilities Scale. The data obtained with content validity also provide important evidence for construct validity (American Educational Research Association, 1998). Linn and Gronlund (1995) stated that it may be sufficient to take expert opinion as the logical way of examining construct validity. In this context, expert opinion on whether each item is sufficient or appropriate to measure defined behaviors within the scope of content validity was accepted as evidence for construct validity.

The Cognitive Abilities Scale was first translated from English to Turkish and opinions of five experts were sought in regards to the translated documents. Translate-back translate procedure was used and the expressions translated into Turkish were translated back to English by other foreign language experts and the original expressions were compared with these expressions. It was observed that there were no semantic differences for all items. This was accepted as proof that language equivalence was ensured for the form obtained by translating the expressions in the source language to Turkish, the target language. Therefore, the Turkish translation of the form was finalized.

The Cognitive Abilities Scale, translated into Turkish, was submitted for expert opinion in line with the requirements of content validity studies. Experts were asked to evaluate each item by using yes-no options for semantic, experiential, conceptual and developmental equality, and to provide constructive criticism to develop the expressions in the scale, if needed. Ten child development and education experts were consulted to test content validity. The items with 90-100% agreement among experts remained the same, the items with 70-80% agreement among experts (Büyüköztürk, 2010) were improved in line with the suggested corrections and were included in the scale. Items with 60% or less agreement among experts were excluded from the scale. Accordingly, in the Cognitive Abilities Scale, items 25 and 30 in verbal language domain were corrected in line with the expert suggestions and item 29 in verbal language domain was removed from the scale based on expert suggestions that it was not appropriate for Turkish language structure and the other items remained the same in line with expert agreement.

Construct validity is the theoretical analysis of whether the items developed to assess a variable really measure it or correlate with the variable. Linn and Gronlund (1995) established three stages for this measure of validity. In the first stage, a number of hypothetical structures are defined to explain scale performance. In the second stage, hypotheses are produced based on the identified structures. In the third stage, these hypotheses are tested by logical methods (expert opinion) or experimental methods. Five subscales (oral language, reading, mathematics, handwriting and enabling behaviors), believed be the basis of the Cognitive Abilities Scale, were developed and standardized in the United States. In order to determine whether these subscales in the measurement tool are also valid for Turkish children, a confirmatory factor analysis was planned to validate a predetermined and constructed structure (Sümer, 2000). However, this procedure was not completed since inadequate sampling, low number of observed variables and data structure resulted in deviation from normality and created values outside the acceptable limits for covariance and correlation matrix values tried to be analyzed in the output file (Anderson and Gerbing, 1984; Lee, Poon and Bentler). 1992; Wothke, 1993; Loehlin, 2004; Şimşek, 2007). Therefore, confirmatory factor analysis could not be performed to validate the subscales of the Cognitive Abilities Scale. The content validity of the scale was regarded as a parallel to the validity of the Cognitive Abilities Scale,

Reliability Study: Reliability refers to the degree to which the measurement tool measures the property accurately. The ratio of actual variance to total variance is reliability. The correlation coefficient (r), calculated as the reliability coefficient of the measurement tool, is used to interpret the accuracy of the measurement results (Anastasia and Urbina, 1997; Büyüköztürk, 2010; Büyüköztürk et al., 2008; Hovardaoğlu, 2007; Tavşancıl, 2006).

Reliability coefficients take into account the group stability of the measured unit. Methods based on item variance and equivalent halves or split-half methods are used in calculating the reliability coefficient by using the scores obtained from applying the measurement tool once to the sample. The Kuder-Richardson (KR-20) and Cronbach Alpha (α) methods, which are based on item variance, were not used in the reliability calculations of the Cognitive Abilities Scale. The fact that some items are scored as zero, one and two and some items scored as zero and one necessitated the use of test-retest correlation and item-total score correlation in calculating the reliability of the Cognitive Abilities Scale. The items with variance values of zero which show how far each value is from the mean distribution (items 1 and 26 in the verbal language subscale; item 6, 7, 8, 9, 10,11, 12, 13, 14, 15, 16, in reading subscale; items 10, 11, 15, 16, 17, 18, 19, 20, 21, 22 in the mathematics subscale and items 1, 2, 3, 4, 5, 6 in enabling behaviors) is not included in the analysis since they do not have any discriminatory value in terms of measurement.

Test-retest reliability: The test-retest method is the re-application of the measurement tool to the same individuals under the same conditions but at different times (Alpar, 2006; Çepni et al., 2009). The correlation coefficient between the scores obtained from both applications is calculated by the correlation technique. The reliability of the Cognitive Abilities Scale was tested by test-retest method. For this purpose, 23 children out of 60 children included in the study in Bolu province were retested three weeks after the first application. Cognitive Abilities Scale subscales had correlation values of .92 or higher during test-retest measurements. Correlation coefficient values of .70-1.00 is defined as high level of relationship (Büyüköztürk, 2010). This result shows that the measuring tool provides consistent measurements over time.

Item- total score correlation: Positive and high correlation between the scores obtained from the test items and the total score of the test indicates that the items exemplify similar behaviors and that the measurement tool has high internal consistency (Büyüköztürk, 2010). Mann-Whitney U Test was used to analyze the differences between the lower and upper 27% item scores according to the total scores of the measurement tool within the scope of item analysis, since the sample was small, the variances were not homogeneous and the data did not show normal distribution (Alpar, 2006; Büyüköztürk, 2010).

The difference between the item mean scores of the upper and lower 27% groups for the oral language subscale of the Cognitive Abilities Scale was found to be significantly meaningful for all items except item 4 ($U = 120, p > .05$) and item 6 ($U = 120, p > .05$). This significant difference indicates that the internal consistency of the test is high (Büyüköztürk, 2010). Items 4 and 6 were excluded from the scale because they did not show any distinguishing features among participants in terms of the measured behavior.

The difference between the item mean scores of the upper and lower 27% groups for the reading subscale of the Cognitive Abilities Scale was found to be significant for all items except item 5 ($U = 96, p > .05$). As it did not show any distinguishing feature among participants in terms of the measured behavior, item 5 was excluded from the scale.

The difference between the item mean scores of the upper and lower 27% groups for mathematics subscale of Cognitive Abilities Scale was found to be statistically significant for all items except item 12 ($U = 128, p > .05$) and 14 ($U = 112, p > .05$). Items 12 and 14 were excluded from the scale because they did not show any distinguishing features among participants in terms of the measured behavior.

The difference between the item mean scores of the upper and lower 27% groups was found to be significant for handwriting subscale of the Cognitive Abilities Scale. Although Item 6 in this subscale was not significant at .01 level, this item was not excluded from the scale based on expert opinions since the item was acceptable at .10 level.

The difference between the item mean scores of the upper and lower 27% groups for enabling behaviors subscale of the Cognitive Abilities Scale was found to be statistically meaningful for all items except item 10 ($U = 88, p > .05$), item 11 ($U = 96, p > .05$).), item 12 ($U = 120, p > .05$), and item 13 ($U = 120, p > .05$). Items 10, 11, 12 and 13 were not found to discriminate between individuals in terms of the measured behavior so it was considered appropriate to remove them from the scale.

According to the Mann-Whitney U test results, the Cognitive Abilities Scale was composed of oral language, reading, mathematics, handwriting and enabling behaviors subscales for the upper and lower 27% groups and items 4 and 6 in the oral language subscale, items 5, 12 and 14 in the mathematics subscale and items 10, 11, 12 and 13 in the enabling behaviors subscale were not included in the measurement tool in this study.

During the adaptation of the Preschool Form to Turkish, it was determined that the scale consisted of five subscales (oral language, reading, mathematics, handwriting and enabling behaviors) and the item numbers of each subscale differed from the original form. In this context, the subscales, the number of items along with the explanations in regards to items characteristics are presented below.

Oral language subscale (25 items); The first ten items of the oral language subscale includes position words and a score of one is given to each correct movement. Items 1, 4 and 6 which evaluate position words by having students place the chain in the box and move the car up and down the ramp are not included in the measurement tool. Ten picture word cards are used in the ten items (Items 11-20) which follow the first ten items. Each card is presented to the child who is asked to name it. Each item correctly identified is given two points and one point is provided in cases where the child shows the unidentified card. The following four items (Items 21-24) include a set of cups, materials with female and male figures and instructions for children to use pronouns when these items are used by the child. Children are given two points when they answer the questions and one point when they follow the directions such as "pour me tea" or "give the female a spoon". The last six items (Items 25-30) include word order questions and therefore non-verbal scoring is not used. Children are evaluated for using noun-verb combinations, plurals, -ing (present participle), possessives and for their ability to describe simple daily tasks in sentences two to three-word sentences. A score is given for each correct answer. Item 29 is not included in the measurement tool because it does not comply with the Turkish language structure.

Reading subscale (4 items); The first four items in the reading subscale are related to the book titled *"Mikey's Favorite Things"*. These items assess the child's ability to look at the book properly, turn the pages of the book, show pictures and explain the event in a few sentences. One point is given for each correct answer. Item 5 where the child is expected to tell a short story and Items 6 through 18 in which sound cards are used are not included in the measurement tool.

Mathematics subscale (10 items); The child is asked to put the rings in the box in the first item of the subscale and one point is given for the correct answer. In the next ten items (Items 2-10), instructions about big-small, empty-full, long-short, little-more and same-different concepts are given to the child for four times in different positions and one point is given for three correct selections. Items 10-11 evaluating the concepts of same-different, Item 15 asking the child to show his/her age by using his/her fingers, Item 14 requiring tower construction with nested containers and Items 15-22 which include sorting, recognition of numbers "2" and "5" and counting rings are not included in the measurement tool.

Handwriting Subscale (6 items); The first two items in this subscale, a score of one is given for the correct posture and pencil grasp while answering items 3 and 6. In the following items (Items 3-6), the child is asked to copy the circle, the vertical line, the horizontal line and the plus sign and two points are given for each figure that the child copies from the picture.

Enabling Behaviors Subscale (4 items); The first three items where child is expected to imitate sounds (Items 1-3) and the next three items where the child is asked to provide physical imitations (items 4-6) and Items 10-13 which require lengthy sentence repetitions are not included in the measurement tool. In Items 7-9, the child is asked to repeat short sentences and each correct repetition is scored as one point. In Item 14, the child is told two lists of words consisting of one word in the first place, two words in the second place, three words in the third place and four words in the fourth place and asked to repeat the words. In any one of the two word lists, children are given points (from one to four points) for the number of words they repeat most.

CONCLUSION

This study, based on Cognitive Abilities Scale-2, used to measure the cognitive abilities of 3-47 month children in order to determine children's current levels and to identify children who need development, aimed to adapt Cognitive Abilities Scale-2 Preschool Form (developed for 24-47 month children) into Turkish by presenting the relevant validity and reliability study for the Preschool Form by focusing on 24-36 month children. Content validity and construct validity were examined in the context of the validity studies for the Cognitive Abilities Scale. The data obtained with content validity also constitute important evidence for construct validity. Linn and Gronlund (1995) stated that it may be sufficient to take expert opinion as the logical way of examining construct validity. In this context, expert opinion on whether each item is sufficient or appropriate to measure defined behaviors within the scope of content validity was accepted as evidence for construct validity. Expert opinions were sought to assess the suitability of the statements/expressions, pictures and materials in the measurement tool to Turkish culture. Necessary arrangements were made in line with the agreement levels of the experts regarding the expressions, pictures and materials in the subscales of the Cognitive Abilities Scale (Büyüköztürk, 2005).

While Items 25 and 30 in the oral language subscale of the Cognitive Abilities Scale were reorganized, Item 29 was excluded from the scale because it was not compatible with the Turkish language structure. Construct validity is the theoretical analysis of whether the items developed to assess a variable really measure it or correlate with the variable. In order to determine whether these five subscales (oral language, reading, mathematics, handwriting and enabling behaviors) in the measurement tool are also valid for Turkish children, a confirmatory factor analysis was conducted. However, this procedure was not completed since inadequate sampling, low number of observed variables and data structure resulted in deviation from normality and created values outside the acceptable limits for covariance and correlation matrix values tried to be analyzed in the output file

Reliability, which is an indicator of how accurately the measurement tool measures the property, is expressed as the ratio of actual variance to total variance. The correlation coefficient (r), calculated as the reliability coefficient of the measurement tool, is used to interpret the accuracy of the measurement results. Reliability coefficients take into account the group stability of the measured unit. The fact that some items are scored as zero, one and two and some items scored as zero and one necessitated the use of test-retest correlation and item-total score correlation in calculating the reliability of the Cognitive Abilities Scale. The items with zero variance values were not included in the analysis as they did not have any discriminatory value in regard to measurement. After excluding the required items from subscales, item-total score correlation and test-retest correlation analyzes were performed. In all subscales of the measurement tool, correlation values were found to be .92 or higher in test-retest measurements. Similarly, Durmuşoğlu (2009) also found the test-retest correlation values to be between .92 and .98 for the subscales of the Cognitive Abilities Scale. The positive and high correlation between the scores obtained from test items and the total score of the test shows that the items exemplify similar behaviors and that the internal consistency of the measurement tool is high (Büyüköztürk, 2010).

Mann-Whitney U Test, one of the nonparametric tests, is used for unrelated measurements when sample size is small, the variances are not homogeneous and the data does not show normal distribution (Alpar, 2006; Büyüköztürk, 2010). For this reason, Mann-Whitney U Test was used to analyze the differences between the lower and upper 27% item scores according to the total scores of the measurement tool within the scope of item analysis. As a result of comparing the lower 27% and upper 27% groups over the total score, the differences in all sub-dimensions were found to be significant in favor of the upper 27% group. In this case, it is concluded that the items in the scale distinguish the upper and lower groups well (Hovardaoğlu, 2007). According to analysis results, the items that did not display any distinguishing characteristics among individuals in terms of the measured behaviors were excluded from the scale and the measurement tool was finalized.

Low level of validity and reliability in some subscales of the measurement tool may related to the fact that the study was limited to a certain interval in terms of age groups in months, the application period for the measurement tool was lengthy and there were difficulties in accessing the implementation group. 30 items in the oral language subscale in the original version of the measuring tool were reduced to 25 items, 16 items in the reading subscale were reduced to 4 items, 22 items in the mathematics subscale were reduced to 10 items, 14 items in the enabling behaviors subscale were reduced to 4 items and 6 items in handwriting sub-scale were remained the same.

When the findings of the study are evaluated in general by keeping the limitations of the study in mind, it is believed that the Cognitive Abilities Scale is a valid and reliable tool that can be used to determine the cognitive abilities of 24-36-month Turkish children. It is thought that bringing the scale to the field will be the basis for early identification of children's developmental needs, the emergence of strengths and weaknesses in children's cognitive abilities and the preparation of educational programs for children, thus providing the necessary cognitive support for them.

The most important limitation of this study is that the preschool form (24-47 months) of the measurement tool was conducted on 24-36 month children of socio-culturally disadvantaged families. Infant and pre-school forms of the measurement tool are recommended to be used for 3-47-month children by reaching the children of families at all month intervals from every socio-cultural level. In addition, standardization of the measurement tool can be undertaken by conducting an adaptation study with a sample representing Turkey.

REFERENCES

Alpar, R. (2006). Spor Bilimlerinde Uygulamalı İstatistik (3. Basım). Ankara: Nobel Yayın Dağıtım.

Anastasi, A., & Urbina, S. (1997). *Psychological Testing* (7th ed.). New Jersey: Prentice-Hall. Inc.

Anderson, J. C., & Gerbing, D. W. (1984). The Effect Of Sampling Error On Convergence, İmproper Solutions, And Goodness-Of-Fit İndices For Maximum Likelihood Confirmatory Factor Analysis. *Psychometrika, 49*(2), 155–173. doi:10.1007/BF02294170

Aral, N. (2011). Bilişsel Gelişim. In N. Aral & G. Baran (Eds.), *Çocuk Gelişimi*. İstanbul: Ya-Pa Yayınları.

Banikowski, A. K., & Mehring, T. A. (1999). Strategies to enhance memory based on brain-research. *Focus on Exceptional Children, 32*(2), 1–16.

Baykul, Y. (2000). *Eğitimde ve Psikolojide Ölçme: Klasik Test Teorisi ve Uygulaması*. Ankara: ÖSYM Yayınları.

Baymur, F. (1994). *Genel Psikoloji*. İstanbul: İnkılap Yayınevi.

Berk, L. E. (2000). *Bebekler ve Çocuklar: Doğum Öncesinden Orta Çocukluğa. (Çev: N. Işıkoğlu-Erdoğan)*. Ankara: Nobel Akademik Yayıncılık.

Bertan, M., Haznedaroğlu, D., Koln, P., Yurdakök, K., & Güçiz, B. D. (2009). Ülkemizde Erken Çocukluk Gelişimine İlişkin Yapılan Çalışmaların Derlenmesi (2000-2007). *Çocuk Sagligi ve Hastaliklari Dergisi*, *52*, 1–8.

Bradley-Johson, S., & Johnson, C. M. (2001). *The Cognitive Abilities Scale* (2nd ed.). Austin, TX: Pro-Ed.

Brockman, J. (1993). *Creativity*. New York: Touchstone.

Bütün-Ayhan, A., & Aral, N. (2007). Bracken temel kavram ölçeği- gözden geçirilmiş formunun altı yaş çocukları için uyarlama çalışması. *Hacettepe Üniversitesi Eğitim Fakültesi Dergisi*, *32*, 42–51.

Büyüköztürk, Ş. (2005). Anket Geliştirme. *Türk Eğitim Bilimleri Dergisi*, *3*(2), 133–151.

Büyüköztürk, Ş. (2010). *Sosyal Bilimler İçin Veri Analizi El Kitabı*. Ankara: Pegem Akademi.

Büyüköztürk, Ş., Kılıç-Çakmak, E., Akgün, Ö. E., Karadeniz, Ş., & Demirel, F. (2008). *Bilimsel Araştırma Yöntemleri*. Ankara: Pegem Akademi.

Çepni, S., Bayrakçeken, S., Yılmaz, A., Yüzel, C., Semerci, Ç., Köse, E., ... Gündoğdu, K. (2009). Geçerlik ve Güvenirlik. In E. Karip (Ed.), *Ölçme ve Değerlendirme* (pp. 89–122). Ankara: Pegem Akademi.

Chein, C., & Hui, A. N. N. (2010). Creativity in early childhood education: Teachers' perception in three Chinese societies. *Thinking Skills and Creativity*, *5*(2), 49–60. doi:10.1016/j.tsc.2010.02.002

Cronbach, L. J. (1951). Coefficient Alpha and The İnternal Structure Of Tests. *Psychometrika*, *16*(3), 297–334. doi:10.1007/BF02310555

Çubukçu, Z. (2004). Öğretmen adaylarının düşünme stillerinin öğrenme biçimlerini tercih etmelerindeki etkisi. XIII. Ulusal Eğitim Bilimleri Kurultayı, Retrieved from http://www.pegema.net/dosya/dokuman/253.pdf

Cummings, L. L. (2000). Parent Training: A Program For Parents Of Two-And Three-Year-Olds. Unpublished Doctoral Dissertation, Central Michigan University, Michigan.

Dağlıoğlu, H. E. (2010). Yaratıcılık, hayal gücü ve zekâ ilişkisi. In E. Çelebi Öncü (Ed.), *Erken Çocukluk Döneminde Yaratıcılık ve Geliştirilmesi*. Ankara: Pegem Akademi.

Domsch, H., Lohaus, A., & Thomas, H. (2009). Prediction Of Childhood Cognitive Abilities From A Set Of Early Indicators Of Information Processing Capabilities. *Infant Behavior and Development*, *32*(1), 91–102. doi:10.1016/j.infbeh.2008.10.006 PMID:19095308

Durmuşoğlu, G. (2009). Item Analysis and Reliability Of The Cognitive Abilities Scale-Second Edition: Turkish Version. Unpublished Doctoral Dissertation, Central Michigan University, Michigan.

Feuerstein, R. S., Hoffman, M. B., Rand, Y., Jensen, M. R., Tzuriel, D., & Hoffmann, D. B. (1985). Learning To Learn: Mediated Learning Experiences And Instrumental Enrichment. *Special Services in the Schools*, *3*(1-2), 49–82. doi:10.1300/J008v03n01_05

Gander, M. J., & Gardiner, H. W. (1995). *Çocuk ve Ergen Gelisimi*. Ankara: İmge Kitabevi.

Hailemariam, A. (2004). Review Of Cognitive Abilities Scale-Second Edition. *Journal of School Psychology*, *42*(2), 171–176. doi:10.1016/j.jsp.2003.10.002

Heppner, P. P., & Krauskorf, C. J. (1987). The integration of personal problem solving process within counseling. *The Counseling Psychologist*, *15*, 371–447. doi:10.1177/0011000087153001

Hovardaoğlu, S. (2007). *Davranış Bilimleri İçin Araştırma Teknikleri*. Ankara: Hatiboğlu Yayınevi.

Karagöz, Y., & Ekici, S. (2004). Sosyal bilimlerde yapılan uygulamalı araştırmalarda kullanılan istatistiksel teknikler ve ölçekler. *Cumhuriyet Üniversitesi İktisadi ve İdari Bilimler Dergisi*, *5*(1), 25–43.

Kavsek, M. (2004). Predicting later IQ from infant visual habituation and dishabituation: A meta-analysis. *Applied Developmental Psychology*, *25*(3), 369–393. doi:10.1016/j.appdev.2004.04.006

Kavsek, M. (2004). Predicting Later Iq From İnfant Visual Habitation And Dishabituation: A Meta-Analysis. *Journal of Applied Developmental Psychology*, *25*(3), 369–393. doi:10.1016/j.appdev.2004.04.006

Keleş, E., & Çepni, S. (2006). Beyin Ve Öğrenme. *Türk Fen Eğitimi Dergisi*, *3*(2), 66–82.

Lee, S.-Y., Poon, W.-Y., & Bentler, P. M. (1992). Structural Equation Models With Continuous And Polytomous Variables. *Psychometrika*, *57*(1), 89–105. doi:10.1007/BF02294660

Linn, R. L., & Gronlund, N. E. (1995). Measurement And Assessment. In *Teaching* (7th ed.). Englewood Cliffs, NJ: Prentice-Hall.

Loehlin, J. C. (2004). Latent Variable Models: An Introduction To Factor, Path And Structural Equation Analysis, (4th ed.). Mahwah, NJ: Lawrence Erlbaum Associates. doi:10.4324/9781410609823

Ömeroğlu, E., & Kandir, A. (2005). *Bilişsel Gelişim*. İstanbul: Morpa Kültür Yayınları.

Ruffin, N. J. (2009). *Understanding Growth and Development Patterns Of Infants*. Retrieved from http://Pubs.Ext.Vt.Edu/350/350-055/350-055.html

Rutter, M., & Rutter, M. (1993). *Developing Minds: Challenge and Continuity Across the Lifespan. Ny.* Basic Books.

Saban, A. (2001). *Çoklu Zekâ Teorisi ve Eğitim*. Ankara: Nobel Yayın Dağıtım.

Samurçay, N. (1983). Zekâ ve yaratıcılık. *Eğitim ve Bilim Dergisi*, *8*(45), 4–12.

San-Bayhan. P ve Artan, İ. (2009). Çocuk Gelişimi ve Eğitimi. İstanbul: Morpa Yayınları.

Senemoğlu, N. (2010). *Gelişim, Öğrenme Ve Öğretim: Kuramdan Uygulamaya. 18. Baskı*. Ankara: Pegem Akademi.

Şimşek, Ö. F. (2007). *Yapısal Eşitlik Modellemesine Giriş, Temel İlkeler ve Lisrel Uygulamaları*. Ankara: Ekinoks.

Sümer, N. (2000). Yapısal eşitlik modelleri: temel kavramlar ve örnek uygulamalar. *Türk Psikoloji Yazilari*, *3*(6), 49–74.

Summak, M. S., & Summak, A. E. (2005). *Akıllı Bebekler Akademisi; Bebekler İçin Çoklu Zekâ Ve Beyin Geliştirme Rehberi*. İstanbul: Sistem Yayıncılık.

Swanson, J. R., Bradley-Johnson, S., Johnson, C. M., & O'dell, A. R. (2009). The Cognitive Abilities Scale-Preschool Form: Studies Of Concurrent Criterion-Related, Construct, And Predictive Criterion-Related Validity. *Journal of Psychoeducational Assessment*, *27*(1), 46–56. doi:10.1177/0734282908322454

Tabachnick, B., & Fidell, L. S. (2007). *Using Multivariate Statistics* (5th ed.). Boston: Pearson Education.

Tavşancıl, E. (2006). *Tutumların Ölçülmesi ve Spss İle Veri Analizi*. Ankara: Nobel Yayın Dağıtım.

Tusing, M. E., Maricle, D. E., & Ford, L. (2003). Assessment With The Woodcock-Johnson Iıı And Young Children. In F. A. Schrank & D. P. Flanagan (Eds.), *WJIII Clinical Use And Interpretation: Scientist-Practitioner Perspectives* (pp. 243–283). Academic Press. doi:10.1016/B978-012628982-4/50009-X

Ülgen, G. (1996). *Kavram Geliştirme Kuramlar ve Uygulamalar. Geliştirilmiş ikinci baskı*. Ankara: Setma Baskı.

Ülke-Kürkçüoğlu, B. (2010). 0-6 Yaş Arası Çocukların Temel Gelişimsel Özellikleri: Bilişsel Gelişim Ve Dil Gelişimi. In I.H. Diken (Ed.). Erken Çocukluk Eğitimi. Ankara: Pegem Akademi.

Üstün, E., & Akman, B. (2003). Üç yaş grubu çocuklarda kavram gelişimi. *Hacettepe Üniversitesi Eğitim Fakültesi Dergisi*, *24*, 137–141.

Waxman, S. R., & Leddon, E. M. (2011). Early Word - Learning And Conceptual Development Everything Had A Name and Each Name Gave Birth To A New Thought. In U. Goswami (Ed.), *Wiley - Blackwell Handbooks Of Developmental Psychology* (2nd ed., pp. 180–208). Ma: Wiley-Blackwell.

Wothke, W. (1993). Nonpositive Definite Matrices İn Structural Modeling. In K. A. Bollen & J. S. Long (Eds.), Testing Structural Equation Models (pp. 256-293). Newbury Park, CA: Sage Publications.

Yeşildere, S., & Türnüklü, E. B. (2007). Öğrencilerin matematiksel düşünme ve akıl yürütme süreçlerinin incelenmesi. *Ankara Üniversitesi Eğitim Bilimleri Fakültesi Dergisi*, *40*(1), 181–213.

KEY TERMS AND DEFINITIONS

Adaptation: The process of changing to suit different conditions.

Cognitive: Connected with thinking or conscious mental processes.

Measurement: A value, discovered by measuring, that corresponds to the size, shape, quality, etc. of something.

Reliability: The quality of being able to be trusted or believed because of working or behaving well.

Scale: A set of numbers, amounts, etc., used to measure or compare the level of something.

Translate: To change written or spoken words from one language to another.

Validity: Based on truth or reason; able to be accepted.

Chapter 21
Yoga for Babies and Children:
Introduction, Definition and Importance of Yoga, Benefits of Yoga, Yoga in Education

Besime Bilmez
Genc Vocational School Child Care and Youth Services, University of Bingöl, Turkey

Neriman Aral
Faculty of Health Sciences, Department of Child Development, Ankara University, Turkey

ABSTRACT

This chapter describes how yoga is a means to relax the body, slowing breathing, and mentally calming down. Yoga in general includes physical postures / poses, breathing practices and movement meditation to improve self-regulation skills within the body and mind. An examination is made in this chapter of how yoga can be taught to babies and young children.

INTRODUCTION

Yoga; is based on the combination of techniques that have an influence on the individual after affecting the body, mind and emotions. Yoga is generally practiced as a mental and physical exercise system in the West and causes relaxation. Yoga can be practiced by everyone without any difference in belief (Höbek Akarsu, 2016) and can be widely used to support the development of children.

Yoga provides children with fun, non-competitive and social experience in the educational environment, physical and motor development, social and emotional development, cognitive development, creativity and discovery (Young et al., 2017). Yoga, in its traditional form, is a holistic practice system that includes multiple techniques including physical postures / exercises, breathing exercises, deep relaxation techniques and meditation / awareness practices. Biomedical and psychosocial research on the clinical effectiveness of yoga for both adults and children has been increasing rapidly in recent years, especially in the last five years (Jeter et al., 2015), and systematic research reviews emphasize that yoga is particularly effective in reducing stress and improving mood (Pascoe and Bauer, 2015; Sharma, 2014).

DOI: 10.4018/978-1-7998-2952-2.ch021

Therapeutic applications of yoga have been shown to be valid in improving both physical and mental health when applied to children and adolescents (Ganpat and Nagendra, 2011; Birdee et al., 2009). Telles (2012), one of the world's leading Yoga researchers, concluded that yoga improves the physical and mental health of children in his study named as Effects of Yoga on Children's Mental Health. Three special skills or features developed by yoga include; self-regulation and physical fitness, along with improvements in mind-body awareness, mental state, health, behavior and performance. Through the practice of meditation and awareness, mental body awareness is increased; this contributes to positive behaviors and outcomes as a result of increased awareness of positive feelings associated with healthy behaviors (Butzer et al., 2016). Khalsa et al. (2012) state that Yoga in schools helps children develop their flexibility, mood and self-regulation skills related to emotion and stress. Yoga is therefore an important life skill tool for children and young people to cope with life-long stress and self-regulation. It is claimed that yoga is effective for maintaining emotional balance and training the mind and body. In the studies, it was found that Yoga is a good tool for listening to the heart (Langøien, 2013) and Yoga practices significantly increase the emotion regulation skills of adolescents (Daly et al., 2015).

In Yoga; integrity of mind, body and soul; harmonious operation of all systems in the body and the elimination of contrasts that disturb feelings and thoughts at the mental level can be achieved. Other studies have shown that physical and cognitive therapy, consciousness awareness and mind-body movements such as Yoga are effective in reducing cortisol secretion and thus improve mood and increase happiness (Sharma, 2014; Chong et al., 2011). It can lead the individual to connect with the universal spirit, thus reaching all parts of the self and improving it (Özdemir, 2018). Considering that the benefits of yoga will also contribute to the development of the child, awareness studies should be carried out. In this section, the definition, importance and benefits of yoga, the use of yoga in education, and yoga in infants and children are discussed.

Definition and Importance of Yoga

Yoga is defined as merging or keeping under control in Sanskritic (Duros & Crowley, 2014). Yoga; means relaxing in the body, slowing down in breathing, and mentally calming down (Dolanay, 2007a). Yoga in general includes; physical postures / poses, breathing practices and movement meditation to improve self-regulation skills within the body and mind (Gard et al., 2014;). In the traditional definition, Yoga; is a holistic application system that includes various techniques such as physical postures / exercises, breathing exercises, deep relaxation techniques and meditation / awareness practices (Khalsa and Butzer, 2016).

Yoga was born in India about five thousand years ago and has been spread all over the world by Yoga scholars. Its origins are based on the Indian philosophical system that promotes the integration of body and mind (Salmon et al., 2009). The most common form of practice is known as Hatha Yoga, and it consists of two branches that aim to have control over the emotions and the mind that make spiritual development possible; pranayama (breathing) and asana (postures) (Kobylińska et al., 2018). Of these, the asana is a physical posture and movement that can be combined in many different ways; through these postures and movements, yoga ensures that the various musculoskeletal systems work together, meeting healing, natural and energetic needs and the needs of the body (Staton and Perron, 2002).

Yoga (Duros and Crowley, 2014; Staton and Perron, 2002), which is thought to be the origin of psychotherapy, produces important physiological changes in the organism, and as a result, provides psychological benefits and may affect the muscles and glands that improve physical health when applied in a appropriate way (Ross and Thomas, 2010; Garfinkel and Schumacher, 2000). Studies show that more or less, short or long-term Yoga interventions have positive effects on outcomes (Govindaraj et al., 2016).

It is emphasized that Yoga should be used in the education system from the early years in order to educate individuals in a healthy way. Biomedical and psychosocial studies on the clinical effects of yoga in both adults and children (Jeter et al., 2015) have increased rapidly in recent years, especially in the last five years. Research shows that yoga is especially effective in reducing stress and increasing mental health and well-being (Pascoe and Bauer, 2015; Sharma, 2014). According to data from the Ministry of Health of Malaysia and the United States; it is seen that the incidence of mental health diseases increases in children. It is reported that mental health disorders are seen in one out of ten people in 2011 in Malaysia and in one out of five people in 2016 in America (Nanthakumar, 2018).

Research shows that anxiety and depression are the most common mental health disorders in children. In a study conducted with adolescents between the ages of thirteen and eighteen (10123 adolescents) in the USA; 31.9% of adolescents stated that they had problems because of anxiety disorder (Merikangas et al., 2010). Likewise, as a result of advances in technology, media technologies such as computers and mobile phones are a source of information for everyone, but intense use is becoming uncomfortable and the pressure from parents and educators about technology cause lack of self-confidence because of stress which is due to the fact that they are raised in a competitive environment both at home and in school.

It is seen that physical activity is gradually decreasing and as a result of lack of movement, obesity, physical and mental health problems may occur (Nanthakumar, 2018; Houser, 2015; Hagen and Nayar, 2014). It is emphasized that Yoga is effective in reducing such problems. Mental awareness is part of Yoga practices (Auty et al., 2017; Salmon et al., 2009); awareness meditation is found to be more effective in reducing negative emotions and symptoms of social anxiety in patients with social anxiety disorder, depression (Ramel et al., 2004) and harshness (Krygier et al., 2013; Carmody and Baer, 2008; Grossman et al., 2004; Brown and Ryan, 2003) than aerobic exercises (Goldin et al., 2016; Goldin et al., 2013). Some researches show that Yoga can have many benefits, including increasing flexibility, promoting weight loss, and improving emotional and psychosocial health (King, 2017; Dai et al., 2015; Houser, 2015; Hagen and Nayar, 2014; Telles et al., 2013; Woodyard, 2011; Ehud et al., 2010; White, 2009; Cordeiro-Drever, 2008). Yoga was found to be effective in revealing the self-regulation skills of children and adolescents in schools (Bergen-Cico et al., 2015; Gard et al., 2014).

Benefits of Yoga

Yoga used to be done only by young men in old times but since the twentieth century, it has also been implemented by women in schools in India. Today, it is stated that yoga is one of the most ideal exercises that allow a person to get to know his body in various forms, sexes, sizes, at any time of his life and at any age, even in infants (Staton and Perron, 2002).

When yoga is done regularly, it can have positive effects on all body systems. Yoga has an important place in the body's support systems (skeletal and muscle structure), control systems (nervous and endocrine systems), metabolism regulation (digestive and respiratory systems), minimizing and improving emotional and behavioral problems (Saraswati, 2007; Dolanay, 2007a; İskit, 2007).

The benefits of yoga to children have also been demonstrated in studies conducted. In a study conducted by the Department of Medicine, Harvard University, it was shown that children between the ages of 6-12 have seen both physical and mental benefits of yoga in the research results. Wei (2016) explains that yoga provides benefits to improve balance, increase strength and endurance, improve aerobic capacity, gain better focus, have stronger memory, build more self-confidence, have high academic performance, decrease negative behaviors in the classroom, reduce anxiety and stress. In a deeper dimension, Yoga teaches you to understand that everyone is same and that they are part of a single whole. With the belief that it is always possible to learn and develop at any time with the belief that the individual is doing the best he can with his skills and with his current awareness, yoga makes the individual to realize that he/she is responsible for his own life and offers opportunities for self-regulation and self-knowledge. After understanding to discover, to love and to respect, to accept and to share; Yoga shows children ways to start with themselves and then to be moderate and in harmony with their environment (Dolanay, 2007a).

Yoga in Education

Nowadays, changes in technology and other fields have affected all aspects of life as well as education. Today, instead of classical education, different approaches and techniques are adopted to education system. As Yoga training with different methods and techniques is defined as "a spontaneous event in the deeper levels of mind" in the process of acquiring knowledge; Yoga symbolically transfers knowledge to the subconscious. To balance yoga training to reveal all the personality traits of the child and to ensure that children can benefit from all yoga activities; children should be allowed to have fun and at the same time a play environment should be created (Saraswati, 2007; İskit, 2007).

Yoga is widely used in healthy life, education and in the treatment of various diseases and disorders. Yoga, which is a complementary therapy for various physiological disorders and psychological disorders, has been widely studied in adults (Field, 2011; Cabral et al., 2011), although studies on the therapeutic use of Yoga in children appear to be very limited (Cabral et al., 2011). Current research shows that there are many benefits in Yoga, including increasing flexibility, promoting weight loss, and improving emotional and psychosocial health (King, 2017; Dai et al., 2015; Houser, 2015; Hagen and Nayar, 2014; Telles et al., 2013; Woodyard, 2011; Ehud et al., 2010; White, 2009; Cordeiro-Drever, 2008).

Yoga practices has also been shown to create positive changes in disorders including autism spectrum disorder (Koenig et al., 2012; Radhakrishma, 2010), down syndrome (Salem et al., 2018; Murray and Ryan-Krause, 2010; Monk-Turner and Turner, 2010; Galantino et al., 2008; Sumar, 1998), cerebral palsy (Reedman et al., 2017; Mak et al., 2017), attention deficit and hyperactivity disorder (Chimiklis et al., 2018; Kaley-Isley et al., 2017; Klatt et al., 2013; Pradhan and Nagendra, 2010), mental retardation (Pise et al., 2018; Cohen et al., 2018; Hawkins et al., 2012).

Child and Baby Yoga

The classic Yoga was classified by Patanjali, known as the founder of Yoga, and was divided into eight Yoga sutras. These sutras were divided into eight branches including; yama (universal ethics, which are the basic behaviors and values to be applied in life; not to harm, not to lie, not to steal, personal control, not to be selfish), niyama (individual ethics which means healthy personal habits that we should have in our lives; cleaning, satisfaction, effort, work, faith), asana (postures; Comfortable and stable posture), pranayama (practice of consciously regulating breath), prathyahara (calming the senses; taste, touch, smell, see and hear), dharana (concentration), dhyana (meditation) and finally samadhi (spiritual development, achieving unlimited happiness). Only the first three can be practiced in children's Yoga (Stiles, 2013; Dolanay, 2007a). In addition, children's yoga does not use the techniques of kriya (cleansing) and mudra (locking) that can be used in classic Yoga (Dolanay, 2007a).

Four main objectives of Children's Yoga are; to create physical, emotional and mental awareness of the child and his / her environment, to be balanced in physical and emotional, mental and social relationships, to provide the right breathing with various breathing techniques and to deal with stress through various relaxation techniques, to slow down and to help the person meet the inner self. Yoga helps children become healthy individuals from a young age in these four areas (Dolanay, 2007a).

Against the increasing speed of the modern world; mothers are forced to give their babies an early stage of skills that will help them to build a life of high welfare. In this sense, Yoga offers babies a resource to help them cope with stress positively and know how to relax themselves. In yoga, mothers provide a safe play environment for babies, allowing a safe hand to turn a challenging movement into a safe game and ensure that play and fun continue over and over again. This fun stimulation through yoga-based exercises is the best way to help babies love life with all their demands as they continue to grow (http://www.harmonyyogaireland.com/index.php/classes/baby-yoga).

Mothers help themselves overcome the astonishment, fatigue, psychological and hormonal imbalance they experience after birth with the mother-baby yoga that they can practice together at home without leaving their baby. This special time for themselves and their babies will restore the old physical strength and flexibility to the mother, relax the mother and her baby, help the baby to solve insomnia, gas and digestive problems, if any, and restore mental and spiritual balance (Dolanay, 2007b).

Baby Yoga provides multiple sensory stimulation for infants, helps all body systems (respiratory, circulatory, digestive, nervous and endocrine) to function, strengthens the attachment of the mother to the infant, is a fun activity and it helps to relieve tension in the back, shoulders and hips as a result of reaching out for the mothers and lifting and carrying the baby (http://www.harmonyyogaireland.com/index.php/classes/baby-yoga).

Baby Yoga is started when the mother feels ready. The period between the second and third months after the baby is born until the baby crawls is the most appropriate period for mother-baby yoga. Baby Yoga:

A: Suitable postures and movements for mothers with 3-6 months old baby.
B: Suitable postures and movements for mothers with 6-12 months old baby.
AB: It is appropriate for everyone to start at appropriate levels and months of action. (Dolanay, 2007b).

Children and young people, like babies, need yoga activities to listen to their bodies, emotions, and ideas sincerely. Therefore, Yoga contributes to healthy development and good mental health; includes encouraging and developing healthy living for children, improving attention, strengthening self-confidence and self-regulation. Children and adolescents need to develop according to their personalities, interpret and achieve the balance between their strengths and social expectations. Yoga can help them empower themselves and contribute to social beings in healthy ways (Hagen and Nayar, 2014).

In addition to the use of Yoga in children in various activities; It is especially used in storytelling activities. Yoga storytelling is practiced in the same way as the typical storytelling technique. Children's yoga story activities include various animal and nature asanas in other words postures. For children, the role of animals, trees, flowers are fun and engaging, and while practicing various animals and nature asanas, it also enhances the strength of their bodies, making them think that they are like the creatures they describe to improve their awareness. In the same way, in the sandwich pose, they fold forward, thinking that there is a cheese between two breads and increase their flexibility and endurance despite difficulty, and help correct many problems such as posture disorders, digestive problems, constipation and lazy eye (Scherrer, 2018; Sürücü, 2017).

CONCLUSION AND SUGGESTIONS

Today, technological advances made to facilitate daily life sometimes take individuals away from their daily lives, put them under intense stress, reduce their personal skills and prevent them from reflecting their emotions. These factors create a number of problems in the mental, physical and physical condition of the person. In this sense, Yoga provides support to individuals in dealing with these rushes in their daily lives, in making life simpler, in taking care of their environment, and in creating awareness about how the environment impacts themselves and their behaviors affect the environment. Yoga allows the person to learn to re-connect with himself and to see his reactions to himself and those around him from a different perspective. It requires an emotional and spiritual power to directly address personal conflicts, concerns, hopes and fears and to understand how it reacts to them and yoga makes a significant contribution to the individual in this manner.

Yoga is the art and science of life, and it has the discipline that tries to ensure that the individual develops in an integrated way in all aspects of development for the integration and development of the mind and body. In order to achieve this integration, it is connected with the physical development part and then the spiritual part is passed later (Saraswati, 2007).

According to the evidence collected through yoga studies; It continues to demonstrate that it provides many benefits to the physical, mental and mental health of all individuals of infants, children and adults. This supports the use of Yoga for individual's personal health as well as in health and education services in many parts of the world. Regular yoga practices increase awareness of pain in the body and help to restore the balance lost, improve body posture, help the brain to get enough oxygen, gain healthy eating habits, improve mental awareness, improve discipline and improve overall quality of life.

In this respect, the following can be recommended for Yoga and its practices:

- Yoga is an art that can be practiced by people from all religions and shows practical, helpful and correct ways of life. By supporting people to develop positive attitudes towards Yoga education and practices; Yoga exercises can be included in training curricula.
- As a result of literature research, there are almost no studies about yoga training and practices in Turkey would be better. In this direction, Yoga training and practices can be applied to all age groups, especially pre-school period, by incorporating them into educational curricula in subsequent education periods.
- Preschool teachers can be informed about Yoga practices and support can be provided for disseminating Yoga practices and raising social awareness.
- As a result of literature researches in the world and in Turkey; Although, yoga is applied to children and individuals with normal development, it can be applied as a preventive and curative treatment for many disorders and diseases such as obesity and cancer.
- When the researches are examined, it is thought that Yoga researches are limited in terms of infancy, preschool and childhood period and it is necessary to diversify and increase the researches in this direction.

- Schools are an ideal place to promote the development of healthy lifestyle skills and behaviors from an early age. The widespread practice of Yoga in schools can have important implications not only for student health but also for society as a whole.

REFERENCES

Auty, K. M., Cope, A., & Liebling, A. (2017). A Systematic Review and Meta-Analysis of Yoga and Mindfulness Meditation in Prison: Effects on Psychological Well-Being and Behavioural Functioning. *International Journal of Offender Therapy and Comparative Criminology*, *61*(6), 689–710. doi:10.1177/0306624X15602514 PMID:26320031

Bergen-Cico, D., Razza, R., & Timmins, A. (2015). Fostering self regulation through curriculum infusion of mindful yoga: A pilot study of efficacy and feasibility. *Journal of Child and Family Studies*, *24*(11), 3448–3461. doi:10.100710826-015-0146-2

Birdee, G. S., Yeh, G. Y., Wayne, P. M., Phillips, R. S., Davis, R. B., & Gardiner, P. (2009). Clinical applications of yoga for the pediatric population: A systematic review. *Academic Pediatrics*, *9*(4), 212–220. doi:10.1016/j.acap.2009.04.002 PMID:19608122

Brown, K. W., & Ryan, R. M. (2003). The benefits of being present: Mindfulness and its role in psychological well-being. *Journal of Personality and Social Psychology*, *84*(4), 822–848. doi:10.1037/0022-3514.84.4.822 PMID:12703651

Butzer, B., Bury, D., Telles, S., & Khalsa, S. (2016). Implementing yoga within the school curriculum: A scientific rationale for improving social–emotional learning and positive student outcomes. *Journal of Children's Services*, *11*(1), 3–24. doi:10.1108/JCS-10-2014-0044

Cabral, P., Meyer, H. B., & Ames, D. (2011). Effectiveness of yoga therapy as a complimentary treatment for major psychiatric disorders: A meta-analysis. *Prim Care Companion*, *13*(4), 1060–1068.

Carmody, J., & Baer, R. A. (2008). Relationships between mindfulness practice and levels of mindfulness, medical and psychological symptoms and well-being in a mindfulness-based stress reduction program. *Journal of Behavioral Medicine*, *31*(1), 23–33. doi:10.100710865-007-9130-7 PMID:17899351

Chimiklis, A. L., Dahl, V., Spears, A. P., Goss, K., Fogarty, K., & Chack, A. (2018). Yoga, Mindfulness, and Meditation Interventions for Youth with ADHD: Systematic Review and Meta-Analysis. *Journal of Child and Family Studies*, *27*(10), 3155–3168. doi:10.100710826-018-1148-7

Chong, C. S., Tsunaka, M., Tsang, H. W., Chan, E. P., & Cheung, W. M. (2011). Effects of yoga on stress management in healthy adults: A systematic review. *Alternative Therapies in Health and Medicine*, *17*(1), 32–38. PMID:21614942

Cohen, S. C. L., Harvey, D. J., Shields, R. H., Shields, G. S., Rashed, R. N., Tancredi, D. J., ... Schweitzer, J. B. (2018). Effects of Yoga on Attention, Impulsivity, and Hyperactivity in Preschool-Aged Children with Attention-Deficit Hyperactivity Disorder Symptoms. *Journal of Developmental and Behavioral Pediatrics*, *39*(3), 200–209. doi:10.1097/DBP.0000000000000552 PMID:29538185

Cordeiro Drever. L. (2008). *Yoga and children ages 5-12 with Down syndrome: an instructional yoga DVD and parent manual* [M.A. dissertation]. California State University, Long Beach, CA.

Dai, C. L., Nabors, L. A., Vidourek, R. A., King, K. A., & Chen, C. C. (2015). Evaluation of an after-school yoga program for children. *International Journal of Yoga*, *8*(2), 160–161. doi:10.4103/0973-6131.158488 PMID:26170600

Daly, L. A., Haden, S. C., Hagins, M., Papouchis, N., & Ramirez, P. M. (2015). Yoga and Emotion Regulation in High School Students: A Randomized Controlled Trial. *Hindawi Publishing Corporation Evidence-Based Complementary and Alternative Medicine*, *2015*, 1–8. doi:10.1155/2015/794928 PMID:26356561

Dolanay, D. B. (2007a). *Lotus Flowers: Yogamini Project*. Istanbul: ALFA Publications.

Dolanay, D. B. (2007b). *Mother-Baby Yoga and Sample Program*. Retrieved from https://yogadivya.net/yoga-dersleri/anne-bebek-yogasi/

Duros, P., & Crowley, D. (2014). The body comes to therapy too. *Clinical Social Work Journal*, *42*(3), 237–246. doi:10.100710615-014-0486-1

Ehud, M., An, B. D., & Avshalom, S. (2010). Here and now: Yoga in Israeli schools. *International Journal of Yoga*, *3*(2), 42–47. doi:10.4103/0973-6131.72629 PMID:21170229

Field, T. (2011). Yoga clinical research review. *Complementary Therapies in Clinical Practice*, *17*(1), 1–8. doi:10.1016/j.ctcp.2010.09.007 PMID:21168106

Galantino, M., Galbavy, R., & Quinn, L. (2008). Therapeutic effects of yoga for children: A systematic review of the literature. *Pediatric Physical Therapy*, *20*(1), 66–80. doi:10.1097/PEP.0b013e31815f1208 PMID:18300936

Ganpa, T. & Nagendra, H. (2011). Yoga for children. *TANG Humanitas Med.*, *1*, 1–4.

Gard, T., Noggle, J. J., Park, C. L., Vago, D. R., & Wilson, A. (2014). Potential self-regulatory mechanisms of yoga for psychological health. *Frontiers in Human Neuroscience*, *8*(770), 1–20. PMID:25368562

Garfinkel, M., & Schumacher, H. R. Jr. (2000). Yoga. *Rheumatic Diseases Clinics of North America*, *26*(1), 125–132. doi:10.1016/S0889-857X(05)70126-5 PMID:10680200

Goldin, P. R., Morrison, A., Jazaieri, H., Brozovich, F., Heimberg, R., & Gross, J. J. (2016). Group CBT versus MBSR for social anxiety disorder: A randomized controlled trial. *Journal of Consulting and Clinical Psychology*, *84*(5), 427–437. doi:10.1037/ccp0000092 PMID:26950097

Goldin, P. R., Ziv, M., Jazaieri, H., Hahn, K., & Gross, J. J. (2013). MBSR versus aerobic exercise in social anxiety: fMRI of emotion regulation of negative self-beliefs. *Social Cognitive and Affective Neuroscience*, *8*(1), 65–72. doi:10.1093can/nss054 PMID:22586252

Govindara, R., Karmani, S., Varambally, S., & Gangadhar, B.N. (2016). Yoga and physical exercise –A review and comparison. *International Review of Psychiatry*, *28*(3), 242–253.

Grossman, P., Niemann, L., Schmidt, S., & Walach, H. (2004). Mindfulness-based stress reduction and health benefits: A meta-analysis. *Journal of Psychosomatic Research, 57*(1), 35–43. doi:10.1016/S0022-3999(03)00573-7 PMID:15256293

Hagen, I., & Nayar, U. S. (2014). Yoga for children and young people's mental health and well-being: Research review and reflections on the mental health potentials of yoga. *Frontiers in Psychiatry, 5*(35), 1–6. PMID:24765080

Hariprasad, V. R., Arasappa, R., Varambally, S., Srinath, S., & Gangadhar, B. N. (2013). Feasibility and efficacy of yoga as an add-on intervention in attention deficit-hyperactivity disorder: An exploratory study. *Indian Journal of Psychiatry, 55*(3), 379–384. PMID:24049203

Harmony Yoga (n.d.). *Baby Yoga.* Retrieved from http://www.harmonyyogaireland.com/index.php/classes/baby-yoga

Hawkins, B. L., Stegall, J. B., Weber, M. F., & Ryan, J. B. (2012). The influence of a yoga exercise program for young adults with intellectual disabilities. *International Journal of Yoga, 5*(2), 151–156. doi:10.4103/0973-6131.98244 PMID:22870001

Höbek Akarsu. R. (2016). *The effect of pregnancy yoga on the pregnant's psychosocial health and prenatal attachment* [doctoral dissertation]. University of Istanbul, Istanbul.

Houser, M. E. (2015). *A Mixed Methods Evaluation of the Effectiveness of a Group Yoga Intervention as an Adjunctive Trauma Therapy for Adolescent Girls* [doctoral dissertation]. University of Denver, Denver, CO.

İskit, N. (2007). 99 Pages of Pregnancy and Child Yoga. Istanbul: Turkey Isbank Cultural Publications.

Jeter, P. E., Slutsky, J., Singh, N., & Khalsa, S. B. S. (2015). *Yoga as a therapeutic intervention: A bibliometric analysis of published* research studies from 1967–2013. *Journal of Alternative and Complementary Medicine (New York, N.Y.), 21*(10), 586–592. doi:10.1089/acm.2015.0057 PMID:26196166

Kaley Isley, L.C., Peterson, J., Fischer, C., & Peterson, E. (2010). Yoga as a Complementary Therapy for Children and Adolescents A Guide for Clinicians. *Psychiatry, 7*(8), 20–32. PMID:20877530

Khalsa, S. B., Hickey-Schultz, L., Cohen, D., Steiner, N., & Cope, S. (2012). Evaluation of the mental health benefits of yoga in a secondary school: A preliminary randomized controlled trial. *The Journal of Behavioral Health Services & Research, 39*(1), 80–90. doi:10.100711414-011-9249-8 PMID:21647811

Khalsa, S. B. S. (2013). Yoga in schools research: improving mental and emotional health. Presented at the *Second International Conference on Yoga for Health and Social Transformation. Academic Press.*

Khalsa, S. B. S., & Butzer, B. (2016). Yoga in school settings: A research review. *Annals of the New York Academy of Sciences, 1373*(1), 45–55. doi:10.1111/nyas.13025 PMID:26919395

King, S. (2017). *A Case Study of a Yoga and Meditation Intervention in an Urban School: A Complex Web of Relationships and Resilience in the Search for Student Well-Being [PhD]. University of California*, Los Angeles.

Klatt, M., Harpster, K., Browne, E., White, S., & Case-Smith, J. (2013). Feasibility and preliminary outcomes for move-into-learning: An arts-based mindfulness classroom intervention. *The Journal of Positive Psychology*, *8*(3), 233–241. doi:10.1080/17439760.2013.779011

Kobylińska, D., Lewczuk, K., Marchlewska, M., & Pietraszek, A. (2018). For body and mind: Practicing yoga and emotion regulation. *Social Psychological Bulletin*, *13*(1), 1–21. doi:10.5964pb.v13i1.25502 PMID:29154556

Koenig, K. P., Buckley-Reen, A., & Garg, S. (2012). Efficacy of the Get Ready to Learn Yoga Program Among Children With Autism Spectrum Disorders: A Pretest–Posttest Control Group Design. *The American Journal of Occupational Therapy*, *66*(5), 538–546. doi:10.5014/ajot.2012.004390 PMID:22917120

Krygier, J. R., Heathers, J. A., Shahrestani, S., Abbott, M., Gross, J. J., & Kemp, A. H. (2013). Mindfulness meditation, well-being, and heart rate variability: A preliminary investigation into the impact of intensive Vipassana meditation. *International Journal of Psychophysiology*, *89*(3), 305–313. doi:10.1016/j.ijpsycho.2013.06.017 PMID:23797150

Langøien, L. J. (2013). *"Pay Attention – Listen to Your Heart!" Unfolding Practice, Change Realities and Awareness of the Embodied Self in Ashtanga Yoga [PhD thesis]*. Norway: University of Norwegian.

Litchke, L. G., Liu, T., & Castro, S. (2018). Effects of Multimodal Mandala Yoga on Social and Emotional Skills for Youth with Autism Spectrum Disorder: An Exploratory Study. *International Journal of Yoga*, *11*(1), 59–65. PMID:29343932

Mak, C., Whittingham, K., Cunnington, R., & Boyd, R. N. (2017). Effect of mindfulness yoga programme MiYoga on attention, behaviour, and physical outcomes in cerebral palsy: A randomized controlled trial. *Developmental Medicine and Child Neurology*, *60*(9), 922–932. doi:10.1111/dmcn.13923 PMID:29869333

Mehta, S., Shah, D., Shah, K., Mehta, S., Mehta, N., Mehta, V., ... Mehta, D. (2012). Peer-Mediated Multimodal Intervention Program for the Treatment of Children with ADHD in India: One-Year Followup. *International Scholarly Research Network*, *2*(4), 1–7. PMID:23316384

Merikangas, K. R., He, J. P., Burstein, M., Swanson, S. A., Avenevoli, S., Cui, L., ... Swendsen, J. (2010). *Lifetime prevalence of mental disorders in U.S. adolescents: Results from the National Comorbidity Study*—Adolescent Supplement (NCS-A). *Journal of the American Academy of Child and Adolescent Psychiatry*, *49*(10), 980–989. doi:10.1016/j.jaac.2010.05.017 PMID:20855043

Monk-Turner, E., & Turner, C. (2010). Does yoga shape body, mind and spiritual health and happiness: Differences between yoga practitioners and college students. *International Journal of Yoga*, *3*(2), 48–54. doi:10.4103/0973-6131.72630 PMID:21170230

Murray, J., & RyanKraus, P. (2010). *Obesity in children with down syndrome: Background and recommendations for management.*

Nanthakumar, C. (2018). The benefits of yoga in children. *Journal of Integrative Medicine*, *16*(1), 14–19. doi:10.1016/j.joim.2017.12.008 PMID:29397087

Özdemir, A. (2018). *The effect of yoga on burn patients self-esteem and body image* [doctoral dissertation]. University of Inönü, Malatya.

Pascoe, M. C., & Bauer, I. E. (2015). A systematic review of randomised control trials on the effects of yoga on stress measures and mood. *Journal of Psychiatric Research*, *68*, 270–282. doi:10.1016/j.jpsychires.2015.07.013 PMID:26228429

Pise, V., Pradhan, B., & Gharote, M. (2018). Effect of yoga practices on psycho-motor abilities among intellectually disabled children. *Journal of Exercise Rehabilitation*, *14*(4), 581–585. doi:10.12965/jer.1836290.145 PMID:30276177

Pradhan, B., & Nagendra, H. (2010). Immediate effect of two yoga-based relaxation techniques on attention in children. *International Journal of Yoga*, *3*(2), 67–69. doi:10.4103/0973-6131.72632 PMID:21170232

Radhakrishma, S. (2010). Application of integrated yoga therapy to increase imitation skills in children with autism spectrum disorder. *International Journal of Yoga*, *3*(1), 26–30. doi:10.4103/0973-6131.66775 PMID:20948898

Ramel, W. I, Goldin, P. R., Carmona, P. E., & McQuaid, J. R. (2004). The effects of mindfulness meditation on cognitive processes and affect in patients with past depression. *Cognitive Therapy and Research*, *28*(4), 433–455. doi:10.1023/B:COTR.0000045557.15923.96

Reedman, S., Boyd, R., & Sakzewskı, L. (2017). Efficacy of a participation-focused therapy intervention on participation in physical activities, health-related quality of life, and behavioural barriers to participation in children with cerebral palsy. *Developmental Medicine and Child Neurology*, *59*(S3), 32–32. doi:10.1111/dmcn.13413

Ross, A., & Thomas, S. (2010). The health benefits of yoga and exercise: A review of comparison studies. *Journal of Alternative and Complementary Medicine*, *16*(1), 3–12. doi:10.1089/acm.2009.0044 PMID:20105062

Salem, Y., Liu, H., Young, A., Tolbert, M., & Elokda, A. (2018). Effects of yoga exercise program on physical function in young children with Down syndrome. *Developmental Medicine and Child Neurology*, *60*(S3), 32–32. doi:10.1111/dmcn.47_14017

Salmon, P., Lush, E., Jablonski, M., & Sephton, S. E. (2009). Yoga and mindfulness: Clinical aspects of and ancient mind/body practice. *Cognitive and Behavioral Practice*, *16*(1), 59–72. doi:10.1016/j.cbpra.2008.07.002

Saraswati, S. P. (2007). *Yoga Education for Children. Trust.* Munger, Bihar: Yoga Publications.

Scherrer, K. (2018). Stories, Songs, Stretches Creating playful story times with yoga and movement. American Libraries Magazine. Retrieved from https://americanlibrariesmagazine.org/2018/01/02/stories-songs-stretches-storytime-yoga/

Scime, M., Cook-Cottone, C., Kane, L., & Watson, T. (2006). Group prevention of eating disorders with fifth-grade females: Impact on body dissatisfaction, drive for thinness, and media influence. *Eating Disorders*, *14*(2), 143–155. doi:10.1080/10640260500403881 PMID:16777811

Sharma, M. (2014). Yoga as an alternative and complementary approach for stress management: A systematic review. *Journal of Evidence-Based Complementary & Alternative Medicine*, *19*(1), 59–67. doi:10.1177/2156587213503344 PMID:24647380

Sifferlin, A. (2013). *The most stressed-out generation? Young Adults*. Time. Retrieved from http://healthland.time.com/2013/02/07/the-most-stressed-outgeneration-young-adults/

Spinazzola, J., Rhodes, A. M., Emerson, D., Earle, E., & Monroe, K. (2011). Application of yoga in residential treatment of traumatized youth. *Journal of the American Psychiatric Nurses Association, 17*(6), 431–444. doi:10.1177/1078390311418359 PMID:21868714

Staton, L., & Perron, S. (2002). *Baby Om: Yoga for Mothers and Babies*. New York: Holt Paperbacks.

Stiles, M. (2013). *Structural yoga therapy: adapting to the individual*. San Francisco: Red Wheel/Weiser.

Sumar, S. (1998). Yoga for the special child: A therapeutic approach for infants and children with Down syndrome, cerebral palsy, and learning disabilities. *Special Yoga, 36*(6), 314–319.

Sumar, S. (1998). *Yoga for the special child: A therapeutic approach for infants and children with Down syndrome, cerebral palsy, and learning disabilities*. Buckingham, VA: Special Yoga.

Sürücü, Ş. (2017). The most entertaining teaching for children: Children's yoga. *Uplifers*. Retrieved from https://www.uplifers.com/cocuklar-icin-en-eglenceli-ogreti-cocuk-yogasi/)

Telles, S. (2012). The Effect of Yoga on Mental Health of Children. In U. Nayar (Ed.), Child and Adolescent Mental Health (pp. 219-227). New Delhi: Sage Publications.

Telles, S., Singh, N., & Bhardwaj, A. K. (2013). Effect of yoga or physical exercise on physical, cognitive and emotional measures in children: A randomized controlled trial. *Child and Adolescent Psychiatry and Mental Health, 7*(37), 1–16. PMID:24199742

Wei, M. (2016). *More than just a game: Yoga for school-age children*. Retrieved from http://www.health.harvard.edu/blog/more-than-just-a-game-yoga-for-school-age-children-201601299055

White, L. S. (2009). Yoga for children. *Pediatric Nursing, 35*, 277–295. PMID:19916343

Woodyard, C. (2011). Exploring the therapeutic effects of yoga and its ability to increase quality of life. *International Journal of Yoga, 4*(2), 49–54. doi 10.4103/0973-6131.85485 PMID:22022122

Young, A. J., Sıllıman-French, L., & Crawford, L. (2017). Yoga For Young Children With Down Syndrome. *Palaestra-Journal, 31*(3), 18–23.

Compilation of References

Abrams, A., Hazen, E., & Penson, R. (2007). Psychosocial issues in adolescents with cancer. *Cancer Treatment Reviews*, *33*(7), 622–630. doi:10.1016/j.ctrv.2006.12.006 PMID:17434265

Acehan, S., Bilen, A., Oğuzhan Ay, M., Gülen, M., Avcı, A., & İçme, F. (2013). Çocuk istismarı ve ihmalinin değerlendirilmesi. *Arşiv Kaynak Tarama Dergisi*, *22*(4), 591–614.

Aghion, P., Akcigit, U., Bergeaud, A., Blundell, R., & Hemous, D. (2019). Innovation and Top Income Inequality. *The Review of Economic Studies*, *86*(1), 1–45. doi:10.1093/restud/rdy027

Ahmad, I., & Shahid, H. (2019). A study of the problems in expressive language learning faced by hearing impaired children. *Academic Research International*, *10*(1).

Ahmetoğlu, E. (2009). Sosyal gelişim. In Y. Fazlıoğlu (Ed.), *Erken çocukluk gelişimi ve eğitimi* (pp. 39–62). İstanbul: Kriter.

Ahmetoğlu, E. (2018). Ahlak Gelişimi. In N. Aral & T. Duman (Eds.), *Eğitim Psikolojisi* (pp. 216–239). Ankara: Pegem.

Aitchison, J. (2012). *Words in the mind: An introduction to the mental lexicon*. John Wiley & Sons.

Akçamete, G. (2010). Children with special needs. In G. Akçamete (Ed.), Students with special needs in general education schools and special education (pp. 31-76). Ankara: Root Spring.

Akçan, A. B. (2013). Turner sendromu. *Konuralp Tip Dergisi*, *5*(2), 53–61.

Akçay, D. & Emiroğlu, O. N. (2016). Medya etkileşimine bağlı okul öncesi dönemdeki çocukta saldırganlık davranışı. *Türkiye Klinikleri*, *2*(1), 116–120.

Akdemir, D.P. (2017). The Effect of Nutritional and Physical Activity Interventions on Nutritional Status and Obesity in Primary School Children: A Cluster Randomized Controlled Study. Retrieved from https://acikerisim.cumhuriyet.edu.tr/xmlui/handle/20.500.12418/6763

Aker, G. (2014). *Pre-service teachers' attitudes towards mainstreaming education*. Unpublished master's thesis, Trakya University, Institute of Social Sciences, Edirne.

Akgül, H., & Yeşilyaprak, B. (2018). Yalnızlığı Azaltma Psiko-Eğitim Programının Yaşlıların Yalnızlık Düzeyine Etkisi. *OPUS Uluslararası Toplum Araştırmaları Dergisi*, *8*(14), 11–52. doi:10.26466/opus.401331

Akoğlu, G., & Şipal, F. (2012). Düşük Sosyo Ekonomik Düzeydeki Çocuklara Yönelik Erken Müdahale Yaklaşımları: Dünya'dan ve Türkiye'den Uygulamalar. *Toplum ve Sosyal Hizmet Dergisi*, *23*(1), 173–184.

Aksan, D. (1995). Her Yönüyle Dil Ana Çizgileriyle Dilbilim. Ankara: TDK.

Akseer, N., Salehi, A. S., Hossain, S. M. M., Mashal, M. T., Rasooly, M. H., & Bhatti, Z. … Bhutta, Z. A. (2016). Achieving maternal and child health gains in Afghanistan: A Countdown to 2015 country case study. *The Lancet. Global Health*. doi:10.1016/S2214-109X(16)30002-X

Akyo, I. T. (2012). Investigation of values in children's picture books. Unpublished masters thesis, Hacettepe University Social Sciences Institute, Ankara, Turkey. (In Turkish) Retrieved from https://tez.yok.gov.tr/UlusalTezMerkezi .

Alakoç-Pirpir, D. & Yıldız-Çiçekler, C. (2018). Fiziksel gelişim. In N. Aral & Z. F. Temel (Ed.), Çocuk gelişimi (pp. 112-139). Ankara: Hedef.

Alat, Z., Akgümüş, Ö., & Cavalı, D. (2012). Okul öncesi eğitimde açık hava etkinliklerine yönelik öğretmen görüş ve uygulamaları. *Mersin Üniversitesi Eğitim Fakültesi Dergisi, 8*(3), 47–62.

Albert, R. R., Schwade, J. A., & Goldstein, M. H. (2018). The social functions of babbling: Acoustic and contextual characteristics that facilitate maternal responsiveness. *Developmental Science, 21*(5), e12641. doi:10.1111/desc.12641 PMID:29250872

Algül, S. (2009). *Erzurum'da yetiştirme yurtlarında kalan korunmaya muhtaç ergenlerde sosyodemografik özellikler ile psikiyatrik belirti ve tanı dağılımı*. Atatürk Üniversitesi Tıp Fakültesi. Tıpta Uzmanlık Tezi.

Alink, L. R., Mesman, J., vanZeijl, J., Stolk, M. N., Juffer, F., Koot, H. M., ... van IJzendoorn, M. H. (2006). The early childhood aggression curve: Development of physical aggression in 10-to 50-month-old children. *Child Development, 77*(4), 954–966. doi:10.1111/j.1467-8624.2006.00912.x PMID:16942499

Alkan Ersoy, Ö., Kurtulmuş, Z., & Çürük Tekin, N. (2014). Aile çocuk eğitim programının annelerin çocuk yetiştirme tutumlarına ve ev ortamını düzenlenmelerine etkisinin incelenmesi. *Kastamonu Eğitim Dergisi, 22*(3), 1077–1090.

Alkaya, E. (2007). Bilingualism and translation of concepts from Russion to Tatar Turkish. *Türkoloji Araştırmaları, 2*(2), 41–53. (In Turkish)

Alpar, R. (2006). Spor Bilimlerinde Uygulamalı İstatistik (3. Basım). Ankara: Nobel Yayın Dağıtım.

Alpaydın, Y., & Canel, A. N. (2011). *Aile eğitimi programları üzerine notlar*. Ankara: Aile ve Sosyal Politikalar Bakanlığı.

Altay, B., Çavuşoğlu, F., & Çal, A. (2016). Yaşlıların sağlık algısı, yaşam kalitesi ve sağlıkla ilgili yaşam kalitesini etkileyen faktörler. *Turk Silahli Kuvvetleri Koruyucu Hekimlik Bulteni, 15*(3), 181–189. doi:10.5455/pmb.1-1438765038

Altıntaş, T. T., & Yıldız-Bıçakçı, M. (2015). Çocukların hastane yaşantısına hazırlanması ve hastaneye hazırlayıcı eğitim. In. A. Bütün-Ayhan (Ed.), Hasta çocukların gelişimi ve eğitimi (pp. 80-109). Eskişehir: Anadolu Üniversitesi.

Altıparmak, S., Yıldırım, G., Yardımcı, F., & Ergin, D. (2010). Annelerden alınan bilgilerle çocuk istismarı ve etkileyen faktörler. *Anadolu Psikiyatri Dergisi, 14*(4), 354–361. doi:10.5455/apd.39819

Altuntaş, O., & Kayıhan, H. (2015). Ev Düzenlemelerinin Yaşlıların Yaşam Kalitesine Etkisi. *Fizyoterapi Rehabilitasyon, 26*(1), 1–13.

American College of Obstetricians and Gynecologists. (2014). ACOG Practice Bulletin No. 1 Safe Prevention of the Primary Cesarean Delivery. *Obstetrics and Gynecology, 114*(1), 2.

American Library Association. (2009). *Early Literacy*. Retrieved from http://www.ala.org/united/products_services/booksforbabies/earlyliteracy

Amerikan Psikiyatri Birliği. (2001). Psikiyatride hastalıkların tanımlanması ve sınıflandırılması el kitabı. Ankara: Hekimler Yayın Birliği.

Ammaniti, M., Lucarelli, L., Cimino, S., D'Olimpio, F., & Chatoor, I. (2012). Feeding disorders of infancy: A longitudinal study to middle childhood. *International Journal of Eating Disorders*, *45*(2), 272–280. doi:10.1002/eat.20925 PMID:21495054

Anastasi, A., & Urbina, S. (1997). *Psychological Testing* (7th ed.). New Jersey: Prentice-Hall. Inc.

Anderson, J. C., & Gerbing, D. W. (1984). The Effect Of Sampling Error On Convergence, İmproper Solutions, And Goodness-Of-Fit İndices For Maximum Likelihood Confirmatory Factor Analysis. *Psychometrika*, *49*(2), 155–173. doi:10.1007/BF02294170

Angı, R. (1999). Yabacı dil öğrenirken-öğretirken. *Okul Aile Dergisi*, *1*(12), 18–20.

Anlar, B., Serdaroğlu, A., & Yakut, A. (2008). *Gelişimsel Cocuk nörolojisi.*. Ankara: Hacettepe Üniversitesi Hastaneleri Basımevi.

Anne Çocuk Eğitim Vakfı (AÇEV). (2007). *Ekonomik ve toplumsal kalkınma için erken çocukluk eğitimi: Önemi, yararları ve yaygınlaştırma önerileri*. İstanbul: Anne Çocuk Eğitim Vakfı Yayınları.

Anne Çocuk Eğitim Vakfı (AÇEV). (2013). *Erken çocukluk eğitimi politikaları.*Ankara:Yaygınlaşma, yönetişim ve yapılar toplantısı" raporu.

Anonim. (2010). Türkiye'de Ergen Profili 2008. Başbakanlık Aile ve Sosyal Araştırmalar Genel Müdürlüğü.

Apadoca, C. (2008). Preventing child malnutrition: Health and agriculture as determinants of child malnutrition. *Journal of Children & Poverty*, *14*(1), 21–40. doi:10.1080/10796120701871264

Arabacı, N. (2017). Aile katılımı. Ayşe Belgin Aksoy (Ed.), Aile eğitimi ve katılımı (s.125-148). Ankara: Hedef CS Basın Yayın.

Aral, N. & Doğan- Keskin, A. (2015). Hastanede oyun ve oyun odaları. In A. Bütün-Ayhan (Ed.), *Hasta çocukların gelişimi ve eğitimi* (pp. 138-159). Eskişehir: Anadolu Üniversitesi.

Aral, N. & Gürsoy, F. (2007). Özel eğitim gerektiren çocuklar ve özel eğitime giriş. İstanbul: Morpa Yayınları.

Aral, N. (2011). Language development. In N. Aral and G. Baran (Eds.), In Çocuk Gelişimi (pp. 163-192). İstanbul: Ya-Pa Publications. (In Turkish).

Aral, N., & Doğan Keskin, A. (2016). Bilişsel gelişim ve dil gelişimi. In N. Aral (Ed.), Çocuk gelişimi. (pp. 96-126). Eskişehir: AÖF Yayınları.

Aral, N., Baran, G., Bulut, Ş., & Çimen, S. (2000). Çocuk gelişimi 1. cilt. İstanbul: Ya-Pa Yayınları.

Aral, N. (2002). *Okul öncesi eğitim ve okul öncesi eğitim programı. Öğretmen rehber kitabı*. İstanbul: YA-PA.

Aral, N. (2011). Bilişsel Gelişim. In N. Aral & G. Baran (Eds.), *Çocuk Gelişimi*. İstanbul: Ya-Pa Yayınları.

Aral, N., Özdoğan-Özbal, E., Gürsoy, F., Çetin-Sultanoğlu, S., Aydos, S., Tosun, S., Karaaslan, T., & Kadan, G. (2016). Poverty definitions through children's eyes. *Journal of Current Researches on Health Sector*, *6*(1), 39–50.

Aral, N., & Baran, G. (2011). *Çocuk gelişimi*. İstanbul: Ya-Pa.

Aral, N., Baran, G., Bulut, Ş., & Çimen, S. (1981). *Çocuk Gelişimi*. İstanbul: YA-PA.

Aral, N., Baran, G., Bulut, Ş., & Çimen, S. (2000). *Çocuk gelişimi 1*. İstanbul: Ya-Pa.

Aral, N., Baran, G., Bulut, Ş., & Çimen, S. (2000). *Eğitimde drama*. İstanbul: YA-PA Yayınları.

Aral, N., Baran, G., Şenay, G., & Çimen, S. (2001). *Çocuk gelişimi I*. Ankara: YA-PA Yayınları.

Aral, N., & Başar, F. (1998). Çocukların kaygı düzeylerinin yaş, cinsiyet, sosyoekonomik düzey ve ailenin parçalanma durumuna göre incelenmesi. *Eğitim ve Bilim Dergisi, 22*(110), 7–11.

Aral, N., & Gürsoy, F. (2011). *Özel eğitim gerektiren çocuklar ve özel eğitime Giriş*. İstanbul: Ya- Pa.

Aral, N., & Kadan, G. (2018a). Zihinsel yetersizliği olan çocuklar. In N. Aral & F. Gürsoy (Eds.), *Özel gereksinimli çocuklar ve eğitimleri: Özel gereksinimli çocuklar* (pp. 14–43). Ankara: Hedef.

Aral, N., & Kadan, G. (2018b). Sosyal gelişim. In N. Aral & Z. F. Temel (Eds.), *Çocuk gelişimi* (pp. 236–262). Ankara: Hedef.

Aram, D., Korat, O., Saiegh-Haddad, E., Arafat, S. H., Khoury, R., & Elhija, J. A. (2013). Early literacy among Arabic-speaking kindergartners: The role of socioeconomic status, home literacy environment and maternal mediation of writing. *Cognitive Development, 28*(3), 193–208. doi:10.1016/j.cogdev.2012.10.003

Arı, M. (1986). Çocukluk döneminde uyum bozuklukları. In Ş. Bilir, (Ed.), Özürlü çocuklar ve eğitimleri (pp. 52-59). Ankara: Ayyıldız Matbaası.

Arı, M. (2018). Çocuklarda uyum bozukluğu, duygu ve davranış sorunları. In E.N. Metin, (Ed.), Özel gereksinimli çocuklar (2. Baskı, pp. 319-356). Ankara: Anı Yayıncılık.

Arıcıoğlu, A., & Gültekin, F. (2017). Zihinsel engelli çocuk annelerinin sosyal destek ve gelecek algıları: Psikolojik danışmanların rolleri. *Mehmet Akif Ersoy Üniversitesi Eğitim Fakültesi Dergisi, 42*, 1–26.

Arı, R. (2005). *Gelişim ve öğrenme*. Ankara: Nobel.

Arı, R., Deniz, M. E., Erişen, Y., & Çeliköz, N. (2006). *Milli Eğitim Bakanlığı projeler koordinasyon merkezi başkanlığı. Çocukların okula hazırlanması (Anne-Çocuk Eğitimi Temel Araştırması)*. Ankara: T.C. Milli Eğitim Bakanlığı Projeler Koordinasyon Merkezi Başkanlığı.

Armbruster, B. B., Lehr, F., Osborn,J. (2006). *Put reading first: The research building blocks for teaching children to read: Kindergarten through grade 3*. Retrieved from https://www.readingrockets.org/guides/put-reading-first-research-building-blocks-teaching-children-read

Arnold, R. D., & Wade, J. P. (2015). A definition of systems thinking: A systems approach. *Procedia Computer Science, 44*(C), 669–678. doi:10.1016/j.procs.2015.03.050

Artan, G. (2003). Engelli çocuk ve gençlerin cinsel eğitimi [Sexual education of children and youth with disabilities]. *Bilim ve Aklın Aydınlığında Eğitim Dergisi [J. Ment. Sci. Educ. Light]*, (4), 18-20.

Aryanti, A., Warsini, S., & Haryanti, F. (2018). Relationship between quality of care of young mothers and socio-emotional development in preschool children. *Belitung Nursing Journal, 4*(1), 31–38. doi:10.33546/bnj.344

Aşık, E. (2006). *Yetiştirme yurdunda kalan gençlerin ergenlik sorunları ve baş etmeleri*. Abant İzzet Baysal Üniversitesi Sağlık Bilimleri Enstitüsü Yüksek Lisans Tezi.

Aslan, D. (2017). Gelişimle ilgili temel konular. In A. Köksal Akyol (Ed.), Erken çocukluk döneminde gelişim (pp. 86-105). Ankara: Anı Yayıncılık.

Aslan, D. (2019). Gelişim ile ilgili temel konular. In A.K. Akyol, (Ed). Erken Çocukluk Döneminde Gelişim I (pp. 82-109). Ankara: Anı Yayıncılık.

Aslan, E., Ergün, N., Duman, B., Bozdağ, F., Karataş, S., & Fakirullahoğlu, A. M. (2018). Yaşlılarda eş kaybı, kayıp sonrası başa çıkma stratejileri ve gelecek beklentileri üzerine nitel bir araştırma. Retrieved from [REMOVED HYPERLINK FIELD]http://acikerisim.artuklu.edu.tr/xmlui/handle/20.500.12514/180#sthash.Vnjj8kT2.dpbs

Atabek, E. (2002). *Erken Büyüyen Çocuklar*. İstanbul: Altın Kitaplar Yayınevi.

Atabey, D. (2018). A study into the effective communication and social skills of preschool children. *Inonu University Journal of the Faculty of Education, 19*(1), 185–199. doi:10.679/inuefd.323598.186

Ataman, A. (2003). Gelişim, kuramlar ve kavramlar. In A. Ataman (Ed.), Gelişim ve Öğrenme. (pp. 9-50). Ankara: Gündüz Eğitim ve Yayıncılık.

Ataman, E., Yılmaz-Güleç, E., Hozan, F., Parıltay, E., Acar, D., Gedikbaşı, A., & Aslan, H. (2016). İkinci trimester-sonografik taramasında multipl yapısal anomaliler gösteren Trizomi 22 olgusu ve literatür derlemesi. *DEÜ Tıp Fakültesi Dergisi, 30*(1), 39-43.

Ataman, A. (1982). How can parents and teachers help gifted children? *Ankara University Journal of the Faculty of Educational Sciences, 15*(1), 335–344.

Ataov, A. (2004). Water as part of children's environment & children's preferences for waterfront settings. *Paper presented at the First International Conference on Inclusive Environments: Open Space-People Space.* Academic Press.

Atasü, T., & Gezer, A. (2000). Gebelikte ilaç kullanımı. In T. Atasü (Ed.), *Gebelikte fetüse ve yeni doğana etkiler* (pp. 373–444). Ankara: Nobel.

Atay, M. (2011). Erken çocukluk döneminde gelişim I. Ankara: Kök Yayıncılık.

Atli, S., & Baran, G. (2017). Gelişimin desteklenmesi, In A. Köksal Akyol (Ed.), *Erken Çocukluk Döneminde Gelişim, 1*, 449–495.

Atmaca Koçak, A. (2004). *Baba destek programını değerlendirme raporu.* İstanbul: Anne-Çocuk Eğitim Vakfı Yayınları.

Aujoulat, I., Simanelli, F. & Deccache, A. (2006). Health promotion needs of children and adolescents in hospitals: A review. *Patient Education and Counseling, 61,* 23-32 doş:. doi:10.1016/j.pec.2005.01.015

Austin, V. L., & Sciarra, D. T. (2013). Çocuk ve ergenlerde duygusal ve davranışsal bozukluklar (M. Özekes, Çev. Ed.). Ankara: Nobel Yayıncılık.

Austin, V. I. V., & Sciarra, D. T. (2015). *Çocuk ve ergenlerde duygusal ve davranışsal bozukluklar.* Ankara: Nobel.

Auty, K. M., Cope, A., & Liebling, A. (2017). A Systematic Review and Meta-Analysis of Yoga and Mindfulness Meditation in Prison: Effects on Psychological Well-Being and Behavioural Functioning. *International Journal of Offender Therapy and Comparative Criminology, 61*(6), 689–710. doi:10.1177/0306624X15602514 PMID:26320031

Avanoğlu-Kızıltepe, Z. (2004). *Teaching and Learning; A contemporary approach to education psychology.* İstanbul: Ofset Publications. (In Turkish)

Ayan, S. (2007). Aile içinde şiddete uğrayan çocukların saldırganlık eğilimleri. *AnatolianJournal of Psychiatry, 8,* 206–214.

Aydın, B. (2013). Eğitim Psikolojisi Gelişim-Öğrenme-Öğretim. In Gelişimin doğası. Baskı, Ankara: Pegem.

Aydın, A. (2008). *Education Psychology.* Ankara: Pegem Publications. (In Turkish)

Aydın, A. (2008). *Sınıf yönetimi (9. Baskı).* Ankara: Pegem Akademi.

Aykanat, B., & Gözen, D. (2014). Çocuk sağlığı hemşireliğinde aile merkezli bakım yaklaşımı. *Gümüşhane Üniversitesi Sağlık Bilimleri Dergisi, 3*(1), 683–695.

Aytaş, G. (2003). The role of children's literature in reading development. *Türklük Bilimi Araştırmaları, 13*, 155–160. (In Turkish)

Aytekin, A. (2005). Meslek seçimini etkileyen sosyo-ekonomik ve kültürel faktörler Isparta örneği, Süleyman Demirel Üniversitesi.

Aytekin, Ç., & Bayhan, P. (2015). Erken Müdahalede Uygulama Basamakları. *Hacettepe University Faculty of Health Sciences Journal, 2*(2), 1–14.

Babaroğlu, A. (2018). *Çocuk psikolojisi ve ruh sağlığı. (5. Baskı)*. Ankara: Vize.

Bacanlı, H. (2002). *Gelişim ve Öğrenme*. Ankara: Nobel Yayınları.

Bağcı Ayrancı, B. (2018). 0-12 yaş dil gelişimi uygulamaları ve yapılması gerekenler. *Kırıkkale Üniversitesi Sosyal Bilimler Dergisi, 8*(1), 13–34.

Bağcıoğlu Ünver, G. (2013). Gelişimle ilgili temel kavramlar, gelişimin temel ilkeleri ve gelişimi etkileyen etmenler. In A. Ulusoy (Ed.), Gelişim ve öğrenme psikolojisi içinde (pp. 1-17). Ankara: Anı.

Baker, A. J., Piotrkowski, C. S., & Brooks-Gunn, J. (1999). The home instruction program for preschool youngsters (HIPPY). *The Future of Children, 9*(1), 116–133. doi:10.2307/1602724 PMID:10414013

Baker, C. (2011). *Foundations of bilingual education and bilingualism* (Vol. 79). Multilingual matters.

Baker, L., Scher, D., & Mackler, K. (1997). Home and Family Influences on Motivations for Reading. *Educational Psychologist, 32*(2), 69–82. doi:10.1207153206985ep3202_2

Bakırcıoğlu, R. (2010). *Çocuk ve Ergende Ruh Sağlığı*. Ankara: Anı.

Bakırcıoğlu, R. (2013). *Çocuk ve ergenlerde ruh sağlığı*. (5. Baskı). Ankara: Anı.

Bakırcıoğlu, R. (2007). *Çocuk ruh sağlığı ve uyum bozuklukları*. Ankara: Anı Yayıncılık.

Bakırcıoğlu, R. (2011). *Çocuk ve ergende ruh sağlığı*. Ankara: Anı Yayıncılık.

Bakır, E., & Kapucu, S. (2017). Çocuk ihmali ve istismarının Türkiye'de yapılan araştırmalara yansıması: Bir literatür incelemesi. *Hacettepe Üniversitesi Hemşirelik Fakültesi Dergisi, 4*(2), 13–24.

Ball, J. W., & Bindler, R. C. (2003). *Pediatric nursing caring for children*. Upper Saddle River, N.J: Pearson Education.

Baloğlu, M. (2016). *Öğrenme psikolojisi*. Nobel.

Baltes, P. B. (1987). Theoritical propositons of life-span developmental psychology on the dynamics between growth and decline. *Developmental Psychology, 23*(5), 611–626. doi:10.1037/0012-1649.23.5.611

Banikowski, A. K., & Mehring, T. A. (1999). Strategies to enhance memory based on brain-research. *Focus on Exceptional Children, 32*(2), 1–16.

Baran, G. (2011). Çocuk gelişimine giriş. In N Aral & G. Baran (Eds.), *Çocuk gelişimi*. (pp. 17-52)İstanbul: Yapa Yayınları.

Barclay, K. H. (1995). Literacy Begins at Birth: What Caregivers Can Learn from Parents of Children Who Read Early. *Young Children, 50*(4), 24–28. doi:10.12691/education-6-6-14

Barclay, K. H. (2010). Using Song Picture Books to Support Early Literacy Development. *Childhood Education, 86*(3), 138–145. doi:10.1080/00094056.2010.10523134

Barlas, Y. (1996). System Dynamics: Systemic Feedback Modeling for Policy Analysis. In *System dynamics review* (Vol. 12, pp. 183–210). Academic Press. Retrieved from http://pisis.unalmed.edu.co/cursos/material/3004639/1/Barlas1996Formal-aspects-of-model-validity.pdf

Baroody, A. E., & Diamond, K. E. (2010). Links Among Home Literacy Environment, Literacy Interest, and Emergent Literacy Skills in Preschoolers At Risk for Reading Difficulties. *Topics in Early Childhood Special Education, 32*(2), 78–87. doi:10.1177/0271121410392803

Baruca, R. & Bialystok, E. (2012). Bilingual effects on cognitive and linguistic development: Role of language, cultural background, and education. *Child Development, 83*(2), 413-422. doi:1 doi:0.1111/j.1467-8624.2011.01707.x

Barut, Y. (2011). Özel gereksinimli çocuk ve ailelerine yönelik psikolojik danışma ve rehberlik hizmetleri. Özel Gereksinimli Çocuklar ve Özel Eğitime Giriş. Edt: Ayşegül Ataman. S:87-101, Ankara: Gündüz Eğitim ve Yayıncılık.

Baş, M., Molu, B., Tuna, H. İ. & Baş, İ. (2017). Göç eden ailelerin sosyo-kültürel ve ekonomik değişiminin kadın ve çocuk yaşamına etkisi. *İnsan ve Toplum Bilimleri Araştırmaları Dergisi, 6*(3), 1680-1693.

Başal, H. A. (2003). *Nasıl mutlu çocuk yetiştirebilirim?* Ankara: Morpa.

Başal, H. A. (2012). *Nasıl mutlu ve başarılı bir çocuk yetiştirebilirim?* Bursa: Ekin Basım Yayın ve Dağıtım.

Başaran, İ. (1996). *Eğitim psikolojisi.* Ankara: Gül Yayınevi.

Başbakkal, Z., Sönmez, S., Şen, N., & Esenay, F. I. (2010). 3-6 yaş grubu çocukların hastaneye yatışa karşı davranışsal tepkileri konusunda verilen eğitimin annelerin anksiyete düzeyi üzerine etkisinin incelenmesi. *Atatürk Üniversitesi Hemşirelik Yüksekokulu Dergisi, 12*(4), 59–65.

Bathea, L. (1999). Primaryprevention of childabuse. *American Family Physician, 59*, 1577–1591. PMID:10193598

Baydar, N., Küntay, A. C., Yagmurlu, B., Aydemir, N., Cankaya, D., Göksen, F., & Cemalcilar, Z. (2014). "It takes a village" to support the vocabulary development of children with multiple risk factors. *Developmental Psychology, 50*(4), 1014–1025. doi:10.1037/a0034785 PMID:24188041

Bayhan, P. (2012). Erken Müdahale Programlarında Aile Merkezli Uygulamalar ve Geçiş: Önemi, Özellikleri, Uygulaması. In N. Metin (Ed.), Özel Gereksinimli Çocuklar İçinde (pp. 29-52). Ankara: Maya Akademi.

Bayhan, P. (2018). Erken Müdahaleye Genel Bakış. In P. Bayhan (Ed.). Erken Müdahale İçinde. Ankara: Hedef Yayıncılık.

Bayhan, S.P. & Artan, İ. (2004). Çocuk gelişimi ve eğitimi. İstanbul: Morpa.

Bayhan, P., & Artan, İ. (2011). *Çocuk gelişimi ve eğitimi.* İstanbul: Morpa Kültür Yayınları.

Bayhan, P., & Sipal, R. F. (2011). Early Intervention Services in Turkey: Perspectives Of South-Eastern Families. *International Social Work, 54*(6), 781–799. doi:10.1177/0020872811404262

Baykara, A., Güvenir, T., & Miral, S. (2011). Hastalık ve hastaneye yatışın çocuk üzerinde etkisi. In A. Ekşi (Ed.), *Ben hasta değilim* (pp. 398–403). Ankara: Nobel.

Baykoç Dönmez, N. (2010). Children with special needs and special education. In N. Baykoç (Ed.), Special education for teaching programs (pp. 13-25). Ankara: Daytime.

Baykoç Dönmez, N. (2000). *Üniversite Çocuk Gelişimi ve Eğitimi Bölümü ve Kız Meslek Lisesi Öğrencileri İçin Oyun Kitabı.* İstanbul: Esin Yayınevi.

Baykoç, N. (2006). *Hastanede çocuk ve genç*. Ankara: Gazi.

Baykoç, N. (2011). Uzun süreli (süreğen) hastalığı olan çocuklar ve eğitimleri. In N. Baykoç (Ed.), *Öğretmenlik programları için özel eğitim* (pp. 276–300). Ankara: Gündüz.

Baykul, Y. (2000). *Eğitimde ve Psikolojide Ölçme: Klasik Test Teorisi ve Uygulaması*. Ankara: ÖSYM Yayınları.

Baymur, F. (1994). *Genel Psikoloji*. İstanbul: İnkılap Yayınevi.

Beard, K. Y., & Sugai, G. (2004). First Step to Success: An Early Intervention For Elementary Children At Risk for Antisocial Behaviour. *Behavioral Disorders*, *29*(4), 396–409. doi:10.1177/019874290402900407

Beck, L., Kumschick, I. R., Eid, M., & Klann-Delius, G. (2012). Relationship between language competence and emotional competence in middle childhood. *Emotion*, *12*(3), 503–514. doi:10.1037/a0026320 PMID:22148995

Bee, H., & Boyd, D. (2009). *Child development psychology* (O. Gündüz, trans.). İstanbul: Kaknüs Publications.

Bee, H., & Boyd, D. (2009). *Çocuk gelişim psikolojisi (O. Gündüz Çev.)*. İstanbul: Kaknüs Yayınları.

Bee, H., & Boyd, D. (2009). *Çocuk Gelişim Psikolojisi*. İstanbul, Turkey: Kaknüs Yayınları.

Bee, H., & Boyd, D. (2009). *Çocuk gelişim psikolojisi*. İstanbul: Kaknüs.

Bekman, S. (2000). *Eşit fırsat: Anne-çocuk eğitim programının değerlendirilmesi*. İstanbul: Anne-Çocuk Eğitim Vakfı Yayınları.

Bekman, S., Aksu-Koç, A., & Erguvanlı-Taylan, E. (2004). *Güneydoğu Anadolu bölgesinde bir erken müdahale modeli: Yaz anaokulu pilot uygulaması*. İstanbul: Anne Çocuk Eğitimi Vakfı Yayınları.

Benati, A. (2018). Grammar-Translation Method. In The TESOL Encyclopedia of English Language Teaching. Academic Press.

Bencik-Kangal, S., & Özkızıklı, S. (2015). Teknoloji ve eğitim. In P. Bayhan (Ed.), Okul öncesi eğitimde teknolojinin rolü (pp. 10-31). Ankara: Hedef.

Benson, R., & Migeon, C. (1975). Physiological and pathological puberty, and human behavior. In B. Eleftheriou & R. Sprott (Eds.), Hormonal Correlates of Behavior (Vol. 1). New York: Plenum Press. doi:10.1007/978-1-4615-7189-6_4

Bento, D., & Dias, G. (2017). The importance of outdoor play for young children's healthy development. *Porto Biomedical Journal*, *2*(5), 157–160. doi:10.1016/j.pbj.2017.03.003

Bergen-Cico, D., Razza, R., & Timmins, A. (2015). Fostering self regulation through curriculum infusion of mindful yoga: A pilot study of efficacy and feasibility. *Journal of Child and Family Studies*, *24*(11), 3448–3461. doi:10.1007/10826-015-0146-2

Bergen, D. (2002). The Role of Pretend Play in Children's Cognitive Development. *Early Childhood Research & Practice*, *4*(1), 1–13.

Bergen, D., Davis, D. R., & Abbitt, J. T. (2015). *Technology play and brain development: Infancy to adolescence and future implications*. Routledge. doi:10.4324/9781315681436

Bergeron, B. S. (1990). What does the term whole language mean? Constructing a definition from the literature. *Journal of Reading Behavior*, *22*(4), 301–329. Retrieved from doi:10.1080/10862969009547716

Berk, L. (2013). Bebekler ve çocuklar, doğum öncesinden orta çocukluğa. Ankara: Nobel.

Berk, L. (2013). Bebekler ve çocuklar. In N. Işıkoğlu Erdoğan (Ed.), Doğum öncesinden orta çocukluğa. Ankara: Nobel.

Berk, E. L. (2007). *Child development*. New York: Pearson.

Berk, E. L. (2013). *Bebekler ve cocuklar, doğum Öncesinden orta cocukluğa. (N. Işıkoğlu Özer, Trans.). Ankara: Nobel.*

Berk, E. L. (2013). *Bebekler ve çocuklar: Doğum öncesinden orta çocukluğa.* Ankara: Nobel.

Berk, L. E. (1999). *Infant children and adolescents.* Boston: Ally and Bacon.

Berk, L. E. (2000). *Bebekler ve Çocuklar: Doğum Öncesinden Orta Çocukluğa. (Çev: N. Işıkoğlu-Erdoğan).* Ankara: Nobel Akademik Yayıncılık.

Berk, L. E. (2013). *Bebekler ve Çocuklar Doğum Öncesinden Orta Çocukluğa.* Ankara: Nobel Akademik Yayıncılık.

Berk, L. E. (2013). *Cognition and language development* (A. Dönmez, trans.). (pp. 272–422). Ankara: İmge Publications.

Berk, L. E., & Meyers, A. B. (2015). *Infants, children, and adolescents.* Pearson.

Bertan, M., Haznedaroğlu, D., Koln, P., Yurdakök, K., & Güçiz, B. D. (2009). Ülkemizde Erken Çocukluk Gelişimine İlişkin Yapılan Çalışmaların Derlenmesi (2000-2007). *Çocuk Sagligi ve Hastaliklari Dergisi, 52,* 1–8.

Beyazıt, U. & Bütün- Ayhan, A. (2015). Hasta çocukların ruhsal özellikleri ve hasta çocuğa yaklaşım. In. A. Bütün-Ayhan (Ed.). *Hasta çocukların gelişimi ve eğitimi* (pp. 58-78). Eskişehir: Anadolu Üniversitesi.

Bialystok, E. (2001). *Bilingualism in development: Language, literacy, and cognition.* Cambridge University Press. doi:10.1017/CBO9780511605963

Bicakci, M. Y., Er, S., & Aral, N. (2017). Views of mothers towards interactive story reading to their children. *Education in Science, 42*(191), 53–68. doi:10.15390/EB.2017.7164

Bicakci, M. Y., Er, S., & Aral, N. (2018). The effects of interactive story reading on children's language development. *Kastamonu Eğitim Dergisi, 26*(1), 201–208. (In Turkish)

Biehl, M. C., Park, M. J., Brindis, C. D., Pantell, R. H., & Irwin, C. E. Jr. (2002). *The health of America's middle childhood population.* San Francisco: Public Policy Analysis and Education Center for Middle Childhood and Adolescent Health.

Biemiller, A. (2006). *Vocabulary Development and Instruction: A Prerequisite for School Learning. In The Handbook of Early Literacy Research.* New York, NY: Guilford Press.

Bildik, T. (1998). *Çalışan ergenlerin sorun alanları ve etkili olan psikososyal faktörler.* Ege Üniversitesi.

Bildik, T., Büküşoğlu, N., & Kesici, H. (2004). Çalışan ergenlerin psikososyal sorunları ve kendiçözüm önerileri. *Ege Tıp Dergisi, 43,* 105–111.

Bilgin, R. (2012). Sokakta çalışan çocukları bekleyen risk ve tehlikeler: Diyarbakır örneği. *ZKÜ Sosyal Bilimler Dergisi, 8*(15).

Bilgin, R. (2009). Diyarbakır'da sokakta çalışan çocuklar üzerine sosyolojik bir araştırma. *Elektronik Sosyal Bilimler Dergisi, 8*(27), 232–244.

Bilgin, R. (2012). Sokakta çalışan çocukları bekleyen risk ve tehlikeler: Diyarbakır örneği. *ZKÜ Sosyal Bilimler Dergisi, 8*(15), 79–96.

Birch, A. (1997). *Developmental psychology: From infancy to adulthood.* Macmillan International Higher Education. doi:10.1007/978-1-349-14204-0

Birdee, G. S., Yeh, G. Y., Wayne, P. M., Phillips, R. S., Davis, R. B., & Gardiner, P. (2009). Clinical applications of yoga for the pediatric population: A systematic review. *Academic Pediatrics*, *9*(4), 212–220. doi:10.1016/j.acap.2009.04.002 PMID:19608122

Birkan, B. (2002). *Küçük Adımlar Kursunun Gelişimsel Geriliği Olan Çocuğa Sahip Annelerin Küçük Adımları Uygulama Becerilerini Kazanmalarına Etkisi*. Eskişehir, Turkey: Anadolu Üniversitesi Yayinlari.

Bishai, D. M., Cohen, R., Alfonso, Y. N., Adam, T., Kuruvilla, S., & Schweitzer, J. (2016). Factors Contributing to Maternal and Child Mortality Reductions in 146 Low- and Middle-Income Countries between 1990 and 2010. *PLoS One*, *11*(1), e0144908. doi:10.1371/journal.pone.0144908 PMID:26783759

Bjorklund, D. F. (1995). *Children's Thinking. Developmental Function and Individual Differences*. USA: Brooks/Cole Publishing Company.

Blackman, J. A. (2002). Early Intervention: A Global Perspective. *Infants and Young Children*, *15*(2), 11–19. doi:10.1097/00001163-200210000-00004

Blair, C., McKinnon, R. D., & Daneri, M. P. (2018). Effect of the tools of the mind kindergarten program on children's social and emotional development. *Early Childhood Research Quarterly*, *43*, 52–61. doi:10.1016/j.ecresq.2018.01.002

Blakemore, S.J. (2010). The developing social brain: Implications for education. *Neuron*, *65*, 744-747.

Bloomberg, S., Rosander, M., & Andersson, G. (2006). Fears, hyperacusis and musicality in Williams Syndrome. *Research in Developmental Disabilities*, *27*(6), 668–680. doi:10.1016/j.ridd.2005.09.002 PMID:16269236

Bloomfield, L. (1933). *Language*. New York: Holt, Rinehart & Winston.

Bloom, L. (1973). *Language Development: Form and Function in Emerging Grammars*. Cambridge: M. J. T. Press.

Bodur, S., Durduran, Y., Küçükkendirci, H., & Doğan, C. (2009). Bebek ölümlerinin prenatal- natal sağlık hizmet kullanımı ve demografik özelliklerle ilişkisi: Vaka kontrol çalışması. *Dicle Tıp Dergisi*, *36*(4), 288–293. doi:10.17517/ksutfd.393180

Bolat, N., Doğangün, B., Yavuz, M., Demir, T., & Kayaalp, L. (2011). Doğuştan tam görme engeli olan ergenlerin depresyon, kaygı düzeyleri ve benlik kavramı özellikleri. *Türk Psikiyatri Dergisi*, *22*(2), 77–62. PMID:21638229

Bonvillain, N. (2019). *Language, culture, and communication: The meaning of messages*. Rowman & Littlefield.

Booij, G. (2012). *The grammar of words: An introduction to linguistic morphology*. Oxford University Press.

Bornstein, M. H., Arterberry, M. E., & Lamb, M. E. (2014). *Development in infancy: A contemporary introduction*. Taylor & Francis Group.

Bornstein, M. H., & Lamb, M. E. (2002). *Development in infancy: An introduction*. Psychology Press. doi:10.4324/9781410602596

Boyden, J., Dawes, A., Dornan, P., & Tredoux, C. (2019). Middle childhood: A key time for healthy development and learning. In *Tracing the Consequences of Child Poverty: Evidence from the Young Lives study in Ethiopia, India, Peru and Vietnam* (pp. 73-100). Bristol, UK: Bristol University Press. Retrieved from http://www.jstor.org/stable/j.ctvkjb390.10

Bozkurt, Ö. S. (2007). Examination of the developmental characteristics of children who are nominated as superior and special talented by their teachers according to their peers. Unpublished Master's Thesis, Hacettepe University Institute of Health Sciences, Ankara.

Bracken, S. S., & Fischel, J. E. (2008). Family reading behavior and early literacy skills in preschool children from low-income backgrounds. *Early Education and Development*, *19*(1), 45–67. doi:10.1080/10409280701838835

Bradley-Johson, S., & Johnson, C. M. (2001). *The Cognitive Abilities Scale* (2nd ed.). Austin, TX: Pro-Ed.

Bredekamp, S. (2014). *Effective practices in early childhood education: building a foundation*. USA: Pearson.

Bredekamp, S. (2015). *Effective applications in early childhood* (H. Z. İnan & T. İnan, trans.). Ankara: Nobel Publications. (In Turkish)

Brock, A., & Rankin, C. (2008). Communication, Language and Literacy from Birth to Five. *Sage (Atlanta, Ga.)*.

Brockman, J. (1993). *Creativity*. New York: Touchstone.

Brookes, G., Berlin, J., & Fuligni, S. (2000). Early Childhood Intervention Programs: What About The Family? In J. P. Shonkoff & S. J. Meisels (Eds.), *Handbook of Early Childhood Interventin* (2nd ed., pp. 549–588). United States of America: Cambridge University Press. doi:10.1017/CBO9780511529320.026

Brown, K. W., & Ryan, R. M. (2003). The benefits of being present: Mindfulness and its role in psychological well-being. *Journal of Personality and Social Psychology*, *84*(4), 822–848. doi:10.1037/0022-3514.84.4.822 PMID:12703651

Brown, L. S. (1993). *The Politics of Individualism: Liberalism, Liberal Feminism, and Anarchism*. Black Rose Books.

Brown, W. H., Pfeiffer, K. A., McIver, K. L., Dowda, M., Addy, L., & Pate, R. R. (2009). Social and environmental factors associated with preschoolers' nonsedentary physical activity. *Child Development*, *80*(1), 45–58. doi:10.1111/j.1467-8624.2008.01245.x PMID:19236392

Bruder, M. B. (2010). Family Centered Early Intervention: Clarifitin Our Values for New Milenium. Topic in Early Childhood. *Special Education*, *20*(2), 105–115.

Brunet, C. & Sarfati, A. C. (2000). 1-7 yaş arası çocuğun eğitimi (K. Bıçakçı, ed.). Ankara: Kapital Medya Hizmetleri A.Ş.

Buckingham, J., Wheldall, K., & Beaman-Wheldall, R. (2013). Why poor children are more likely to become poor readers: The school years. *Australian Journal of Education*, *57*(3), 190–213. doi:10.1177/0004944113495500

Bügü, B. Ö. (2001). *Gebelik ve laktasyon döneminde ilaç kullanımı*. Ankara: Nobel.

Bukatko, D., & Daehler, M. W. (2012). *Child development: A thematic approach*. Nelson Education.

Bülbül, A., Okan, F., Şahin, S., & Nuhoğlu, A. (2006). Düşük doğum ağırlıklı erken doğmuş bebeklerde erken dönem hastalık ve ölüm oranı sonuçları. *Türk Pediatri Arşivi*, *43*, 94–98.

Bülbül, S., & Doğan, S. (2016). Suça sürüklenen çocukların durumu ve çözüm önerileri. *Çocuk Sagligi ve Hastaliklari Dergisi*, *59*(1), 31–36.

Bulut, Ş. (2000). *Altı yaş çocuklarında davranış problemleri ile anne ve öğretmenlerin uyum düzeyleri arasındaki ilişkinin incelenmesi*. Yayınlanmamış Yüksek Lisans Tezi, Ankara Üniversitesi.

Burdette, H. L., & Whitaker, R. C. (2005). A national study of neighborhood safety, outdoor play, television viewing, and obesity in preschool children. *Pediatrics*, *116*(3), 657–662. doi:10.1542/peds.2004-2443 PMID:16140705

Burke, M. D., Hagan-Burke, S., Kwok, O., & Parker, R. (2009). Predictive validity of early literacy indicators from the middle of kindergarten to second grade. *The Journal of Special Education*, *42*(4), 209–226. doi:10.1177/0022466907313347

Bus, A.G., Belsky, J., Van Ijzendoorn, M.H., & Crnic, K. (1995). Attachment and Bookreading Patterns: A Study of Mothers, Fathers, and Their Toddlers. *Early Childhood Research Quarterly*, 81-98.

Bütün-Ayhan, A. (2017). Bilişsel gelişim. In N. Aral & T. Duman (Ed.), Eğitim psikolojisi (pp. 104-128). Ankara: Pegem.

Bütün-Ayhan, A. (2017). Bilişsel gelişim. In N. Aral & T. Duman (Eds.), *Eğitim psikolojisi* (pp. 104–128). Ankara: Pegem.

Bütün-Ayhan, A., & Aral, N. (2007). Bracken temel kavram ölçeği- gözden geçirilmiş formunun altı yaş çocukları için uyarlama çalışması. *Hacettepe Üniversitesi Eğitim Fakültesi Dergisi*, *32*, 42–51.

Butzer, B., Bury, D., Telles, S., & Khalsa, S. (2016). Implementing yoga within the school curriculum: A scientific rationale for improving social–emotional learning and positive student outcomes. *Journal of Children's Services*, *11*(1), 3–24. doi:10.1108/JCS-10-2014-0044

Büyüköztürk, Ş. (2005). Anket Geliştirme. *Türk Eğitim Bilimleri Dergisi*, *3*(2), 133–151.

Büyüköztürk, Ş. (2010). *Sosyal Bilimler İçin Veri Analizi El Kitabı*. Ankara: Pegem Akademi.

Büyüköztürk, Ş., Kılıç-Çakmak, E., Akgün, Ö. E., Karadeniz, Ş., & Demirel, F. (2008). *Bilimsel Araştırma Yöntemleri*. Ankara: Pegem Akademi.

Cabral, P., Meyer, H. B., & Ames, D. (2011). Effectiveness of yoga therapy as a complimentary treatment for major psychiatric disorders: A meta-analysis. *Prim Care Companion*, *13*(4), 1060–1068.

Çağdaş, A. & Şahin Seçer, Z. (2015). Anne-baba eğitimi. Ankara: Eğiten Kitap.

Çağdaş, A. (2008). *Anne baba çocuk iletişimi*. Ankara: Kök.

Çağdaş, A., & Şahin-Seçer, Z. (2015). *Anne- baba eğitimi*. Ankara: Eğiten.

Çağdaş, A. (2012). *Anne-baba- çocuk iletişimi*. Ankara: Eğiten.

Çağlar, D. (1981). Uyumsuz çocuklar ve eğitimleri. Ankara: Ankara Üniversitesi Eğitim Fakültesi Yayınları No: 103.

Caglar, D. (1972). Characteristics of gifted children. *Ankara University Journal of the Faculty of Educational Sciences.*, *5*(3), 95–110.

Çakırer Özservet, Y. (2015). Geleneksel çocuk oyunları için sokak temelli geleneksel oyun alanlarının yeniden keşfi, I. In *Uluslararası Türk Dünyası Çocuk Oyun ve Oyuncakları Kurultayı*. Academic Press.

Cakmak Gulec, H. & Gecgel, H. (2006). Children's literature. Ankara: Kökpublications (in Turkish)

Çakmak, A., & Pektaş, D. (2018). Duygusal gelişim. In N. Aral & T. Duman (Eds.), *Çocuk gelişimi* (pp. 264–290). Ankara: Hedef.

Çamdeviren, Ş. (2014). Difficulties faced by parents of gifted children attending Science and Art Center (Bilsem) (case of Sakarya province). Unpublished Master's Thesis, Sakarya University Institute of Educational Sciences, Sakarya.

Cameron, L. (2001). Teaching languages to young learners. United Kingdom: Cambridge. doi:10.1017/CBO9780511733109

Camkurt, M. Z. (2013). 2022 Sayılı kanuna göre engelliliğin tespiti ve engelliliğe dayalı aylıkların bağlanması işlemleri. *Mali Çözüm Dergisi*, *23*, 79–118.

Campbell-Hicks, R. (2016). Early literacy programmes in public libraries: Best practice. *The Australian Library Journal*, *65*(2), 121–129. doi:10.1080/00049670.2016.1121770

Campbell, S. B. (1995). Behavior problems in preschool children: A review of recent research. *Journal of Child Psychology and Psychiatry, and Allied Disciplines*, *36*(1), 113–149. doi:10.1111/j.1469-7610.1995.tb01657.x PMID:7714027

Can, A. (2006). *Çocuk ve çizgi film*. Konya: Öz Eğitim.

Canadian Council on Learning. (2006). How parents foster early literacy. Lessons in Learning. Retrieved from http://www.child-encyclopedia.com/sites/default/files/docs/suggestions/how-parents-foster-early-literacy.pdf

Cansız, R. (2004). *Çocuk yetiştirme modellerinde cinsiyetçi yaklaşımlar.* Ankara Üniversitesi Sosyal Bilimler Enstitüsü, Kadın Çalışmaları Anabilim Dalı.

Cardon, G. M., & De Bourdeaudhuij, M. M. (2008). Are preschool children active enough? Objectively measured physical activity levels. *Research Quarterly for Exercise and Sport, 79*(3), 326–332. doi:10.1080/02701367.2008.10599496 PMID:18816944

Carmody, J., & Baer, R. A. (2008). Relationships between mindfulness practice and levels of mindfulness, medical and psychological symptoms and well-being in a mindfulness-based stress reduction program. *Journal of Behavioral Medicine, 31*(1), 23–33. doi:10.100710865-007-9130-7 PMID:17899351

Carr, A. (2011). Social and emotional development in middle childhood. *Child Psychology and Psychiatry*, 56-61.

Carroll, M. (2018). *Clinical Reproductive Science.* John Wiley & Sons. doi:10.1002/9781118977231

Caspi, A., & Roberts, B. W. (2009). Personality development across the life course: The argument for change and continuity. *Psychological Inquiry, 12*(2), 49–66. doi:10.1207/S15327965PLI1202_01

Catts, H. W., Fey, M. E., Tomblin, J. B., & Zhang, X. (2002). A longitudinal investigation of reading outcomes in children with language impairments. *Journal of Speech, Language, and Hearing Research: JSLHR, 45*(6), 1142–115. doi:10.1044/1092-4388(2002/093) PMID:12546484

Çavuşoğlu, H. (2013). *Çocuk sağlığı hemşireliği 1.* Ankara: Sistem.

Cedar Mill Community Libraries. (2017). *Early Literacy.* Retrieved from https://library.cedarmill.org/kids/early-literacy/

Çelikkaleli, Ö. (2014). Bilim, eğitim, psikolojisi ve gelişimle ilgili temel kavramlar. In R. Çeçen Eroğul & F. Yurtal (Eds.), Eğitim Psikolojisi El Kitabı (pp. 1-46). Ankara: Anı Yayıncılık.

Çepni, S., Bayrakçeken, S., Yılmaz, A., Yüzel, C., Semerci, Ç., Köse, E., ... Gündoğdu, K. (2009). Geçerlik ve Güvenirlik. In E. Karip (Ed.), *Ölçme ve Değerlendirme* (pp. 89–122). Ankara: Pegem Akademi.

Cer, E. (2016). Reality and relativity of child in 0-6 year children's books. Ankara: Eğiten Kitap Publications. (in Turkish)

Çetin Gündüz, H., & Nalbantoğlu Yılmaz, F. (2016). Lise Öğrencilerinin Kariyer Kaygılarını Belirlemeye Yönelik Ölçek Geliştirme Çalışması. *Mersin Üniversitesi Eğitim Fakültesi Dergisi, 12*(3), 1008–1022.

Çetin, S., & Özözen, M. (2015). Doğum öncesi gelişim. In N. Baysal-Metin (Ed.), Doğum öncesinden ergenliğe çocuk gelişimi (pp. 23-54). Ankara: Pegem.

Çetingöz, D., & Yıldırım Doğru, S. (2016). Gelişimi risk altında olan okul öncesi dönem çocuklarının sosyal beceri düzeylerinin değerlendirilmesi. *MediterraneanJournal of Humanities., VI*(2), 151–172. doi:10.13114/MJH.2016.291

Çetintaş, S. (2015). *5 yaş çocuklarında görülen davranış problemleri ile annelerin çocuklarına uyguladıkları disiplin yöntemleri arasındaki ilişkinin incelenmesi.* Yayınlanmamış Yüksek Lisans Tezi, Gazi Üniversitesi.

Çevik, C. (2017). Yaşlıların Sağlık Hizmetine Erişimi. *Turkiye Klinikleri J Public Health Nurs-Special Topics, 3*(3), 165–171.

Ceylan, Ş., & Yalman, F. (2018). Erken Müdahale Tarihçesi. In P. Bayhan (Ed.). Erken Müdahale İçinde (pp. 38-61). Ankara: Hedef Yayıncılık.

Ceylan, Ş., Gözün Kahraman, Ö., & Ülker, P. (2016). Annelerin erken çocukluk dönemine bakış açısı. *İnsan ve Toplum Bilimleri Araştırmaları Dergisi, 5*(5), 1333-1356.

Ceylan, G., Erol, D., & Yüce, H. (2009). DeNovo 21/21 Translokasyonu olan Down sendromlu iki olgu: Literatür araştırması. *Firat Tip Dergisi, 14*(4), 280–282.

Ceylan, R. (2017). Fiziksel gelişim. In N. Aral & T. Duman (Eds.), *Eğitim psikolojisi* (pp. 56–73). Ankara: Pegem.

Charlesworth, R. (1993). Understanding child development. Albany, NewYork: Delmar publishers Inc.

Chein, C., & Hui, A. N. N. (2010). Creativity in early childhood education: Teachers' perception in three Chinese societies. *Thinking Skills and Creativity, 5*(2), 49–60. doi:10.1016/j.tsc.2010.02.002

Chetty, R., Hendren, N., Kline, P., & Saez, E. (2014). Where is the land of opportunity? The geography of intergenerational mobility in the United States. *The Quarterly Journal of Economics, 129*(4), 1553–1623. doi:10.1093/qje/qju022

Chimiklis, A. L., Dahl, V., Spears, A. P., Goss, K., Fogarty, K., & Chack, A. (2018). Yoga, Mindfulness, and Meditation Interventions for Youth with ADHD: Systematic Review and Meta-Analysis. *Journal of Child and Family Studies, 27*(10), 3155–3168. doi:10.100710826-018-1148-7

Chomsky, A.N. (2002). *Syntactic Structures.* Walter de Gruyter.

Chomsky, N. (2004). Knowledge of Language as a Focus of Inquiry. In B. Lust & C. Foley (Eds.), First language acquisition: The essential readings (pp. 15 24). Malden: MA: Blackwell Pub.

Chomsky, N. (1975). *Reflections on language.* NY: Pantheon.

Chomsky, N. (1986). *Knowledge of language: its nature, origin and use.* New York: Praeger.

Chong, C. S., Tsunaka, M., Tsang, H. W., Chan, E. P., & Cheung, W. M. (2011). Effects of yoga on stress management in healthy adults: A systematic review. *Alternative Therapies in Health and Medicine, 17*(1), 32–38. PMID:21614942

Chopra, M., Bhutta, Z. A., Mason, E., Preserr, C., Stenberg, K., Sheehan, P., ... Ostergren, M. (2013). Advancing social and economic development by investing in women's and children's health: A new Global Investment Framework. *Lancet, 383*(9925), 1333–1354. doi:10.10160140-6736(13)62231-x PMID:24263249

Cicek, A. (2002). Language development of 0-6 year children. *Erzincan Eğitim Fakültesi Dergisi, 4*(1), 115–124.

Çifci-Tekinarslan, İ. (2012). Zihinsel yetersizliği olan öğrenciler. In İ. H. Diken (Ed.), Özel eğitime gereksinimi olan çocuklar ve özel eğitim (pp. 137- 163). Ankara: Pegem.

Cihangir-Altay, N. (2008). Çocuklarda ameliyat öncesi hazırlık. *Sağlık Bilimleri Fakültesi Hemşirelik Dergisi,* 68-75. doi:10.14527/9786052410240.03

Citil, M., & Ataman, A. (2018). Reflection of behavioral characteristics of gifted students in primary school age on educational environments and problems that may arise. *Gazi University Journal of the Faculty of Education, 38*(1), 185–231.

Claessens, A., Duncan, G., & Engel, M. (2009). Kindergarten skills and fifth-grade achievement: Evidence from the ECLS-K. *Economics of Education Review, 28*(4), 415–427. doi:10.1016/j.econedurev.2008.09.003

Clark, S. M., Ghulmiyyah, L. M., & Hankins, G. D. (2008). Antenatal, antecedents and the impact of obstetric care in the etiology of cerebral palsy. *Clinical Obstetrics and Gynecology, 51*(4), 775–786. doi:10.1097/GRF.0b013e3181870994 PMID:18981802

Cliff, D. P., Hesketh, K. D., Vella, S. A., Hinkley, T., Tsiros, M. D., Ridgers, N. D. ... Plotnikoff, R. C. (2016). Objectively measured sedentary behaviour and health and development in children and adolescents: systematic review and meta-analysis. *Obesity Reviews*, *17*(4), 330-344. Retrieved from https://onlinelibrary.wiley.com/doi/full/10.1111/obr.12371

Çoban, A. E. (2010). Mesleki rehberlik. In M. Güven (Ed.), Psikolojik Danışma ve Rehberlik (pp. 132-159). Ankara: Anı.

Çoban, S. (2015). Türkiye'de risk altındaki çocuklar ve çocuk suçluluğu üzerine bir değerlendirme. *Sosyoloji Konferansları*, *52*(2), 791–810.

Cohen, S. C. L., Harvey, D. J., Shields, R. H., Shields, G. S., Rashed, R. N., Tancredi, D. J., ... Schweitzer, J. B. (2018). Effects of Yoga on Attention, Impulsivity, and Hyperactivity in Preschool-Aged Children with Attention-Deficit Hyperactivity Disorder Symptoms. *Journal of Developmental and Behavioral Pediatrics*, *39*(3), 200–209. doi:10.1097/DBP.0000000000000552 PMID:29538185

Cömert, D., & Erdem, E. (2010). *Erken çocukluk döneminde aile katılımı etkinlikleri.* Ankara: Eğiten Kitabevi.

Cömert, D., & Erdem, E. (2016). *Erken çocukluk eğitiminde aile eğitimi.* Ankara: Eğiten Yayıncılık.

Cömertpay, B. (2006). Dramanın 5-6 yaş grubu çocuklarının dil edinimine etkisi, Sosyal Bilimler Enstitüsü Okul Öncesi Eğitimi Ana Bilim Dalı. Unpublished master thesis, Çukurova Üniversitesi.

Condon, E., Crepinek, M. K., & Fox, M. K. (2009). School meals: Types of foods offered to and consumed by children at lunch and breakfast. *Journal of the American Dietetic Association*, *109*(2), 67–78. doi:10.1016/j.jada.2008.10.062 PMID:19166674

Cook, J. L., & Cook, G. (2005). *Child development principles and perspectives.* Boston, Mass: Pearson.

Çopur, Z., Erkal, S., & Şafak, Ş. (2006). Annelerin 12 yaş üzeri çocuklarıyla ilişkilerinde yaşadıkları çatışmaların incelenmesi, I. In Ulusal ev ekonomisi kongresi. Academic Press.

Cordeiro Drever. L. (2008). *Yoga and children ages 5-12 with Down syndrome: an instructional yoga DVD and parent manual* [M.A. dissertation]. California State University, Long Beach, CA.

Cosco, N. G. (2007). Environmental interventions for healthy development of young children in the outdoors. In C. Ward-Thompson & P. Travlou (Eds.), *Open space people space* (pp. 125–135). London, England: Taylor and Francis.

Coşkun, A. (2008). Yüksek riskli gebeliklerin fetüs/yeni doğana etkisi ve hemşirelik yaklaşımı. In T. Daloğlu & G. Görak (Eds.), Temel neonatoloji ve hemşirelik ilkeleri içinde (pp. 57-94). İstanbul: Nobel Tıp Kitabevi.

Couch, A., & DeCocker-Geist, K. (2014). Physiological alterations during the postnatal period. In G. Robin (Ed.), *Prenatal and postnatal care a woman-centered approach* (pp. 419–422). Iowa: John Wiley & Sons, Inc.

Crisp, A. (1999). Anoreksia nervoza: Psikanalitik bakış. In A. Ekşi (Ed.), Ben hasta değilim (pp. 223-228). İstanbul: Nobel Tıp Kitabevi Ltd. Şti.

Cronbach, L. J. (1951). Coefficient Alpha and The İnternal Structure Of Tests. *Psychometrika*, *16*(3), 297–334. doi:10.1007/BF02310555

Çubukçu, Z. (2004). Öğretmen adaylarının düşünme stillerinin öğrenme biçimlerini tercih etmelerindeki etkisi. XIII. Ulusal Eğitim Bilimleri Kurultayı, Retrieved from http://www.pegema.net/dosya/dokuman/253.pdf

Cüceloğlu, D. (2000). *İnsan ve davranışı.* İstanbul: Remzi Kitabevi.

Çuhadaroğlu, F., Canat, S., Kılıç, E., Şenol, S., Ruganci, N., Öncü, B., ... Avcı, A. (2004). *Ergen ve Ruhsal Sorunlar. Durum Saptama Çalışması. Türkiye Bilimler Akademisi (TÜBA).* Ankara: Raporları.

Çukur, D. (2009). Çocuk dinlenimi açısından oyunun önemi ve konut yakın çevresinde oyun değerini artırıcı mekansal düzenlemeler. *Ege Mimarlık*, 2(69).

Cummings, L. L. (2000). Parent Training: A Program For Parents Of Two-And Three-Year-Olds. Unpublished Doctoral Dissertation, Central Michigan University, Michigan.

Cunningham, J. N., Kliewer, W., & Garner, P. W. (2009). Emotion socialization, child emotion understanding and regulation, and adjustment in urban African American families: Differential associations across child gender. *Development and Psychopathology*, *21*(1), 261–283. doi:10.1017/S0954579409000157 PMID:19144233

da Cunha, A. J. L. A., Leite, Á. J. M., & de Almeida, I. S. (2015). The pediatrician's role in the first thousand days of the child: The pursuit of healthy nutrition and development. [Versão Em Português]. *Jornal de Pediatria*, *91*(6), S44–S51. doi:10.1016/j.jped.2015.07.002 PMID:26351759

Daelmans, B., Darmstadt, G. L., Lombardi, J., Black, M. M., Britto, P. R., & Lye, S. … Richter, L. M. (2017). Early childhood development: The foundation of sustainable development. *Lancet*. doi:10.1016/S0140-6736(16)31659-2

Dağ, İ., Öktem, F., Yazıcı, K., Güvenç, G., Rezaki, M., Özer, Ö., & Tunçel, M. (2007). Televizyon programlarındaki şiddet içeriğinin, müstehcenliğin ve mahremiyet ihlallerinin izleyicilerin ruh sağlığı üzerindeki olumsuz etkileri. Ankara: RTUK.

Dagabakan, O., & Dagabakan, F. (2007). Language and theories of language development for children. *Milli Eğitim Dergisi*, *174*, 155–161. (in Turkish)

Dağlıoğlu, H. E. (2006). *Üstün yetenekli çocuklar ve televizyon. M.Ü. Atatürk Eğitim Fakültesi 1. Uluslararası Okul Öncesi Eğitim Kongresi Bildiriler Kitabı*. İstanbul: Ya- Pa.

Dağlıoğlu, H. E. (2010). Yaratıcılık, hayal gücü ve zekâ ilişkisi. In E. Çelebi Öncü (Ed.), *Erken Çocukluk Döneminde Yaratıcılık ve Geliştirilmesi*. Ankara: Pegem Akademi.

Dahl, M. (1987). Early feeding problems in an affluent society. III. Follow-up at two years: Natural course, health, behaviour, and development. *Acta Paediatrica (Oslo, Norway)*, *76*(6), 872–880. doi:10.1111/j.1651-2227.1987.tb17257.x PMID:3480683

Dahl, M., & Sundelin, C. (1992). Feeding problems in an affluent society. Follow-up at four years of age in children with early refusal to eat. *Acta Paediatrica (Oslo, Norway)*, *81*(8), 575–579. doi:10.1111/j.1651-2227.1992.tb12303.x PMID:1392380

Dai, C. L., Nabors, L. A., Vidourek, R. A., King, K. A., & Chen, C. C. (2015). Evaluation of an afterschool yoga program for children. *International Journal of Yoga*, *8*(2), 160–161. doi:10.4103/0973-6131.158488 PMID:26170600

Dalgalı, H. (2014). *Televizyonda yayınlanan çocuk programlarının okul öncesi eğitim becerilerinin gelişimine olan etkisinin ebeveyn görüşlerine göre incelenmesi*. Kayseri: Erciyes Üniversitesi /Eğitim Bilimleri Enstitüsü.

Dalgar, G., & Kaytez, N. (2018). Doğum öncesi gelişim ve doğum. In P. Ünüvar (Ed.), Bebeklik döneminde gelişim ve eğitim (pp. 24-68). Ankara: Anı.

Daly, L. A., Haden, S. C., Hagins, M., Papouchis, N., & Ramirez, P. M. (2015). Yoga and Emotion Regulation in High School Students: A Randomized Controlled Trial. *Hindawi Publishing Corporation Evidence-Based Complementary and Alternative Medicine*, *2015*, 1–8. doi:10.1155/2015/794928 PMID:26356561

Damon, W., & Lerner, R. M. (2008). *Child and adolescent development: An advanced course*. Canada: John Wiley & Sons, Inc.

Daskalovska, N. (2018). Historical Perspectives on Teaching Speaking. In The TESOL Encyclopedia of English Language Teaching. Academic Press. doi:10.1002/9781118784235.eelt0710

Davidse, N. J., de Jong, M. T., Bus, A. G., Huijbregts, S. C., & Swaab, H. (2011). Cognitive and environmental predictors of early literacy skills. *Reading and Writing*, *24*(4), 395–412. doi:10.100711145-010-9233-3 PMID:21475735

Davies, D. (2010). *Child Development*. New York, NY: TheGuilford Pres.

Dawkins, M. P., Fullilove, C., & Dawkins, M. (1995). Early assessment of problem behavior among young children in high-risk environments. *Family Therapy*, *22*(3), 133–141.

De Houwer Annick, D. (1999). *Two or more languages in early childhood: some general points and practical recommendations*. Retrieved from http://www.cal.org/resources/digest/earlychild.html

De Saussure, F. (1985). Genel dilbilim dersleri. Ankara: Birey ve Toplum Yayınları.

Dean, S. V., Lassi, Z. S., Imam, A. M., & Bhutta, Z. A. (2014). Preconception care: Closing the gap in the continuum of care to accelerate improvements in maternal, newborn and child health. *Reproductive Health*, *11*(Suppl. 3), S1. doi:10.1186/1742-4755-11-S3-S1 PMID:25414942

Decasper, A., & Fifer, W. P. (1980). Of human bonding: Newborns prefer their mothers voice, *Science*. *Sayı*, *208*, 1174–1176. PMID:7375928

Dede, H., Kandemir, Ö., Yalvaç, S. & Altay, M. (2017). İleri anne yaşı nedeniyle yapılan ikinci trimester amniyosentez çalışmalarımız: Üç yıllık deneyim. *The Journal of Gynecology- Obstetrics and Neonatology, 10*(38), 1586-1588.

Değirmencioğlu, H. (2018). Kişilik gelişimi. In N. Aral & Z. F. Temel (Ed.), Çocuk gelişimi (pp. 292-313). Ankara: Hedef CS Basın Yayın.

Demetrioua, A., Merrell, C., & Tymms, P. (2017). Mapping and predicting literacy and reasoning skills from early to later primary school. *Learning and Individual Differences*, *54*, 217–225. doi:10.1016/j.lindif.2017.01.023

Demir, İ. & Yavuzer, H. (2016). Yeni kuşak anne-babalar ve çocukları. İstanbul: Remzi.

Demircan, Ö. (2005). Yabancı dil öğretim yöntemleri. İstanbul: Der Yayınları.

Demirci, K. (2011). On the importance of phoneme theory. *Türkoloji Araştırmaları*, *6*(2), 359–366. (in Turkish)

Demircioğlu, H. (2012). *Altı yaşında çocuğu olan annelere uygulanan aile eğitimi programının aile işlevleri ve anne-baba tutumları ile çocuğun gelişimi üzerindeki etkisinin incelenmesi. Yayımlanmamış doktora tezi*. Ankara: Gazi Üniversitesi.

Demirel, Ö. (2002). *Teaching Turkish*. Ankara: Pegem A Publications. (in Turkish)

Demiriz, S., Karadağ, A., & Ulutaş, I. (2003). *Okulöncesi eğitim kurumlarında eğitim ortamı ve donanımı*. Ankara: Anı Yayıncılık.

Demir-Lira, Ö. E., Applebaum, L. R., Goldin-Meadow, S., & Levine, S. C. (2019). Parents' early book reading to children: Relation to children's later language and literacy outcomes controlling for other parent language input. *Developmental Science*, *22*, e12764. doi:10.1111/desc.12764 PMID:30325107

Demir, T. (2010). An essay on egocentric speaking in speech therapy. *Kastamonu Eğitim Dergisi*, *188*(2), 415–430. (in Turkish)

Demir, V., & Yıldırım, B. (2017). Sanatla Terapi Programının Üniversite Sınavına Hazırlanan Öğrencilerin Depresyon, Anksiyete ve Stres Belirti Düzeylerine Etkililiği. *Ege Eğitim Dergisi*, *18*(1), 311–344. doi:10.12984/egeefd.280267

Denham, S. A., Bassett, H. H., & Wyatt, T. (2007). The socialization of emotional competence. In J. E. Grusec & P. D. Hastings (Eds.), *Handbook of socialization. New York*: Guilford.

Denham, S., Blaır, K., Demlılder, E., Levitas, J., Sawyer, K., Auerbach, S., & Queenan, P. (2003). Preschool Emotional Competence: Pathway to Social Competence. *Child Development*, *74*(1), 238–256. doi:10.1111/1467-8624.00533 PMID:12625448

Deniz, Ü. (2015). Üreme sistemi anatomisi ve fizyolojisi. In Ü. Deniz & Ö. R. Önder (Eds.), *Okul öncesi eğitimi öğretmen adayları ve öğretmenler için anne çocuk sağlığı ve ilk yardım* (pp. 11–22). Ankara: Nobel.

Deniz, Ü. (2017). Gelişim. In N. Aral & T. Duman (Eds.), *Eğitim psikolojisi* (pp. 32–53). Ankara: Pegem. doi:10.14527/9786052410240.02

Deniz, Ü. (2018). Gelişim. In N. Aral & T. Duman (Eds.), *Eğitim Psikolojisi* (pp. 32–53). Ankara: Pegem. doi:10.14527/9786052410486

Dere Çiftçi, H., & Aksoy, B. A. (2017). Aile eğitim programları. In A.B. Aksoy (Ed.), Aile eğitimi ve katılımı (pp. 43-70). Ankara: Hedef CS Basın Yayın.

Deretarla Gül, E. (2013). Okul öncesi eğitimde babaların katılımını sağlama. In Y.A. Arnas (Ed.), Aile eğitimi ve okul öncesinde aile katılımı (pp. 223-237). Ankara: Vize Yayıncılık.

Deretarla Gul, E. (2015). Adopt reading habit to preschoolers. In M. Gönen (Ed.), Child Development (pp. 305-314). Ankara: Eğiten Kitap Publications (in Turkish)

Deretarla-Gül, E. (2012). Ailelerin çocuk bahçelerine ve çocuk bahçelerindeki materyallere bakış açılarının incelenmesi. *Ç.Ü. Sosyal Bilimler Enstitüsü Dergisi*, *21*(3), 261–274.

Deveci, E., & Açık, Y. (2003). Çocuk istismarı nedenlerinin incelenmesi. *Arşiv*, *12*, 396–405.

Dewey, J. (1930). *Individualism Old and New*.

Dighe, J. (1993). Children and the earth. *Young Children*, *48*(3), 58–63.

Diken, İ., Bayhan, P., Turan, F., Sipal, F., Sucuoglu, B., Ceber-Bakkaloglu, H., ... Kara, O. K. (2012). Early Childhood Intervention and Early Childhood Special Education in Turkey within the Scope of the Developmental System Approach. *Infants and Young Children*, *25*(4), 346–353. doi:10.1097/IYC.0b013e318268541d

Dikici Sığırtmaç, A. (2013). Family participation in the education of children with special needs. In Y. Aktaş Arnas (Ed.), Family education and pre-school family participation (2nd ed., pp. 174-176). Adana: Visa.

Dikici Sığırtmaç, A., & Deretarla Gül, E. (2010). *Preschool special education*. Ankara: Visa.

Dilsiz, H., & Mağden, D. (2015). Öğretmenlerin çocuk istismar ve ihmali konusunda bilgi ve risk tanıma düzeylerinin tespit edilmesi. *Hacettepe Üniversitesi Sağlık Bilimleri Dergisi*, *1*(2), 678–694.

Dinçel, E. (2006). *Ergenlik dönemi gelişimsel ödevleri ve psikolojik problemler*. Ankara Üniversitesi Sosyal Bilimler Enstitüsü Yüksek Lisans Tezi.

Dixon-Krauss, L., Januszka, C.M. & Chae, C. (2010). Development of the dialogic reading inventory of parent- child book reading. *Journal of Research in Childhood Education*, *24*(3), 266–277.

Dodici, B. J., Draper, D. C., & Peterson, C. A. (2003). Early parent—Child interactions and early literacy development. *Topics in Early Childhood Special Education*, *23*(3), 124–136. doi:10.1177/02711214030230030301

Doğan, B. (2010). *Hastanede yatan çocuğun bakımına ebeveynlerin katılımı konusunda hemşirelerin, doktorların ve ebeveynlerin görüşleri ve beklentileri.* Ankara: Başkent Üniversitesi/Sağlık Bilimleri Enstitüsü.

Dokuyan, M. (2016). 12. Sınıf öğrencilerinde algılanan anne- baba tutumları ile benlik saygısı arasındaki ilişkinin incelenmesi. *Ahi Evran Üniversitesi Sosyal Bilimler Enstitüsü Dergisi, 2*(2), 1–21.

Dol, A. (1999). Illustrations of 0-6 year children's books. Unpublished masters thesis, Onsekiz Mart University, Social Sciences Institute, Çanakkale. (In Turkish)

Dolanay, D. B. (2007b). *Mother-Baby Yoga and Sample Program.* Retrieved from https://yogadivya.net/yoga-dersleri/anne-bebek-yogasi/

Dolanay, D. B. (2007a). *Lotus Flowers: Yogamini Project.* Istanbul: ALFA Publications.

Dombrowski, S. C., Noonan, K., & Martin, R. P. (2007). Low birth weight and cognitive outcomes: Evidence for a gradient relationship in an urban, poor, African American birth cohort. *School Psychology Quarterly, 22*(1), 26–43. doi:10.1037/1045-3830.22.1.26

Domitrovich, C. E., Durlak, J. A., Staley, K. C., & Weissberg, R. P. (2017). Social-Emotional Competence: An Essential Factor for Promoting Positive Adjustment and Reducing Risk in School Children. *Child Development, 88*(2), 408–416. doi:10.1111/cdev.12739 PMID:28213889

Domsch, H., Lohaus, A., & Thomas, H. (2009). Prediction Of Childhood Cognitive Abilities From A Set Of Early Indicators Of Information Processing Capabilities. *Infant Behavior and Development, 32*(1), 91–102. doi:10.1016/j.infbeh.2008.10.006 PMID:19095308

Donmez, N., Abidioglu, U., Dinçer, C., & Erdenir, N. ve Gumuscu, S. (2000). Language development activities in early years. İstanbul: Yapa Publications (in Turkish).

Dönmez, B. N., Abidoğlu, Ü., Dinçer, Ç., Erdemir, N., & Gümüşçü, Ş. (1997). *Okul öncesi dönemde dil gelişimi etkinlikleri.* Ankara: Sim Yayıncılık.

Dowdell, K., Gray, T., & Malone, K. (2011). Nature and its influence on children's outdoor play. *Australian Journal of Outdoor Education, 15*(2), 24–35. doi:10.1007/BF03400925

Driscoll, K., & Pianta, R. C. (2011). Mothers' and fathers' perceptions of conflict and closeness in parent-child relationships during early childhood. *Journal of Early Childhood and Infant Psychology, 7*, 1–24.

Drossos, N. T. (2004). *Execute function and externalizing behavior problems in preschool children* [Doctoral Dissertation]. The Graduate College Of The Illinois Institute Of Technology.

Dudley, R. (2011). Overview of System Dynamics Modeling. *Earth01.Net*, 1–8. Retrieved from http://earth01.net/RG-Dudley/PDF/Dudley - Overview of System Dynamics Modeling.pdf

Duncan, G. J., Dowsett, C. J., Claessens, A., Magnuson, K., Huston, A. C., Klebanov, P., ... Japel, C. (2007). School readiness and later achievement. *Developmental Psychology, 43*(6), 1428–1446. doi:10.1037/0012-1649.43.6.1428 PMID:18020822

Duncan, G. J., Ziol-Guest, K. M., & Kalil, A. (2010). Early-childhood poverty and adult attainment, behavior, and health. *Child Development, 81*(1), 306–325. doi:10.1111/j.1467-8624.2009.01396.x PMID:20331669

Dunlap, L. D. (2009). *An introduction to early childhood special education.* Upper Saddle River, NJ: Pearson Merrill Prentice Hall.

Durand, V. M., & Mindell, J. A. (1990). Behavioral treatment of multiple childhood sleep disorders: Effects on child and family. *Behavior Modification*, *14*(1), 37–49. doi:10.1177/01454455900141003 PMID:2294900

Durgun, Ö. (2011). Türkiye'de yoksulluk ve çocuk yoksulluğu üzerine bir inceleme. *Bilgi Ekonomisi ve Yönetimi Dergisi*, *6*(1), 143–154.

Durmuşoğlu, G. (2009). Item Analysis and Reliability Of The Cognitive Abilities Scale-Second Edition: Turkish Version. Unpublished Doctoral Dissertation, Central Michigan University, Michigan.

Durmuşoğlu, M. C. (2018). Çocukta gelişim ve öğrenme. In F. Temel & N. Aral (Eds.), Gelişimsel Destek Programları (pp. 135-176). Ankara: Hedef CS Yayıncılık ve Mühendislik.

Duros, P., & Crowley, D. (2014). The body comes to therapy too. *Clinical Social Work Journal*, *42*(3), 237–246. doi:10.100710615-014-0486-1

Dursun, A. (2010). *Okul öncesi dönemdeki çocukların davranış problemleriyle anne- baba tutumları arasındaki ilişkinin incelenmesi*. Yayınlanmamış Yüksek Lisans Tezi, Dokuz Eylül Üniversitesi.

Dursunoğlu, H. (2013). Child and literature. In Ö. Yılar and L. Turan (Eds.), children's literature for faculty of education (pp. 27-30). Ankara: Pegem Academy Publications. (in Turkish)

Durualp, E. (2017). Doğum öncesi gelişim ve doğum. In A. Köksal-Akyol (Ed.), Erken çocukluk döneminde gelişim (pp. 114-149). Ankara: Anı.

Durualp, E. (2019). Doğum öncesi gelişim ve doğum. In A. Köksal Akyol (Ed.), Erken cocukluk döneminde gelişim-I (pp. 114-144). Ankara: Anı Yayıncılık.

Durualp, E. (2019). Doğum öncesi gelişim ve doğum. In A.K. Akyol (Ed.), Erken Çocukluk Döneminde Gelişim I (pp. 11-149). Ankara: Anı Yayıncılık.

Durualp, E. (2015). Çocukların ve ailelerin tanı, tedavi ve ameliyat sürecine hazırlanması. In A. Bütün-Ayhan (Ed.), *Hasta çocukların gelişimi ve eğitimi* (pp. 112–136). Eskişehir: Anadolu Üniversitesi.

Durualp, E. (2017). Doğum, doğum sonrası. In A. K. Akyol (Ed.), *Erken çocukluk döneminde gelişim I*. Ankara Anı Press.

Durualp, E., Kaytez, N., & Kadan, G. (2016). Okul öncesi dönemdeki çocukların annelerine verilen aile eğitiminin annelerin bilgi düzeyleri ile çocukların gelişimlerine olan etkisinin incelenmesi. *Abant İzzet Baysal Üniversitesi Eğitim Fakültesi Dergisi*, *16*(4), 1706–1722.

Duursma, E. (2014). Parental book reading practices among families in the Netherlands. *Journal of Early Childhood Literacy*, *14*(4), 435–458. doi:10.1177/1468798414523025

Duursma, E. (2016). Who does the reading, who the talking? Low-income fathers and mothers in the US interacting with their young children around a picture book. *First Language*, *36*(5), 465–484. doi:10.1177/0142723716648849

Dwyer, K. P., Osher, D., & Hoffman, C. C. (2000). Creating responsive schools: Contextualizing early warning, timely response. *Exceptional Children*, *66*(3), 347–365. doi:10.1177/001440290006600306

Dyment, J. E., & Bell, A. C. (2008). Grounds for movement: Green school grounds as sites for promoting physical activity. *Health Education Research*, *23*(6), 952–962. doi:10.1093/her/cym059 PMID:17956885

Eccles, J. S. (1999). The development of children ages 6 to 14. *The Future of Children*, *9*(2), 30–44. doi:10.2307/1602703 PMID:10646256

Edmond, K. M., Strobel, N. A., Adams, C., & McAullay, D. (2019). Effect of early childhood development interventions implemented by primary care providers commencing in the neonatal period to improve cognitive outcomes in children aged 0–23 months: Protocol for a systematic review and meta-analysis. *Systematic Reviews*, *8*(1), 224. doi:10.118613643-019-1142-1 PMID:31470911

Education Encyclopedia. (n.d.). Stages of growth in child development. Retrieved from http://education.stateuniversity.com/pages/1826/Child-Development-Stages-Growth.html#ixzz0j0jMHgRB

Efe, E. (2013). *Üniversite Öğrencilerinin Eş Seçme Eğilimleri*. Hacettepe Üniversitesi Sosyal Bilimler Enstitüsü Yüksek Lisans Tezi.

Efe, H., & Uluoğlu, S. A. (2015). Dünyada çocuk işçiliğiyle mücadelede gelinen nokta ve geleceğe dair bazı öngörüler. *Eğitim Bilim Toplum Dergisi*, *13*(51), 46–72.

Ege, P. (2006). Language and concept development. In S. Topbaş (Ed.), Literacy development in childhood (pp. 140-157). Ankara: Kök Publications. (In Turkish).

Ehud, M., An, B. D., & Avshalom, S. (2010). Here and now: Yoga in Israeli schools. *International Journal of Yoga*, *3*(2), 42–47. doi:10.4103/0973-6131.72629 PMID:21170229

Ekmekçi, Ö. (1995). *Psikodilbilim*. Çukurova Üniversitesi, Yadim.

Elibol, F. (2016). 0-18 yaş çocuk gelişimi ve gelişim ilkeleri ile çocuk gelişiminde tanı ve değerlendirmenin yeri, önemi ve amaçları. In Ç. Dinçer (Ed.), Gelişimsel Tanı ve Değerlendirme Yöntemleri (pp. 26-58). Eskişehir: AÖF Yayınları.

Elibol, F. (2007). *12-36 aylar arasında çocukları olan annelere verilen grup eğitiminin annelerin anne-babalık görevlerinde öz yeterliliklerine yönelik katkısının incelenmesi, Yayınlanmamış yüksek lisans tezi*. Ankara: Hacettepe Üniversitesi.

Ellis, N. C. (2019). Essentials of a theory of language cognition. *Modern Language Journal*, *103*, 39–60. doi:10.1111/modl.12532

Elmlinger, S. L., Schwade, J. A., & Goldstein, M. H. (2019). The ecology of prelinguistic vocal learning: Parents simplify the structure of their speech in response to babbling. *Journal of Child Language*, *46*(5), 998–1011. doi:10.1017/S0305000919000291 PMID:31307565

Emery, R. C. (2009). Stay for the children? Husband violence, marital stability, and children's behavior problems. *Journal of Marriage and the Family*, *71*(4), 905–916. doi:10.1111/j.1741-3737.2009.00643.x

Emmorey, K. (2001). *Language, cognition, and the brain: Insights from sign language research*. USA: Psychology Press. doi:10.4324/9781410603982

Enderby, P., Judge, S., Creer, S., & John, A. (2013). Examining the need for and provision of AAC methods in the UK. *Advances in Clinical Neuroscience & Rehabilitation*, *13*, 20-23. Retrieved from https://www.thecommunicationtrust.org.uk/media/3414/other_ways_of_speaking_final.pdf

Erdem, A. Y. & Bilge, F. (2016). Lise öğrencileri için meslek seçimine ilişkin akılcı olmayan inançlar ölçeği'nin geliştirilmesi. *Türk Psikolojik Danışma ve Rehberlik Dergisi*, *3*(29), 95-114

Erden, M. & Akman, Y. (2004). Gelişim ve öğrenme. Ankara: Arkadaş Yayınları.

Erdener, E. (2009). Views of Vygotsky on language and thought; A criticism on Vygotsky. *Türk Eğitim Bilimleri Dergisi*, *7*(1), 85–103. (in Turkish)

Erden, M., & Akman, Y. (1998). *Gelişim-öğrenme-öğretme: eğitim psikolojisi*. Ankara: Arkadaş.

Erden, M., & Akman, Y. (2006). *Eğitim psikolojisi. Gelişim-öğrenme-öğretme. Geliştirilmiş on beşinci baskı* (15th ed.) Ankara: Arkadaş Yayınevi.

Erden, M., & Akman, Y. (2009). *Educational Psychology*. Ankara: Arkadaş Publications. (in Turkish)

Erdoğan, N. I., Atan, A., Asar, H., Mumcular, F., Yüce, A., Kıraç, M., & Kilimlioğlu, Ç. (2016). Ebeveyn ve öğretmenlerin birlikte hikâye okuma etkinliklerinin incelenmesi. *İlköğretim Online, 15*(1). doi:10.17051/io.2016.05211

Erermiş, S. & Aydın, C. (1999). Çocuk ruh sağlığı yönünden enürezis. In A. Ekşi (Ed.), Ben hasta değilim. İstanbul: Nobel Tıp Kitabevi Ltd. Şti.

Ereş, F. (2009). Tolumsal bir sorun: Suçlu çocuklar ve ailenin önemi. *Aile ve Toplum Eğitim Kültür ve Araştırma Dergisi, 17*, 88–96.

Ergenç, İ. (1983). Yabancı dil öğretimi ve olumsuz aktarım. *Türk Dili Aylık Dil ve Yazın Dergisi, Dil Öğretimi Özel Sayısı, 32*, 379–380.

Ergin, H., Yıldız, A., Erdoğan, G., Mercan, S.Ç., Çelikkol, K.A. … Gazioğlu, G., E. (2010). Gelişim psikolojisi. Academic Press.

Ergül, C., Karaman, G., Akoğlu, G., Tufan, M., Sarıca, A. D., & Kudret, Z. B. (2014). Okul öncesi öğretmenlerinin "Erken Okuryazarlık" kavramına ilişkin bilgi düzeyleri ve sınıf uygulamaları. *İlköğretim Online, 13*(4). doi:10.17051/io.2014.71858

Ergül, C. (2014). Sık rastlanan yetersizlikler. In İ. H. Diken (Ed.), *Erken çocukluk eğitimi* (pp. 258–262). Ankara: Pegem.

Ergül, C., Akoğlu, G., Sarica, A., Tufan, M., & Karaman, G. (2015). Ana sınıflarında gerçekleştirilen birlikte kitap okuma etkinliklerinin" etkileşimli kitap okuma" bağlamında incelenmesi. *Mersin Üniversitesi Eğitim Fakültesi Dergisi, 11*(3). doi:10.17860/efd.88429

Ergül, C., Sarıca, A. D., & Akoğlu, G. (2016). Etkileşimli kitap okuma: Dil ve erken okuryazarlık becerilerinin geliştirilmesinde etkili bir yöntem. *Ankara Üniversitesi Eğitim Bilimleri Fakültesi Özel Eğitim Dergisi, 17*(2), 193–206. doi:10.21565/ozelegitimdergisi.246307

Ergül, Ş. & Kalkım, A. (. (2011). Önemli bir kronik hastalık: Çocukluk ve ergenlik döneminde obezite. *Turk Silahli Kuvvetleri Koruyucu Hekimlik Bulteni, 10*(2), 223–230. doi:10.5455/pmb.20110119104521

Eripek, S. (2002). Uyumsuz çocuklar ve eğitimleri. Y. Özsoy, M. Özyürek ve S. Eripek (Eds.), Özel eğitime muhtaç çocuklar özel eğitime giriş (pp. 199-223). Ankara: Karatepe Yayınları.

Eripek, S., & Vuran, S. (2010). Education of children with intellectual disabilities. In G. Akçamete (Ed.), Students with special needs in special education schools and special education (pp. 245-282). Ankara: Kök Publishing.

Eripek, S. (1992). Disabled and their education. *Anadolu University Journal of Education Faculty, 3*(1-2), 79–84.

Eripek, S., & Vuran, S. (2008). *Education of children with intellectual disabilities. In G. Akçamete (Ed.), Special education* (pp. 245–278). Ankara: Kök Publishing.

Erkan, R. & Erdoğdu, M. Y. (2006). Göç ve çocuk suçluluğu. *Aile ve Toplum, 3*(9), 79-90.

Erkan, S. (2010). Aile ve aile eğitimi ile ilgili temel kavramlar. In Z.F. Temel (Ed.), Aile Eğitimi ve Erken Çocukluk Eğitiminde Aile Katılımı (pp. 3-43). Ankara: Anı Yayıncılık.

Er, M. (2006). Çocuk, hastalık, anne-babalar ve kardeşler. *Çocuk Sagligi ve Hastaliklari Dergisi, 49*, 155–168.

Er, S. (2016). The importance of parents' interactive story reading. *Başkent University Journal of Education, 2*(3), 156–160.

Er, S., Aral, N., & Yıldız Bıçakçı, M. (2013). Identifying the relationship between children's language skills and parents' story reading self-efficacy. *Procedia: Social and Behavioral Sciences*, *93*, 1269–1274. doi:10.1016/j.sbspro.2013.10.027

Er-Sabuncuoğlu, M. (2015). Genetik, çevre ve gelişim. In F. Turan & A.İ. Yükselen (Eds.), Her Yönüyle Okul Öncesi Eğitim: Çocuk Gelişimi 1 (pp. 73-98). Ankara: Hedef CS Yayıncılık ve Mühendislik.

Ersay, E. (2013). Aile eğitimi programlarının planlanması ve değerlendirilmesi. In F. Temel (Ed.), Aile eğitimi ve erken çocukluk eğitiminde aile katılım çalışmaları (pp. 274-324). Ankara: Anı yayıncılık.

Ersoy, Ö. & Avcı, N. (2001). Özel gereksinimi olan çocuklar ve eğitimleri-özel eğitim. İstanbul: Ya-Pa Yayınları.

Eryılmaz, A. (2011). Yaşam boyu gelişim yaklaşımı. *Psikiyatride Güncel Yaklaşımlar-Current Approaches in Psychiatry*, *3*(1), 49–66.

Evirgen, Ş. (2002). *Okul öncesi eğitim çerçevesinde okul destekli anne eğitim programının anneler üzerindeki etkilerinin incelenmesi, Yayınlanmamış yüksek lisans tezi.* İstanbul: Marmara Üniversitesi.

Ewing, R., & Maher, M. (2014). *Phonics: its place in the literacy story.* ALEA. Retrieved from https://www.alea.edu.au/documents/item/943

Fabes, R., & Martin, C. (2000). *Exploring Child Development* (pp. 175–195). Boston: Pearson Education, Inc.

Fair, D. & Schlaggar, B. L. (2008). Brain development. In M.M. Haith, J.B Benson (Eds.), Encyclopedia of infantvand early childhood development. Oxford, UK: Elsevier.

Family Strengthening Policy Center. (2007). *Home visiting: Strengthening families by promoting parenting success. National human services assembly*, Policy Brief, No. 23.

Fazlıoğlu, Y. (2009). *Erken çocukluk gelişim ve eğitimi. 1.* Baskı, Krite.

Fazlıoğlu, Y. (2017). Sosyal gelişim. In N. Aral & T. Duman (Eds.), *Eğitim psikolojisi* (pp. 154–176). Ankara: Pegem. doi:10.14527/9786052410240.07

Feldman, R. S. (2017). *Development across the life span* (8th ed.). Pearson.

Fenichel, E., & Mann, T. L. (2001). Early Head Start for low-income families with infants and toddlers. *The Future of Children*, *11*(1), 135–141. doi:10.2307/1602817 PMID:11712450

Fernandez, E. M., & Smith Cairns, H. (2011). *Fundamentals of psycholinguistics*. Malden, MA: WileyBlackwell.

Ferrer-Mico, T., Prats-Fernàndezb, M. A., & Redo-Sanchez, A. (2012). Impact of Scratch programming on students' understanding of their own learning process. *Procedia: Social and Behavioral Sciences*, *46*, 1219–1223. doi:10.1016/j.sbspro.2012.05.278

Feuerstein, R. S., Hoffman, M. B., Rand, Y., Jensen, M. R., Tzuriel, D., & Hoffmann, D. B. (1985). Learning To Learn: Mediated Learning Experiences And Instrumental Enrichment. *Special Services in the Schools*, *3*(1-2), 49–82. doi:10.1300/J008v03n01_05

Fields, M.V., & Boesser, C. (2002). *Constructive guidance and discipline: Preschool and primary education.* Upper Saddle River, NJ: Merrill Prentice Hall.

Field, T. (2011). Yoga clinical research review. *Complementary Therapies in Clinical Practice*, *17*(1), 1–8. doi:10.1016/j.ctcp.2010.09.007 PMID:21168106

Figlio, D., Guryan, J., Karbownik, K., & Roth, J. (2014). The effects of poor neonatal health on children's cognitive development? *The American Economic Review*, *104*(12), 3921–3955. doi:10.1257/aer.104.12.3921 PMID:29533575

Filburn, A. G., Lahiri, T., & Ren, C. L. (2016). *Handbook o cystic fibrozis*. Springer. doi:10.1007/978-3-319-32504-0

Flavell, J. H., Millir, P. H., & Millir, S. S. (1993). Cognitive development. New Jersey: Prentice-Hall.

Fonagy, P. (2001). *Introduction to attachment theory. In attachment theory and psychoanalysis*. New York: Other Press.

Forcada-Guex, M., Pierrehumbert, B., Borghini, A., Moessinger, A., & Muller-Nix, C. (2006). Early dyadic patterns of mother-infant interactions and outcomes of prematurity at 18 months. *Pediatrics*, *118*(1), 107–114. doi:10.1542/peds.2005-1145 PMID:16818525

Foster, R., & Park, J. (2012). An integrative review of literature examining psychometric properties of instruments measuring anxiety or fear in hospitalized children. *Pain Management Nursing*, *13*(2), 94–106. doi:10.1016/j.pmn.2011.06.006 PMID:22652282

Foster, S. H. (1990). *The communicative competence of young children: A modular approach*. London: Longman.

Freeman, J. L. (2007). Using a balanced literacy approach in elementary schools. (Master Dissertation). Available from UNI Scholar Works. (UMI No. 690)

Friberg, I. K., Kinney, M. V., Lawn, J. E., Kerber, K. J., Odubanjo, M. O., Bergh, A. M., ... Black, R. E. (2010). Sub-Saharan Africa's mothers, newborns, and children: How many lives could be saved with targeted health interventions? *PLoS Medicine*, *7*(6), e1000295. doi:10.1371/journal.pmed.1000295 PMID:20574515

Froiland, J. M., Powell, D. R., Diamond, K. E., & Son, S. H. C. (2013). Neighborhood socioeconomic well-being, home literacy, and early literacy skills of at-risk preschoolers. *Psychology in the Schools*, *50*(8), 755–769. doi:10.1002/pits.21711

Galantino, M., Galbavy, R., & Quinn, L. (2008). Therapeutic effects of yoga for children: A systematic review of the literature. *Pediatric Physical Therapy*, *20*(1), 66–80. doi:10.1097/PEP.0b013e31815f1208 PMID:18300936

Galda, L., Cullinan, B. E., & Strickland, D. S. (1997). Language, literacy and child. Orlando: Harcourt Brace & Company Puplishing.

Gallahue, D. L., Ozmun, J. C., & Goodway, J. D. (2014). Motor gelişimi anlamak, bebekler, çocuklar, ergenler ve yetişkinler, (D. Sevimay Özer & A. Aktop, Trans.). Ankara: Nobel Yayıncılık.

Gallahue, D. L., Ozmun, J. C., & Goodway, J. D. (2014). *Motor gelişimi anlamak. Bebekler, çocuklar, ergenler, yetişkinler.* Ankara: Nobel.

Gamez, P. B., Griskell, H. L., Sobrevilla, Y. N., & Vazquez, M. (2019). Dual Language and English-Only Learners' Expressive and Receptive Language Skills and Exposure to Peers' Language. *Child Development*, *90*(2), 471–479. doi:10.1111/cdev.13197 PMID:30566240

Gander, M. J. & Gardiner, H. W. (2004). Child and Adolescent Development. Ankara: İmge Kitabevi Yayınları. (in Turkish)

Gander, M. J., & Gardiner, H. W. (2004). Çccuk ve ergen gelişimi. Ankara: İmge Kitabevi.

Gander, M. J., & Gardiner, H. W. (1993). *Çocuk ve Ergen Gelişimi* (B. Onur, Ed.). Ankara: İmge Yayınevi.

Gander, M. J., & Gardiner, H. W. (1995). *Çocuk ve Ergen Gelisimi*. Ankara: İmge Kitabevi.

Gander, M. J., & Gardiner, H. W. (2010). *Çocuk ve ergen gelişimi*. Baskı, İmge.

Gander, M. J., & Gardiner, H. W. (2015). *Çocuk ve ergen gelişimi.* Ankara: İmge.

Ganpa, T. & Nagendra, H. (2011). Yoga for children. *TANG Humanitas Med.*, *1*, 1–4.

Garces, E., Thomas, D., & Currie, J. (2002). Longer-term effects of Head Start. *The American Economic Review, 92*(4), 999–1012. doi:10.1257/00028280260344560

Garcia, O. (2011). *Bilingual education in the 21st century: A global perspective.* John Wiley & Sons.

Gardner, F. & Shaw, D. S. (2008). Behavioral problems of infancy and preschool children (0–5). In M. Rutter, D. Bishop, D. Pine et al. (Eds.), Rutter's child and adolescent psychiatry (5th ed., pp. 882-894). London: Blackwell Press.

Gard, T., Noggle, J. J., Park, C. L., Vago, D. R., & Wilson, A. (2014). Potential self-regulatory mechanisms of yoga for psychological health. *Frontiers in Human Neuroscience, 8*(770), 1–20. PMID:25368562

Garfinkel, M., & Schumacher, H. R. Jr. (2000). Yoga. *Rheumatic Diseases Clinics of North America, 26*(1), 125–132. doi:10.1016/S0889-857X(05)70126-5 PMID:10680200

Gass, S. M., & Selinker, L. (2008). *Second Language Acquisition: An Introductory Course.* United States: Routledge. doi:10.4324/9780203932841

Gauavin, M., & Parke, R. D. (2010). Socialization. In M. H. Bornstein (Ed.), *Handbook of cultural development science (pp. 239-258).* New York: Psychology Press.

Genç, A. (2018). *Okul öncesi ve ilkokul döneminde olan işitme kayıplı çocukların davranış problemlerinin incelenmesi.* Yayınlanmamış Yüksek Lisans Tezi, Anadolu Üniversitesi.

Gencer, T. E. (2017). Göç ve eğitim ilişkisi üzerine bir değerlendirme: Suriyeli çocukların eğitim gereksinimi ve okullaşma süreçlerinde karşılaştıkları güçlükler. *Uluslararası Sosyal Araştırmalar, 10*(54), 839–851.

Gesell, A., Halverson, H. M., Thomson, H., Ilg, F. L., Castner, B. M., Ames, L. B., & Amatruda, C. S. (1940). *The first five years of life, the preschool years.* Oxford, England: Harpers.

Ghaffarzadegan, N., Lyneis, J., & Richardson, G. P. (2011). How small system dynamics models can help the public policy process. *System Dynamics Review, 27*, 22–44. doi:10.1002dr.442

Gibbs, L.J. & Earley, E.J. (1994). *Using children's literature to develop core values.* Bloomington, IN: Phi Delta Kappa Educational Foundation. Retrieval https://files.eric.ed.gov/fulltext/ED366992.pdf

Gilkerson, J., Richards, J. A., & Topping, K. J. (2017). The impact of book reading in the early years on parent–child language interaction. *Journal of Early Childhood Literacy, 17*(1), 92–110. doi:10.1177/1468798415608907

Gill, P. (2004). *Halk Kütüphanesi Hizmeti: Gelişim için IFLA/UNESCO İlkeleri* (B. Yılmaz, Trans.). İstanbul, Turkey: Pandora. (Original work published 2001)

Gleason, J. B., & Ratner, N. B. (2005). *The development of language.* Pearson/Allyn and Bacon.

Gökçay, G., Köklük, S., Kayadibi, F., Eraslan, E., & Çalışkan, M. (2000). Çocuklarda ilk iki yılda gelişimi etkileyen faktörler. *Istanbul Tip Fakültesi Mecmuasi, 63*(4), 395- 405.

Goldin, P. R., Morrison, A., Jazaieri, H., Brozovich, F., Heimberg, R., & Gross, J. J. (2016). Group CBT versus MBSR for social anxiety disorder: A randomized controlled trial. *Journal of Consulting and Clinical Psychology, 84*(5), 427–437. doi:10.1037/ccp0000092 PMID:26950097

Goldin, P. R., Ziv, M., Jazaieri, H., Hahn, K., & Gross, J. J. (2013). MBSR versus aerobic exercise in social anxiety: fMRI of emotion regulation of negative self-beliefs. *Social Cognitive and Affective Neuroscience, 8*(1), 65–72. doi:10.1093can/nss054 PMID:22586252

Goldstein, M. H., King, A. P., & West, M. J. (2003). Social Interaction Shapes Babbling: Testing Parallels between Birdsong and Speech. *Proceedings of the National Academy of Sciences of the United States of America*, *100*(13), 8030–8035. doi:10.1073/pnas.1332441100 PMID:12808137

Gomby, D. S., Culross, P. L., & Behrman, R. E. (1999). Home visiting: Recent program evaluations: Analysis and recommendations. *The Future of Children*, *9*(1), 4–26. doi:10.2307/1602719 PMID:10414008

Gonen, M. (2000). Investigation of thesis on last 10 years' children's books about the content, graphics, physical properties. Ankara University, Ankara. (In Turkish)

Gönener, D., & Görak, G. (2009). Okul yaş grubu çocukların hastane ve hastalığı ile ilgili Bilgilendirme Durumlarının Endişe Kaynakları ile Etkileşimi. *Gaziantep Tıp Dergisi*, *15*(1), 41–48.

Gonen, M. & Arı, M. (1989). Investigation of the effects of picture books on language development of 4-5 year old children. *Education in Science*, *13*(72), 21–30. (in Turkish)

Gönen, M., Durmuşoğlu, M., & Severcan, S. (2009). Examining the views of the preschool education teachers on the content, illustrations and physical characteristics of the picture story books used in education. *Procedia: Social and Behavioral Sciences*, *1*(1), 753–759. doi:10.1016/j.sbspro.2009.01.133

Gonzalez, J. E., Taylor, A. B., Davis, M. J., & Kim, M. (2013). Exploring the underlying factor structure of the Parent Reading Belief Inventory (PRBI): Some caveats. *Early Education and Development*, *24*(2), 123–137. doi:10.1080/10409289.2011.623651

Gordon, T. (2014). *Etkili anne baba eğitimi.* İstanbul: Profil.

Gove, A., & Black, M. M. (2016). Measurement of Early Childhood Development and Learning under the Sustainable Development Goals. *Journal of Human Development and Capabilities*, *17*(4), 599–605. doi:10.1080/19452829.2016.1243520

Govindara, R., Karmani, S., Varambally, S., & Gangadhar, B.N. (2016). Yoga and physical exercise –A review and comparison. *International Review of Psychiatry*, *28*(3), 242–253.

Gözün Kahraman, Ö., & Yuvacı, M. (2017). Baba katılımı. In A.B. Aksoy (Ed.), Aile eğitimi ve katılımı (pp. 197-234). Ankara: Hedef CS Basın Yayın.

Grant, N., Strevens, H., & Thornton, J. (2015). Physiology of labor. In *Epidural Labor Analgesia* (pp. 1–10). Cham: Springer.

Greenfield, C. (2007). A case study of children's and adults' perceptions of 'being outside' in one early childhood centre. Unpublished Qualitative Masters of Education Thesis, Massey University, Palmerston North, New Zealand.

Greenfield, P. M., & Bruner, J. S. (1966). Culture and Cognitive Growth. *International Journal of Psychology*, *1*(2), 89–107. doi:10.1080/00207596608247117

Greenman, J. (2005). Caring spaces, learning places, Environments that work. Redmond, WA: Exchange Press.

Greenman, J., & Stonehouse, A. (1996). *Prime times a handbook forexcellence in infant and toddler care*. Redleaf Press.

Grossman, P., Niemann, L., Schmidt, S., & Walach, H. (2004). Mindfulness-based stress reduction and health benefits: A meta-analysis. *Journal of Psychosomatic Research*, *57*(1), 35–43. doi:10.1016/S0022-3999(03)00573-7 PMID:15256293

Gui, L. (2019). Research on Teaching Chinese as a Foreign Language from the Perspective of Ecological Linguistics. In *Proceedings of the 5th International Workshop on Education, Development and Social Sciences (IWEDSS 2019)*. Academic Press. Retrieved from https://webofproceedings.org/proceedings_series/ESSP/IWEDSS%202019/IWEDSS19334.pdf

Güler, T. (2010). Ailenin tanımı, tarihsel gelişimi ve aile çeşitleri. In T. Güler (Ed.), Anne baba eğitimi (pp. 1-14). Ankara: Pegem Akademi.

Güler, G., Güler, N., Kocataş, S., & Akgül, N. (2009). Yaşlıların sağlık bakım gereksinimleri. *Cumhuriyet Medical Journal, 31*(4), 367–373.

Güler, N., Uzun, S., Boztas, Z., & Aydoğan, S. (2002). Anneleri tarafından çocuklara uygulanan duygusal ve fiziksel istismar/ihmal davranışı ve bunu etkileyen faktörler. *Cumhuriyet Üniversitesi Tıp Fakültesi Dergisi, 24*(3), 128–134.

Gültekin Akduman, G., & Baran, G. (2011). Suça karışan ve karışmayan 12-15 yaş grubundaki çocuklarda akran istismarının çeşitli ailesel özelliklerle ilişkisi. *Anadolu Psikiyatri Dergisi, 12*, 76–83.

Gültekin-Akduman, G., Günindi, Y., & Türkoğlu, D. (2015). Okul öncesi dönem çocuklarının sosyal beceri düzeyleri ile davranış problemleri arasındaki ilişkinin incelenmesi. *Uluslararası Sosyal Araştırmalar Dergisi, 8*(37), 673–683.

Gültekin, G., & Baran, G. (2005). Hastalık ve çocuk. *Aile ve Toplum, 7*(2), 61–68.

Gültekin, G., & Baran, G. (2013). 9-14 yaş grubu akut ve kronik hastalığı olan çocukların denetim odağı düzeylerinin incelenmesi. *Türk Pediatri Arşivi, 40*, 211–220.

Gültekin, M. (2006). *Okul öncesinde duygu ve davranış sorunları*. İstanbul: Nesil.

Gündoğdu, R. (2013). Öğrenmenin doğası ve temel kavramlar. In B. Çapri & B. Gündüz (Eds.), Eğitim psikolojisi (pp. 169-192). Adana: Karahan Kitabevi.

Gündüz, C., Çoğulu, Ö., Özkınay, C., Tütüncüoğlu, S., Sapmaz, G., Kösem, M., & Özkınay, F. (2000). İdiyopatik mental retardasyonda Fragil X sendromu araştırması. *Ege Tıp Dergisi, 39*(2), 97–104.

Güneri Yöyen, E. (2001). Şiddet türleri ve kişilik özellikleri. *Yaşam Becerileri Psikoloji Dergisi, 1*(1), 35–50. doi:10.31461/ybpd.316724

Güngör Aytar, A. (2015). Toplumsal ve Duygusal Gelişim. In A. Ulusoy (Ed.), Gelişim ve Öğrenme Psikolojisi (pp. 93-126). Ankara: Anı.

Güngör Aytar, A., & Kurtoğlu Karataş, B. (2018). Gelişim psikolojisine giriş. In A. Güngör Aytar (Ed.), Gelişim psikolojisi (pp. 12-44). Ankara: Hedef CS BasınYayın.

Güngör, M. (2009). Eğitim hakkı ve sokakta yaşayan/çalışan çocuklar: Mersin ili örneği. *Mersin Üniversitesi Eğitim Fakültesi Dergisi, 5*(1), 28–42.

Güngör, M. (2013). Risk altındaki çocukların aile yapıları ve suça yönelimleri (Mersin İli Örneği). *Mersin Üniversitesi Eğitim Fakültesi Dergisi, 9*(2), 421–434.

Günşen İçli, T. (2009). *Çocuk Suç ve Sokak*. Başbakanlık Aile ve Sosyal Araştırmalar Genel Müdürlüğü.

Gunter, P. L., Coutinho, M. J., & Cade, T. (2002). Classroom factors linked with academic gains among students with emotional and behavioral problems. *Preventing School Failure, 46*(3), 126–132. doi:10.1080/10459880209603357

Guralnick, M. J. (2011). Why Early İntervention Works: A Systems Perspective. *Infants and Young Children, 24*(1), 6–28. doi:10.1097/IYC.0b013e3182002cfe PMID:21532932

Gür, S. B., & Kurt, T. (2011). Türkiye'de ailelerin eğitim ihtiyaçları. *Aile ve Toplum, 7*(27), 33–62.

Gürsoy, F. & Özaslan, H. (2017). Çocuk gelişimine giriş. In N. Aral (Ed.), Çocuk gelişimi. Eskişehir: Açık Öğretim Fakültesi Yayını.

Gürsoy, F., & Özaslan, H. (2016). Çocuk gelişimine giriş. In N. Aral (Ed.), Çocuk gelişimi (pp. 2-24). Eskişehir AÖF Yayınları.

Gürsoy, F., Aral, N., Yıldız Bıçakçı, M. (2005). Yetiştirme yurdunda kalan ve kalmayan ergenlerin benlik tasarım düzeylerinin incelenmesi. *Eğitim Araştırmaları Dergisi*, 24-34

Hadadian, A., & Koch, K. R. (2013). Issues in Labeling Young Children with Developmental Delay: Whose Responsibility is it? *International Journal of Early Childhood Special Education*, *5*(2), 187–199.

Hagen, I., & Nayar, U. S. (2014). Yoga for children and young people's mental health and well-being: Research review and reflections on the mental health potentials of yoga. *Frontiers in Psychiatry*, *5*(35), 1–6. PMID:24765080

Hailemariam, A. (2004). Review Of Cognitive Abilities Scale-Second Edition. *Journal of School Psychology*, *42*(2), 171–176. doi:10.1016/j.jsp.2003.10.002

Hale, L., & Guan, S. (2015). Screen time and sleep among school-aged children and adolescents: A systematic literature review. *Sleep Medicine Reviews*, *21*, 50–58. doi:10.1016/j.smrv.2014.07.007 PMID:25193149

Halfon, H., Forrest, C. B., Lerner, R. M., & Faustman, E. M. (2018). *Handbook of Life Course Health Development*. Springer. doi:10.1007/978-3-319-47143-3

Halfon, N. (2009). *Life course health development: A new approach for addressing upstream determinants of health and spending*. Washington: Expert Voices, National Institute for Health Care Management Foundation. Retrieved from http://www.nihcm.org/pdf/ExpertVoices_Halfon_FINAL.pdf

Halıcı, P. (2004). Çocuklarda yabancı dil eğitimi [Doktora semineri]. Ankara Üniversitesi.

Hamamcı, Z. & Hamurlu, M. (2005). Anne babaların meslek gelişimine yardımcı olmaya yönelik tutumları ve bilgi düzeylerinin çocuklarının mesleki kararsızlıkları ile ilişkisi. *İnönü Üniversitesi Eğitim Fakültesi Dergisi*, *6*(10), 55-69.

Hammer, C. S., Farkas, G., & Macguza, S. (2010). Language and Literacy Development of Head Start Children: A Study Using the Family and Child Experiences Survey Database. *Language, Speech, and Hearing Services in Schools*, *41*(1), 70–83. doi:10.1044/0161-1461(2009/08-0050) PMID:19948772

Hargie, O., & Dickson, D. (2004). *Skilled Interpersonal Communication Research, Theory and Practice* (4th ed.). London: Routledge Publishing. doi:10.4324/9780203427880

Hariprasad, V. R., Arasappa, R., Varambally, S., Srinath, S., & Gangadhar, B. N. (2013). Feasibility and efficacy of yoga as an add-on intervention in attention deficit-hyperactivity disorder: An exploratory study. *Indian Journal of Psychiatry*, *55*(3), 379–384. PMID:24049203

Harmony Yoga (n.d.). *Baby Yoga*. Retrieved from http://www.harmonyyogaireland.com/index.php/classes/baby-yoga

Harris, R. J. (1992). *Cognitive processes in bilingual*. Amsterdam, Netherland: Elsevier Science Publishers

Harris, S. S. (2000). Readiness to participate in sports. In J. A. Sullivan & S. J. Anderson (Eds.), *Care of the Young Athlete* (pp. 19–34). Rosemont, IL: American Academy of Orthopedic Surgeons and American Academy of Pediatrics.

Harunoğulları, M. (2016). Suriyeli sığınmacı çocuk işçiler ve sorunları: Kilis örneği. *Göç Dergisi*, *3*(1), 29–63.

Hatun, Ş. (2002). Çocuk hakları sözleşmesinin 13. yüzyılında yoksulluk ve çocuklar üzerine etkileri. *Türk Tabipleri Birliği Merkez Konseyi:5-37*. Retrieved from http://www.ttb.org.tr/eweb/yoksulluk_ve_cocuklar.pdf

Haugen, E. I. (1966). *Language conflict and language planning: The case of modern Norwegian*. Harvard University Press. doi:10.4159/harvard.9780674498709

Hawkins, B. L., Stegall, J. B., Weber, M. F., & Ryan, J. B. (2012). The influence of a yoga exercise program for young adults with intellectual disabilities. *International Journal of Yoga*, *5*(2), 151–156. doi:10.4103/0973-6131.98244 PMID:22870001

Heckman, J. J. (2012). The Heckman Equation Invest in early childhood development : Reduce deficits, strengthen the economy . The Heckman Equation Make greater investments in young children to see greater returns in education, health and productivity. *The Heckman Equation*, 1–2.

Hendon, C., & Bohan, L. M. (2008). Hospitalized children's mood differences during play and music therapy. *Child: Care, Health and Development*, *34*(2), 141–144. doi:10.1111/j.1365-2214.2007.00746.x PMID:18257785

Hengirmen, M. (2001). *Yabancı dil öğretim yöntemleri ve tömer yöntemi.* Ankara: Engin Yayın Evi.

Henry, G. T., Gordon, C. S., & Rickman, D. K. (2006). Early education policy alternatives: Comparing quality and outcomes of Head Start and state prekindergarten. *Educational Evaluation and Policy Analysis*, *28*(1), 77–99. doi:10.3102/01623737028001077

Heppner, P. P., & Krauskorf, C. J. (1987). The integration of personal problem solving process within counseling. *The Counseling Psychologist*, *15*, 371–447. doi:10.1177/0011000087153001

Hernandez, D. J. (2011). *Double Jeopardy: How Third-Grade Reading Skills and Poverty Influence High School Graduation.* Retrieved from https://eric.ed.gov/?id=ED518818

Hess, Y. S., Lonnerdal, B., Hatz, C., Rivers, A. J., & Brown, H. K. (2009). Recent advances in knowledge of zinc nutritional. *Human Health Food Bulletin*, *30*(1), 5–11.

Hindman, A. H., Connor, C. M., Jewkes, A. M., & Morrison, F. J. (2008). Untangling the effects of shared book reading: Multiple factors and their associations with preschool literacy outcomes. *Early Childhood Research Quarterly*, *23*(3), 330–350. doi:10.1016/j.ecresq.2008.01.005

Hindman, A. H., Skibbe, L. E., & Foster, T. D. (2014). Exploring the variety of parental talk during shared book reading and its contributions to preschool language and literacy: Evidence from the Early Childhood Longitudinal Study-Birth Cohort. *Reading and Writing*, *27*(2), 287–313. doi:10.100711145-013-9445-4

Hjorth, P., & Bagheri, A. (2006). Navigating towards sustainable development: A system dynamics approach. *Futures*, *38*(1), 74–92. doi:10.1016/j.futures.2005.04.005

Höbek Akarsu. R. (2016). *The effect of pregnancy yoga on the pregnant's psychosocial health and prenatal attachment* [doctoral dissertation]. University of Istanbul, Istanbul.

Hodnett, E. D., Downe, S., & Walsh, D. (2012). Alternative versus conventional institutional settings for birth. *Cochrane Database of Systematic Reviews*, (8): CD000012. doi:10.1002/14651858.CD000012.pub4 PMID:22895914

Holzman, L. (2009). *Vygotsky at work and play.* Clifton, NJ: Psychology Press.

Homer, J. B., & Hirsch, G. B. (2006). System dynamics modeling for public health: Background and opportunities. *American Journal of Public Health*, *96*(3), 452–458. doi:10.2105/AJPH.2005.062059 PMID:16449591

Hopwood Stephens. (2015). Outdoor Play for 1–3 Year Olds. New York: Routledge.

Houser, M. E. (2015). *A Mixed Methods Evaluation of the Effectiveness of a Group Yoga Intervention as an Adjunctive Trauma Therapy for Adolescent Girls* [doctoral dissertation]. University of Denver, Denver, CO.

Hovardaoğlu, S. (2007). *Davranış Bilimleri İçin Araştırma Teknikleri.* Ankara: Hatiboğlu Yayınevi.

Howard, V. F., Williams, B., & Lapper, E. C. (2011). *Özel Gereksimi Olan Küçük Çocuklar*. Ankara: Nobel Yayınevi.

Howard, V. F., Williams, B., & Lepper, C. E. (2011). *Özel gereksinimi olan küçük çocuklar*. Ankara: Nobel.

Hughes, E. (2003). Learners with emotional or behavioral disorders. In D. R. Hallahan & J. M. Kauffman (Eds.), *Exceptional learners introduction to special education* (9th ed., pp. 221–261). Boston: Allyn & Bacon.

Hullinger-Sirken, H., & Staley, L. (2016). Preschool Through Grade 3: Understanding Writing Development: Catie's Continuum. *YC Young Children, 71*(5), 74–78. Retrieved from http://www.jstor.org/stable/ycyoungchildren.71.5.74

Hunter, C. A., & Hurst, A. (2016). *Understanding doulas and childbirth: Women, love, and advocacy*. Springer. doi:10.1057/978-1-137-48536-6

Hutagalung, S. I. S. (2018). *Using Berlitz Method to Improve Students' Vocabulary at Happy Holy Kids Cinere Preschool* [Doctoral dissertation]. Universitas Kristen Indonesia.

Hutchins, T., & Sims, M. (1999). *Programme planning for infants and toddlers: An ecological approach*. Sydney: Prentice Hall.

Hutchison, E. D. (2015). *Dimensions of Human Behavior: The Changing Life Course Los Angeles*. SAGE Publishing.

Hutton, J. S., Phelan, K., Horowitz-Kraus, T., Dudley, J., Altaye, M., & Holland, S. K. (2017). Shared reading quality and brain activation during story listening in preschool-age children. *The Journal of Pediatrics, 191*, 204–212. doi:10.1016/j.jpeds.2017.08.037 PMID:29173308

IISD. (2019). *Modelling for Sustainable Development New Decisions for a New Age*. 85.

İnal-Emiroğlu, F.N. & Pekcanlar- Akay, A. (2008). Kronik hastalıklar, hastaneye yatış ve çocuk. *DEU Tıp Fakültesi Dergisi, 22*(2), 99-105.

İnan, M. (2004). 3-9 yaş çocukları için uygulamalı hareket eğitimi öğretmen el kitabı. İstanbul: Morpa

İnanç, B. Y., Bilgin, M., & Kılıç Atıcı, M. (2007). *Gelişim psikolojisi, çocuk ve ergen gelişimi*. Ankara: Pegem.

Ingram, D. (1989). *First language acquisition: Method, description, and explanation*. Cambridge, UK: Cambridge University Press.

International Reading Association. (1999). *Using Multiple Methods of Begining Reading Instruction*. Retrieved from https://www.literacyworldwide.org/docs/default-source/where-we-stand/multiple-methods-position-statement.pdf?sfvrsn=d04ea18e_6

İrge, E., Timur, S., Zincir, H., Oltuoğlu, H., & Dursun, S. (2005). Gebelikte beslenmenin değerlendirilmesi. *Sted, 14*(7), 157–160.

Işık, H. K. (2018). *Çalışan/çalışmayan ebeveynlerin çocuklarına karşı tutumları ve bu tutumların 5-6 yaş grubu çocukların davranış problemleri ile ilişkisinin incelenmesi*. Yayınlanmamış Yüksek Lisans Tezi, Maltepe Üniversitesi.

Işık, M. (2007). *Televizyon ve çocuk*. Konya: Eğitim.

Işikoglu Erdogan, N. (2016). Investigation of the effects of parent-child interactive story reading activities *Kastamonu Eğitim Dergisi, 24*(3), 1071-1086. (in Turkish)

İskit, N. (2007). 99 Pages of Pregnancy and Child Yoga. Istanbul: Turkey Isbank Cultural Publications.

Israel, S. E. (2008). *Early reading first and beyond*. London: Corwin.

Jackson, S., Pretti-Frontczak, K., Harjusola-Webb, S., Grisham-Brown, J., & Romani, J. M. (2009). Response to İntervention: Implications For Early Childhood Professionals. *Language, Speech, and Hearing Services in Schools*, *40*(4), 424–434. doi:10.1044/0161-1461(2009/08-0027) PMID:19801404

Jansen, P. W., Verlinden, M., Dommisse-van Berkel, A., Mieloo, C., van der Ende, J., Veenstra, R., ... & Tiemeier, H. (2012). Prevalence of bullying and victimization among children in early elementary school: Do family and school neighbourhood socioeconomic status matter?. BMC public health, 12(1), 494. doi:10.100712552-016-9182-9

Jantscher-Krenn, E., Marx, C., & Bode, L. (2013). Human milk oligosaccharides are differentially metabolised in neonatal rats. *British Journal of Nutrition*, *110*(4), 640–650. doi:10.1017/S0007114512005727 PMID:23312292

Jeter, P. E., Slutsky, J., Singh, N., & Khalsa, S. B. S. (2015). *Yoga as a therapeutic intervention: A bibliometric analysis of published* research studies from 1967–2013. *Journal of Alternative and Complementary Medicine (New York, N.Y.)*, *21*(10), 586–592. doi:10.1089/acm.2015.0057 PMID:26196166

Jimenez Soto, E., La Vincente, S., Clark, A., Firth, S., Morgan, A., Dettrick, Z., ... Widiati, Y. (2012). Developing and Costing Local Strategies to Improve Maternal and Child Health: The Investment Case Framework. *PLoS Medicine*, *9*(8), e1001282. doi:10.1371/journal.pmed.1001282 PMID:22879817

Jirasek, J. E. (2001). *An atlas of the human embryo and fetus: A phofografik rewiew of human prenatal development*. NY: The Parthenon Publishing Group.

Johnson, A. D., Martin, A., Brooks-Gunn, J., & Petrill, S. A. (2008). Order in the house! Associations among household chaos, the home literacy environment, maternal reading ability, and children's early reading. *Merrill-Palmer Quarterly Journal of Developmental Psychology*, *54*(4), 445–472. doi:10.1353/mpq.0.0009 PMID:19526070

Jong-Wha, L. (2015). *Why we must invest more in education*. World Economic Forum.

Jordan, G. E., Snow, C. E., & Porche, M. V. (2000). Project EASE: The effect of a family literacy project on kindergarten students' early literacy skills. *Reading Research Quarterly*, *35*(4), 524–546. doi:10.1598/RRQ.35.4.5

Juel, C. (1988). Learning to read and write: A longitudinal study of 54 children from first through fourth grades. *Journal of Educational Psychology*, *80*(4), 437–447. doi:10.1037/0022-0663.80.4.437

Jurimae, T., & Jurimae, J. (2001). *Growth, physical activity, and motor development in prepubertal children*. USA: Taylor & Francis Group. doi:10.1201/b14219

Justice, L. M., & Kaderavek, J. N. (2002). Using shared storybook reading to promote emergent literacy. *Teaching Exceptional Children*, *34*(4), 8–13. Retrieved from doi:10.1177/004005990203400401

Justice, L. M., & Kaderavek, J. N. (2004). Embedded-explicit emergent literacy intervention I. *Language, Speech, and Hearing Services in Schools*, *35*(3), 201–211. doi:10.1044/0161-1461(2004/020) PMID:15248791

Justice, L. M., & Pence, K. L. (2005). *Scaffolding with storybooks: A guide for enhancing young children's language and literacy achievement*. Newark, DE: International Reading Association.

Kabasakal, U. (2007). Ergen ve genç intiharında aile faktörü. *Doğu Anadolu bölgesi araştırma ve uygulama merkezi*, *6*(1), 32-37.

Kadayıfçı, O., Ürünsak, İ. F., Ürünsak, M., & Atay, Y. (2001). Gebelikte enfeksiyon hastalıkları. In S. M. Beksaç, N. Demir, A. Koç, & A. Yüksel (Eds.), *Obstetrik maternal fetal tıp ve perinatoloji* (pp. 878–939). Ankara: Nobel.

Kagan, J. (2010). Emotions and temperament. In M. H. Bornstein (Ed.), *Handbook of cultural developmental science*. New York: Psychology Press.

Kağıtçıbaşı, Ç., Sunar, D., Bekman, S., & Cemalcılar, Z. (2005). *Erken Müdahalenin Erişkinlikte Süren Etkileri: Erken Destek Projesinin İkinci Takip Araştırmasının Ön Bulguları*. İstanbul: Anne Çocuk Eğitimi Vakfı Yayınlar.

Kahraman, F., & Kahraman, A. B. (2017). Dezavantajlılığın çocuk halleri: Dünyada ve Türkiye'de dezavantajlı gruplarda yer alan çocukların durumu. *Gümüşhane Üniversitesi Sosyal Bilimler Enstitüsü Elektronik Dergisi, 8*(20), 255–268.

Kaley Isley, L.C., Peterson, J., Fischer, C., & Peterson, E. (2010). Yoga as a Complementary Therapy for Children and Adolescents A Guide for Clinicians. *Psychiatry, 7*(8), 20–32. PMID:20877530

Kalınkara, V., & Başıbüyük, G., & Ay, F. (2016). Yaşlıların geronteknolojik ürünleri kabule yönelik tutumları. *Yaşlı Sorunları Araştırma Dergisi, 9*(2), 1–19.

Kalkan, E. (2010). Dünyada anne baba eğitimi ve aile katılımı: Amerika Birleşik Devletleri örneği. In T. Güler (Ed.), Anne baba eğitimi (pp. 166-181). Ankara: Pegem Akademi.

Kandır, A & Alpan, Y. (2008). Okul öncesi dönemde sosyal-duygusal gelişime anne-baba davranışlarının etkisi. *Aile ve Toplum Eğitim Kültür ve Araştırma Dergisi, 4*(14), 33-38

Kandır, A. (2003). My 3-6 year old child is growing. İstanbul: Morpa Publications. (in Turkish)

Kandır, A. (2007). Gaining language abilities in cognitive development. In E. Omeroglu and A. Kandır (Eds.), Cognitive Development (pp. 131-149). İstanbul: Morpa Kültür Yayınları. (in Turkish)

Kandır, A., & Alpan, Y. (2008). Okul öncesi dönemde sosyal-duygusal gelişime anne-baba davranışlarının etkisi. *Sosyal Politika Çalışmaları Dergisi, 14*(14), 33–38.

Kaner, S. (2009). Duygusal ve davranış bozuklukları olan çocuklar. In A Ataman, (Ed.), Özel gereksinimli çocuklar ve özel eğitime giriş (pp. 171-194). Ankara: Gündüz Eğitim ve Yayıncılık.

Kapçı, E. G. (2005). Erken çocukluk döneminde sıklıkla karşılaşılan duygusal davranışsal sorunlar. In Y. Uzuner (Ed.), Çocukta ruh sağlığı, uyum bozukluğu (pp. 74-92). Eskişehir: Anadolu Üniversitesi Yayınları.

Karaalp, T., Dinç, P., & Teke, A. (2015). Sağlık yapılarında iyileştirme önerilerinin geliştirilmesi: Çocuk psikiyatrisi kliniği örneği. *Anatolian Journal of Clinical Investigation, 9*(3), 121–138.

Karaaslan, T., Beyazıt, U., & Bütün-Ayhan, A. (2015). Gelişim dönemler açısından hasta çocuklar ve hastane ortamının düzenlenmesi. In A. Bütün-Ayhan (Ed.), *Hasta çocukların gelişimi ve eğitimi* (pp. 26–56). Eskişehir: Anadolu Üniversitesi.

Karabekiroğlu, K. (2011). *Çocuğuma nasıl davranmalıyım?* İstanbul: Say.

Karabulut, R. (2018). Examining the effect of education on problem solving skills of gifted children attending primary school. Unpublished PhD Thesis, Gazi University Institute of Educational Sciences, Ankara.

Karacan, E. (2000). Bebeklerde ve çocuklarda dil gelişimi. *Klinik Psikiyatri Dergisi, 3*, 263–268.

Karacan, E. (2000). Factors affecting language development. *Sürekli Tıp Eğitim Dergisi, 7*, 254–258. (in Turkish)

Karagöz, Y., & Ekici, S. (2004). Sosyal bilimlerde yapılan uygulamalı araştırmalarda kullanılan istatistiksel teknikler ve ölçekler. *Cumhuriyet Üniversitesi İktisadi ve İdari Bilimler Dergisi, 5*(1), 25–43.

Karahan, F. (2005). Bilingualism in Turkey. In *Proceedings of the 4th international symposium on bilingualism*. Cascadilla Press.

Karakaş, B., & Çevik, Ö. C. (2016). Çocuk refahı: Çocuk hakları perspektifinden bir değerlendirme. *Gazi Üniversitesi İktisadi ve İdari Bilimler Fakültesi Dergisi, 18*(3), 887–906.

Karakaya, Z. (2008). Dil edinimi "Okul öncesi dil ve oyun eğitimi." Samsun: Eflatun Matbaacılık.

Karakuş, I. Ş. (2006). The effects of children's literature on reading development. Unpublished masters thesis, Gazi University Educational Sciences Institute, Ankara. (in Turkish)

Karakus, F. (2010). Difficulties faced by gifted parents. *Mersin University Journal of the Faculty of Education*, *6*(1), 127–144.

Karaman, G. (2014). Erken Okuryazarlık Becerilerinin Tarihsel Gelişimi, Kuramsal Temelleri ve Kapsami. In F. Temel (Ed.), *Her Yönüyle Okul Öncesi Eğitim 4 –Dil ve Erken Okuryazarlık* (pp. 12–24). Ankara, Turkey: Hedef Press.

Karataş, K. (1993). Çocuk işgücü sorunu: Nedenleri, sonuçları ve çözüm önerileri. *Hacettepe Üniversitesi Sosyal Hizmetler Yüksekokulu Dergisi*, *11*, 85–102.

Karoly, A. L., Kilburn, R. M., & Cannon, J. (2005). *Early Childhood Interventions, Proven Results, Future Promise*. Santa Monica: Rand Corporation. doi:10.7249/MG341

Kartal, H. (2013). Erken eğitim ve destek programları. In F. Temel (Ed.), Aile eğitimi ve erken çocukluk eğitiminde aile katılım çalışmaları (pp. 159-221). Ankara: Anı yayıncılık.

Kasatura, İ. (2002). Sokaktaki çocuklar sokakta çalışan çocuklar. *İstanbul Ticaret Üniversitesi Dergisi*, 211-227.

Katar, S., & Devecioğlu, C. (2006). Dicle üniversitesi yeni doğan bakım ünitesinde izlenen çok düşük doğum ağırlıklı bebeklerin değerlendirilmesi. *Dicle Tıp Dergisi*, *33*(4), 248–251.

Kauffman, J. M. (2001). *Characteristics of emotional and behavioral disorders of children and youth* (7th ed.). Upper Saddle River, NJ: Merrill/Prentice Hall.

Kavaklı, U., & Özkara, E. (2012). Kişisel, sosyal ve iş hayatında özürlü bireylere tanınan haklar ve özürlü olguların medikolegal değerlendirmesi. *DEÜ Tıp Fakültesi Dergisi*, *26*(1), 65–74.

Kavsek, M. (2004). Predicting later IQ from infant visual habituation and dishabituation: A meta-analysis. *Applied Developmental Psychology*, *25*(3), 369–393. doi:10.1016/j.appdev.2004.04.006

Kaya, A. (2016). Views of Turkish teachers about mainstreaming and special needs students. Unpublished Master's Thesis, Gazi University, Institute of Educational Sciences, Ankara.

Kaya, Y., & Şahin, N. (2013). Kadınlarda madde kullanımı ve hemşirenin rolü. *Hemşirelikte Eğitim ve Araştırma Dergisi*, *10*(1), 3–7.

Kaytez, N. (2018). Motor gelişim. In N. Aral & Z. F. Temel (Eds.), *Çocuk gelişimi* (pp. 142–170). Ankara: Hedef.

Kazdin, A. E. (1998). Conduct disorder. In R. J. Morris & T. R. Kratochwill (Eds.), *The practice of child theraphy* (3rd ed., pp. 199–230). Boston: Allyn & Bacon.

Keenan, T. (2002). *An Introduction to Child Development*. USA: Sage Publication.

Keleş, E., & Çepni, S. (2006). Beyin Ve Öğrenme. *Türk Fen Eğitimi Dergisi*, *3*(2), 66–82.

Kernan, M. (2010). Space and place as a source of belonging and participation in urban environments: Considering the role of early childhood education and care settings. *European Early Childhood Education Research Journal*, *18*(2), 199–213. doi:10.1080/13502931003784420

Keyou, G., Jianbin, J., & Hang, L. (2007). Food-based dietary guidelines in China. *Annals of Nutrition & Metabolism*, *51*(2), 26–31. doi:10.1159/000103564

Khalsa, S. B. S. (2013). Yoga in schools research: improving mental and emotional health. Presented at the *Second International Conference on Yoga for Health and Social Transformation. Academic Press.*

Khalsa, S. B. S., & Butzer, B. (2016). Yoga in school settings: A research review. *Annals of the New York Academy of Sciences, 1373*(1), 45–55. doi:10.1111/nyas.13025 PMID:26919395

Khalsa, S. B., Hickey-Schultz, L., Cohen, D., Steiner, N., & Cope, S. (2012). Evaluation of the mental health benefits of yoga in a secondary school: A preliminary randomized controlled trial. *The Journal of Behavioral Health Services & Research, 39*(1), 80–90. doi:10.100711414-011-9249-8 PMID:21647811

Kıldan, O. (2010). Aile içi ilişkiler ve özel durumlarda aile. In T. Güler (Ed.), Anne baba eğitimi (pp. 56-71). Ankara: Pegem Akademi.

Kılıç, Ç. (2010). Aile Eğitim Programları ve Türkiye'deki örnekleri. *Ab ant İzzet Baysal Üniversitesi Dergisi. 10*(1), 100-111.

Kılıç, K. M. (2018). Gelişimin genetik temelleri. In A. Güngör-Aytar (Ed.), Gelişim psikolojisi (pp. 46-71). Ankara: Hedef.

Kılıç, Ş. (2018). Orta çocukluk dönemi. In. F. A. Güngör-Aytar (Ed.), Gelişim psikolojisi (pp. 118-158). Ankara: Hedef.

Kilicoglu, M. (2007). *Effective teacher for active learning*. Istanbul: Morpa Culture Publications.

Kılınç, B., & Kılınç, E. P. (2014). Yeni medya ortamında çocuk birey: Yeni iletişim teknolojileri ve medya pedagojisinin önemi. *Akdeniz İletişim Dergisi, 22*, 9–23. doi:10.31123/akil.441825

Kılınç, E. F., & Aral, N. (2015). Erken Müdahale İçin Bir Model: Anne Çocuk Etkileşim Programı. *Hacettepe Üniversitesi Sağlık Bilimleri Fakültesi Dergisi, 1*(2), 63–76.

King, S. (2017). *A Case Study of a Yoga and Meditation Intervention in an Urban School: A Complex Web of Relationships and Resilience in the Search for Student Well-Being [PhD]. University of California*, Los Angeles.

King, S., & Saxton, M. (2010). Opportunities for language development: Small group conversations in the nursery class. *Educational and Child Psychology, 27*(4), 31–44.

Kır, E. (2013). Çocuklara yönelik cinsel taciz ve istismara karşı önleyici eğitim çalışmaları. *İÜHFM C. LXXI. 1*(786), 785-800.

Kıran, Z. & Kıran, A. (2002). Dilbilime giriş. Ankara: Seçkin Yayıncılık.

Kıran, A., & Kıran, Z. (2002). *Dilbilime giriş*. Ankara: Seçkin.

Kircaali-Iftar, G. (2000). *Küçük Adımlar Gelişimsel Geriligi Olan Cocuklara Yonelik Erken Egitim Programinin Aile Goruslerine Dayali Olarak Değerlendirilmesi.* Eskişehir, Turkey: Anadolu Universitesi Yayinlari.

Klatt, M., Harpster, K., Browne, E., White, S., & Case-Smith, J. (2013). Feasibility and preliminary outcomes for move-into-learning: An arts-based mindfulness classroom intervention. *The Journal of Positive Psychology, 8*(3), 233–241. doi:10.1080/17439760.2013.779011

Klerman, L. (2000). Family planning: Needs and opportunities. In M. C. McCormick & J. E. Siegel (Eds.), *Prenatal care: effectiveness and implementation* (pp. 271–284). Cambridge: Cambridge University Press.

Knapp, M. L., & Hall, J. A. (2007). Nonverbal Communication in Human Interaction (5th ed.) Wadsworth: Thomas Learning.

Knitzer, J., & Lefkowitz, J. (2006). *Helping The Most Vulnerable Infants, Toddlers, and Their Families.* New York: National Center for Children in Poverty, Columbia University.

Kobylińska, D., Lewczuk, K., Marchlewska, M., & Pietraszek, A. (2018). For body and mind: Practicing yoga and emotion regulation. *Social Psychological Bulletin, 13*(1), 1–21. doi:10.5964pb.v13i1.25502 PMID:29154556

Kocacık, F. (2001). Şiddet olgusu üzerine. C.Ü. *İktisadi ve İdari Bilimler Dergisi*, 2(1), 3-7.

Koçak, Ü., Albayrak, M., Erol, R., & Şanlı, C. (2012). Evaluation of sleep patterns and sleep disturbances in children: a preliminary study in Kırıkkale. *Türkiye Çocuk Hastalıkları Dergisi, 6*(2), 81-87.

Koç, S. (2011). Hastane palyaçolarının ameliyat öncesi dönemdeki çocuklar ve ebeveynlerinin anksiyete düzeylerine etkisinin belirlenmesi. *Hemşirelikte Eğitim ve Araştırma Dergisi, 8*(3), 26–31.

Koçtürk, N. (2018). Çocuk ihmalini ve istismarını önlemede okul çalışanlarının sorumlulukları. *MSKU Eğitim Fakültesi Dergisi, 5*(1), 38–47.

Koçtürk, N., & Yılmaz, N. (2018). Çocuk istismarı ve ihmali için risk altındaki çocukları belirlemeye ve müdahale etmeye yönelik model/veri tabanı önerisi. *Kastamonu EducationJournal, 26*(6), 1919–1927.

Koçyiğit, S., Tuğluk, M.N., & Kök, M. (2007). Çocuğun gelişiminde eğitsel bir etkinlik olarak oyun. *KKEFDI- OKKEF, 16*, 325-342.

Koenig, K. P., Buckley-Reen, A., & Garg, S. (2012). Efficacy of the Get Ready to Learn Yoga Program Among Children With Autism Spectrum Disorders: A Pretest–Posttest Control Group Design. *The American Journal of Occupational Therapy, 66*(5), 538–546. doi:10.5014/ajot.2012.004390 PMID:22917120

Kolac, E., & Demir, T. & Karadag, R. (2012). Views of pre service teachers on using children's literature in language education. *Türkiye Sosyal Araştırmalar Dergisi, 161*, 195–213. (in Turkish)

Kormi-Nouri, R., Moradi, A. R., Moradi, S., Akbari-Zardkhaneh, S., & Zahedian, H. (2012). The effect of bilingualism on letter and category fluency tasks in primary school children: Advantage or disadvantage? *Bilingualism: Language and Cognition, 15*(2), 351–364. doi:10.1017/S1366728910000192

Kotaman, H. (2007). Impacts of dialogical storybook reading on young children's reading attitudes and vocabulary development. *Reading Improvement, 45*(2), 55–61.

Kotaman, H. (2009). Parents' story reading self efficacy scale. *Türk Eğitim Bilimleri Dergisi, 7*(4), 767–780. (in Turkish)

Kotiadis, K. (2007). Using soft systems methodology to determine the simulation study objectives. *Journal of Simulation, 1*(3), 215–222. doi:10.1057/palgrave.jos.4250025

Kotil, Ç. (2002). *Yabancı dil eğitimi*. Nisan: Çocuk ve Aile.

Kraayenoord, C. E. V., & Paris, S. G. (1996). Story construction from a picture book: An assessment activity for young learners. *Early Childhood Research Quarterly, 11*(1), 41–61. doi:10.1016/S0885-2006(96)90028-9

Krashen, S. D., & Terrel, D. T. (1983). *The natural approach: language acquisition in the classroom*. Hayward, CA: Alemany Press.

Krygier, J. R., Heathers, J. A., Shahrestani, S., Abbott, M., Gross, J. J., & Kemp, A. H. (2013). Mindfulness meditation, well-being, and heart rate variability: A preliminary investigation into the impact of intensive Vipassana meditation. *International Journal of Psychophysiology, 89*(3), 305–313. doi:10.1016/j.ijpsycho.2013.06.017 PMID:23797150

Küçükkaragöz, H. (2011). Cognitive development and language development. In B. Yeşilyaprak (Ed.), Development and learning psychology (pp. 37-45). Ankara: Pegema Publications. (in Turkish)

Kuday, F. S. (2007). *Aile destekli kurum merkezli eğitim alan ve hiç okul öncesi eğitim almayan 3–6 yaş çocukların bilişsel gelişimlerinin karşılaştırılması,* (unpublished master thesis) Marmara Üniversitesi, Eğitim Bilimleri Enstitüsü, Eğitim Bilimleri Ana Bilim Dalı, Okul Öncesi Öğretmenliği Bilim Dalı, İstanbul

Kuebli, J. (1994). Young children's understanding of everyday emotions. *Young Children*, 36–48.

Kuhn, T. (2006). *Grammatik im englisch unterricht der primarstufe: theoretische grundlagen und praktische unterrichtsvorschläge*. Springer.

Kulaksızoğlu, A. (1998). *Ergenlik psikolojisi*. İstanbul: Remzi Yayınevi.

Kurar, İ., & Baltacı, F. (2015). Halkın Boş Zaman Değerlendirme Alışkanlıkları: Alanya Örneği. *International Journal of SportCultureandScience*, *2*(2), 39–52.

Kurt, B. (2008). The effects of children's literature products on basic language skills. Unpublished masters thesis, Gazi University Educational Sciences Institute, Ankara. (in Turkish)

Kurtay, D. (2018). Çocuk ihmal ve istismarının önlenmesi. Retrieved from http://ahto.meb.k12.tr/ meb_iys_dosyalar/06/25/733770/dosyalar/2018_01/03144225_cocukihmalveistismarınınönlenmesi.doc

Kurt, G., Yücel Beyaztaş, F., & Erkol, Z. (2010). Yaşlıların sorunları ve yaşam memnuniyeti. *TurkishJournal of ForensicMedicine*, *24*(2), 32–39.

Kuruvilla, S., Schweitzer, J., Bishai, D., Chowdhury, S., Caramani, D., & Frost, L. … Bustreo, F. (2014). Success factors for reducing maternal and child mortality. *Bulletin of the World Health Organization*. doi:10.2471/BLT.14.138131

Kuther, T. L. (2016). *Physical and Cognitive Development in Middle Childhood in Lifespan Development: Lives in Context*. Canada: SAGE Publicatiions.

Kuz, T. (2001). Esi Examination of Attitudes towards Inclusive Education. Unpublished Thesis, Prime Ministry Administration for Disabled People Publications, Ankara.

Kuzgun, Y. (2000). *Meslek Danışmanlığı Kuramlar ve Uygulamalar*. Ankara: Nobel.

Kyle, T. (2008). *Essentials of pediatric nursing*. Philadelphia, PA: LWW.

Laboo, L. (2005). Books and computer response activities that support literacy development. *The Reading Teacher*, *59*(3), 288–292. doi:10.1598/RT.59.3.10

Ladewig, P. A. W., London, M. L., & Davidson, M. R. (2014). *Contemporary maternal-newborn nursing care* (8th ed.). Upper Saddle River, NJ: Pearson Education, Inc.

Lago-Delello, E. (1998). Classroom dynamics and the development of serious emotional disturbance. *Exceptional Children*, *64*(4), 479–492. doi:10.1177/001440299806400404

Langford, K. S. (2002). Infectious disease and pregnancy. *Current Obstetrics & Gynaecology*, *12*(3), 125–130. doi:10.1054/cuog.2001.0247

Langøien, L. J. (2013). *"Pay Attention – Listen to Your Heart!" Unfolding Practice, Change Realities and Awareness of the Embodied Self in Ashtanga Yoga [PhD thesis]*. Norway: University of Norwegian.

Larsen, F. D., & Long, M. H. (1991). An introduction to second language acquisition research. New York: Longman Inc.

Lawn, J. E., Blencowe, H., Oza, S., You, D., Lee, A. C. C., Waiswa, P., ... Cousens, S. N. (2014). Every newborn: Progress, priorities, and potential beyond survival. *Lancet*, *384*(9938), 189–205. doi:10.1016/S0140-6736(14)60496-7 PMID:24853593

Lee, S.-Y., Poon, W.-Y., & Bentler, P. M. (1992). Structural Equation Models With Continuous And Polytomous Variables. *Psychometrika, 57*(1), 89–105. doi:10.1007/BF02294660

Lenroot, R. K., & Giedd, J. N. (2006). Brain development in children and adolescents: Insights fromanatomical magnetic resonance imaging. *Neuroscience and Biobehavioral Reviews, 30*(6), 718–729. doi:10.1016/j.neubiorev.2006.06.001 PMID:16887188

Levine, L. E., & Munsch, J. (2016). *Child development from infancy to adolescence: An active learning approach.* Sage Publications.

Levine, L. E., & Munsch, J. (2016). *Social and Emotianal Development in Middle Childhood in book Child Development From Infancy to Adolescence.* Canada: SAGE Publications, Inc.

Levine, L. E., & Munsch, J. A. (2015). *Child development: An active learning approach.* Canada: Sage Publication.

Levin, I., & Aram, D. (2012). Mother-child joint writing and storybook reading and their effects on kindergartners' literacy: An intervention study. *Reading and Writing, 25*(1), 217–249. doi:10.100711145-010-9254-y

Lewis, G. (1982). Human migration: A geographical perspective. London: Canberra.

Lewis, M., & Goldberg, S. (1969). Perceptual-cognitive development in infancy: A generalized expectancy model as a function of the mother-infant interaction. *Merrill-Palmer Quarterly of Behavior and Development, 15*(1), 81–100. Retrieved from http://www.jstor.org/stable/23082954

Lieberman, A., & Holt, L. H. (2005). *Nine months and a day: A pregnancy and birth companion.* Harvard Common Press.

Liew, J. (2012). Effortful control, executive functions, and education: Bringing self-regulatory and social-emotional competencies to the table. *Child Development Perspectives, 6*(2), 105–111. doi:10.1111/j.1750-8606.2011.00196.x

Lindberg, L., Bohlin, G., Hagekull, B., & Thunström, M. (1994). Early food refusal: Infant and family characteristics. *Infant Mental Health Journal, 15*(3), 262–277. doi:10.1002/1097-0355(199423)15:3<262::AID-IMHJ2280150303>3.0.CO;2-Q

Lindon, J. (2010). Understanding child development 2nd edition linking theory and practice. London: Hachette UK Company.

Linn, R. L., & Gronlund, N. E. (1995). Measurement And Assessment. In *Teaching* (7th ed.). Englewood Cliffs, NJ: Prentice-Hall.

Litchke, L. G., Liu, T., & Castro, S. (2018). Effects of Multimodal Mandala Yoga on Social and Emotional Skills for Youth with Autism Spectrum Disorder: An Exploratory Study. *International Journal of Yoga, 11*(1), 59–65. PMID:29343932

Lloyd, D. N. (1978). Prediction of School Failure From Thirdgrade Data. Educational and Psychological Measurement. *Mental Health Study Center National Institute of Mental Health, 38*(4), 1193–1200. doi:10.1177/001316447803800442

Loehlin, J. C. (2004). Latent Variable Models: An Introduction To Factor, Path And Structural Equation Analysis, (4th ed.). Mahwah, NJ: Lawrence Erlbaum Associates. doi:10.4324/9781410609823

Lordoglu, K. & Aslan, M. (2018). Görünmeyen göçmen çocukların işçiliği: Türkiye'deki Suriye'li çocuklar. *Çalışma ve Toplum, 2*, 715-732.

Lou, C., Higgins, K., Gelfer, J., Hong, E., & Miller, S. (2005). The effects of teacher facilitation on the social interactions of young children during computer activities. *Topics in Early Childhood Special Education, 25*(4), 208–217. doi:10.1177/02711214050250040201

Louw, D. A. (2002). *Human Development. Characteristics of Language of Preschooler.* South Africa: Perskor.

Love, J. M., Kisker, E. E., Raikes, H. H., & Tarullo, L. (1997). Overview of the Early Head Start research and evaluation project: February 1998. *NHSA Dialog, 1*(4), 181–192. doi:10.120719309325nhsa0104_8

Luby, J., Belden, A., Botteron, K., Marrus, N., Harms, M. P., Babb, C., ... Barch, D. (2013). The effects of poverty on childhood brain development: The mediating effect of caregiving and stressful life events. *JAMA Pediatrics, 167*(12), 1135–1142. doi:10.1001/jamapediatrics.2013.3139 PMID:24165922

Lyytinen, P., Laakso, M. L., & Poikkeus, A. M. (1998). Parental contribution to child's early language and interest in books. *European Journal of Psychology of Education, 13*(3), 297–308. doi:10.1007/BF03172946

Mackenzie, R. J. (2014). *Çocuğunuza sınır koyma*. İstanbul: Yakamoz.

Madazi, R., & Atasü, T. (2000). Fetüs ve yeni doğana etkili enfeksiyon hastalıkları. In T. Atasü (Ed.), *Gebelikte fetüs ve yeni doğana zararlı etkiler* (pp. 319–372). Ankara: Nobel.

Mağden, D., Şahin, S., & Çetinkaya, Ş. (2001). *Üç-dört yaş grubu çocukların fotoğraf ve çizimlerdeki insanların bakış yönlerini algılamaları ve görsel dikkatin incelenmesi*. Milli Eğitim.

Mak, C., Whıttıngham, K., Cunnıngton, R., & Boyd, R. N. (2017). Effect of mindfulness yoga programme MiYoga on attention, behaviour, and physical outcomes in cerebral palsy: A randomized controlled trial. *Developmental Medicine and Child Neurology, 60*(9), 922–932. doi:10.1111/dmcn.13923 PMID:29869333

Mandan Sürücü, S. (2009). 11-17 yaş orta düzeyde zihin engelli kız ergenlerin temel cinsel bilgi ve cinsel istismarı algılamalarının belirlenmesi [Yüksek Lisans Tezi]. Marmara Üniversitesi Eğitim Bilimleri Enstitüsü.

Maraşuna, O., & Eroğlu, O. (2013). Ortaokul öğrencilerinin tıbbi işlem korkuları ve etkileyen faktörler. *Güncel Pediatri, 11*(1), 13–22. doi:10.4274/Jcp.11.03

Maree, J. (2018). Promoting career development and life design in the early years of a person's life. *Early Child Development and Care, 188*(4), 425–436. doi:10.1080/03004430.2017.1345892

Mariscal, S., & Benavides, A. A. (2017). Typical Language Development of Monolingual Spanish-Speaking Children. In *Language Development and Disorders in Spanish-speaking Children* (pp. 3–36). Cham: Springer. doi:10.1007/978-3-319-53646-0_1

Marjanovič-Umek, L., Hacin, K., & Fekonja, U. (2017). The quality of mother–child shared reading: Its relations to child's storytelling and home literacy environment. *Early Child Development and Care*. doi:10.1080/03004430.2017.1369975

Marotz, L. R., & Allen, K. E. (2016). *Developmental Profiles: Pre-Birth Through Adolescence*. United States: Cengage Learning.

Mash, E., & Wolfe, D. (2007). *Abnormal child psychology* (3rd ed.). USA: Wodsworth Learning.

Maviş, A. (2005). *Anlayarak hızlı okuma ve öğrenme teknikleri*. İstanbul: Hayat Yayınları Dergisi.

Mayer, G. R. (1995). Preventing antisocial behavior in schools. *Journal of Applied Behavior Analysis, 28*(4), 467–478. doi:10.1901/jaba.1995.28-467 PMID:16795877

Mayes, L. C., & Cohen, D. J. (2002). *Çocuğunuzu anlama rehberi*. İstanbul: Beyaz Balina.

Maynard, T., & Waters, J. (2007). Learning in the outdoor environment: A missed opportunity? *Early Years, 27*(3), 255–265. doi:10.1080/09575140701594400

McAfee, O., Leong, D., & Bodrova, E. (2004). *Basics of Assessment: A Primer for Early Childhood Professionals*. Washington, D.C.: National Association for the Education of Young Children.

McDevitt, T. M., & Ormrod, J. E. (2002). *Child development: educating and working with children and adolescents*. Upper Saddle River, NJ: Pearson Prentice Hall.Merrill.

Mcdevitt, T. M., & Ormrod, J. E. (2010). *Child development and education* (7th ed.). Pearson.

McDonagh, M. S., Osterweil, P., & Guise, J. (2005). The benefits and risks of inducing labour in patient with prior cesarean delivery: A systematic review. *BJOG*, *112*(8), 1007–1015. doi:10.1111/j.1471-0528.2005.00623.x PMID:16045510

McKinney, S. S. (2008). *Yanlış çocuk: Çocuk yetiştirmede yapılan hatalar*. İstanbul: Ekinoks.

McLeod, S. A. (2008). Psychosexual stages. Simply Psychology. Retrieved from www.simplypsychology.org/psychosexual.html

MEB / Milli Eğitim Bakanlığı. (2012). *Gebelik ve fetüs fizyolojisi*. Ankara: MEB yayınları.

MEB. (2013). 0-36 Ay Çocuklar İçin Eğitim Programı, Ankara, Turkey: MEB-UNİCEF.

Meece, J. L. (1997). *The communicative competence of youngchildren*. New York: Addison Wesley Longman.

Meggitt, C. (2006). *Child Development: An Illustrated Guide*. Oxford: Heinemann.

Mehta, S., Shah, D., Shah, K., Mehta, S., Mehta, N., Mehta, V., ... Mehta, D. (2012). Peer-Mediated Multimodal Intervention Program for the Treatment of Children with ADHD in India: One-Year Followup. *International Scholarly Research Network*, *2*(4), 1–7. PMID:23316384

Mendive, S., Lissi, M. R., Bakeman, R., & Reyes, A. (2017). Beyond mother education: Maternal practices as predictors of early literacy development in Chilean children from low-SES households. *Early Education and Development*, *28*(2), 167–181. doi:10.1080/10409289.2016.1197014

Mengütay, S. (2005). *Çocuklarda hareket gelişimi ve spor*. İstanbul: Morpa.

Meraki, Z., & Yıldız Bıçakçı, M. (2015). Bebeklik ve İlk Çocukluk Döneminde Erken Müdahale, Bebeklik ve İlk Çocukluk Döneminde (0-36 Ay) Gelişim. In M. Yıldız Bıçakçı (Ed.), Duyuların Gelişimi ve Desteklenmesi (pp. 179-210). Ankara: Eğiten Kitap.

Meral, B. F., Özbey, F., & Sola, C. (2011). Küçük adımlar erken eğitim programının aile görüşlerine göre değerlendirilmesi. *Sakarya Üniversitesi Eğitim Fakültesi Dergisi*, *22*, 25–40.

Merikangas, K. R., He, J. P., Burstein, M., Swanson, S. A., Avenevoli, S., Cui, L., ... Swendsen, J. (2010). *Lifetime prevalence of mental disorders in U.S. adolescents: Results from the National Comorbidity Study*—Adolescent Supplement (NCS-A). *Journal of the American Academy of Child and Adolescent Psychiatry*, *49*(10), 980–989. doi:10.1016/j.jaac.2010.05.017 PMID:20855043

Metin, E. N. (2018). Zihinsel engelli çocuklar. In N. Metin (Ed.), Özel gereksinimli çocuklar (pp. 74-115). Ankara: Anı.

Metindoğan Wise, A. (2015). Gelişim bilimi: Tanımı, kapsamı, tarihsel temelleri, araştırma konu ve yöntemleri. In F. Turan & A.İ. Yükselen (Eds.), Her Yönüyle Okul Öncesi Eğitim: Çocukluk Gelişimi 1 (pp. 18-44), Ankara: Hedef CS Yayıncılık ve Mühendislik.

Michalow, J., Chola, L., McGee, S., Tugendhaft, A., Pattinson, R., Kerber, K., & Hofman, K. (2015). Triple return on investment: The cost and impact of 13 interventions that could prevent stillbirths and save the lives of mothers and babies in South Africa. *BMC Pregnancy and Childbirth*, *15*(1), 1–13. doi:10.118612884-015-0456-9 PMID:25879579

Millî Eğitim Bakanlığı Hayat Boyu Öğrenme Genel Müdürlüğü. (2014). Aile eğitimi kurs programı (0-18 yaş çocuğu olan ailelere).

Milli Eğitim Bakanlığı. (1997). Özel Eğitim Hakkında Kanun Hükmünde Kararname. 06.06.1997 tarih ve 23011 sayılı Resmî Gazete, Ankara.

Milli Eğitim Bakanlığı. (2006). Özel Eğitim Hizmetleri Yönetmeliği. 31.05.2006 tarih ve 26184 sayılı Resmî Gazete, Ankara.

Ministry of Education. (2018). *Ministry of National Education Special Education Services Regulation*. Ankara: Official Gazette.

Mishra, G. (2006). When child becomes a teacher - the child to child programme. *Indian Journal of Community Medicine*, *31*(4), 277–278.

Monk-Turner, E., & Turner, C. (2010). Does yoga shape body, mind and spiritual health and happiness: Differences between yoga practitioners and college students. *International Journal of Yoga*, *3*(2), 48–54. doi:10.4103/0973-6131.72630 PMID:21170230

Moore, D. W., Bean, T. W., Birdyshaw, D., & Rycik, J. A. (1999). Adolescent literacy: A position statement. *Journal of Adolescent & Adult Literacy*, *43*(1), 97–112. Retrieved from https://eric.ed.gov/?id=ED437640

Moore, K. L., Persaud, T. V. N., & Torchia, M. G. (2013). *The developing human: cilinically oriented embroyology*. Philadelphia: Elsevier Saunders.

Morgan, C. T. (1991). *Psikolojiye giriş*. Ankara: Hacettepe Üniversitesi, Meteksan A.Ş.

Morgan, C. T. (2013). *Psikolojiye Giriş. (20th ed.) Yirminci Basım*. Konya: Eğitim Kitabevi.

Morrison, G. (2007). *Fundemantals of Early Childhood Education. United States: Coral Graphic Service*. United States: Pearson Merill Prentince Hall.

Morrow, L. M. (2005). *Literacy Development in the Early Years Helping Children Read and Write* (5th ed.). Boston: Allyn & Bacon: Pearson.

Morrow, L., & Gambrell, L. (2005). *Using children's literature in preschool: Comprehending and enjoying books*. Newark, DE: International Reading Association.

Muis, K. R., Ranelluci, J., Trevors, G., & Duffy, M. C. (2015). The effects of technology mediated immediate feedback on kindergarten students' attitudes, emotions, engagement and learning outcomes during literacy skills Development. *Learning and Instruction*, *38*, 1–13. doi:10.1016/j.learninstruc.2015.02.001

Mundy, L. K., Canterford, L., Kosola, S., Degenhardt, L., Allen, N. B., & Patton, G. C. (2017). Peer Victimization and Academic Performance in Primary School Children. *Academic Pediatrics*, *17*(8), 830–836. doi:10.1016/j.acap.2017.06.012 PMID:28652070

Murray, J., & RyanKraus, P. (2010). *Obesity in children with down syndrome: Background and recommendations for management.*

Murray, S. S., & McKinney, E. S. (2013). *Foundations of maternal-newborn and women's health nursing-e-book*. Elsevier Health Sciences.

Mustard, J. F. (2002). Early child development and the brain – the base for health, learning and behavior throughout life. *Early Child Development to Human Development*, *10*, 23–62.

Mutlu, H. (2018). *Baba Destek Eğitim Programının aile işlevleri üzerindeki etkisinin incelenmesi. Yayınlanmamış yüksek lisans tezi*. İzmir: Dokuz Eylül Üniversitesi.

Myers, R. (1996). *Hayatta kalan on iki*. İstanbul: Varol Matbaası.

NAEYC. (2009). *Position Statement (PDF) Developmentally Approptiate Practice in Early Childhood Programs Serving Children from Birth through Age 8, adapted in 2009*. National Association for the Education of Young Children. Retrieved from https://www.naeyc.org/files/naeyc/file/positions/PSDAP.pdf

Nagtalon-Ramos, J. (2014). *Maternal- newborn nursing care: best evıdence-based practıces*. Philadelphia: F. A. Davis Company.

Nalbant, A., & Babaoğlan, E. (2016). Risk altındaki öğrenciler: Yozgat örneği. *E-Uluslararası Eğitim Araştırmaları Dergisi, 7*(3), 11–30.

Nanthakumar, C. (2018). The benefits of yoga in children. *Journal of Integrative Medicine, 16*(1), 14–19. doi:10.1016/j.joim.2017.12.008 PMID:29397087

Nathan, L., Stackhouse, J., Goulandris, N., & Snowling, M. J. (2004). The development of early literacy skills among children with speech difficulties. *Journal of Speech, Language, and Hearing Research: JSLHR, 47*(2), 377–391. doi:10.1044/1092-4388(2004/031) PMID:15157138

National Association for the Education of Young Children. (1996). *Technology and Young Children Ages 3 Through 8*. Washington, DC: NAEYC.

National Research Council. (1984). *Development During Middle Childhood: The Years From Six to Twelve*. Washington, DC: The National Academies Press.

NCAC. (2009). Putting Children First. *The Magazine Of the National Childcare Accreditation Council (NCAC), 30*(6), 3-5.

Nedovic, S., & Morrissey, A. (2013). Calm active and focused: Children's responses to an organic outdoor learning environment. *Learning Environments Research, 16*(2), 281–295. doi:10.100710984-013-9127-9

Neilson, C., Bharadwaj, P., & Løken, K. V. (2013). Early life health interventions and academic achievement. *The American Economic Review, 103*(5), 1862–1891. doi:10.1257/aer.103.5.1862 PMID:29533048

Nelsen, J., Lott, L., & Glenn, S. (2002). *A'dan Z'ye pozitif disiplin (M. Ersin, Çev.)*. İstanbul: Hayat.

Nelsen, J., Lott, L., & Glenn, S. (2002). *Çocuk Gelişiminde a'dan z'ye pozitif disiplin (M. Ersin Çev.)*. İstanbul: Hayay Yayıncılık.

Ng, K., Jarvinen, A., & Bellugi, U. (2014). Toward a deeper characterization of the social phenotype of Williams Syndrome. *The Association Between Personality and Social Drive Research in Developmental Disabilities, 35*(8), 1838–1849. PMID:24794322

Nicoladis, E. (1998). First clues to the existence of two input language: Pragmatic And Lexical Differentiation In Bilingual Child. *Bilingualism: Language and Cognition, 1*(2), 105–116. doi:10.1017/S1366728998000236

Northrop, L., & Killeen, E. (2013). A Framework for using iPads to build early literacy skills. *The Reading Teacher, 66*(7), 531–537. doi:10.1002/TRTR.1155

OBADER. (2013). *Okul öncesi eğitim programı ile bütünleştirilmiş aile destek eğitim rehberi*. Ankara: Millî Eğitim Bakanlığı Temel Eğitim Genel Müdürlüğü.

Ocak, Ş., & Arda, T. B. (2014). Okul Öncesi Dönemde Önleyici Müdahale Edici Programların Karşılaştırılması. *Hacettepe Üniversitesi Eğitim Fakültesi Dergisi, 29*(4), 171–188.

Odom, S. L., Boyd, B. A., Hall, L. J., & Hume, K. (2010). Evaluation of comprehensive treatment models for individuals with autism spectrum disorders. *Journal of Autism and Developmental Disorders, 40*(4), 425–443. doi:10.100710803-009-0825-1 PMID:19633939

Odom, S. L., & Wolery, M. (2003). A Unified Theory of Practice in Early İntervention/Early Childhood Special Education: Evidence-Based Practices. *The Journal of Special Education, 37*(3), 164–173. doi:10.1177/00224669030370030601

Oebel, G. (2008). *So-called" Alternative FLL-Approaches*. GRIN Verlag.

Oğuzkan, Ş., & Oral, G. (1997). *Okulöncesi eğitimi*. İstanbul: Milli Eğitim Basımevi.

Oktay, A., & Unutkan, Ö. P. (2005). Erken Çocuklukta Gelişim ve Eğitimde Yeni Yaklaşımlar. In M. Sevinç (Ed.), İlköğretime hazır oluş ve okul öncesi eğitimle ilköğretimin karşılaştırılması (pp, 145-155). Istanbul, Turkey: Morpa Kültür Yayınları.

Oktay, A. (2000). *Magic years of life: Preschool period.* İstanbul: Epsilon Publications. (in Turkish)

Oktik, N., Çukur, C. Ş., Ünal, H., Gökalp-Yılmaz, G., Özbek, Ç., Doğan, V., & Hasırcı, D. (2010). Sosyal risk analizi: Mardin proje raporu örneği. Muğla: Muğla Üniversitesi.

Okumuş, E. (2009). Sokak çocuklarının sosyolojisi-Diyarbakır örneği. *Dinbilimleri Akademik Araştırma Dergisi, IX*(1), 9–37.

Okutan, E., Akbaş, M. G. (2019). 15-24 Yaş arası Öğrencilerin Kariyer Kaygılarını İncelemeye Yönelik Literatür Araştırması. *Siyaset, Ekonomi ve Yönetim Araştırmaları Dergisi, 7*(1), 33-41.

Olaya, B., Moneta, M. V., Pez, O., Bitfoi, A., Carta, M. G., Eke, C., ... Kovess, V. (2015). Country-level and individual correlates of overweight and obesity among primary school children: A cross-sectional study in seven European countries. *BMC Public Health, 15*(475), 1–12. PMID:25952506

Olds, A. (2001). *Child care design guide*. New York: McGraw Hill.

Oller, D. K. (2000). *The emergence of the speech capacity*. Mahwah, NJ: Erlbaum. doi:10.4324/9781410602565

Olsen, O., & Clausen, J. A. (2012). Planned hospital birth versus planned home birth. *Cochrane Database of Systematic Reviews, 12*(9), 1–26. PMID:22972043

Ömeroğlu, E. (2007). Cognitive and language development in school children. In E. Ömeroğlu & İ. Ulutaş (Eds.), Child and adolescent development (pp. 35-61). İstanbul: Morpa Publications (in Turkish)

Ömeroğlu, E., & Kandir, A. (2005). *Bilişsel Gelişim*. İstanbul: Morpa Kültür Yayınları.

Ömeroğlu, E., & Kandır, A. (2005). *Bilişsel gelişim*. İstanbul: Morpa.

Ömeroğlu-Turan, E., Tezel Şahin, F., Turla, A., & Can, M. (1999). *Anne çocuk eğitimi programının annelerin çocuk gelişimi ile ilgili bilgilerine ve ev ortamlarının düzenlenmesine etkisi. In Ulusal Eğitim Bilimleri Kongresi Bildirgeleri* (pp. 306–325). Eskişehir: Anadolu Üniversitesi Yayınları.

Öncü, E., Kurt, Ö.A., & Esenay, F.I., & Özer, F. (2012). Çalışan çocukların ailede istismarı. *Turkish Journal of Public Health, 10*(3), 128–140.

Önder, A. (2003). *Ailede iletişim: Konuşarak ve dinleyerek anlaşalım*. İstanbul: Morpa.

Orhan, S., & Genç, K. G. (2015). Engellilere yönelik ülkemizdeki özel eğitim hizmet uygulamaları ve örnek ülke karşılaştırması. *Sosyal Politika Çalışmaları Dergisi, 35*(2), 115–146.

Ortega, L. (2014). *Understanding second language acquisition*. Routledge. doi:10.4324/9780203777282

Oswalt, A. (2019). Infancy Physical Development: Gross Motor Skills. Grace Point Wellness. Retrieved from https://www.gracepointwellness.org/461-child-development-parenting-infants-0-2/article/10109-infancy-physical-development-gross-motor-skills

Otto, B. (2006). *Language development in early childhood*. Upper Saddle River, NJ: Pearson.

Ou, S. R., & Reynolds, A. J. (2006). Early Childhood Intervention And Educational Attainment: Age 22 findings from the Chicago Longitudinal Study. *Journal of Education for Students Placed at Risk*, *11*(2), 175–198. doi:10.120715327671espr1102_4

Owens, E. R. (1984). *Language development: An introduction*. Ohio: Merrill Publications.

Owens, R. (1999). *Language disorders: A functional approach to assessment and intervention*. Boston: Allyn & Bacon.

Öz, İ. (2002). Çocuğum neden çalıyor? *Çoluk Çocuk Anne-Baba-Eğitimci Dergisi, 21*, 20-21.

Özakar, S. ve Gözen, D. (2012). Erken çocukluk döneminde masturbasyon sorunu olan çocuğa hemşirelik yaklaşımı. *Ege Üniversitesi Hemşirelik Fakültesi Dergisi*, *28*(1), 113–122.

Özay, H. (2004). Ailenin çocuğa ilişkin kabul ve beklenti düzeyleri. In *Özel Eğitim Kongresi Bildirileri Özel Eğitimden Yansımalar*. Ankara: Kök.

Özbar N. (2007). *Hareket eğitimi programının 4-6 yaş grubu çocuklarda motor beceri ve vücut kompozisyonu üzerine etkisinin incelenmesi* [Doktora TeziM]. Ü. Sağlık Bilimleri Enstitüsü.

Özbay, H, Şahin, N, Hınçal, G, Güngör, S., Öztürk, E, Mavili-Aktaş, A., Aybaş, M., Göka, E. (1991). Ergenlikte Benlik İmajı: Çalışan ve öğrenci ergenler arasında karşılaştırmalı bir çalışma. *3P Dergisi*, *2*(2), 82-95.

Özbay, Y. (2003). Gelişim ve öğrenme psikolojisi. Trabzon: Akademi Kitabevi.

Özbay, H., & Öztürk, E. (1992). *Gençlik*. İstanbul: İletişim Yayıncılık A.Ş.

Özbey, S. (2015). Çocuklarda uyum ve davranış problemleri. In Y. Aydoğan & G. Gültekin Akduman (Eds.), Çocuk ruh sağlığı (pp. 173-188). Ankara: Eğiten Kitap.

Özcan, M. (2018). Comparison of attitudes of gifted and non-gifted students towards physical education and sports lesson. Unpublished Master's Thesis, Health Sciences Institute, Ondokuz Mayıs University, Samsun.

Özçerezci-Güner, Ö., & Kavlak, O. (2016). Doğum. In Ü. Sevil & G. Ertem (Eds.), Perinatoloji ve Bakım (pp. 299-372). Azmir: Ankara Nobel Tıp Kitabevi.

Özdal, F. (2003). İlköğretim dört ve beşinci sınıfa devam eden anne-babası ile yaşayan ve baba yoksunu olan çocukların kaygı düzeylerinin incelenmesi [Yüksek Lisans tezi]. Ankara Üniversitesi.

Özdemir, A. (2018). *The effect of yoga on burn patients self-esteem and body image* [doctoral dissertation]. University of Inönü, Malatya.

Özdemir, Y. (2009). Ergenlik döneminde benlik kurgusu gelişiminin kültür ve aile bağlamında incelenmesi [Doktora tezi]. Ankara Üniversitesi.

Özdemir, Y., & Çok, F. (2011). Autonomy Development in Adolescence. *Türk Psikolojik Danışma ve Rehberlik Dergisi*, *4*(36), 152–164.

Özel, E., & Zelyurt, H. (2016). Anne baba eğitiminin aile çocuk ilişkilerine etkisi. *Sosyal Politika Çalışmaları Dergisi, 36*, 9–34.

Özen, A. (2011). Aile eğitimi. In İ.H. Diken (Eds.), Özel Eğitime Gereksinimi Olan Öğrenciler ve Özel Eğitim (pp. 89-107). Ankara: Pegem.

Özenoğlu, A., Özdemir, A., Tecer, Ö., Balcı, H., Uğurlu, S., & Tüzün, Ş. ve Uğur M. (2008). Anoreksiya nevroza tedavisine ekip yaklaşımı. *Gözde Tıp Dergisi, 23*(2), 66–69.

Özenoğlu, A., & Ünal, G. (2015). Açık ve yoksulluğun çocuklarda saldırganlık ve şiddet davranışları ve ilişkisi. *Gümüşhane Üniversitesi Sağlık Bilimleri Dergisi, 4*(1), 162–179.

Özer, D. S. & Özer, K. (2000). *Çocuklarda Motor Gelişim*. İstanbul: Kazancı

Özer, D. S., Özer, K. (2000). *Çocuklarda Motor Gelişim*. İstanbul: Kazancı.

Özer, Y.E., & Çolak, M. (2015). Yoksulluk, dezavantajlı grup ve suç ilişkisine yönelik bir alan araştırması: İzmir kadın sığınma evleri örneği. *İktisadi ve İdari Bilimler Dergisi, 29*(3), 479-491.

Özer, D. S., & Özer, K. (2009). *Çocuklarda motor gelişim*. Ankara: Nobel Yayınları.

Özgen-Macon, B. (2010). Fötal santral sinir sisteminin manyetik rezonans görüntülemesi. *Hacettepe Tıp Dergisi, 41*, 155–161.

Özgün, Ö. (2008). Sınıfta istenmeyen davranışlar, nedenleri ve yönetimi. In Y. Aktaş Arnas & F. Sadık, (Eds.), Okul öncesi eğitimde sınıf yönetimi (pp. 243-276). Ankara: Kök Yayıncılık.

Özgün, Ö. (2011). Ahlak gelişimi. In M. E. Deniz (Ed.), Erken cocukluk döneminde gelişim. (pp.285-304). Ankara: Ertem Basım Ltd. Şti.

Öz, İ. (2003). *Çocukta uyum ve davranış bozuklukları*. Ankara: Kök Yayıncılık.

Özkan, Y. Ö. (2009). Effect of inclusion preparatory activities of students with normal development in providing social acceptance of individuals with special needs. Unpublished Master's Thesis, Institute of Educational Sciences, Anadolu University, Eskisehir.

Ozokcu, O. (2013). Mental disability, visual impairment, hearing impairment. In I. Diken (Ed.), Inclusion in primary education (p. 56-87). Ankara: Pegem Academy.

Özsoy, S. (2014). Adölesan annelerin emzirmeye yönelik düşünce ve uygulamaları. *F.N. Hemşirelik Dergisi, 22*(2), 84–93.

Öztabak, M., & Erdoğan, İ. (2017). Lise Öğrencilerinin Düşünme Stilleri ile Karar Verme Stilleri Arasındaki İlişkinin İncelenmesi. *HAYEF Journal of Education, 14*(2), 33–57.

Öztürk Çopur, E., Ulutaşdemir, N., & Balsak, H. (2015). Çocuk ve suç. *Hacettepe Üniversitesi Sağlık Bilimleri Dergisi, 1*(2), 120–124.

Özyürek, A., Çetin, A. & Yıldırım, R. (2018). Aile hekimi ve öğretmenlerin çocuk ihmal ve istismarı konusundaki görüş ve tutumları. *İnsan ve Toplum Bilimleri Araştırmaları Dergisi, 7*(1), 436-453.

Özyürek, M. (1985). Orthopedic defective placement in educational environments. *Ankara University Journal of the Faculty of Educational Sciences, 17*(1).

Özyürek, A., Bedge, Z., & Yavuz, N. F. (2014). Okul öncesi çocuklarının sosyal becerileri ile yakın çevredeki yetişkin etkileşimleri arasındaki ilişki. *Sosyal Bilimler Dergisi, 16*(2), 115–134.

Ozyürek, A., Ozkan, I., Begde, Z., & Yavuz, N. F. (2015). Okul öncesi dönem çocuklar için hareket ve sosyal becerilerin gelişiminin desteklenmesine yönelik eğitim programı. *International Journal of Science Culture and Sport*, *3*(11), 2148–1148. doi:10.14486/IJSCS261

Özyürek, M. (2006). *Changing attitudes towards people with disabilities*. Ankara: Kök Publishing.

Özyurt, B. E. (2012). Gelişim konularına genel bakış. In A. Kaya (Ed.). Eğitim psikolojisi (pp. 1-27). Ankara: Pegem A Yayıncılık.

Ozyurt, B. E. (2012). Gelişim konularına genel bakış. In A. Kaya (Ed.), *Eğitim psikolojisi* (pp. 1–27). Ankara: Pegem A Yayıncılık.

Paap, K. R., Johnson, H. A., & Sawi, O. (2015). Bilingual advantages in executive functioning either do not exist or are restricted to very specific and undetermined circumstances. *Cortex*, *69*, 265–278. doi:10.1016/j.cortex.2015.04.014 PMID:26048659

Paktuna Keskin, S. (2012). *Çocuk davranışlarındaki korkuyu tanımak ve başetmek*. İstanbul: Boyut Yayınları.

Pamukcu, M. (2016). Perceptions and competencies of primary school principals in the education of children with special needs bir araştırma A research in Maltepe district of Istanbul Province. Unpublished Master's Thesis, Institute of Social Sciences, Maltepe University, Istanbul.

Papalia, D. E., Olds, S. W., & Feldman, R. D. (2004). *Human development* (9th ed.). New York: McGraw Hill.

Parke, R., & Gauvain, M. (2009). *Child psychology: A contemporary viewpoint* (8th ed.). USA: McGraw Hill.

Park, J., Vincent, D., & Hastings-Tolsma, M. (2007). Disparity in prenatal care among women of colour in the USA. *Midwifery*, *23*(1), 28–37. doi:10.1016/j.midw.2005.08.002 PMID:16842895

Pascoe, M. C., & Bauer, I. E. (2015). A systematic review of randomised control trials on the effects of yoga on stress measures and mood. *Journal of Psychiatric Research*, *68*, 270–282. doi:10.1016/j.jpsychires.2015.07.013 PMID:26228429

Paul, S. T. (1996). Search for semantic inhibition failure during sentence comprehension by younger and older adults. *Psychology and Aging*, *11*(1), 10–20. doi:10.1037/0882-7974.11.1.10 PMID:8726366

Pehlivan, D., Çefle, K., Öztürk, Ş., Akbulut, F.M., & Palandüz, Ş. (2013). 47 Xİ(Xq)Y karyotipi saptanan bir Kleinfelter sendromu olgusu. *İstanbul Tıp Fakültesi Dergisi, 71*, 91-93.

Pehlivan, H. (2005). *Oyun ve Öğrenme*. Ankara: Anı Yayıncılık.

Pekşen Süslü, D. (2002). *Çalışan ve öğrenci ergenlerde kimlik duygusu kazanımının araştırılması*. Ankata Üniv. Eğitim Bilimleri Enstitüsü. Yüksek Lisans Tezi.

Pelendecioğlu, B., & Bulut, S. (2009). Çocuğa yönelik aile içi fiziksel istismar. *Abant İzzet Baysal Üniversitesi Dergisi*, *9*(1), 49–62.

Pelham, S. D., & Abrams, L. (2014). Cognitive advantages and disadvantages in early and late bilinguals. *Journal of Experimental Psychology. Learning, Memory, and Cognition*, *40*(2), 313–325. doi:10.1037/a0035224 PMID:24294916

Perdahlı Fiş, N., Arman, A., Ay, P., Topuzoğlu, A., Güler, A. S., Gökçe İmren, S., & Ersu, R., & Berkem, M. (2010). Çocuk uyku alışkanlıkları anketinin Türkçe geçerliliği ve güvenilirliği. *Anadolu Psikiyatri Dergisi*, *11*, 151–160.

Perrotta, D. (2015). Neonatal care in labor and delivery room. In C. Capogna (Ed.). Epidural labor analgesia: Childbirth without pain (pp. 221-235). Springer. doi:10.1007/978-3-319-13890-9_17

Perry, J. P. (2001). *Outdoor play: Teaching strategies with young children*. New York: Teachers College Press.

Petropoulou, H., Stratigos, J., & Katsambas, A. D. (2006). Human immunodeficiency virus infection and pregnancy. *Clinics in Dermatology*, *24*(6), 536–542. doi:10.1016/j.clindermatol.2006.05.003 PMID:17113971

Peykerli, G. (2003). Ölümcül hastalıklara psikolojik yaklaşım. *C.Ü. Tıp Fakültesi Dergisi*, *25*(4), 62–65.

Peynirci, H., & Ertürk, E. (2012). Kleinfelter syndrome. *Turk Jern*, *17*, 63–70.

Phillips, L. M., Norris, S. P., & Anderson, J. (2008). Unlocking the door: Is parents' reading to children the key to early literacy development? *Canadian Psychology*, *49*(2), 82–88. doi:10.1037/0708-5591.49.2.82

Piaget, J. (2007). Çocukta dil ve düşünme. Ankara: Palme Yayıncılık.

Piaget, J. (1956). *The language and thought of a child*. New York, NY: Harcourt Brace and Co.

Pierce, C. D., Reid, R., & Epstein, M. H. (2004). Teacher-mediated interventions for children with EBD and their academic outcomes: A review. *Remedial and Special Education*, *25*(3), 175–188. doi:10.1177/07419325040250030501

Pignot Shahov, V. (2012). Measuring L2 Receptive and Productive Vocabulary Knowledge. *University of Reading Language studies working papers*, *4*, 37-45

Pınar, S., & Erkut, O. (2000). *Artistik Jimnastik Yaş Grupları Gelişim Programı*. Ankara: G.S.G.M. Basımevi.

Pise, V., Pradhan, B., & Gharote, M. (2018). Effect of yoga practices on psycho-motor abilities among intellectually disabled children. *Journal of Exercise Rehabilitation*, *14*(4), 581–585. doi:10.12965/jer.1836290.145 PMID:30276177

Plak, R. D., Merkelbach, I., Kegel, C. A. T., Von-Ijzendoorn, M. H., & Bus, A. G. (2016). Brief computer interventions emergent academic skills is susceptible children: A gene by environment experiment. *Learning and Instruction*, *45*, 1–8. doi:10.1016/j.learninstruc.2016.06.002

Plaza, M., & Cohen, H. (2006). The Contribution of Phonological Awareness and Visual Attention in Early Reading and Spelling. *Dyslexia*, *13*(1), 67–76. doi:10.1002/dys.330 PMID:17330736

Pletcher, C. L., & Younggren, N. (2011). Early intervention-IDEA Part C: Service Delivery Approaches And Practices. In C. Groark & S. P. Maude (Eds.), Early Childhood İntervention: Shaping The Future For Children With Special Needs And Their Families (pp. 1-39). USA: ABC-CLIO, LLC.

Polat Uluocak, G. (2009). İç göç yaşamış ve yaşamamış çocukların okulda uyumu. *Dokuz Eylül Üniversitesi Buca Eğitim Fakültesi Dergisi*, *26*, 35–44.

Polat, O. (2016). Şiddet. *MÜHF - HAD*, *22*(1), 15-34.

Polat, G. (2007). İç göçün çocuk ruh sağlığına etkisi ve sosyal hizmet müdahalesi. *Toplum ve Sosyal Hizmet*, *18*(1), 89–106.

Posada, G. (2009). Attachment. In J. B. Benson & M. M. Haith (Eds.), *Social and emotional development in infancy and early childhood* (pp. 30–41). Academic Press.

Potter, M., Spence, J. C., Boulé, N. G., Stearns, J. A., & Carson, V. (2017). Associations between physical activity, screen time, and fitness among 6- to 10-year-old children living in Edmonton, Canada. *Applied Physiology, Nutrition, and Metabolism*, *42*(5), 487–494. doi:10.1139/apnm-2016-0419 PMID:28177727

Powell, L. H., & Cassidy, D. (2001). *Family life education. An introduction*. London: Mayfield Publishing.

Pradhan, B., & Nagendra, H. (2010). Immediate effect of two yoga-based relaxation techniques on attention in children. *International Journal of Yoga*, *3*(2), 67–69. doi:10.4103/0973-6131.72632 PMID:21170232

Prado, E. L., & Dewey, K. G. (2014). Nutrition and brain development in early life. *Nutrition Reviews*, *72*(4), 267–284. doi:10.1111/nure.12102 PMID:24684384

Prechtl, H. F. R. (2004). Principles Of Early Motor Development In The Human. In A.F. Kalverboer, B. Hopkıns, & R. Geuze (Eds.), Motor Development in Early And Later Childhood: Longitudinal Approaches (pp. 35-51). New York: Cambridge.

Prendergast, T. (2013). Growing readers: A critical analysis of early literacy content for parents on Canadian public library websites. *Journal of Library Administration*, *53*(4), 234–254. doi:10.1080/01930826.2013.865389

Puckering, C., Pickles, A., Skuse, D., Heptinstall, E., Dowdney, L., & Zur-Szpiro, S. (1995). Mother-child interaction and the cognitive and behavioural development of four-year-old children with poor growth. *Journal of Child Psychology and Psychiatry, and Allied Disciplines*, *36*(4), 573–595. doi:10.1111/j.1469-7610.1995.tb02315.x PMID:7650084

Purutçuoğlu, E. (2015). Hasta çocuk hakları: Hasta çocuğa ve ailesine yönelik sosyal destek. In A. Bütün-Ayhan (Ed.), Hasta çocukların gelişimi ve eğitimi (pp. 184-213). Eskişehir: Anadolu Üniversitesi.

Radhakrishma, S. (2010). Application of integrated yoga therapy to increase imitation skills in children with autism spectrum disorder. *International Journal of Yoga*, *3*(1), 26–30. doi:10.4103/0973-6131.66775 PMID:20948898

Rahman, S., Salameh, K., Bener, A., & El Ansari, W. (2010). Socioeconomic associations of improved maternal, neonatal, and perinatal survival in Qatar. *International Journal of Women's Health*, *2*(1), 311–318. doi:10.2147/IJWH.S12426 PMID:21151678

Raikes, H., Pan, B. A., & Luze, G. (2006). Mother-child book reading in low income families: Correlates and outcomes during the first three years of life. *Child Development*, *77*(4), 924–953. doi:10.1111/j.1467-8624.2006.00911.x PMID:16942498

Ramel, W. I, Goldin, P. R., Carmona, P. E., & McQuaid, J. R. (2004). The effects of mindfulness meditation on cognitive processes and affect in patients with past depression. *Cognitive Therapy and Research*, *28*(4), 433–455. doi:10.1023/B:COTR.0000045557.15923.96

Ranweiler, L. (2004). *Preschool Readers and Writers: Early Literacy Strategies for Teachers* (N. Brickman & H. Barton, Eds.). Ypsilanti, Michigan: High/Scope Press.

Rathus, S. A. (2006). *Childhood voyage in development (2nd ed.)*. New York: Thomson Wadsworth.

Rathus, S. A. (2006). *Childhood: voyages in development*. Belmont, CA: Thomson/Wadsworth.

Rauh, H. (2005). At-risk concept. In B. Hopkins, R. G. Barr, G. F. Micheland, & P. Rochat (Eds.), Encyclopedia of child development (pp. 101-106). New York: Cambridge University.

Raver, C. (1997). Poor Children Gain Social Competence From Sensitive İnteraction With Parent. *The Brown University Child and Adolescent Behavior Letter*, *13*(7), 1–2.

Raynor, P., & Rudolf, M. C. J. (1996). What do we know about children who fail to thrive? *Child: Care, Health and Development*, *22*(4), 241–250. doi:10.1111/j.1365-2214.1996.tb00427.x PMID:8818427

Reach Out and Read. (2013). *Milestones of Early Literacy*. Retrieved from http://www.reachoutandread.org/FileRepository/RORmilestones_English.pdf

Real Play Coalition (RPC). (2019). Retrieved from https://www.realplaycoalition.com/

Reçber, D., & Özen, S. (2005). Trizomi 13, Patou syndrome: Bir olgu sunumu. *Van Tıp Dergisi*, *12*(1), 29–31.

Reedman, S., Boyd, R., & Sakzewskı, L. (2017). Efficacy of a participation-focused therapy intervention on participation in physical activities, health-related quality of life, and behavioural barriers to participation in children with cerebral palsy. *Developmental Medicine and Child Neurology*, *59*(S3), 32–32. doi:10.1111/dmcn.13413

Reich, P. A. (1986). *Language development*. Englewood Cliffs, NJ: Prentice-Hall.

Revelle, G., & Bowman, J. (2017). Parent child dialogue with ebooks. In *Proceedings of the 10th International Conference on Interaction Design and Children* (pp. 346-351). New York: ACM. doi:10.1145/3078072.3079753

Reynolds, A. J., Temple, J. A., Robertson, D. L., & Mann, E. A. (2001). Longterm Effects Of An Early Childhood İntervention on Educational Achievement And Juvenile Arrest: A 15-Year Follow Up Of Low-İncome Children in Public Schools. *Journal of the American Medical Association*, *285*(18), 2339–2347. doi:10.1001/jama.285.18.2339 PMID:11343481

Richards, J. C., & Rodgers, T. S. (2014). *Approaches and methods in language teaching*. Cambridge university press.

Richmond, B., & Peterson, S. (2001). *An introduction to systems thinking*. High Performance Systems., Incorporated.

Riehl, C. M. (2011). *Çocuklarda ikinci dilin önemi.* Retrieved from http://parkakademi.com/haber.html

Rodriguez, E. T., & Tamis-LeMonda, C. S. (2011). Trajectories of the home learning environment across the first 5 years: Associations with children's vocabulary and literacy skills at prekindergarten. *Child Development*, *82*(4), 1058–1075. doi:10.1111/j.1467-8624.2011.01614.x PMID:21679179

Rogoff, B., Sellers, M., Pirrotta, S., Fox, N., & White, S. (1975). Age of assignment of roles and responsibilities in children: A cross-cultural survey. *Human Development*. *18*(5), 353–369. doi:10.1159/000271496

Rolffs, S. (2009). Yine iki dillilik üzerine. *Die Gaste*, *5*(Ocak-Şubat).

Romaine, S. (1995). Bilingualism (2nd ed.). Malden.

Ross, A., & Thomas, S. (2010). The health benefits of yoga and exercise: A review of comparison studies. *Journal of Alternative and Complementary Medicine*, *16*(1), 3–12. doi:10.1089/acm.2009.0044 PMID:20105062

Roth, F. P., Paul, D. R., & Pierotti, A. M. (2006). *Let's Talk For People With Special Communication Needs. Emergent Literacy: Early Reading and Writing Development.* ASHA. Retrieved from https://www.asha.org/public/speech/emergent-literacy.htm

Rothbart, M. K., & Gartstein, M. A. (2009). Temperament. In J. B. Benson & M. M. Haith (Eds.), *Social and emotional development in infancy and early childhood* (pp. 480–495). Academic Press.

Rowland, C. F., & Noble, C. L. (2010). The role of syntactic structure in children's sentence comprehension: Evidence from the dative. *Language Learning and Development*, *7*(1), 55–75. doi:10.1080/15475441003769411

Rowland, T. (1981). Physical fitness in children: Implications for the prevention of coronary artery disease. *Current Problems in Pediatrics*, 9. PMID:7026178

Rubin, K., Coplan, R., Fox, N., & Calkıns, S. (1995). Emotionaltiy, Emotion Regulation and Preschoolers Social Adaptation. Development and Psychopathology, (7), 49-62.

Rudan, I., Kapiriri, L., Tomlinson, M., Balliet, M., Cohen, B., & Chopra, M. (2010). Evidence-based priority setting for health care and research: Tools to support policy in maternal, neonatal, and child health in Africa. *PLoS Medicine*, *7*(7), e1000308. doi:10.1371/journal.pmed.1000308 PMID:20644640

Ruffin, N. J. (2009). *Understanding Growth and Development Patterns Of Infants*. Retrieved from http://Pubs.Ext.Vt.Edu/350/350-055/350-055.html

Rutter, M., & Rutter, M. (1993). *Developing Minds: Challenge and Continuity Across the Lifespan. Ny*. Basic Books.

Rwashana, A. S., Nakubulwa, S., Nakakeeto-Kijjambu, M., & Adam, T. (2014). Advancing the application of systems thinking in health: Understanding the dynamics of neonatal mortality in Uganda. *Health Research Policy and Systems, 12*(1), 1–14. doi:10.1186/1478-4505-12-36 PMID:25104047

Saarnı, C. (2001). Cognition, Context and Goals: Significant Components in Social-Emotional Efffectiveness. *Social Development, 10*(1), 125–127. doi:10.1111/1467-9507.00152

Saban, A. (2001). *Çoklu Zekâ Teorisi ve Eğitim*. Ankara: Nobel Yayın Dağıtım.

Sadıkoğlu, G., Gökdağ, R., Alper, Z., Özçakır, A., & Özdemir, H. & Uncu, Y. (2007). Aile psikolojisi ve eğitimi. Anadolu Üniversitesi Yayınları.

Sáez-López, J. M., Román-González, M., & Vázquez-Cano, E. (2016). Visual programming languages integrated across the curriculum in elementary school: A two year case study using "Scratch" in five schools. *Computers & Education, 97*, 129–141. doi:10.1016/j.compedu.2016.03.003

Sağıroglu, N. (2006). Expectations of families with special needs from special education and rehabilitation centers where their children continue. Unpublished Master's Thesis, University of Abant İzzet Baysal.

Sağlam, M. (2017). Aile içi iletişim. In N. Aral (Ed.), *Çocuk ve iletişim* (pp. 308–324). Ankara: Vize.

Sağlam, M., Ulutaş, A., & Çalışkan, Z. (2015). Children's rights in the pediod of infancy]. *Hacettepe University Faculty of Health Sciences Journal, 1*(2), 502–510.

Sağlık Bakanlığı. (2014). *Doğum öncesi bakım yönetim rehberi*. Ankara: Sağlık Bakanlığı Yayınları.

Şahin, S. (2015). Okul Öncesi Dönemde Motor Gelişim. In F. Turan, A.İ. Yükselen (Eds.), Çocuk Gelişimi 2 Okul Öncesi Döneminde Gelişim (pp. 24-35). Ankara:Hedef

Şahin, S. (2017). Portage Erken Çocukluk Dönemi Eğitim Programı. In Z.F. Temel (Ed.), Erken Çocukluk Eğitiminde Yaklaşımlar ve Programlar İçinde (pp. 327-353). Ankara: Vize Yayıncılık.

Şahin, S. (2018). Çocuk gelişimine giriş. In N. Aral & Z. F. Temel (Eds.), Çocuk gelişimi (pp. 16-31). Ankara: Hedef CS Basın Yayın.

Şahiner, Ü. M., Yurdakök, K., Kavak, U. S., & Çetin, İ. (2001). Tıbbi açıdan çocuk istismarı. *Katkı Pediatri Dergisi, 22*, 276.

Şahin, F. (2001). Çocuk istismarı ve ihmalinin önlenmesinde hekimin rolü. *Sürekli Tıp Eğitimi Dergisi, 10*, 246.

Şahin, S., & Aral, N. (2012). Aile içi iletişim. *Ankara Sağlık Bilimleri Dergisi, 1*(3), 55–66. doi:10.1501/Asbd_0000000029

Şahin, S., Işıtan, S., & Gündüz, S. (2011). *Özel gereksinimli çocuklar ve eğitimleri*. Ankara: Eğiten Kitap.

Sak, E. M., Özkul, Ö., Evsen, S. M., Sak, S., & Eviroğlu, O. (2009). Gebelik anemisinin perinatal sonuçlara etkisi. *Dicle Tıp Dergisi, 36*(1), 39–42.

Salem, Y., Liu, H., Young, A., Tolbert, M., & Elokda, A. (2018). Effects of yoga exercise program on physical function in young children with Down syndrome. *Developmental Medicine and Child Neurology, 60*(S3), 32–32. doi:10.1111/dmcn.47_14017

Salmon, P., Lush, E., Jablonski, M., & Sephton, S. E. (2009). Yoga and mindfulness: Clinical aspects of and ancient mind/body practice. *Cognitive and Behavioral Practice*, *16*(1), 59–72. doi:10.1016/j.cbpra.2008.07.002

Sameroff, A. J., & Fiese, B. H. (2000). Models of Development And Developmental Risk. In C. H. Zeanah Jr., (Ed.), *Handbook Of İnfant Mental Health* (pp. 3–19). New York: The Guilford Press.

Samovar, L. A., McDaniel, E. R., Porter, R. E., & Roy, C. S. (2015). *Communication between cultures*. Nelson Education.

Sampson, M. R., & Briggs, L. (1981). What Does Research Say About Beginning Reading? *Reading Horizons: A Journal of Literacy and Language Arts*, *21*(2). Retrieved from https://scholarworks.wmich.edu/reading_horizons/vol21/iss2/6

Samurçay, N. (1983). Zekâ ve yaratıcılık. *Eğitim ve Bilim Dergisi*, *8*(45), 4–12.

San Bayhan, P., & Artan, İ. (2011). *Çocuk Gelişimi ve Eğitimi*. Istanbul, Turkey: Morpa Yayınları, İstanbul.

San-Bayhan, P., & Artan, İ. (2009). Çocuk gelişimi ve eğitimi. İstanbul: Morpa.

San-Bayhan. P ve Artan, İ. (2009). Çocuk Gelişimi ve Eğitimi. İstanbul: Morpa Yayınları.

Sandall, S. R., Smith, B. J., McLean, M. E., & Alison, B. R. (2002). Qualitative research in early intervention/early childhood special education. *Journal of Early Intervention*, *25*(2), 129–136. doi:10.1177/105381510202500210

Santrock, J. W. (2012). Yaşam boyu gelişim. Ankara: Nobel.

Santrock, J. W. (2013). Life-span development (14th ed.). McGraw-Hill.

Santrock, J. W. (2011). *Lifelong development*. Mc Graw Hill.

Santrock, J. W. (2011). *Life-span development* (13th ed.). Ankara: Nobel Yayın.

Santrock, J. W. (2016). *Yaşam boyu gelişim (13. Baskı)*. Ankara: Nobel.

Santrock, J. W. (2016). *Yaşam boyu gelişim*. Ankara: Nobel.

Santrock, J. W. (2017). *Educational psychology* (6th ed.). New York: McGraw-Hill Education.

Santrock, J. W. (2018). *Lıfe-span development* (17th ed.). New York: McGraw-Hill Education.

Sapsağlam, Ö., & Ömeroğlu, E. (2016). Examining the effect of social values education program being applied to nursery school students upon acquiring skills. *Educational Research Review*, *11*(3), 1261–1271.

Sarac, S. (2019). Examination of attitudes of special gifted 5th, 6th, 7th grade students towards individuals with special needs affected by disability. Unpublished Master's Thesis, Institute of Educational Sciences, Bahcesehir University, Istanbul.

Saracho, O. N. (2017). Bullying young children's roles, social status and prevention programmes. *Early Child Development*, *187*(1), 68–79. doi:10.1080/03004430.2016.1150275 PMID:28042899

Saracho, O. N. (2017). Parents' shared storybook reading – learning to read. *Early Child Development and Care*, *187*(3-4), 554–567. doi:10.1080/03004430.2016.1261514

Saraswati, S. P. (2007). *Yoga Education for Children. Trust*. Munger, Bihar: Yoga Publications.

Şarbat, G. & Demirkol, M. (1999). Yeme bozuklukları-nutrisyonel bakış. In A. Ekşi (Ed.), Ben hasta değilim (pp. 212-222). İstanbul: Nobel Tıp Kitabevi Ltd. Şti.

Sawyer, W. F. (2009). *Growing up with literature* (5th ed.). USA: Delmar.

Sax, K. (1918). The Behavior of the Chromosomes in Fertilization. *Genetics*, *3*(4), 309–327. PMID:17245907

Sayan, H. (2016). Okul öncesi eğitimde teknoloji kullanımı. *Yüzyılda Eğitim ve Toplum*, *5*(3), 67–83.

Saygılı, S. (2013). *Çocuklarda davranış bozuklukları*. İstanbul: Elit.

Sayın, Y., Usanmaz, A., & Aslangiri, F. (2016). Uluslararası göç olgusu ve yol açtığı etkiler: Suriye göçü örneği. *KMÜ Sosyal ve Ekonomik Araştırmalar Dergisi*, *18*(31), 1–13.

Schelletter, C. (2002). The effect of form similarity on bilingual children's lexical development. bilingualism: Language and cognition. *Language and Cognition*, *5*, 93–107.

Scherrer, K. (2018). Stories, Songs, Stretches Creating playful story times with yoga and movement. American Libraries Magazine. Retrieved from https://americanlibrariesmagazine.org/2018/01/02/stories-songs-stretches-storytime-yoga/

Schmid, R. F., Miodrag, N., & DiFrancesco, N. (2008). A human computer technology analysis and recommendations. *The Future of Children*, *10*(2), 4–30.

Schmidt, H., Gostin, L. O., & Emanuel, E. J. (2015). Public health, universal health coverage, and Sustainable Development Goals: Can they coexist? *Lancet*, *386*(9996), 928–930. doi:10.1016/S0140-6736(15)60244-6 PMID:26138141

Schore, A. N. (1999). *Affect regulation and the origin of the self: The neurobiology of emotional development*. Florence, KY: Psychology Press.

Schunk, D. H. (2009). *Öğrenme Teorileri: Eğitimsel Bir Bakış*. Ankara: Nobel Akademik Yayıncılık, Ankara.

Scime, M., Cook-Cottone, C., Kane, L., & Watson, T. (2006). Group prevention of eating disorders with fifth-grade females: Impact on body dissatisfaction, drive for thinness, and media influence. *Eating Disorders*, *14*(2), 143–155. doi:10.1080/10640260500403881 PMID:16777811

Scribner, E. (2013). Exploring Different Factors of Language Development. 2013 Awards for Excellence in Student Research & Creative Activity - Documents. The Keep. Retrieved from http://thekeep.eiu.edu/lib_awards_2013_docs/1

Şeker, G., & Kaya, A. (2019). Mesleki karar verme programının onuncu sınıf öğrencilerinin mesleki kararsızlık düzeyleri üzerinde etkililiğinin sınanması. *Gazi Üniversitesi Eğitim Fakültesi Dergisi*, *39*(1), 457–476.

Selçuk, Z. (1997). Eğitim psikolojisi. Ankara: Pegema Yayınları.

Selçuk, Z. (2000). *Gelişim psikolojisi.* Ankara: Nobel.

Semerci, B. (2015). *Birlikte büyütelim*. İstanbul: Alfa.

Semwanga, A. R., Nakubulwa, S., & Adam, T. (2016). Applying a system dynamics modelling approach to explore policy options for improving neonatal health in Uganda. *Health Research Policy and Systems*, *14*(1), 35. doi:10.118612961-016-0101-8 PMID:27146327

Şen-Beytut, D., Bolışık, B., Solak, U. & Seyfioğlu, U. (2009). Çocuklarda hastaneye yatma etkilerinin projektif yöntem olan resim çizdirme yoluyla incelenmesi. *Maltepe Üniversitesi Hemşirelik Bilim ve Sanatı Dergisi*, *2*(3), 35–44.

Senemoğlu, N. (2005). *Gelişim öğrenme ve öğretim kuramdan uygulamaya (12. Baskı)*. Ankara: Gazi Kitabevi.

Senemoğlu, N. (2007). *Gelişim öğrenme ve öğretim- kuramdan uygulamaya*. İstanbul: Gönül Yayınevi.

Senemoğlu, N. (2007). *Gelişim Öğrenme ve Öğretim*. Ankara: Gönül.

Senemoğlu, N. (2007). *Gelişim öğrenme ve öğretim: Kuramdan uygulamaya*. Gönül Yayıncılık.

Senemoğlu, N. (2010). *Gelişim, Öğrenme Ve Öğretim: Kuramdan Uygulamaya. 18. Baskı*. Ankara: Pegem Akademi.

Senemoğlu, N. (2013). *Development, learning and teaching. From theory to practice*. Ankara: Gönül Publications. (in Turkish)

Senemoğlu, N. (2013). *Gelişim, öğrenme ve öğretim: Kuramdan uygulamaya*. Ankara: Yargı.

Senemoğlu, N. (2018). *Gelişim, öğrenme ve öğretim*. Ankara: Anı Yayıncılık.

Şener, D. K., & Ocakçı, A. F. (2012). Yoksulluğun çocuk sağlığı üzerine çok boyutlu etkileri. *Ankara Sağlık Hizmetleri Dergisi*, *13*(1), 57–68. doi:10.1501/Ashd_0000000098

Şentürk, Ü. (2012). Parçalanmış aile çocuklarının eğitimdeki başarı/başarısızlık durumu (Malatya örneği 2006). *Sosyal Politika Çalışmaları*, *7*(29), 105–126.

Seven, S. (2018). *Çocuk ruh sağlığı (5. Baskı)*. Ankara: Pegem Akademi.

Sever, S. (2017). *Teaching Turkish and learning* (6th ed.). Ankara: Anı Publications. (in Turkish)

Sevinç, M. (2003). Erken çocuklukta gelişim ve eğitimde yeni yaklaşımlar. İstanbul: Morpa Kültür Yayınları.

Sevinç, M. (2004). *Erken çocukluk gelişimi ve eğitiminde oyun*. Ankara: Morpa Kültür Yayınları.

Sezer, Ş. (2018). Öğretmenlerin sınıf yönetimi tutumlarının öğrencilerin gelişimi üzerindeki ekileri: Fenomenolojik bir çözümleme. *Hacettepe Üniversitesi Eğitim Fakültesi Dergisi*, *33*(2), 534–549.

Shaffer, D., & Kipp, K. (2014). *Developmental psychology: Childhood and adolescence*. USA: Wodsworth Cengage Learning.

Shahaeian, A., Wang, C., Tucker-Drob, E., Geiger, V., Bus, A. G., & Harrison, L. J. (2018). Early shared reading, socio-economic status, and children's cognitive and school competencies: Six years of longitudinal evidence. *Scientific Studies of Reading*, *22*(6), 485–502. doi:10.1080/10888438.2018.1482901

Sharma, H. D. (2004). *Dengeli çocuk yetiştirme*. Eskişehir: Platform.

Sharma, M. (2014). Yoga as an alternative and complementary approach for stress management: A systematic review. *Journal of Evidence-Based Complementary & Alternative Medicine*, *19*(1), 59–67. doi:10.1177/2156587213503344 PMID:24647380

Sheehan, E. (1999). Kaygı bozuklukları: Kaygı, fobiler ve panik ataklar (M. Sağlam, Çev.). İstanbul: Alfa Yayınları.

Shelov, S. P., & Altmann, T. R. (2015). Bebek ve Küçük Çocuk Bakımı (0-5 Yaş). Ankara: İmge Kitapevi.

Shelton, R. C. (2013). *Anksiyete bozuklukları (Çev. İ. Ekmekçi Ertek). In Psikiyatri tanı ve tedavi* (pp. 351–365). Ankara: Güneş Yayınları.

Shonkoff, J. P., & Phillips, D. A. (2002). (Eds.). From neurons to neighborhood: The science of early childhood development. Washington, DC: National Academy Press. Retrieved from http://files.eric.ed.gov/fulltext/ED446866.pdf

Shonkoff, J. P., Boyce, W. T., & McEwen, B. S. (2009). Neuroscience, molecular biology, and the childhood roots of health disparities: Building a new framework for health promotion and disease prevention. *Journal of the American Medical Association*, *301*(21), 2252–2259. doi:10.1001/jama.2009.754 PMID:19491187

Shonkoff, J. P., & Phillips, D. A. (2001). From neurons to neighborhoods: the science of early childhood development: an introduction. *Zero to Three*, *21*, 4–7.

Shrier, C. (2013). *ABC's of Early Literacy: The importance of developing early literacy skills*. Retrieved from https://www.canr.msu.edu/news/abcs_of_early_literacy_the_importance_of_developing_early_literacy_skills

Sifferlin, A. (2013). *The most stressed-out generation? Young Adults.* Time. Retrieved from http://healthland.time.com/2013/02/07/the-most-stressed-outgeneration-young-adults/

Sigelman, C.K. & Rider, E.A. (2015). *Life span human development.* USA: Cengage Learning.

Sigelman, C. K., & Rider, E. A. (2017). *Human development across the life span* (9th ed.). Cengage Learning.

Simsek, T., & Yakar, Y. M. (2014). Child and Literature. In T. Şimşek (Ed.), Children's literature from theory to practice (pp. 13- 44). Ankara: Grafiker. (in Turkish)

Şimşek, Ö. F. (2007). *Yapısal Eşitlik Modellemesine Giriş, Temel İlkeler ve Lisrel Uygulamaları.* Ankara: Ekinoks.

Sınar, A. (2006). Children's literature studies in Turkey. *Türkiye Araştırmaları Literatür Dergisi, 4*(7), 175–225.(in Turkish)

Sinclair, D. (1978). *Human Growth After Birth* (3rd ed.). London: Oxford University Press.

Şirin, M. R. (2000). *Children's literature in 99 questions.* İstanbul: Çocuk Vakfı Publications. (in Turkish)

Sirin, M. R. (2007). *A critical view towards children's literature.* Ankara: Kök Publications. (in Turkish)

Skinner, B. F. (1957). *Verbal behavior.* New York: Appleton-Century-Crofts. doi:10.1037/11256-000

Slee, P. T., & Shute, R. (2014). *Child development: Thinking about theories texts in developmental psychology.* New York: Routledge.

Sless, D. (1981). *Learning and visual communication.* Routledge.

Smith, D. E., & Chapel, J. R. (1970). Reading readiness. *Reading Horizons: A Journal of Literacy and Language Arts, 10*(2), 3. Retrieved from https://scholarworks.wmich.edu/reading_horizons/vol10/iss2/3

Smith, C. B. (1994). *Whole language: The debate* (W. Lewis, Ed.). Bloomington, Indiana: EDINFO Press and ERIC Clearinghouse on Reading, English, and Communication.

Smith, D. (1977). *Growth and Its Disorders.* Philadelphia: W.B. Saunders.

Snow, C. E., Burns, M. S., & Griffin, P. (Eds.). (1998). *Preventing Reading Difficulties in Young Children.* Washington, DC: National Academies Press.

Solidtaş, S. (2019). Principal, teacher and parent views on the roles of school principals in providing children with special needs. Unpublished doctoral dissertation, Ankara University, Institute of Educational Sciences, Ankara.

Sørlie, M. A., Idsoe, T., Ogden, T., Olseth, A. R., & Torsheim, T. (2018). Behavioral Trajectories During Middle Childhood: Differential Effects of the School-Wide Positive Behavior Support Model. *Prevention Science, 19*(8), 1055–1065. doi:10.100711121-018-0938-x PMID:30121877

Sperber, D., & Wilson, D. (1986). *Relevance: Communication and cognition* (Vol. 142). Cambridge, MA: Harvard University Press.

Spinazzola, J., Rhodes, A. M., Emerson, D., Earle, E., & Monroe, K. (2011). Application of yoga in residential treatment of traumatized youth. *Journal of the American Psychiatric Nurses Association, 17*(6), 431–444. doi:10.1177/1078390311418359 PMID:21868714

Spira, E. G., Bracken, S. S., & Fischel, J. E. (2005). Predicting improvement after first-grade reading difficulties: The effects of early language, emergent literacy, and behavior skills. *Developmental Psychology, 41*(1), 225–234. doi:10.1037/0012-1649.41.1.225 PMID:15656751

Sprague, J., & Walker, H. (2000). Early identification and intervention for youth with antisocial and violent behavior. *Exceptional Children, 66*(3), 367–379. doi:10.1177/001440290006600307

Staats, A. W. (1971). Linguistic-mentalistic theory versus an explanatory S-R learning theory of language development. In D. I. Slobin (Ed.), *The ontogenesis of grammar* (pp. 103–152). New York: Academic Press.

Staempfli, M. B. (2009). Reintroducing adventure into children's outdoor play environments. *Environment and Behavior, 41*(2), 268–280. doi:10.1177/0013916508315000

Stark, R. E. (1986). Prespeech segmental feature development. In P. Fletcher & M. Garman (Eds.), *Language acquisition* (2nd ed., pp. 149–173). Cambridge, UK: Cambridge University Press. doi:10.1017/CBO9780511620683.010

Staton, L., & Perron, S. (2002). *Baby Om: Yoga for Mothers and Babies.* New York: Holt Paperbacks.

Steinberg, L., Bornstein, M. H., & Vandell, D. L. (2010). *Life-span development: Infancy through adulthood.* Cengage Learning.

Stephenson, A. (1998). Opening up the outdoors: A reappraisal of young children's outdoor experiences. Unpublished Master Thesis, Wellington.

Stephenson, A. (2002). Opening up the outdoors:exploring there lation ship between the in door and outdoor environments of acentre. *European Early Childhood Education Research Journal, 10*(1), 29–38. doi:10.1080/13502930285208821

Stiles, M. (2013). *Structural yoga therapy: adapting to the individual.* San Francisco: Red Wheel/Weiser.

Stillerman, K. P., Mattison, D. R., Giudice, L. C., & Woodruff, T. J. (2008). Environmental exposures and adverse pregnancy outcomes: A review of the science. *Reproductive Sciences (Thousand Oaks, Calif.), 15*(7), 631–650. doi:10.1177/1933719108322436 PMID:18836129

Stirparo, S. (2015). Maternal expectational, information, and satisfaction with labor analgesia. In C. Capogna (Ed.). Epidural labor analgesia: Childbirth without pain (pp. 53-69). Springer.

Strickland, D. S., & Schickedanz, J. A. (2004). *Learning about Print in Preschool: Working with Letters, Words, and Beginning Links with Phonemic Awareness.* Newark, DE: International Reading Association.

Subaşıoğlu, F. (2008). Üniversitelerin bilgi ve belge yönetimi bölümlerinin "engellilik farkındalığı" üzerine bir araştırma. *Bilgi Dünyası, 9*(2), 399–430.

Sucuoğlu, B. (2006). *İlköğretimde Kaynaştırma Uygulamaları-Yaklaşımlar, Yöntemler, Teknikler.* Morpa Kültür Yayınları.

Sucuoglu, B., & Kargin, T. (2006). *Ilkögretimde kaynastirma uygulamalari: Yaklasimlar, yöntemler, teknikler.* Istanbul: Morpa Yayincilik.

Sumar, S. (1998). *Yoga for the special child: A therapeutic approach for infants and children with Down syndrome, cerebral palsy, and learning disabilities.* Buckingham, VA: Special Yoga.

Sumar, S. (1998). Yoga for the special child: A therapeutic approach for infants and children with Down syndrome, cerebral palsy, and learning disabilities. *Special Yoga, 36*(6), 314–319.

Sümer, N. (2000). Yapısal eşitlik modelleri: temel kavramlar ve örnek uygulamalar. *Türk Psikoloji Yazilari. 3*(6), 49–74.

Sümer, N., Gündoğdu-Aktürk, E., & Helvacı, F. (2010). Anne-baba tutum ve davranışlarının psikolojik etkileri: Türkiye'de yapılan çalışmalara toplu bakış. *Türk Psikoloji Yazilari, 13*(25), 42–59.

Summak, M. S., & Summak, A. E. (2005). *Akıllı Bebekler Akademisi; Bebekler İçin Çoklu Zekâ Ve Beyin Geliştirme Rehberi.* İstanbul: Sistem Yayıncılık.

Sürücü, Ş. (2017). The most entertaining teaching for children: Children's yoga. *Uplifers*. Retrieved from https://www.uplifers.com/cocuklar-icin-en-eglenceli-ogreti-cocuk-yogasi/)

Swanson, J. R., Bradley-Johnson, S., Johnson, C. M., & O'dell, A. R. (2009). The Cognitive Abilities Scale-Preschool Form: Studies Of Concurrent Criterion-Related, Construct, And Predictive Criterion-Related Validity. *Journal of Psychoeducational Assessment*, *27*(1), 46–56. doi:10.1177/0734282908322454

Tabachnick, B., & Fidell, L. S. (2007). *Using Multivariate Statistics* (5th ed.). Boston: Pearson Education.

Tahiroğlu, A. Y., Fırat, S., & Diler, R. S. ve Avcı, A. (2005). Erkek çocuklarda yeme bozuklukları; bir anoreksiya nervosa vakası. *Çocuk Sagligi ve Hastaliklari Dergisi*, *48*, 151–157.

Talim, B. (2015). Doğum öncesi gelişim, doğum ve yeni doğan. In F. Turan & A. İ. Yükselen (Eds.), *Çocuk gelişimi 1: Bebeklik döneminde gelişim* (pp. 99–112). Ankara: Hedef.

Tamayo, M., Rebolledo, J., & Besoaín-Saldaña, A. (2017). Monitoring inclusive education in Chile: Differences between urban and rural areas. *International Journal of Educational Development*, *53*, 110–116. doi:10.1016/j.ijedudev.2017.01.002

Tambalis, K. D., Panagiotakos, D. B., Psarra, G., & Sidossis, L. S. (2019). Association of cardiorespiratory fitness levels with dietary habits and lifestyle factors in schoolchildren. *Applied Physiology, Nutrition, and Metabolism*, *44*(5), 539–545. doi:10.1139/apnm-2018-0407 PMID:30321490

Taner Derman, M. & Başal, H. A. (2013). Okul öncesi çocuklarında gözlenen davranış problemleri ile ailelerinin anne-baba tutumları arasındaki ilişki. *Amasya Üniversitesi Eğitim Fakültesi Dergisi*, *2*(1), 115–144.

Tanju, E. H. (2011). Okul öncesi dönemde davranış problemleri ve alınacak önlemler. In G.U. Balat & H. Bilgin (Eds.), Okul öncesi eğitimde sınıf yönetimi (pp. 45-92). Ankara: Eğiten Kitap.

Tanyeri-Bayraktar, B., Bayraktar, S., Hepokur, M., & Kardaş, M. (2015). Hidrosefali ile birliktelik gösteren fetal alkol sendromu. *Bezmialem Science*, *3*, 48–50.

Tarı-Cömert, I., & Kayıran, S.M. (2010). Çocuk ve ergenlerde internet kullanımı. *Çocuk Dergisi, 10*(4), 166-170 doi:10.5222/j.child.2010.166

Tasci, K. N. (2017). Feelings and thoughts of parents and siblings in families with special needs. Unpublished Master's Thesis, Institute of Social Sciences, Cukurova University, Adana.

Taşkın, L. (2016). *Doğum ve kadın sağlığı hemşireliği*. Ankara: Özyurt.

Tavşancıl, E. (2006). *Tutumların Ölçülmesi ve Spss İle Veri Analizi*. Ankara: Nobel Yayın Dağıtım.

Teale, W. H., & Sulzby, E. (1986). *Emergent Literacy: Writing and Reading. Writing Research: Multidisciplinary Inquiries into the Nature of Writing Series*. Norwood, NJ: Ablex Publishing Corporation.

Tekin-Ersan, D., & Kizir, M. (2016). Williams sendromu olan bireylerin ailelerinin deneyimlerinin ve karşılaştıkları güçlüklerin belirlenmesi. *Ankara Üniversitesi Eğitim Bilimleri Fakültesi Özel Eğitim Dergisi*, *17*(3), 299–316. doi:10.21565/ozeleegitinmdergisi.268557

Teksöz, E., & Ocakçı, F. A. (2014). Çocuk hemşireliğinde sanat uygulamaları. *Dokuz Eylül Üniversitesi Hemşirelik Yüksekokulu Dergisi*, *7*(2), 119–123.

Telles, S. (2012). The Effect of Yoga on Mental Health of Children. In U. Nayar (Ed.), Child and Adolescent Mental Health (pp. 219-227). New Delhi: Sage Publications.

Telles, S., Singh, N., & Bhardwaj, A. K. (2013). Effect of yoga or physical exercise on physical, cognitive and emotional measures in children: A randomized controlled trial. *Child and Adolescent Psychiatry and Mental Health*, *7*(37), 1–16. PMID:24199742

Temel, F. (2003). *Aile eğitim modeli: Dünya'da ve Türkiye'deki uygulamalar. Erken çocukluk eğitimi politikaları: yaygınlaşma, yönetişim ve yapılar toplantısı raporu*. Ankara: Anne-Çocuk Eğitim Vakfı.

Temel, Z. F. (2001). *Okul Öncesi Eğitime Aile Katılımı. Gazi Üniversitesi Anasınıfı-Anaokulu Öğretmen El Kitabı*. İstanbul: YA-PA Yayınları.

Temel, Z. F., Bekir, H., & Gül Yazıcı, Z. (2014). *Erken çocuklukta dil edinimi*. Ankara: Vize.

Temizyurek, F. (2003). The importance of children's literature in teaching Turkish. *Türklük Bilimi Araştırmaları*, *13*, 161–167. (in Turkish)

Ten Hoope-Bender, P., De Bernis, L., Campbell, J., Downe, S., Fauveau, V., Fogstad, H., ... Van Lerberghe, W. (2014). Improvement of maternal and newborn health through midwifery. *Lancet*, *384*(9949), 1226–1235. doi:10.1016/S0140-6736(14)60930-2 PMID:24965818

Tercanlı Metin, G. and Gokcay, G. (2014). Story reading in infancy and early childhood *Çocuk Dergisi*, *14*(3), 89-94. (in Turkish)

Tezcan, S., & Eryurt, M. A. (2006). Türkiye'de iç göç ve çocuk sağlığı. *Nufusbilim Dergisi*, *28-29*, 15–28.

Tezel Şahin, F., & Özyürek, A. (2010b). Anne-baba eğitimi ve Türkiye'de uygulanan aile eğitim programları In T. Güler (Ed.), Anne baba eğitimi (pp. 142-158). Ankara: Pegem Akademi.

Tezel Şahin, F., & Özyürek, A. (2010a). *Anne baba eğitimi ve okul öncesinde aile katılımı*. İstanbul: Morpa Kültür Yayınları.

Tezel, A. (2002). Çocuğa yönelik şiddet. *Atatürk Üniversitesi Hemşirelik Yüksekokulu Dergisi*, *5*(1), 93–100.

Thompson, R. A. (2009). Emotional development. In R. A. Schweder (Ed.), *The Chicago companion to the child*. Chicago: University of Chicago Press.

Tipton, L. A. (2014). The parent's role: Shared book reading and the child with asd. Unpublished doctorate dissertation, University of California, Riverside.

Tolani, N., Brooks-Gunn, J., & Kagan, L. S. (2006). Parenting education programs for poor young children: A cross-national exploration. In *Proceedings of the UNICEF/New School International Conference*. Academic Press.

Tompkins, G. E. (2006). *Literacy for the 21st century* (4th ed.). New Jersey: Pearson Prentice Hall.

Topbaş, S. (2006). The components of language. In S. Topbaş (Ed.), Language and concept development (pp. 21-30). Ankara: Kök Publications. (in Turkish)

Topbaş, S. (2010). Dil ve kavram gelişimi. Ankara: Kök Yayıncılık.

Topbaş, S., & Mavis, İ. (2006). Pragmatics development. In S. Topbaş (Ed.), Language and concept development (pp. 123-139). Ankara: Kök Publications. (in Turkish)

Topçuoğlu, T. (2014). Çocuk suçluluğu ve gelişimsel (risk-odaklı) suç önleme. *KMÜ Sosyal ve Ekonomik Araştırmalar Dergisi*, *16*, 217-226.

Toprak, F. (2018). Comparison of life satisfaction, quality of life and social support perceptions of families with mentally retarded and gifted children. Unpublished Master's Thesis, Hacettepe University Institute of Social Sciences, Ankara.

Toran, M. (2010). *Yoksul çocukların kaderi çalışmak: Türkiye'de çocuk emeği*. Ankara: Ütopya.

Torr, J. (2005). Child Development and Teaching Young Children. In A. Talay-Ongan & E. Ap (Eds.), *Language Development in Early Childhood* (pp. 79–98). Canada: Thomson Learning Nelson.

Törüner, E. K., & Büyükgönenç, L. (2011). *Çocuk Sağlığı*. Ankara: Gökçe.

Törüner, E. K., & Büyükgönenç, L. (2012). *Çocuk sağlığı temel hemşirelik yaklaşımları*. Ankara: Göktuğ.

Tovey, H. (201 1). Achieving the balance. In J. White (Ed.), *Outdoor provision in the early years* (pp. 12–22). London: Sage Publication.

Transforming the Workforce for Children Birth Through Age 8. (2015). In *Transforming the Workforce for Children Birth Through Age 8*. doi:10.17226/19401

Traş, Z., & Gökce, Z. (2017). Lise Öğrencilerinin Kariyer Karar Verme Güçlükleri ve Ego Kimlik Statülerinin Bazı Değişkenlere Göre İncelenmesi. *Türk Eğitim Bilimleri Dergisi*, *15*(1), 58–69.

Trawich Swith, J. (2013). *Erken Çocukluk Döneminde Gelişim*. Ankara: Nobel Akademik Yayıncılık.

Trawick Swith, J. (2013). Erken çocukluk döneminde gelişim (Çok kültürlü bir bakış açısı). Ankara: Nobel.

Trawick-Swith, J. (2013). Erken çocukluk döneminde gelişim. Ankara: Nobel.

Tucker-Drob, E. M., Briley, D. A., & Harden, K. P. (2013). Genetic and Environmental Influences on Cognition Across Development and Context. *Current Directions in Psychological Science*, *22*(5), 349–355. doi:10.1177/0963721413485087 PMID:24799770

Tuğrul, B. (2002). Oyun çocuğun en ciddi işidir, *Çoluk Çocuk Dergisi*, 30-31.

Tümkaya, S. (2012). Okul öncesi eğitim döneminde ailenin rolü ve önemi. In S. Tümkaya & F. Gülaçtı (Eds.), *Erken çocukluk eğitimi* (pp. 336–367). Ankara: Pegem.

Tümkaya, S., Aybek, B., & Çelik, M. (2010). Yoksul ailelerden gelen ergenlerde psikososyal bir olgu olarak umutsuzluk ve boyun eğici davranışların incelenmesi. *Uluslararası İnsan Bilimleri Dergisi*, *7*(1), 970–984.

Tuna, M. D., Sümer, H. M., & Diken, İ. H. (2018). Erken müdahale modelleri ve temel kuramlar. In P. Bayhan (Ed.), Erken müdahale (pp. 64-81). Ankara: Hedef CS Basın Yayın.

Tuna, M., Sümer, M., & Diken, İ. H. (2018). Erken Eğitsel Müdahale Modelleri ve Temel Kuramlar. In P. Bayhan (Ed.) Erken Müdahale içinde (pp. 64-81). Ankara: Hedef.

Tur, G., & Turla, A. (2005). *Child, literature and book in early childhood*. İstanbul: Yapa. (in Turkish)

Türkaslan, N. (2007). Boşanmanın çocuklar üzerine olumsuz etkileri ve bunlarla başetme yolları. *Aile ve Toplum*, *3*(11), 99–108.

Türkmen, B. (2005). Sosyal risk altındaki çocuklar, gençler ve yaşlılar. *Gazi Üniversitesi Kastamonu Eğitim Dergisi.*, *13*(1), 11–22.

Tusing, M. E., Maricle, D. E., & Ford, L. (2003). Assessment With The Woodcock-Johnson Iıı And Young Children. In F. A. Schrank & D. P. Flanagan (Eds.), *WJIII Clinical Use And Interpretation: Scientist-Practitioner Perspectives* (pp. 243–283). Academic Press. doi:10.1016/B978-012628982-4/50009-X

Ülgen, G. (1996). *Kavram Geliştirme Kuramlar ve Uygulamalar. Geliştirilmiş ikinci baskı*. Ankara: Setma Baskı.

Ulke Kürkçüoğlu, B. (2010). 0-6 Yaş Arası Çocukların Temel Gelişimsel Özellikleri: Bilişsel gelişim ve dil gelişimi, In İ. H. Diken (Ed.), Erken çocukluk eğitimi. Ankara, Turkey: Pegem Akademi.

Ülke-Kürkçüoğlu, B. (2010). 0-6 Yaş Arası Çocukların Temel Gelişimsel Özellikleri: Bilişsel Gelişim Ve Dil Gelişimi. In I.H. Diken (Ed.). Erken Çocukluk Eğitimi. Ankara: Pegem Akademi.

Ulu, Y. (2018). *Okul öncesi 3-6 yaş grubu çocuklarının davranış problemleri ve anne-baba tutumlarının incelenmesi.* Yayınlanmamış Yüksek Lisans Tezi, Çağ Üniversitesi.

Uluç, G. (2003). *Küreselleşen medya.* İstanbul: Anahtar Kitaplar.

Uluğ, H. (2007). Kuzey Adana'daki çocuk oyun alanlarının bitki seçimi yönünden irdelenmesi. Retrieved from http://library.cu.edu.tr/tezler/6587.pdf

Ünal, F. (1999). Çocukluk mastürbasyonu. In A. Ekşi (Ed.), Ben hasta değilim (pp. 72-84). İstanbul: Nobel Tıp Kitabevi Ltd.

Ünal, F. (2014). Aile eğitimi. A.B. Aksoy (Ed.). Her Yönüyle okul öncesi eğitim içinde (pp. 273-303). Ankara: Hedef Yayıncılık.

Ünal, F. (2017). Aile eğitimi. Ayşe Belgin Aksoy (Ed.), Aile eğitimi ve katılımı (s.11-42). Ankara: Hedef CS Basın Yayın.

Unal, M. (2009). The place and importance of playgrounds in child development. *Inonu University Journal of Faculty of Education, 10*(2), 95–109.

UNICEF. (2015). *Education the most powerful investment in our future - UNICEF Connect.*

United Nations Educational Scientific and Cultural Organization. (1997). *50 Years for Education.* Retrieved from http://unesdoc.unesco.org/images/0011/001102/110264eb.pdf

United States Department of Education Office of Communications and Outreach. (2005). *Helping Your Child Become a Reader.* Washington, DC: U.S. Department of Education.

Ünlüer, E. (2010). Okul öncesi eğitimde ailenin yeri ve önemi. In T. Güler (Ed.), Anne baba eğitimi (pp. 93-101). Ankara: Pegem Akademi.

Ünver, G. (2007). Gelişimle ilgili temel kavramlar, gelişimin temel ilkeleri ve gelişimi etkileyen etmenler. In A. Ulusoy (Ed.), Eğitim Psikolojisi. Ankara: Anı Yayıncılık

Ünver, S., & Yıldırım, M. (2013). Ameliyat sürecinde çocuk hastaya yaklaşım. *Güncel Pediatri, 11*, 128–133 doi:10.4274/Jcp.11.35229

Üstün, E., & Akman, B. (2003). Üç yaş grubu çocuklarda kavram gelişimi. *Hacettepe Üniversitesi Eğitim Fakültesi Dergisi, 24*, 137–141.

Utchell, L. A., Schmitt, A. J., Mccallum, E., Mcgoey, K. E., & Piselli, K. (2016). Ability of Early Literacy Measures to Predict Future State Assessment Performance. *Journal of Psychoeducational Assessment, 34*(6), 511–523. doi:10.1177/0734282915621221

Van Itallie, T. (1979). Obesity: Adverse effects on health and longevity. *The American Journal of Clinical Nutrition, 32*(12), 2723–2733. doi:10.1093/ajcn/32.12.2723 PMID:389027

Vander Zanden, J. W. (1985). *Human Development.* New York: Random House.

Vandermaas-Peeler, M., Sassine, B., Price, C., & Brilhart, C. (2011). Mothers 'and fathers' guidance behaviours during storybook reading. *Journal of Early Childhood Literacy, 12*(4), 415–442. doi:10.1177/1468798411417331

Vardar, B. (1998). Açıklamalı dilbilim terimleri sözlüğü. İstanbul: ABC Kitabevi.

Veziroglu, M. (2009). *Investigation of the harmony of picture books and MoNE early childhood gains*. Ankara: Hacettepe University, Social Sciences Institute. (in Turkish)

Vitaro, F., Boivin, M., & Bukowski, W. M. (2009). The role of friendship in child and adolescent psychosocial development. In K. H. Rubin, W. M. Bukowski, & B. Laursen (Eds.), *Social, emotional, and personality development in context. Handbook of peer interactions, relationships, and groups* (pp. 568–585). New York: The Guilford Press.

Voogt, J., & Mckenney, S. (2016). TPACK in teacher education: Are we preparing teachers to use technology for early literacy? *Technology, Pedagogy and Education*, *26*(1), 69–83. doi:10.1080/1475939X.2016.1174730

Vygotsky, L. (1962). *Thought and language*. Cambridge, MA: MIT Press. doi:10.1037/11193-000

Vygotsky, L. S. (1978). *Mind in Society: The Development of Higher Psychological Processes*. Cambridge, MA: Harvard University Press.

Wagner, M. M., & Clayton, S. L. (1999). The parents as teachers program:Results from two demonstrations. *The Future of Children*, *9*(1), 91–115. doi:10.2307/1602723 PMID:10414012

Walker, F., Christian, L., & Black, E. R. (2007). Functional indicator for assessing zinc deficiency. *Food and Bulletin*, *28*(3), 454–479. doi:10.1177/15648265070283S305

Walker, H. M., & Golly, A. (1999, November/December). Developing behavioral alternatives for antisocial children at the point of school entry. *ERIC Clearing House*, *73*(2), 104–107. doi:10.1080/00098659909600159

Walker, S. P., Wachs, T. D., Meeks Gardner, J., Lozoff, B., Wasserman, G. A., Pollitt, E., & Carter, J. A. (2007). Child development: Risk factors for adverse outcomes in developing countries. *Lancet*, *369*(9556), 145–157. doi:10.1016/S0140-6736(07)60076-2 PMID:17223478

Walter-Laager, C., Brandenberg, K., Tinguely, L., Schwarz, J., Pfiffner, M. R., & Moschner, B. (2017). Media assisted language learning for young children: Effects of a world- learning app on the vocabulary acquisition of two year olds. *British Journal of Educational Technology*, *48*(4), 1062–1072. doi:10.1111/bjet.12472

Wang, C. (2006). *Designing communicative tasks for college English courses: a dessertation submitted as a partial fulfillment for the degree of m.a. in English language and literature, school of foreign languages and literature*. Chongqing University & Yangtze Normal University, China. Retrieved from www.asian-efl-journal.com/thesis_Wang_Cheng_jun.pdf

Wasserman, G. A., Liu, X., Parvez, F., Ahsan, H., Factor-Litvak, P., van Geen, A., ... Graziano, J. H. (2004). Water arsenic exposure and children's intellectual function in Araihazar, Bangladesh. *Environmental Health Perspectives*, *112*(13), 1329–1333. doi:10.1289/ehp.6964 PMID:15345348

Watanabe, N. (2017). Accelerated Cognitive Development—Piaget's Conservation Concept. *Journal of Educational and Developmental Psychology*, *7*(2), 68–74. doi:10.5539/jedp.v7n2p68

Watson, C. (1995). *One Language or two? helping families from other one language or two? helping families from othercultures decide on: how to talk to their child with language delays*. Spanish Workshop for Children. Retrieved from http://www.spanishworkshopforchildren.com/resources/resources.htm

Waxman, S. R., & Leddon, E. M. (2011). Early Word - Learning And Conceptual Development Everything Had A Name and Each Name Gave Birth To A New Thought. In U. Goswami (Ed.), *Wiley - Blackwell Handbooks Of Developmental Psychology* (2nd ed., pp. 180–208). Ma: Wiley-Blackwell.

Weaver, C., Shaw, D., Dishion, T., & Wilson, M. (2008). Parenting self-efficacy and problem behavior in children at high risk early conduct problems: The mediating role of maternal depression. *National Institutes of Health, 31*(4), 594-605.

Wei, M. (2016). *More than just a game: Yoga for school-age children.* Retrieved from http://www.health.harvard.edu/blog/more-than-just-a-game-yoga-for-school-age-children-201601299055

Wei, L. (2000). *The bilingualism reader*. UK: Routled.

Weilburger, L. S. (2008). *Çocuk ve disiplin: Çocuğumuza davranış yöntemleri (İ. Özbaş, Çev.).* İstanbul: Ekinoks.

Weilburger, S. L. (2008). *Baba sakın bize vurma: Çocuk ve aile içi şiddet.* İstanbul: Ekinoks.

Weinberg, R., Tenenbaum, G., McKenzie, A., Jackson. S., Anshel, M., Grove, R., & Fogarty, G. (2000). Motivation for youth participation in sport and physical activity: Relationships to culture, self-reported activity levels, and gender. *International Journal of Sport Psychology, 31*, 321–346.

Wei, S., Wo, B. L., Qi, H. P., Xu, H., Luo, Z. C., Roy, C., & Fraser, W. D. (2012). Early amniotomy and early oxytocin for prevention of, or therapy for, delay in first stage spontaneous labour compared with routine care. *Cochrane Database of Systematic Reviews*, 9. PMID:22972098

West, Z. (2008). *Acupuncture in pregnancy and childbirth*. Elsevier Health Sciences.

White, H. (2005). *Developing early literacy skills in the early years*. London: Paul Chapman.

Whitehurst, G. J. (2003). *Read Together, Talk Together: A Dialogic Reading Program for Young Children*. New York, NY: Pearson Early Learning Group.

Whitehurst, G. J., Falco, F. L., Lonigan, C. J., Fischel, J. E., Debaryshe, B. D., Valdez-Menchaca, M. C., & Caulfield, M. (1988). Accelerating language development through picture-book reading. *Developmental Psychology, 24*(4), 552–558. doi:10.1037/0012-1649.24.4.552

Whitehurst, G. J., & Lonigan, C. J. (1998). Child development and emergent literacy. *Child Development, 69*(3), 848–872. doi:10.1111/j.1467-8624.1998.tb06247.x PMID:9680688

White, J. (2008). *Playing and learning out doors. Making provision forhigh-quality experiences in the outdoor environment*. Oxon: Routledge.

White, L. S. (2009). Yoga for children. *Pediatric Nursing, 35*, 277–295. PMID:19916343

Whitworth, J., Sewankambo, N. K., & Snewin, V. A. (2010). Improving implementation: Building research capacity in maternal, neonatal, and child health in Africa. *PLoS Medicine, 7*(7), 2–4. doi:10.1371/journal.pmed.1000299 PMID:20625547

WHO, & UNICEF. (2014). Every Newborn: an action plan to end preventable deaths: Executive summary. *Geneva: World Health Organization.* Retrieved from http://www.everynewborn.org/Documents/Executive-summary_EN.pdf%0Awww.who.int/about/licensing/copyright_form/en/index.html

WHO. (2002). *Health Research on Arsenic Poisoning (Scientific Debate)* http://apps.who.int/iris/bitstream/10665/127108/1/SEA-ACHR-27-10.pdf?ua=1adresindenerişildi

WHO. (2010). *WHO Investing in our common future Joint Action Plan for Women's and Children's Health.*

WHO. (2016) Standards for Improving Quality of Maternal and Newborn Care in Health Facilities. Retrieved from https://apps.who.int/iris/bitstream/handle/10665/249155/9789241511216-eng.pdf;jsessionid=B32399A6F963AA391 7AE1E001D053E71?sequence=1

Wijnhoven, T. M. A. (2015). Overweight and obesity in primary-school children: a surveillance system for policy-making in Europe from 2007 onwards [PhD Thesis]. Wageningen University.

Wilensky, D. S., Ginsberg, G., Altman, M., Tulchinsky, T. H., Ben Yishay, F., & Auerbach, J. (1996). A community based study of failure to thrive in Israel. *Archives of Disease in Childhood, 75*(2), 145–148. doi:10.1136/adc.75.2.145 PMID:8869197

Willoughby, C., & Polatajko, H. J. (1995). Motor problems in children with developmental coordination disorder: Review of the literature. *The American Journal of Occupational Therapy, 49*(8), 787–794. doi:10.5014/ajot.49.8.787 PMID:8526224

Wilson, R. A. (1996). Starting Early: Environmental Education During the Early Childhood Years. ERIC Clearing house for Science, Mathematics and Environmental Education. (ED402147)

Wilson, R. (2008). *Nature and young children: Encouraging creative play and learning in natural environments.* New York, NY: Routledge.

Wilson, R. A., Kilmer, S. J., & Knauerhase, V. (1996). Developing an environmental outdoor play space. *Young Children, 51*(6), 56–61.

Wolstenholme, E. (2005). The potential of system dynamics. *Leading Edge,* (October). Retrieved from http://www.iseesystems.com/resources/Articles/leading_edge_10.pdf

Woodyard, C. (2011). Exploring the therapeutic effects of yoga and its ability to increase quality of life. *International Journal of Yoga, 4*(2), 49–54. doi:10.4103/0973-6131.85485 PMID:22022122

World Bank. (2010). *Child Youth Development.*

Wortham, S. C. (1989). Outdoor play environments for infants and toddlers. *Day Care and Early Education, 16*(4), 28–30. doi:10.1007/BF01619376

Wothke, W. (1993). Nonpositive Definite Matrices İn Structural Modeling. In K. A. Bollen & J. S. Long (Eds.), Testing Structural Equation Models (pp. 256-293). Newbury Park, CA: Sage Publications.

Wu, C. C., & Honig, A. S. (2010). Taiwanese mothers' beliefs about reading aloud with preschoolers: Findings from the parent reading belief inventory. *Early Child Development and Care, 180*(5), 647–669. doi:10.1080/03004430802221449

Yalçın, H., Bay, M., & Ekmekçioğlu, E. (2014). Türkiye ve Türk dünyasında çocuk hakları ile ilgili çalışmaların yararları. Retrieved from http://www.bilgesam.org/Images/Dokumanlar/0-279-2014070118sosbilkongre33.pdf

Yalçın, A., & Aytaş, G. (2014). *Children's literature.* Ankara: Akçağ Publications. (in Turkish)

Yan, J. H., & McCullagh, P. (2004). Cultural Influence on Youth's Motivation of Participation in Physical Activity. *Journal of Sport Behavior, 27*(4), 378–390.

Yaşar, M. (2015). Okul Öncesi Dönemde Sosyal-Duygusal Gelişim. Ankara: Hedef.

Yaşar-Ekici, F. (2015). Okul öncesi eğitime devam eden çocukların sosyal beceri ile aile özellikleri arasındaki ilişkinin incelenmesi. *The Black Sea Journal of Social Sciences, 7*(1), 223–259.

Yaşar, F. (1996). *Elazığ ve Malatya illerindeki çocuk yuvası ve yetiştirme yurtlarında kalan korunmaya muhtaç çocukların fiziksel ve psiko-sosyal sorunlarının araştırılması.* Elazığ: Tıpta Uzmanlık Tezi, Fırat Üniversitesi Tıp Fakültesi Çocuk Sağlığı ve Hastalıkları.

Yates, S. (2010). *Pregnancy and Childbirth E-Book: A holistic approach to massage and bodywork.* Elsevier Health Sciences.

Yavuzer, H. (1995). *Child Psychology*. İstanbul: Remzi Publications. (in Turkish)

Yavuzer, H. (1999). *Çocuk Psikolojisi*. İstanbul: Remzi.

Yavuzer, H. (2011). *Ana-baba ve çocuk*. İstanbul: Remzi.

Yavuzer, H. (2011). *Okul Çağı Çocuğu*. İstanbul, Turkey: Remzi Kitapevi.

Yavuzer, H. (2012). *Çocuk psikolojisi*. İstanbul: Remzi.

Yavuzer, H. (2012a). *Çocuk psikolojisi (34. Baskı)*. İstanbul: Remzi Kitabevi.

Yavuzer, H. (2012b). *Anne-baba ve çocuk (23. Baskı)*. İstanbul: Remzi Kitabevi.

Yavuz, O., & Yavuz, Y. (2018). Huzurevindeki Yaşlı Bireylere Oynatılan Zekâ Oyununun Yaşlıların Bilişsel Becerilerine, Yalnızlık ve Psikolojik İyi Oluş Düzeylerine Etkisi. *Yaşam Becerileri Psikoloji Dergisi*, *2*(3), 127–141. doi:10.31461/ybpd.427039

Yazgan İnanç, B., Bilgin, M., & Kılıç Atıcı, M. (2007). *Gelişim Psikolojisi. Çocuk ve ergen gelişimi*. Ankara: Pegem A Yayıncılık.

Yazgan İnanç, B., Bilgin, M., & Kılıç Atıcı, M. (2010). *Gelişim Psikolojisi*. Ankara: Pegem.

Yazıcı, Z. (2003). Okul öncesi eğitimin okul olgunluğu üzerine etkisinin incelenmesi. *M. E. B. Dergisi*, 155-156.

Yazıcı, E. (2012). Korunmaya muhtaç çocuklar ve çocuk evleri. *Mustafa Kemal Üniversitesi Sosyal Bilimler Enstitüsü Dergisi*, *9*(18), 499–525.

Yazıcı, E., & Gençer, E. (2016). Okul öncesi çocuklarının bilgi ve iletişim teknolojileriyle etkileşimlerinin bazı değişkenler yönünden incelenmesi. *Kastamonu Eğitim Dergisi*, *24*(5), 2235–2252.

Yazıcı, Z. ve Can Yaşar, M. (2006). Erken çocukluk döneminde dil kazanımı. *Mesleki Eğitim Dergisi Sayı*, *8*(15), 126–138.

Yeşilçiçek-Çalık, K., & Aktaş, S. (2011). Gebelikte depresyon, sıklık ve risk faktörleri ve tedavisi. *Psikiyatride Güncel Yaklasimlar*, *3*(1), 142–162.

Yeşildere, S., & Türnüklü, E. B. (2007). Öğrencilerin matematiksel düşünme ve akıl yürütme süreçlerinin incelenmesi. *Ankara Üniversitesi Eğitim Bilimleri Fakültesi Dergisi*, *40*(1), 181–213.

Yeşilyaprak, B. (2013). *Eğitimde Rehberlik Hizmetleri Gelişimsel Yaklaşım*. Ankara: Nobel.

Yıldız Akkuş, S. & Bütün Ayhan, A. (2018). Kronik hastalığı olan çocukların davranışlarının ve yaşam kalitelerinin incelenmesi. *Türkiye Çocuk Hastalıkları Dergisi*, 1-7.

Yıldız Bıçakçı, M., & Aral, N. (2009). Language development. In N. Aral & T. Duman (Eds.), Educational psychology (pp. 115- 130). İstanbul: Kriter Publications. (in Turkish)

Yıldız Bıçakçı, M., & Durualp, E. (2016). Çocuklarda sosyal ve duygusal gelişim. In N. Aral (Ed.), Çocuk gelişimi (pp. 128-156). Eskişehir: AÖF Yayınları.

Yıldız, M. A. & Duy, B. (2015). Sokakta çalışmak ve gelir düzeyi erinlerin benlik saygısı, yaşam doyumu ve depresyon düzeylerine etki eder mi? (Diyarbakır ili örneği). *İlköğretim Online*, *14*(2), 522-537.

Yıldız, A., Baran, E., Akdur, R., Ocaktan, E., & Kanyılmaz, O. (2008). Bir sağlık ocağı bölgesinde 0- 11 aylık bebekleri olan annelerin emzirme durumları ve etkileyen faktörler. *Ankara Üniversitesi Tıp Fakültesi Mecmuası*, *61*(2), 61–67.

Yıldız-Bıçakçı, M. & Aral, N. (2017). Dil gelişimi. In. N. Aral & T. Duman (Eds.), Eğitim psikolojisi (pp. 132-150). Ankara: Pegem.

Yıldız-Bıçakçı, M., & Aral, N. (2017). Dil gelişimi. In N. Aral & T. Duman (Eds.), Eğitim psikolojisi (pp. 132-150). Ankara: Pegem.

Yıldız-Bıçakcı, M., Dursun, Ş. S., & Bolatbaş, D. E. (2018). Erken Müdahale Hizmetleri. In P. Bayhan (Ed.), Erken Müdahale içinde (pp. 84-102). Ankara: Hedef.

Yılmaz Bolat, E., & Özgün, Ö. (2013). Erken çocuklukta aile eğitimi ve aile eğitimine yönelik uygulanan programlar. In Y.A. Arnas (Ed.), Aile eğitimi ve okul öncesinde aile katılımı (pp. 21-51). Ankara: Vize Yayıncılık.

Yılmaz Ünal, S. & Deniz, Ü. (2018). Davranış ve uyum problemleri olan çocuklar. In N. Aral & F. Gürsoy (Eds.), Özel gereksinimli çocuklar ve eğitimleri-Özel gereksinimli çocuklar (pp. 85-111). Ankara: Hedef CS Basın Yayın.

Yılmaz, M. T. (2008). Cinsel gelişim. In M. E. Deniz (Ed.), Erken cocukluk döneminde gelişim (pp. 291-323). Ankara: Maya Akademi.

Yılmaz, E., Aydın-Tosun, Ö., Tarhan, N., Ayaz, R., Turgut, A., & Karateke, A. (2016). İleri anne yaşı gebeliklerinde perinatal sonuçlar. *Journal of Clinical and Experimental Investigations*, *7*(2), 157–162.

Yılmazer, Y., Yıldız, S., & Şahin-Öz, A. (2015). Hasta çocuğa yönelik eğitim düzenlemeleri. In A. Bütün-Ayhan (Ed.), *Hasta çocukların gelişimi ve eğitimi* (pp. 162–183). Eskişehir: Anadolu Üniversitesi.

Yolcuoğlu, G. İ. (2009). Türkiye'de çocuk koruma sisteminin genel olarak değerlendirilmesi. *Aile ve Toplum*, *5*(18), 43–57.

Yolcuoğlu, İ. G. (2009). Türkiye'de çocuk koruma sisteminin genel olarak değerlendirilmesi. *Aile ve Toplum Eğitim Kültür ve Araştırma Dergisi*, *18*, 43–58.

Yörükoğlu, A. (2011). *Çocuk ruh sağlığı*. İstanbul: Özgür.

Yörükoğlu, A. (2011). *Çocuk ruh sağlığı: Çocuğun kişilik gelişimi, eğitimi ve ruhsal sorunları (31. Baskı)*. İstanbul: Özgür Yayınları.

Yorulmaz, M. (2009). *Göreve dayalı öğretim yönteminin Türkçenin yabancı dil olarak öğretiminde dinleme becerisinde kullanılmasının avantajları*. Unpublished master thesis, Trakya Üniversitesi, Edirne.

Young, A. J., Sıllıman-French, L., & Crawford, L. (2017). Yoga For Young Children With Down Syndrome. *Palaestra-Journal*, *31*(3), 18–23.

Yücel, M. K. (2015). *Doğum ve kadın sağlığı*. Ankara: Palme.

Yücelyiğit, S., & Aral, N. (2018). Technology Addiction in Children. In R. Efe (Ed.), *Educational Sciences Research in the Globalizing World*. Sofia: St. Kliment Ohridski University Press.

Yule, G. (2016). *The study of language*. Cambridge university press.

Yumuş, M. (2013). *Okul öncesi eğitimcilerinin 36-72 ay aralığındaki çocukların davranış problemleri ile ilgili görüşlerinin incelenmesi ve başa çıkma stratejilerinin belirlenmesi*. Yayınlanmamış Yüksek Lisans Tezi, Hacettepe Üniversitesi.

Yurtseven, L. (2011). Investigation of mothers' story reading effects on children's social skills development. Unpublished masters thesis, İnönü University, Educational Sciences Institute, Malatya. (in Turkish)

Yüzer, S., Yiğit, R., & Taşdelen, B. (2006). Çocuğu hastanede yatan annelerin aldığı sosyal destek ile depresyon düzeyleri arasındaki ilişki. *Atatürk Üniversitesi Hemşirelik Yüksekokulu Dergisi*, *9*(4), 54–62.

Zembat, R. (1999). *Okul öncesi eğitimde temel konular*. İstanbul: Turan Ofset.

Zengin, D. (2003). Sıfatların başka sözcük türlerine kayması, *Ankara Üniversitesi Dil ve Tarih Coğrafya Fakültesi Dergisi Sayı*, *43*, 83–99.

Zevenbergen, A. A., & Whitehurst, G. J. (2003). Dialogic reading: A shared picture book reading intervention for preschoolers. In A. Van Kleeck, S.A. Stahl, & E.B. Bauer (Eds.), *On reading books to children: Parents and teachers* (pp. 177-200). Mahwah, NJ: Lawrence Erlbaum Associates. Retrieved from http://citeseerx.ist.psu.edu/viewdoc/download?doi=10.1.1.31.5935&rep=rep1&type=pdf

Zhang, C., Schwartz, B., & Lee, H. (2006). Collaborative Services for Infants and Toddlers with Disabilities: Perspectives From Professionals in an Urban Setting. *Early Child Development and Care*, *176*(3-4), 299–311. doi:10.1080/03004430500092282

Ziegler Delahunt, J. (2002). Motor Development Child Development. USA: Macmillan.

Zwisler, R. (2006). Sprachentwicklung. Retrieved from http://www.zwisler.de/scripts/Sprachentwicklung

About the Contributors

Neriman Aral, Prof. Dr., was born in Ankara. She completed her higher education in 1983 at the Department of Child Development and Education of Hacettepe University. She gained her master's degree in 1986 from Hacettepe University. She has two doctorate degree one of them from Ankara University (1990) and the other one from Hacettepe University (1992). She has worked in a center for the children with Cerebral Palsy in 1983-1986. She was assigned as research assistant in 1987 at the University of Ankara. She became an associate professor at Ankara University in 1995. She has assigned as professor in 2001. Prof. Dr. Neriman Aral has held many administrative positions. She was the Head of the Department of Child Development and Education in Ankara University Agriculture Faculty School of Home Economics between 1998 and 2000, Head of the Department of Child Development and Education from 2004 to 2007, Vice Principal in the School of Home Economics between 2000-2007 and Coordinator of Ankara University Children's Science Center between 2012-2017. After establishing the Department of Child Development at Ankara University Faculty of Health Sciences, in 2009, her professor title was transferred to the Department of Child Development. She was Head of the Department of Child Development between 2009 and 2017 in Ankara University Faculty of Health Sciences. She has worked as a representative of the Council of Higher Education in Committee of Prime Ministry for Protection Young Ones from Harmful Publications, between 2001 and 2007. In 2011, she has made observations for three months at the University at Buffalo the State University of New York, Institute of Education Department of Learning and Teaching Early Years and Childhood as guest researcher. Prof. Dr. Neriman Aral took part in the projects supported by European Union, Tubitak and University resources. She has submitted papers in many international and national congresses and also organized the scientific congress. And she has many articles in international and national journals, many books and books chapters on the field. Professor Dr. Neriman Aral is currently a lecturer in the Department of Child Development at the Health Sciences Faculty of Ankara University and also she is the Dean of same faculty. She is married and mother of a child.

Devlet Alakoç Pirpir completed the doctorate program at Selcuk University Graduate School of Social Sciences Department of Child Development and Home Economics Teaching in 2011. Her areas of study cover early childhood education, parental education, school readiness, approaches in early childhood education. Currently, she is an assistant professor at Selcuk University Faculty of Health Sciences Department of Child Development.

Sibel Atli, Asst. Prof., was born in Van / Baskale. In 2005, she completed her undergraduate education at the Department of Preschool Teaching at the Department of Primary Education in the Faculty of Education at Van Yüzüncü Yıl University. In 2011, she completed her master's degree in Educational Programs and Teaching Department of the Institute of Educational Sciences at Yüzüncü Yıl University. She completed her PhD at Ankara University, Institute of Health Sciences, Department of Child Development in 2019. Between 2005-2012, she worked as a pre-school teacher in elementary and kindergartens of Van Ministry of National Education. Between 2012 to 2019, she worked as an instructor at Yüzüncü Yıl University Vocational School, Child Development Program and Faculty of Education Preschool Teaching Departments. Van Yüzüncü Yıl University, Faculty of Health Sciences, Department of Child Development she continues to work as a Faculty Member.

Burçin Aysu was born in Kadıkoy. She graduated from Gazi University, Faculty of Education, Department of Preschool Education and completed a doctorate degree in child development and education at Ankara University, Institute of Health Science, Department of Child Development. She worked as a teacher and a director at private and public preschools. She currently continues works at Ankara Yıldırım Beyazıt University, Faculty of Health Science, Department of Child Development as an Assistant Profesor.

Burçin Aysu, M. Sc., was born in Kadıkoy. She graduated from Gazi University, Faculty of Education, Department of Preschool Education.

Gülen Baran, Professor Dr., was born in Ankara. She completed her undergraduate education in 1986 at Ankara University School of Home Economics. In 1987, she entered Ankara University, School of Home Economics, Department of Child Development as a research assistant. In 1989, she completed her master's degree and in 1995, she completed her doctorate. She became an Assistant Professor in 1997 and Associate Professor in 1998 at the Department of Child Development. In 2008, she was appointed as Professor. She served at Ankara University Faculty of Health Sciences Department of Child Development as academic member and the head of the department. She has many national and international books, book chapters, article publications and project and journal referees. She is a variety of undergraduate and graduate levels conducts lessons. She has the advisor of many graduate and doctoral students in the Child Development and Family Counseling Departments. She is currently a faculty member in the Faculty of Health Sciences, Department of Child Development.

Adalet Biltekin graduated from İ.D. Bilkent University, Department of Psychology in 2014. In 2019, she completed her master's degree at Ankara University, Department of Child Development. She wrote her master's thesis on early literacy.

Joyce Cheptum, Dr., has over 15 years of experience in Nursing and Midwifery practice. She holds a PhD in Nursing from the University of Nairobi and a Master of Science in Public Reproductive Health from Kenyatta University. She holds a Bachelor of Science in Nursing from Moi University. She has a wealth of experience in midwifery clinical practice and training. She has participated in various projects aimed at improving maternal and neonatal health. She is a trainer in emergency obstetrics and newborn care. She has played key roles in training, designing and supervising health research. She has published several articles in peer reviewed journals. She spearheaded the neonatal nursing conference which saw the launch of the Association of the neonatal nurses of Kenya. She has successfully organized midwifery scientific conferences. She has mentored many nursing students and supervised several undergraduate and postgraduate research successfully. She has participated in development and review of various curricula for nursing and midwifery training and is a reviewer of various journals. She is also involved in various community service activities.

Sühendan Er got her undergraduate degree in English Language Teaching at Gazi University in 1990. She worked at TED Ankara College Primary school as an English teacher between 1990 and 1997. In 1997 she started working at Gazi University as an English language instructor. She got her MS degree from the same university in the field of English Language Teaching. In 2011, she got PhD from Ankara University, Child Development and Education Department with the thesis on children's foreign language learning and belief. While she was working at Gazi University, she completed teacher training programme. Besides, she took place as Tester, Academic Coordinator, Teacher Trainer and lastly the Vice Director of the school of foreign languages. Since 2012 she has been working at TED University, Education Faculty, Department of Early Childhood Education. Dr. Er has been having her academic and scientific studies on children's language development in early childhood, teaching foreign languages to children, values education, school readiness and constructive approach in education. She is the founder member of the World Council for Curriculum and Instruction and she has been the vice president of the institution's Turkish chapter since 2011.

Figen Gürsoy was born in 1967. She graduated from School of Home Economics at University of Ankara. She received her master's degree in 1992 and PhD in 1996 from Ankara University, Faculty of Home Economics (Child Development). She became research assistant in 1992, assistant professor in 2000, associate professor in 2001 and full professor in 2008. She continues her academic carrier at University of Ankara, Department of Child of Child Development. Her research interests include socially disadvantaged children, children drifting into crime, broken families' children, school-age children, adolescence and characteristics, children with disabilities, children with behavior problems. She conducted many courses and supervised graduate thesis on adolescence, children with disabilities, developmental psychology, communication skills, cognitive and language development in children and education of children with special needs, and developmental disorders in children.

Gül Kadan is a PhD student at Ankara University, Faculty of Health Sciences, Department of Child Development. She is working as a lecturer at Çankırı Kartekin University, Faculty of Health Sciences, Department of Child Development.

Rıdvan Karabulut was born in 1980 in Kayseri. He completed his primary, secondary, and high school education in Kayseri. He graduated from Kayseri Sumer High School. He completed his undergraduate education in Erciyes University, his master's degree in Abant İzzet Baysal University, and his doctorate in preschool education at Gazi University. He has worked in many private and public institutions. In the private sector, she worked as the coordinator for the education of special gifted children at Tekden Private Primary School. He worked as a classroom teacher in Istanbul and Sivas in institutions affiliated to the Ministry of National Education. In 2010, he started to work as a lecturer in Erciyes University Child Development Department. She started working at Ahi Evran University in 2017 and is currently working as a faculty member in the Department of Child Development at the School of Health. He is married and has one child.

Nezahat Karaca was born in Afyonkarahisar. She graduated from Celal Bayar University, Faculty of Education, Class Education Department in 2004. In 2005, she started to work as a research assistant in Afyon Kocatepe University, Faculty of Education, Department of Primary Education. In 2008, she completed her master's degree at Afyon Kocatepe University, Department of Elementary Education and received her Master of Science degree. She worked as an instructor between 2011-2015. In 2015, she completed her PhD at Ankara University, Department of Child Development and Education. In 2015, he became a faculty member at the Department of Preschool Education at Afyon Kocatepe University. He has articles in national and international journals, author of book chapters and books about the field. He is currently working in Afyon Kocatepe University Faculty of Education Preschool Education Department

Fatma Elif Kılınç completed her undergraduate degree from Selçuk University Faculty of Vocational Education, Child Development and Education Department in 1997, Kılınç completed her master's degree at the same university with a thesis on emotional intelligence. Between 2001-2005, she studied mother-child interactions, alternative approaches in education, children's books and library services for children at various schools and institutions in the United States of America. She was granted the title of doctor with her thesis on mother-child interaction in 2011. In 2012, she was appointed to Abant İzzet Baysal University Faculty of Education Department of Early Childhood Education as assistant professor. Between 2010-2012, she participated in the founding of the AİBÜ Early Childhood Education Research and Application Center and served as administrator. Between 2012-2014, she served as an advisor to the President of the Council of Higher Education. She served as a member of the National Ministry of Education Board of Education between 2014-2016. Since 2016, she has been working as a lecturer and department chair at Ankara Yıldırım Beyazıt University Faculty of Health Sciences Department of Child Development. She became an associate professor in 2019. She conducted many projects supported by the resources of the European Union, TÜBİTAK and universities. She is currently conducting work related to in-service trainings offered by the National Ministry of Education, teacher training, training Syrian educators working at Temporary Education Centers and training Turkish teachers working with Syrian students. She has studies on alternative approaches, the Montessori method, cognitive and language development, parent education, mother-child interaction, children's publications, interactive reading, environmental education, children's rights, values education and differentiated instruction. She has had articles published in national and international journals and has presented at various conferences.

Lucy Kivuti-Bitok, Dr., is a Senior Lecturer at the University of Nairobi. Acknowledging the inroads system dynamics modelling and Mathematics have gained in every discipline and the modern view of management as a system of mathematical models and processes, Dr Kivuti-Bitok obtained her PhD in Mathematical modelling (Health Systems Engineering) from the University of Nairobi and a Post-Doctoral fellowship from the University of Otago where she worked on System Dynamics modelling of Health impact of Bicycling in New Zealand's major cities as well as One Health approach in Antimicrobial resistance.. She obtained her Master in Health Sector Management from Roskilde University / Galilee College and BScN from the University of Nairobi.She is consultant in System Dynamics Modelling. With over 20 years of experience as a researcher and academician, Lucy takes pride in the infinite nature and versatility of the nursing profession which enabled her to venture into the world of mathematical modelling, health systems engineering and system dynamics modelling. She is championing the application of System Dynamics modelling in UHC. Prior to working at the University of Nairobi, Lucy worked with the Ministry of Health as a nurse Manager

Ebru Tanju Aslışen graduated first from Hacettepe University, School of Home Economics, Department of Child Health and Education in 1992 with a Science Promotion Award and the İhsan Doğramacı Outstanding Achievement Award. Between 1992 and 1994, she volunteered to work with mentally disabled children and their families. In 1994, she started to work as a research assistant in the department she graduated. She received her master's degree in 1995 and a PhD degree in 2004 from Hacettepe University Institute of Health Sciences, Department of Child Development and Education. In 1999, she established a special education institution for the education and rehabilitation of mentally disabled children and continued to be active with children and their families. Between 2005 - 2017, she worked as a faculty member at Başkent University Faculty of Education and retired. She worked at Hasan Kalyoncu University Faculty of Education as a faculty member between February 2017 and August 2019. She received the title of Associate Professor on November 15, 2017. She has been working in the Department of Child Development in Ankara Hacı Bayram Veli University since October 2019.

Halil Uzun was born in 1978 in Ceyhan, Adana, and attended primary and high school education in Ceyhan. He graduated from Atatürk University, Kazım Karabekir Faculty of Education, Department of Elementary Education, Inonu University Institute of Social Sciences, Preschool Teacher Education Graduate Program and Ankara University, Institute of Science, Department of Child Development and Education Doctorate Program. He is studying on mother-father training, technology effects and language development in early childhood. Uzun is married and has three children.

Müdriye Yildiz Biçakçi started her academic career at Ankara University in 2002 and is currently working as a professor. During this period, between 2010 and 2015, he worked as vice dean in the Ankara University Faculty of Health Sciences and in certain intervals he was in charge of education in Ankara University Practice Kindergartens. He also served as a member of the Advisory Board of TUBITAK for Early Childhood Periodical Education / Curious Tiny and Books. Since 2016, he has been supporting the Child Development and Educators Association, which he has been supporting since 2015 as a member of the board of directors, as vice president of the association. In 2017, he was appointed to Jauma I University (Castellon) Department of Education on values education in early childhood.

Canan Yıldız Çiçekler completed the doctorate program at Selcuk University Graduate School of Social Sciences Department of Child Development and Teaching in 2016. Her areas of study include creativity, art and education, and visual arts. Currently, she is an assistant professor in the Department of Preschool Education at the education of Faculty of Necmettin Erbakan University.

Secil Yucelyigit, Dr., graduated from Hacettepe University, Department of Mathematics in Ankara, Turkey, and worked as a software specialist for five years. Due to her interest areas, she decided to continue her education on social sciences and studied child development and education during her PhD at Ankara University Graduate School of Natural and Applied Sciences-Child Development and Education. She worked as a manager in the Department of Educational Material Development for a publishing and producer company. Currently, she is working at TED University Faculty of Education, Department of Elementary Education-Early Childhood Education.

Index

A

B

C

D

E

F

G

H

I

S

T

V

Y

www.ingramcontent.com/pod-product-compliance
Lightning Source LLC
LaVergne TN
LVHW062106300926
849161LV00010B/2383